Experimental and Theoretical Advances in Biological Pattern Formation

NATO ASI Series

Advanced Science Institutes Series

A series presenting the results of activities sponsored by the NATO Science Committee, which aims at the dissemination of advanced scientific and technological knowledge, with a view to strengthening links between scientific communities.

The series is published by an international board of publishers in conjunction with the NATO Scientific Affairs Division

A	**Life Sciences**	Plenum Publishing Corporation
B	**Physics**	New York and London
C	**Mathematical and Physical Sciences**	Kluwer Academic Publishers
D	**Behavioral and Social Sciences**	Dordrecht, Boston, and London
E	**Applied Sciences**	
F	**Computer and Systems Sciences**	Springer-Verlag
G	**Ecological Sciences**	Berlin, Heidelberg, New York, London,
H	**Cell Biology**	Paris, Tokyo, Hong Kong, and Barcelona
I	**Global Environmental Change**	

Recent Volumes in this Series

Series A: Life Sciences

Experimental and Theoretical Advances in Biological Pattern Formation

Edited by

Hans G. Othmer

University of Utah
Salt Lake City, Utah

Philip K. Maini

University of Oxford
Oxford, England, United Kingdom

and

James D. Murray

University of Washington
Seattle, Washington

Springer Science+Business Media, LLC

Proceedings of a NATO Advanced Research Workshop on
Biological Pattern Formation,
held August 27–31, 1992,
in Oxford, England, United Kingdom

NATO-PCO-DATA BASE

The electronic index to the NATO ASI Series provides full bibliographical references (with keywords and/or abstracts) to more than 30,000 contributions from international scientists published in all sections of the NATO ASI Series. Access to the NATO-PCO-DATA BASE is possible in two ways:

—via online FILE 128 (NATO-PCO-DATA BASE) hosted by ESRIN, Via Galileo Galilei, I-00044 Frascati, Italy

—via CD-ROM "NATO Science and Technology Disk" with user-friendly retrieval software in English, French, and German (©WTV GmbH and DATAWARE Technologies, Inc. 1989). The CD-ROM also contains the AGARD Aerospace Database.

The CD-ROM can be ordered through any member of the Board of Publishers or through NATO-PCO, Overijse, Belgium.

Library of Congress Cataloging-in-Publication Data

NATO Advanced Research Workshop on Biological Pattern Formation (1992
 : Oxford, England)
   Experimental and theoretical advances in biological pattern
 formation / edited by Hans G. Othmer, Philip K. Maini,. and James D.
 Murray.
      p.   cm. -- (NATO ASI series. Series A, Life sciences ; vol.
 259)
   "Proceedings of a NATO Advanced Research Workshop on Biological
 Pattern Formation, held August 27 - 31, 1992, in Oxford,
 England, United Kingdom"--T.p. verso.
   "Published in cooperation with NATO Scientific Affairs Division."
   Includes bibliographical references and index.
   ISBN 978-0-306-44661-0   ISBN 978-1-4615-2433-5 (eBook)
   DOI 10.1007/978-1-4615-2433-5
   1. Pattern formation (Biology)--Congresses. 2. Morphogenesis-
 -Congresses. 3. Developmental biology--Congresses.   I. Othmer, H.
 G. (Hans G.), 1943-   . II. Maini, Philip K. III. Murray, J. D.
 (James Dickson) IV. North Atlantic Treaty Organization. Scientific
 Affairs Division. V. Title. VI. Series: NATO ASI series. Series
 A, Life sciences ; v. 259.
 QH491.N38 1992   93-39363
 574.3--dc20   CIP

ISBN 978-0-306-44661-0

©1993 Springer Science+Business Media New York
Originally published by Plenum Press, New York in 1993

PREFACE

This volume contains the proceedings of the NATO ARW on 'Biological Pattern Formation' held at Merton College, University of Oxford, on 27-31 August, 1992.

The objective of the workshop was to bring together a select group of theoreticians and experimental biologists to present the latest results in the area of biological pattern formation and to foster interaction across disciplines. The workshop was divided into 5 main areas: (i) limb development, (ii) Dictyostelium discoideum, (iii) Drosophila, (iv) cell movement, (v) general pattern formation.

We thank all the participants for their contributions, enthusiasm, and willingness to collaborate. There was a genuine, open, and extremely fruitful interaction between the experimentalists and theoreticians which made the workshop a success. We also thank The Welcome Trust for providing additional funding.

The local organization fell mainly on Denise McKittrick and Beverley Bhaskhare at the Mathematical Institute, Oxford, and Jeanette Hudson and the staff of Merton College. We greatly appreciate their help and patience. We also thank Jonathan Sherratt, Wendy Brandts and Debbie Benson for helping out in the conference and for providing a happy welcome to participants on a typically cold, wet and windy English summer day.

Hans G. Othmer
Philip K. Maini
James D. Murray

CONTENTS

vii

1. GENERAL MODELS OF PATTERN FORMATION: SOME USES, PROBLEMS AND SUCCESSES

J. D. Murray

Applied Mathematics FS-20
University of Washington
Seattle, WA 98195, USA

1.1 INTRODUCTION

Understanding the evolution of spatial patterns and the mechanisms which create them are among the most crucial issues in developmental biology. Whether the process is the generation of cartilage patterns in the developing limb, the formation of feathers and scales, the myriad of colour patterns on butterfly wings, regeneration in *Hydra*, the segmentation in the egg of a developing *Drosophila*, the patterns on the coats of animals, the formation of the grex in *Dictyostelium* and so on, a key question is "How are these patterns formed and what is the mechanism (or mechanisms) that creates them?" Enormous progress has been made in understanding some of the basic principles that any mechanism must possess to be able to generate spatial patterns. In spite of this we still do not know, with any certainty, definitive details of a single pattern formation mechanism which is involved in development. Model mechanisms - morphogenetic models - for biological pattern generation can suggest to the embryologist possible scenarios as to how, and sometimes when, pattern is laid down and how the embryonic form might be created. Although genes control pattern formation, genetics says nothing about the actual mechanisms involved nor how the vast range of pattern and form that we see evolves from a homogeneous mass of dividing cells. However, with the enormous strides made in our understanding of a wide variety of developmental situations there is room for optimism that in the not too distant future we shall have isolated not only the biochemical and physical elements involved but also have discovered some of the actual mechanisms. There has, in the past few years, been an increasing recognition among experimentalists and theoreticians that the most dramatic progress in biology will come about through a genuine interdisciplinary approach in which biomedical scientists, mathematicians and physical scientists all play a role. The book by Murray (1989) discusses in detail many successful case studies of such an interdisciplinary approach.

Embryogenesis depends on a series of sequential processes which generate specific patterns at each stage in development. Most of these processes depend on the outcome of one or more previous patterning stages. For example, chondrogenic patterns and formation of feather and hair primordia both involve major symmetry-breaking and are essential for subsequent developmental stages. In animal development the basic body plan is laid down very early, of the order of the first few weeks in a gestation period of many months. Answering key questions about what the mechanisms are and which genes turn them on is an important

Experimental and Theorietcal Advances in Biological Pattern Formation,
Edited by H.G. Othmer *et al.*, Plenum Press, New York, 1993

step, but arguably the first is learning exactly when in development the patterning processes take place. Without such knowledge it is very difficult to carry out experiments to try and determine the mechanism, since after it has produced the requisite pattern its role is usually finished and in general it will simply cease to exist. We have a fairly clear picture in the case of *Dictyostelium*, increasingly so for *Drosophila* and to a certain extent for limb cartilage development, but even here, there is no general agreement, as is clear, for example, from the proceedings of the NATO workshop on Developmental Patterning in Development (Hinchliffe *et al.* 1991). In the case of animal integumental patterns we do not, in general, even know within order of days when in development the pattern is actually formed. This was the key question asked and studied by Murray *et al.* (1990), which is a theoretical and experimental paper on alligator (*Alligator mississippiensis*) stripes and when the basic pattern is formed. Depending on the sex of the alligator the time of activation of the stripe patterning mechanism (as yet unknown of course) was determined to within 24 hours, approximately the time we believe it takes to form the stripe pattern.

The evidence is now clear that tissue interaction can also be crucial in the development of many patterns, complex patterns in particular. Here the pattern formation process in both tissues may take place simultaneously, but the interaction is crucial from the start. In the models of Nagorcka *et al.* (1987), (1986), (see also Nagorcka *et al.* this volume), Shaw & Murray (1990), Cruywagen & Murray (1991) two pattern generators were coupled to model such tissue interaction.

When mechanisms are coupled, as in these tissue interaction models, it might appear reasonable to suppose that the pattern spectrum is larger than that which can be created by either mechanism on its own. Numerical simulation of the coupled systems interestingly shows unequivocally that the spectrum of patterns found is almost always restricted to a subset of the theoretically possible. The conclusion is that the coupling of pattern generators, each with its own basin of attraction, instead of introducing more complexity actually has the effect of *reducing* the number of possible patterns. One exception to the reduction in the number is the case when neither mechanism can produce pattern on its own but coupling them together results in a pattern (Shaw & Murray 1990; Cruywagen & Murray 1991). Equally important is the fact that these simulations suggest that the patterns generated by these coupled systems are highly robust, considerably more so than the patterned solutions generated by either mechanism on its own. That tissue interaction can have both a simplifying effect, by reducing the possible patterns, and a stabilising effect could have a profound affect on our thinking of how pattern development is effected during embryogenesis.

Although with spatially homogeneous oscillators, the problem of period locking is well known, spatial pattern locking is an interesting and new phenomenon. The concept of basins of attraction in spatial pattern generators was introduced and discussed in more detail by Murray (1991) at the 1st European Conference on Mathematics Applied to Biology and Medicine (Grenoble, 1991). It is also discussed in a wider context by Goodwin *et al.* (1993).

1.2   PATTERN FORMATION MECHANISMS

Broadly speaking, there are two prevailing views of pattern generation that have influenced the thinking of embryologists in the past ten years. One is the long standing and well known Turing (1952) chemical prepattern approach, and the other is the more recent continuum mechanochemical approach developed by G. F. Oster and J. D. Murray and their colleagues (see, for example, Murray *et al.* 1983; Oster *et al.* 1983; Murray & Oster

1984a, 1984). General descriptions and overview of pattern formation mechanisms are given in the book by Murray (1989).

Turing's (1952) theory of morphogenesis involves hypothetical chemicals - morphogens - which react and diffuse in such a way that if the chemical kinetics and the diffusion coefficients have certain properties, steady state heterogeneous spatially patterned solutions in chemical concentrations can evolve. Morphogenesis then proceeds by the cells reading and reacting to the chemical prepattern and differentiating according to some 'bauplan', such as Wolpert's (1981) 'positional information' concept. This considers the cells to have been pre-programmed to differentiate according to the underlying morphogenetic prepattern. This views morphogenesis as essentially a slave process, since once the chemical prepattern is established all else follows. Turing's theory has stimulated a vast amount of research, both mathematical and experimental. Such reaction diffusion models have been widely studied and applied to a variety of biological problems; see, for example, Meinhardt (1983); Murray (1989); Murray & Myerscough (1991) and papers in this volume. Recent experimental work by de Kepper *et al.* (1991) has dramatically demonstrated these spatial structures in the context of chemical reactions. Interestingly the scale of the patterns displayed is commensurate with what is required in developmental situations. A major unsolved problem with this approach, as it applies to limb development, for example, is that the form-shaping events which are required as the limb evolves are not addressed. Clear demonstration of the existence and identification of morphogens is still a major problem, although retinoic acid is now definitively implicated in cartilage formation in the chick limb (see, Summerbell, this volume). There is, however, no disagreement that chemicals play important roles in embryogenesis and the experimental work by Wolpert & Hornbruch (1990) on limb cartilage formation suggests that there may indeed be a chemical prepattern. Interestingly this latter research was "provoked" by the mechanochemical theory of limb cartilage formation! Diffusion plays a crucial role here and how morphogens might control pattern is important (see, for example, Dillon and Othmer, this volume).

The mechanochemical models of morphogenesis take a quite different approach by directly bringing mechanical forces and known properties of cells and biological tissue into the process of morphogenetic pattern formation. Here pattern formation and morphogenesis go on simultaneously as a single process. The form-shaping movements of the cells and the embryological tissue interact continuously to produce the observed pattern. An important aspect of this approach is that the models are formulated in terms of measurable quantities such as cell densities, elastic forces, cell traction forces, tissue deformation, known chemicals and so on. This focuses attention on the morphogenetic process itself and is more amenable to experimental investigation.

Another class of models rely on chemotaxis and the response of cells to gradients in the chemoattractant. Theories here have been applied with considerable success to various developmental situations, the most widely studied of which is the life cycle of the slime mould *Dictyostelium* (see, for example, Soll *et al.* , Schaap & Wang and Williams & Joss, this volume).

The principal use of any theory is in its predictions and, even though each theory might be able to create similar patterns, they are mainly distinguished by the different experiments they suggest. A major point in favour of simultaneous development is that such mechanisms have the potential for self correction. Embryonic development, which proceeds sequentially, is usually a stable process with the embryo capable of adjusting to many outside disturbances. The process whereby a prepattern exists and then morphogenesis takes place is effectively an open loop system. These are potentially unstable processes and make it difficult for the embryo to make the necessary corrective adjustments as development proceeds.

1.3   GENERAL PROPERTIES OF PATTERN FORMATION MODELS

Although there are major differences in the assumptions made in each of these mechanisms for pattern formation, they all share similar characteristics. They all require what is in effect some form of local activation and lateral inhibition. If we group the various variables, morphogens, cells, extracellular matrix, and so on as "morphogenetic variables", denoted generally by **m**, then from a modelling viewpoint we have for most pattern generator mechanisms

| rate of change morphogen concentrations | = | local dynamics of the morphogen interactions | + | spatial effects (long and short range diffusion, convection, chemotaxis, galvanotaxis, haptotaxis...) |

When these local and spatial effects are acting simultaneously they must conspire to give some kind of local activation of **m** and spatial lateral inhibition as sketched in Figure 1.1.

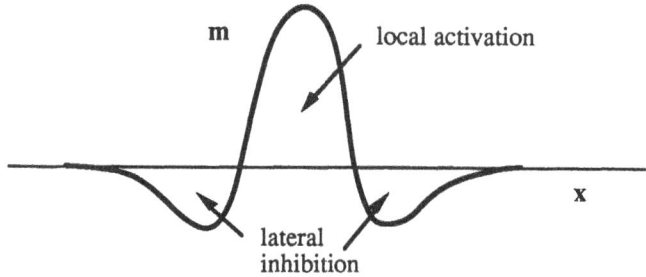

Figure 1.1. Concept of spatial local activation and lateral inhibition which produces spatial patterns in the concentration of morphogenetic variables **m**.

To see how this can produce spatial patterning let us consider a specific mechanism, say one involving cells which produce their own chemoattractant, which is a model used, for example, by Oster & Murray (1989) in their discussion of developmental constraints and by Maini *et al.* (1991) in their study of complex patterns formed on growing domains. Let us suppose that an increase in cell concentration results in more chemoattractant produced, that cells tend to move up a concentration gradient, that both the chemoattractant and cells also diffuse, a process which tends to reduce spatial heterogeneities and that cells and chemoattractant decay according to first order kinetics. Suppose we start with a spatially uniform state and let us suppose that a small heterogeneity in cell density occurs. Once a region of increased cell density appears it produces more chemoattractant, thereby creating a local concentration gradient, and thus induces cells to move up this concentration gradient: this tends to enhance the spatial cell aggregation. Diffusion, of course, tends to decrease the spatial heterogeneity. If there is no cell mitosis such a process decreases the cell density laterally in the vicinity of the aggregation. Far enough away from the aggregation of cells another aggregation can be initiated. It is only if the aggregative (chemotaxis) effect is greater than the dispersal effect that a pattern will form. This introduces the idea of a threshold. It turns out that if the various parameters associated with the model mechanism are in a certain relation and exceed some threshold value, patterns will be produced. Such a process also introduces the concept of a pattern wavelength - the distance between aggregations.

1.4 CARTILAGE PATTERNING IN THE DEVELOPING LIMB: MORPHOGENETIC LAWS

Mathematical models can pose highly relevant biological questions and, *vice versa*, biological problems can pose fascinating mathematical questions. Here I describe one example which shows how the study of model mechanisms for pattern formation can suggest possible and real developmental scenarios for specific developmental processes and hence, in conjunction with experimentalists who confirmed the predictions, further our biological understanding of a developmental process.

As mentioned above, an important feature of morphogenetic patterns is that they are frequently laid down sequentially. In the developing forelimb of the salamander (which is a typical vertebrate limb skeletal structure) the humerus, marked H in Figure 1.2, is laid down first, then the radius (R) and ulna (U) and so on but with considerable increasing complexity. Reaction-diffusion, cell-chemotaxis and mechanochemical models, all of which have been

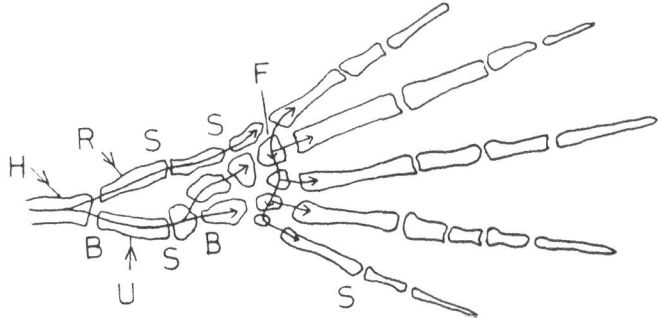

Figure 1.2. Schematic diagram of the forelimb of a salamander. The lines (with arrows) passing through the bones show how the cartilage patterns are built up from sequences of focal (F), branching (B) and segmental (S) bifurcations illustrated in Figure 1.3.

suggested as the primary pre-cartilage patterning mechanism, can display such a sequence of patterns. Limb cartilage patterning is a major area of research for experimentalists and theoreticians alike (see several papers in this volume). However, we still do not yet know enough about the processes involved to say which of the various model mechanisms is the more likely. What we describe briefly here is another aspect of modelling which has not been exploited as much as it perhaps could and should, namely the discovery of fundamental rules or laws which govern certain patterning situations. It crucially requires a close collaboration between experimentalists and theoreticians. We shall demonstrate not only why certain morphologies are unlikely but what the possible sequential scenarios are for constructing the complex architecture in the developing limb. This is intimately related to the important concept of 'developmental constraints' (Oster & Murray 1989) of which there are singularly few accepted extant examples: one occurs in animal coat patterns (Murray 1981). The work, which we briefly describe below, is discussed in detail in Oster *et al.* (1988) and Murray (1989).

Let us suppose that geometry and scale are the parameters which vary in the developing limb bud and see how these can effect pattern variation. The following discussion is based on a detailed mathematical analysis of the model equations for all of the above mentioned mechanisms. Consider the developing limb bud shown schematically in Figure 1.3. We can

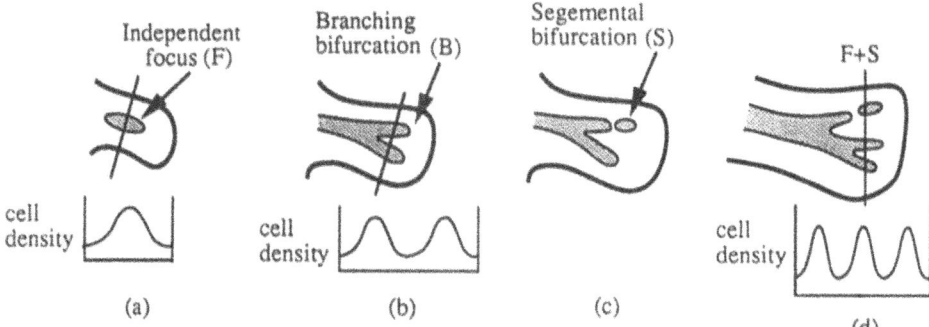

Figure 1.3. The three types of cell condensations that lead to cartilage formation. In (a) the limb width (at the cross section at the line crossing the limb bud) is sufficient for only one cell aggregation - a focal condensation (F) - to occur. As the limb bud widens with growth, a branching bifurcation (B) can be fitted in as in (b). With further elongation a segmental bifurcation (S) becomes possible as in (c). In (d) we see how a cross-section with three patterns is formed with the set of morphogenetic rules based on only the three bifurcations F, B and S.

think of cell aggregations as a 'wave' pattern in cell density which can fit into the domain. The region of higher density of cells become chondrocytes and eventually cartilage. The cross-sectional domain size at AB in Figure 1.3(a) is just sufficient to fit in one wave. We call this a *focal* condensation (F). It is simply an aggregation of cells which forms when there is sufficient space and enough cells to create it: if the cross-section is too small no pattern can be initiated. This condensation recruits cells as the limb grows and eventually becomes the humerus: see Figure 1.2.

With further limb bud growth, a stage is reached when the cross-sectional area increases sufficiently so that two waves can fit into the region as shown in Figure 1.3(b). (With the mechanical theory such a deformation can directly be effected by the aggregative process itself.) The change from a focal aggregation to this Y-form we call a *branching bifurcation* (B). The gaps for joint formation form at a later stage, a precursor of which is the separation of the arms of the Y. The mathematical analysis indicates that in practice it is not possible to have a limb pattern development which goes from two aggregations to one: this is an example of a developmental constraint which is imposed by geometry, scale and the pattern formation mechanism. Another important pattern initiation possibility comes from limb growth without increase in width. Here a focal pattern or one of the legs of a branching pattern can break off longitudinally as shown in Figure 1.3(c): we call this a *segmental bifurcation* (S).

At this stage we might think that the spatial bifurcations simply increase in complexity as the limb bud grows. It should theoretically be possible, for example, to have a transition from one aggregation to three. The model analyses predict, however, that the combination of parameters necessary to produce a trifurcation would have to lie within a very narrow range of values or is simply not possible with a continuous change of shape. Since biological pattern formation is a robust process, such sensitivity is highly unlikely. Instead the model suggests that when three aggregations are formed as in Figure 1.3(d) they arise from a combination of a branching bifurcation and a segmental bifurcation.

1.4.1 Bifurcation pattern sequences suggest a theory of limb architecture morphogenesis

Although there is enormous diversity in limb morphology cartilage formation, in most vertebrate limbs there is a certain similarity of structural organisation. From the mathematical study of the two principal theories of pattern formation we reach the same conclusion regarding the sequence of bifurcating patterns and which encapsulates one of our major theoretical hypotheses. We postulate in Oster *et al.* (1988) that all vertebrate limbs could form via a combination of only the three bifurcation possibilities described above, namely focal, branching and segmental bifurcations. Experiments (see Oster *et al.* 1988) on a variety of limbs indicate that, at least with the quite different species studied in detail, this bifurcation scenario is borne out. The result therefore of this study was the discovery of general *morphogenetic rules* or *laws* for limb architecture formation.

If the early limb, for example that of the salamander (*Ambystoma*) shown in Figure 1.3, is treated with the chemical colchicine which reduces the dermal cell number, the final limb sometimes looks like the paedomorphic, or early embryonic, form *Proteus* which has fewer digits. This suggests that *Proteus* and *Ambystoma* shared a common developmental mechanism. It is thus neither a relevant nor answerable question to ask which specific digits are lost by a reduction in cell density or as a result of evolution. The *developmental constraints* imposed by a decrease in cell density simply limits the number of aggregation centres possible. This has significant evolutionary implications (Oster *et al.* 1988).

1.5 FUTURE NEEDS AND POSSIBLE DIRECTIONS

There are numerous other examples where theory and experiment have influenced each other in major ways, as many of the articles in this volume make clear. One important point which arose as a consequence of the mathematical study of models is that there is no unique effect (like mitosis, chemotaxis and so on) which is responsible for the patterns obtained. Effects are a direct result of the combination of effects and are reflected in values of dimensionless groups of parameters. Thus models can suggest the probable outcomes of varying given physical parameters and hence indicate what type of experiment would be useful in helping to discover the basis for the underlying mechanism.

Modelling in morphogenesis has reached the stage where several different mechanisms, based on quite different biological assumptions, can generate the observed patterns in many biological situations of interest. The question is how to distinguish between them so as to determine which may be the relevant mechanism *in vivo*. These different models, or explanations, for how pattern arises suggest quite different experiments and will lead to a greater understanding of the biological processes involved. The final arbiter of a model's correctness is not so much in what patterns it generates (although a first necessary condition for any such model is that it must be able to produce biologically observed patterns) but in how consistent it appears in the light of subsequent experiments and observations.

In the case of reaction diffusion models the unequivocal identification of morphogens is crucial. Until this has been done it is difficult to test reaction diffusion models experimentally. On the other hand, with mechanical models, which involve cell densities, real tissue and visible deformations it is possible to change several of the real parameters. For example, it is possible to reduce the number of cells (by radiation or appropriate chemical treatment). In the case of the formation of feather germ primordia the mechanical model predicts that spacing between patterns will increase as cell density is decreased and this has been borne out by experiments (Davidson 1983).

Mechanical and mechanochemical models lend themselves to experimental scrutiny more readily than reaction diffusion models. It is likely that both types of models are involved in development, but until more is known about the morphogens involved, it seems that, at this stage, mechanical models can indicate experimental activity to elucidate the underlying mechanisms involved in morphogenesis in a more productive way. The case with cell-chemotaxis models is also encouraging since one of the key chemicals (cAMP) is known. With recent experiments on *E.coli* (Berg & Budrene 1991) there is some hope from recent collaboration between theoreticians and experimentalists that a realistic model mechanism can be produced which reflects the underlying biological process.

The encouraging results described above (and many others not described here) obtained from a mathematical approach to pattern formation suggests that it might be useful and informative to investigate other areas where known mechanism elements, such as cell traction in the mechanical theory, play a key role. One of these, for example, is wound healing. In the case of burns epidermal cells at the wound site appear to adopt dermal cell characteristics and are capable of exerting large traction forces, which exert forces at the wound edges. These large traction forces can cause puckering of the skin and lead to severe scarring and disfigurement. This process could reasonably be modelled using the mechanochemical approach, with a view to trying to minimise the traction-caused puckering, either by artificial or other means, suggested by the model. Mathematically this is a formidable problem but the potential practical rewards justify a detailed study. Considerable research is currently ongoing in this area by the authour and his group. Some first attempts have been reported (Tranquillo & Murray 1992).

In conclusion it should be clearly stated that a crucially important aspect of this research is the interdisciplinary content. Also there is absolutely no way mathematical modelling could ever solve such biological problems on its own. On the other hand I believe it is unlikely that even a reasonably complete understanding could come solely from experiment.

ACKNOWLEDGMENTS This work was in part supported by Grant DMS-9106848 from the U.S. National Science Foundation.

REFERENCES

Berg, H., & Budrene, E. 1991. Complex patterns formed by motile cells in E. coli. *Nature*, **349**, 630–633.

Cruywagen, G. C., & Murray, J. D. 1991. A new tissue interaction model for epidermal-dermal spatial patterns. In *Proc. 1st European Conference on the Applications of Mathematics to Medicine & Biology*, Demongeot, J., & Capasso, V. (eds), Heidelberg, 1991. Springer-Verlag. (In press).

Davidson, D. 1983. The mechanism of feather pattern formation in the chick. I The time of determination of feather position. II Control of the sequence of pattern formation. *J. Embryol. Exp. Morph.*, **74**, 245–273.

de Kepper, P., Castets, V., Dulos, E., & Boissonade, J. 1991. Turing-type chemical patterns in the chlorite-iodide-malonic acid reaction. *Physica*, **D 49**, 161–169.

Goodwin, B. C., Kauffman, S., & Murray, J. D. 1993. Is morphogenesis an intrinsically robust process? *J. Theor. Biol.* (In press).

Hinchliffe, J. R., Hurle, J. M., & Summerbell, D. 1991. *Developmental patterning of the vertebrate limb*. New York: Plenum.

Maini, P. K., Myerscough, M. R., Winters, K. H., & Murray, J. D. 1991. Bifurcating spatially heterogeneous solutions in a chemotaxis model for biological pattern formation. *Bull. Math. Biol.*, **53**, 701–719.

Meinhardt, H. 1983. *Models of biological pattern formation*. London: Academic Press.

Murray, J. D. 1981. On pattern formation mechanisms for lepidopteran wing patterns and mammalian coat markings. *Phil. Trans. Roy. Soc. (Lond.)*, **B295**, 473–496. (Also in the book of the Proceedings of the Royal Sociey Meeting on Theories of Biological Pattern Formation. March 1981.).

Murray, J. D. 1989. *Mathematical biology*. New York.: Springer-Verlag.

Murray, J. D. 1991. Complex pattern formation and tissue interaction. In *Proc. 1st European Conference on the Applications of Mathematics to Medicine & Biology*, Demongeot, J., & Capasso, V. (eds), Heidelberg, 1991. Springer-Verlag.

Murray, J. D., & Myerscough, M. R. 1991. Pigmentation pattern formation on snakes. *J. Theor. Biol.*, **149**, 339–360.

Murray, J. D., & Oster, G. F. 1984a. Cell traction models for generating pattern and form in morphogenesis. *J. Math. Appl. in Medic. & Biol.*, **1**, 51–75.

Murray, J. D., & Oster, G. F. 1984b. Generation of biological pattern and form. *IMA J. Math. Biol.*, **19**, 265–279.

Murray, J. D., Oster, G. F., & Harris, A. K. 1983. A mechanical model for mesenchymal morphogenesis. *J. Math. Biol.*, **17**, 125–129.

Murray, J. D., Deeming, D. C., & Ferguson, M. W. J. 1990. Size dependent pigmentation pattern formation in embryos of Alligator mississippiensis : time of initiation of pattern generation mechanism. *Proc. Roy. Soc. (Lond.)*, **B239**, 279–293.

Nagorcka, B. N. 1986. The role of reaction-diffusion system in the initiation of skin organ primordia. I. the first wave of initiation. *J. Theor. Biol.*, **121**, 449–475.

Nagorcka, B. N., Manoranjan, V. S., & Murray, J. D. 1987. Complex spatial patterns from tissue interactions — an illustrative example. *J. Theor. Biol.*, **128**, 359–374.

Oster, G. F., & Murray, J. D. 1989. Pattern formation models and developmental constraints. *J. Exp. Zool.*, **251**, 186–202.

Oster, G. F., Murray, J. D., & Harris, A. K. 1983. Mechanical aspects of mesenchymal morphogenesis. *J. Embryol. Exp. Morph.*, **78**, 83–125.

Oster, G. F., Shubin, N., Murray, J. D., & Alberch, P. 1988. Evolution and morphogenetic rules. The shape of the vertebrate limb in ontogeny and phylogeny. *Evolution*, **42**, 862–884.

Shaw, L. J., & Murray, J. D. 1990. Analysis of a model for complex skin patterns. *SIAM J. Appl. Math.*, **50**, 628–648.

Tranquillo, R. T., & Murray, J. D. 1992. Continuum model of fibroblast-driven wound contraction: inflammation mediation. *J. Theor. Biol.*, **158**, 135–172.

Turing, A. M. 1952. The chemical basis for morphogenesis. *Phil. Trans. Roy. Soc. Lond.*, **B237**, 37–72.

Wolpert, L. 1981. Positional information and pattern formation. *Phil. Trans. R. Soc. Lond.*, **B295**, 441–450.

Wolpert, L., & Hornbruch, A. 1990. Double anterior chick limb buds and models for cartilage rudiment specification. *Development*, **109**, 961–966.

2. PROCESS AND OUTCOME: EVOLUTIONARY ANALYSIS OF MORPHOLOGICAL PATTERNS

P. Alberch and M. J. Blanco

Museo Nacional de Ciencias Naturales (CSIC)
J. Gutierrez Abascal, 2
28006 Madrid Spain

2.1 INTRODUCTION

Mathematical biologists often construct models to describe the emergence of a *specific global* pattern as a result of a series of local dynamical interactions. The pattern to be explained is the one observed in nature and the interactions postulated need to be plausible, *i. e.* in agreement with known physico-chemical processes. In general, the model must be "robust", in the sense that the resultant pattern should not be too sensitive to perturbations in parameter values.

This approach clearly makes sense, since invariance is one of the key features of biological systems in general, and of developmental ones in particular. For example, within a given species, an overwhelming majority of embryos end up with the same basic morphology in spite of differences in genetic make-up - which are translated into variation of parameter values (*e. g.* enzymatic activities, diffusion rates, membrane adhesiveness, etc.) - or even of small differences in environmental conditions (*e. g.* temperature of incubation of the egg). In fact, one of the most striking features of developmental systems is their resilience against environmental insult.

The counterpart of invariance is change; the latter being of critical importance in an evolutionary context. Developmental systems when subjected to genetic mutation or environmental perturbation sometimes do generate outcomes qualitatively different from the norm. In mathematical jargon, this would correspond to a bifurcation event in a dynamical system. Alberch (1980; 1989) and Oster & Alberch (1982) have reviewed the properties of systems of morphological evolution and their relationship to pattern formation models. They argued that the morphological variation generated within a population, or within a species, is highly ordered and that such ordered structure is a reflection of the properties of the underlying pattern-generating system.

The vertebrate limb has traditionally been a favorite subject for mathematical modeling of processes of pattern formation due to their relatively complex morphology. At the same time, the possibility to reduce the system to two dimensions has made it more amenable to analysis (see review by Hinchliffe & Johnson 1980).

Here, we will briefly review some recent, and mostly unpublished research done in collaboration with G. Wagner. We introduce empirical data on the intraspecific patterns of

Experimental and Theorietcal Advances in Biological Pattern Formation,
Edited by H.G. Othmer *et al.*, Plenum Press, New York, 1993

11

carpal variation in several species of newts of the genus *Triturus*. We integrate this analysis of patterns of variability with a developmental interpretation and within a phylogenetic context. Our goal is to illustrate that one must not only deal with the nature of processes controlling the emergence of patterns but must also integrate the issue of pattern formation with the genesis of variability.

2.2   CARPAL VARIATION IN THE GENUS *TRITURUS*

To illustrate the properties of the variability found in morphological patterns, we have used the newts of the genus *Triturus*. This group has recently been subject to a phylogenetic study based on an integration of data sets, that in addition to morphology, include molecular, chromosomal, or even, behavioural information (Macgregor *et al.* 1990). This work provides us with a reliable hypothesis of cladistic relationships within which analysis of the observed patterns of variability can be carried out.

2.2.1   Developmental Information

Shubin & Alberch (1986) emphasized that the pattern of the vertebrate limb develops sequentially. Chondrogenetic condensations appear in a characteristic spatial and temporal pattern. In agreement with the mechanistic models of Oster *et al.* (1985), Shubin & Alberch (1986) observed three basic spatial patterns of chondrogenesis: 1. *"de novo"* formation of cellular condensations; 2. "branching" of an existing chondrogenetic foci; or 3. "segmentation" of a chondrogenetic foci into two or more elements. Blanco & Alberch (1992) have recently published a detailed analysis of the development of the skeletal elements of the carpus in the marbled newt, *Triturus marmoratus*. Figure 2.1 illustrates a sequence of states in the carpal development of this species.

Such developmental information can be abstracted and depicted using the method proposed by Shubin & Alberch (1986) to describe the patterns of "embryonic connectivity" (Figure 2.2). While in the proximal part the humerus branches into the radius and ulna, in the distal region the first chondrogenic condensation observed is the basale commune (Figure 2.2A). The radius and ulna undergo subsequent segmentations in a proximo-distal direction. The radius splits into the radiale and element Y while the ulna gives rise to the ulnare and distal carpal 4 (Figure 2.2D, E). The basale commune is connected proximally with a central rod that segments into the centrale and intermedium (Figure 2.2E), and laterally with the distal carpal 3 (Figure 2.2C). In summary, developmental patterns in the carpus involve a well-established sequence of branching, segmentation and *de novo* condensations, schematically depicted in Figure 2.2 (see Shubin & Alberch (1986) and Blanco & Alberch (1992) for more detailed descriptions).

2.2.2   Adults Patterns of Variation

Wagner *et al.* (1993) have studied the carpal morphology of 678 forelimbs from eight species and subspecies within the genus *Triturus*.

In the genus *Triturus*, the adult carpus consists of seven elements: the basale commune (= distal carpal 1+2), distal carpals 3 and 4, radiale, element Y, centrale, and intermedium + ulnare fused in a single element (Figure 2.3A).

The analysis of basipodial variation revealed changes in the typical pattern based on the appearance of 'fusion' between adjacent elements, separation of usually fused elements

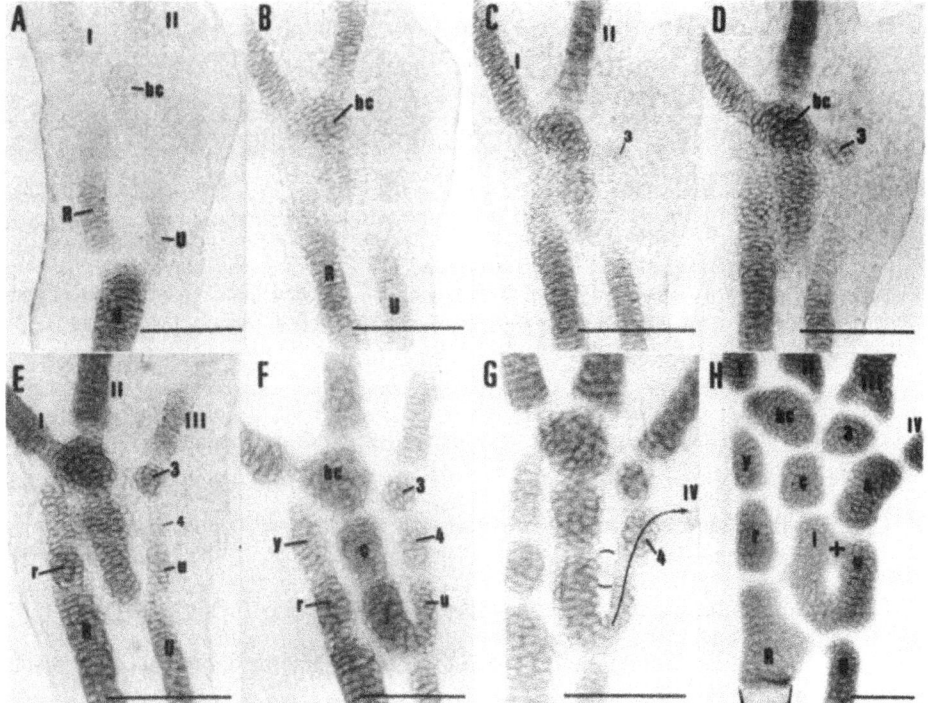

Figure 2.1. Limb buds of *Triturus marmoratus* that have been cleared and stained with Alcian blue to show the distribution of condensations of chondrocytes (cartilage cells) at various stages of development. **A**. Earliest stage illustrated, shows the condensation of the basale commune (**bc**) and metacarpals I and II in the distal region while in the proximal area the radius (**R**) and ulna (**U**) differentiate. **B**. At this stage the **bc** is connected to a chondrogenic condensation that develops from a distal to proximal direction. **C**. The **bc** shows a connection with the differentiating third distal carpal element (**3**). **D**. Close up the same stage showing this connection, so called, "embryonic connection". **E**. The precartilaginous condensation distal to **R** gives rise to the radiale (**r**) and the element y (**y**). Note that the ulnare (**u**) and the carpal distal 4 (**4**) first appear as a single cartilaginous rod connected to **U**; **F**. Close up the same stage showing the condensation proximal to the **bc**, which will give rise to the centrale (**c**) and intermedium (**i**). **G**. A secondary fusion between the element **i** and **u** becomes evident as well as the close relation of the element **4** with the metacarpal 4 (**IV**). **H**. Later stage illustrating a fully formed carpus with all the elements well differentiated. Bars = 0.2 mm.

and supernumerary elements (Table 2.1). Eight types of variation were identified and described as follows (we have used the terminology proposed by Shubin & Alberch 1986):

- f(1+2+3) = distal carpals 1, 2 and 3 form a single element (Figure 2.3B)

- f(3+4) = distal carpals 3 and 4 form one element (Figure 2.3C)

- (r+Y) = the prechondrogenic rod that splits to from the radiale and the element Y fails to segment, thus giving rise to a single element (Figure 2.3D)

- f(ui+c) = the ulnare + intermedium and the centrale appear as a single element (Figure 2.3E)

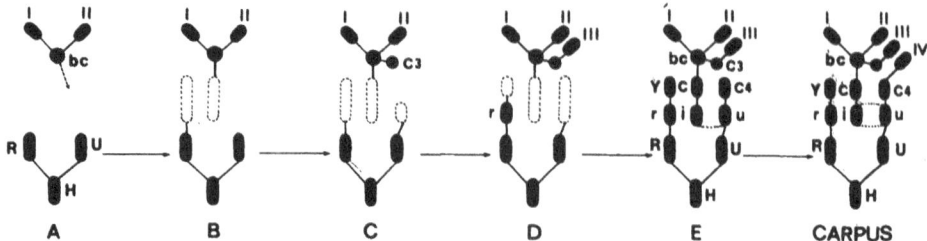

Figure 2.2. Schematic depiction of the branching and segmentation events characteristic of carpus development in *Triturus marmoratus* based on Figure 2.1. Dotted lines depict prechondrogenic condensations, and the line linking **i + u** (intermedium + ulnare) in **E** indicates secondary fusion (from Blanco & Alberch 1992). Abbreviations as in Figure 2.1.

- f(ui + 4) = the distal carpal 4 and the ulnare + intermedium appear as a single element (Figure 2.3F)

- s(ui) = the ulnare and intermedium fail to secondarily fuse, thus appearing as two distinct elements (Figure 2.3G)

- f(ui+r) = the ulnare + intermedium is fused with the radiale (Figure 2.3H)

- sn = an extra element appears between the ulnare + intermedium and carpal distal 4 (Figure 2.3I)

2.2.3   Statistical Analysis of Variability

Our analysis was based on variation in individual limbs rather than whole animals. The reason was the lack of correlation between variations in the left and right limb.

Comparison of the frequencies reported in Table 2.1 was performed by likelihood ratio tests (G - test; (Sokal & Rohlf 1981), see Wagner *et al.* (1993) for further details).

Of the 678 carpi from 8 (sub-)species included in this study, 8.7% deviate from the typical pattern. The percentage of variant morphologies ranges from 0% in *T. vulgaris* (N-32) to 15.8% in *T. marmoratus marmoratus* (N=166). Excluding the small samples of *T. vulgaris* and *T. montadoni* the remaining differences in the level of variation, however, are not significant (G=8.5835, dgf=5, n.s.). Hence the samples from six taxa are homogeneous with respect to the overall level of variation.

The most common carpal variant, which has been found in samples from all taxa is the fusion between distal carpals three and four, f(3+4). The overall frequency of f(3+4) is 5.5%. The f(3+4) frequency is inhomogeneous among the taxa (G=19.638, dgf=5, a<0.005). However, the inhomogeneity is only due to a very high frequency of f(3+4) found in one local population of *T. m. marmoratus* from Santander (northern Spain; Rienesl & Wagner 1992). Excluding this sample from the data set leads to an homogeneous distribution of f(3+4) frequency among samples from all taxa (G=7.730, dgf=5, n.s.). Hence, with the exception of one local sample from *T. m. marmoratus*, the frequency of f(3+4) is the same in all the taxa up to the resolution of the test.

Another commonly found variation is the fusion between the radial and the element Y, f(r+Y). This fusion is found in all taxa except *T. boscai*. In *T. boscai*, the absence of f(r+Y) can be explained by sample size (P(¬f(r+Y)/N = 79) =0.142). Among the other taxa, where f(r+Y) has been found, the frequency of f(r+Y) is homogeneous (G=3.183, dgf=4, n.s.).

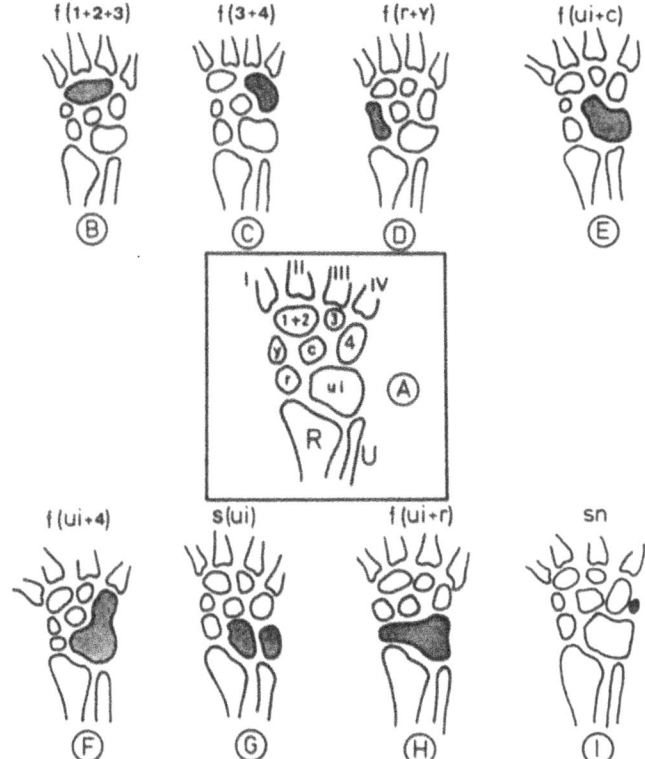

Figure 2.3. Normal carpal morphology characteristic of all species studies (**A**). We encountered the following variants: **B**: "fusion" of distal carpals 1,2 and 3; **C**: "fusion" of distal carpals 3 and 4; **D**: "fusion" of the radiale and element Y; **E**: "fusion" of the intermedium + ulnare and centrale; **F**: "fusion" of intermedium + ulnare and carpal distal 4; **G**: "separation" of intermedium and ulnare; **H**: "fusion" of intermedium + ulnare and radiale; **I**: an extra element appears. Abbreviations as in Figure 2.1.

The most conservative interpretation of the data is that there is no significant difference in the tendency to form a fusion between r and Y.

Fusions between the carpal elements, which consist of the usually single element composed by the ulnare and intermedium, ui, and the radiale, f(ui + r), have been observed only on *T. cristatus* and *T. m. marmoratus*. The frequency of this fusion in these samples is low (p(f(ui+r)) = 0.017) and the absence off the fusion from the other samples is explained by the restricted sample size (P(¬f(ui+r)/N)) is between 0.056 for *T. m. pygmaeus* and 0.25 in *T. boscai*.

Significant differences between the taxa have been found with regard to the fusion between ui and the centrale c, f(ui+c). In *T. m. pygmaeus* and *T. helveticus* the frequency of f(ui + c) is 0.0332. Between these taxa this frequency is homogeneous (G=0.1505, dgf=1, n.s.). f(ui+c) has also been found in *T. boscai* once and could not be included into the test. The absence of f(ui+c) from *T. cristatus* samples is not significant (P=0.0903), but is significant in *T. m. marmoratus* (P=0.033) and *T. alpestris* (P=0.048). Hence, the variation patterns in *T. m. marmoratus* and *T. alpestris* is significantly different from all the taxa where f(ui +c) has been found. In the *T. cristatus* sample the absence of f(ui+c) is not significant. This statistical analysis is summarized in Table 2.2.

Table 2.1. Distribution of types of carpal variation in the genus *Triturus*

Type of variation	T.alp. Nc=90	T.helv. Nc=107	T.bosc. Nc=79	T.m.marm. Nc=101	T.m.pygm. Nc=166	T.cirs. Nc=71	T.mon.* Nc=32
f(3+4)	2	1	3	14	10	2	3
f(r+Y)	2	5	-	1	2	2	-
f(ui+c)	-	3	1	-	6	-	1
s(ui)	-	2	-	-	1	-	1
f(ui+4)	1	1	-	-	-	-	-
f(ui+r)	-	-	-	1	-	1	-
f(1+2+3)	-	-	-	-	1	-	-
sn	-	-	1	-	-	-	-

* Note that *Triturus vulgaris* is absent from the table because none of the limbs analyzed showed any type of variation. Abbreviations: NC = number of carpi; T.alp. = *Triturus alpestris*; T.helv. = *Triturus helveticus*; T.bosc. = *Triturus boscai*; T.m.marm. = *Triturus marmoratus marmoratus*; T.m.pygm. = *Triturus marmoratus pygmaeus*; T.cris. = *Triturus cristatus* and T. mon. = *Triturus montandoni*.

Table 2.2. Summary statistics of carpal variation from the six species with larger sample size.

SPECIES	N	P(T)	f(3+4)	f(r+Y)	f(ui+r)	f(ui+c)
T. cristatus	71	0.92	+	+	+	(-) n.s.
T. m. marmoratus	101	0.89	+	+	+	-, a < 5%
T. m. pygmaeus	164	0.92	+	+	(-) n.s.	+
T. alpestris	90	0.96	+	+	(-) n.s.	-, a < 5%
T. boscai	80	0.94	+	(-) n.s.	(-) n.s.	+
T. helveticus	107	0.90	+	+	(-) n.s.	+

N is the number of carpi examined; P(T) is the relative frequency of typical carpi in the sample: **f (a+b)** is the fusion between elements a and b; + means that this fusion has been found in the respective sample; (-) **n.s.** means that the fusion not been found, but its absence can be due to the limited sample size given the frequency of its occurrence in other samples; **-, a < 5%** means that the fusion is absent from the sample and its absence can not be explained by the limited sample size at the 5% level of significance assuming that its occurs in the population with about the same or higher frequency as in other samples.

2.2.4   Phylogenetic Analysis

A phylogenetic analysis of this difference in variation patterns (see Table 2.1) was done by superimposing our data on the cladogram of Macgregor and coworkers (Macgregor

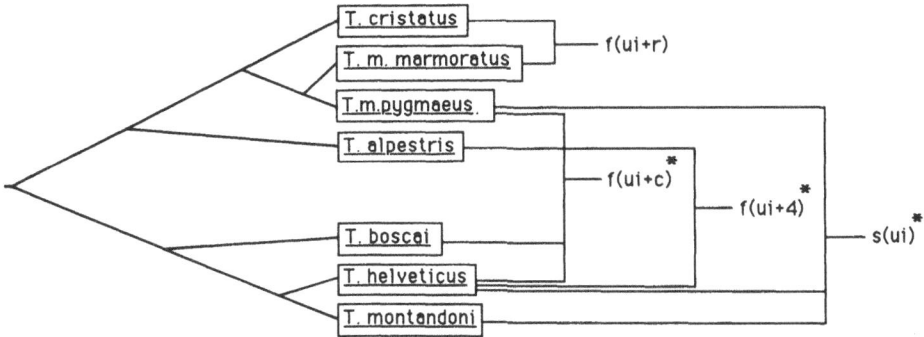

Figure 2.4. Phylogenetic relationships among the species studied (adapted from Macgregor *et al.* 1990). We superimpose on the cladogram the types of morphological variation encountered. **f(r-Y)** and **f(3+4)** were not included because they are ubiquitous. The variants **sn** and **f(1+2+3)** appear only in one species each and have also been excluded. **f(ui +r)** links the large sized, and phylogenetically related *T. marmoratus* and *T. cristatus*. We emphasize that *T. marmoratus pygmaeus* exhibits variants **s(ui)**, **f(ui+c)** and **f(ui+4)** which are only found in the small-bodied species: *T. boscai, T. helveticus* and *T. montandoni*. These data suggest that patterns of variability in *T. marmoratus pygmaeus* are correlated with small size indicative of paedomorphosis (all paedomorphic morphologies are emphasized with asterisks).

et al. 1990). We assumed that the two subspecies of *T. marmoratus* are more closely related to each other than to any other taxon in our analysis (Garcia-Paris, pers. comm.) (Figure 2.4).

Four different cases were considered, because of the unknown pattern in *T. cristatus* and the unknown plesiomorphic state of the variation pattern. In all possible scenarios a convergent origin of different variation patterns is inferred. This result is mainly due to the assumed close relationship between the two *T. marmoratus* subspecies and the different variation patterns found in them.

2.2.5   Developmental Interpretation of Patterns of Intragroup Variability

The analysis of the developmental patterns observed during the carpal and tarsal ontogeny of *T. marmoratus marmoratus* (Figure 2.2) (Blanco & Alberch 1992) allows us to make certain predictions on the likelihood of appearance of a given variant (Figure 2.5). For example, we can define intrapopulational paedomorphic variants resulting from failure to undergo "separation events" of chondrogenic condensation at the terminal stages of the ontogenetic sequence (Figure 2.2). *A priori*, one would expect these morphs to be common in appearance given the simple underlying developmental control mechanism, but, also, *i. e.* most likely, to be paedomorphic (see Alberch *et al.* (1979) for a review of heterochrony and, in particular, paedomorphosis).

The fusion f(y+Y) is relatively common and it appears in all of the species studied. These data suggest that all species of *Triturus*, like all urodeles, are paedomorphic to a certain degree and, therefore, express a tendency towards generating paedomorphic traits. The fusion f(ui+c), however, presents a differential, and more insightful, pattern. It represents a highly paedomorphic morphology and it is, therefore, quite striking to have it found in the species characterized by small body size: *T. marmoratus pygmaeus, T. boscai* and *T. helveticus* (Table 2.3). In fact, the pattern is independent of phylogenetic relationship. This is particularly evident when comparing *T. marmoratus marmoratus* and *T. marmoratus pygmaeus*. The small subspecies (Table 2.3) exhibits the variability characteristic of the distantly related small-size newts (*T. boscai* and *T. helveticus*), instead of the one found in the closel related *T. marmoratus marmoratus* (Figure 2.4). The same point is illustrated by another paedomorphic pattern, the failure of the ulnare and intermedium to secondarily fuse, s(ui), which appears in *T. marmoratus pygmaeus* and small-size species such as *T. helveticus* and *T. montadoni*. These data show that developmental information can be used as a predictor of expected patterns of variability which are amenable to empirical testing.

Table 2.3. Approximate adult size of species compared in this study.

SPECIES	N	SVL (MM) (x± s.e)
T. cristatus	N/A	(91-80)*
T. marmoratus marmoratus	38	74.07 ± 7.82
T. marmoratus pygmaeus	49	59.66 ± 5.95
T. alpestris	14	44.37 ± 7.10
T. montandoni	26	40.14 ± 4.22
T. boscai	11	34.55 ± 2.35
T. helveticus	53	35.69 ± 4.33
T. vulgaris	24	28.19 ± 2.55

* The size range for *T. cristatus* is taken from the literature, based on the total length published and a relative SVL of 57%.

One fusion, f(3+4), has been found in homogeneous frequencies in all samples large enough to be tested statistically, and its probability is thus independent of the size of the subspecies. This fusion is not a paedomorphic variant (see Figures 2.2 and 2.5). Therefore, one should not expect it to correlate with the size, as, in fact, is the case.

2.3 CONCLUSION

The results reviewed above convincingly show that not all possible patterns are generated in nature. In fact, some morphologies are much more likely to occur that others. Such structured pattern of variability can be most reliably interpreted from a developmental perspective. That is, the generation of new variation - and the evolution that such variability enables - results from perturbation within a conserved developmental system. This property is what has been referred to as "developmental constraint" on evolutionary processes (Alberch 1982).

From the perspective of the mathematical modeller, the data discussed here suggests that besides focusing on invariance it is also useful to study the behaviour of dynamical systems under perturbation. It is not sufficient to construct a model that is able to generate a particular pattern. One also needs to focus on the kind of variability that such model can generate under parameter perturbation (*e. g.* see Murray (1981) for such kind of approach). A successful model should be able to make testable predictions on what kinds of morphologies are more likely to appear in nature. If evolutionary biology pretends to be something more than

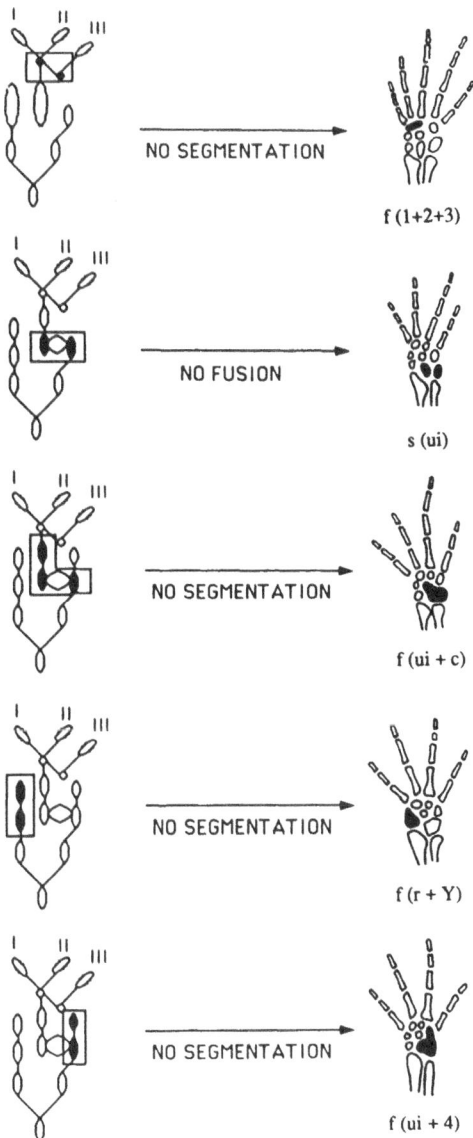

Figure 2.5. Predicted paedomorphic morphologies resulting from the failure of some branching and segmentation events to occur. Also, new patterns may emerge from the absence of secondary fusions. Compare these patterns predicted on the basis of developmental knowledge with the types of variation found in nature (Figure 2.3).

historical reconstruction it needs the incorporation of adequate models of pattern formation to become a full-fledged predictive science.

ACKNOWLEDGMENTS Günter Wagner collaborated with us in this research project. This work was supported by DGICYT # PB 89-0045 grant to P.A. and a postdoctoral fellowship from the Consejo Superior de Investigaciones Científicas (CSIC) to M.J.B.

REFERENCES

Alberch, P. 1980. Ontogenesis and morphological diversification. *Amer. Zoologist*, **20**, 653.

Alberch, P. 1982. Developmental constraints in evolutionary processes. In *Evolution and Development*, Bonner, J. T. (ed). Springer-Verlag, New York: Springer-Verlag. Dahlen Konferenzen.

Alberch, P. 1989. The logic of monsters: Evidence for internal constraint in development and evolution. *Geobios. Mémoire spécial*, **12**, 21.

Alberch, P., Oster, G., Gould, S. J., & Wake, D. B. 1979. Size and shape in ontogeny and phylogeny. *Paleobiology*.

Blanco, J. J., & Alberch, P. 1992. Caenogenesis, developmental variability and evolution in the carpus and tarsus of the marbled newt *Triturus marmoratus*. *Evolution*, **46**, 677.

Hinchliffe, J. R., & Johnson, D. R. 1980. *The development of the vertebrate limb*. Oxford: Oxford University Press.

Macgregor, H. C., Sessions, S. K., & Arntzen, J. W. 1990. An integrative analysis of phylogenetic relationships among newts of the genus *Triturus* (family *Salamandridae*), using comparative biochemistry, cytogenetics and reproductive interactions. *J. Evol. Biol.*, **3**, 329.

Murray, J. 1981. On pattern formation mechanisms for lepidopteran wing patterns and mammalian coat patterns. *Philos. Trans. R. Soc. Lond. (Biol.)*, **295**, 473–496.

Oster, G., Murray, J., & Maini, P. 1985. A model for chondrogenic condensations in the developing limb: The role of extracellular matrix and cell tractions. *J. Embryol. Exp. Morphol.*, **89**, 93.

Oster, G. F., & Alberch, P. 1982. Evolution and bifurcation of developmental programs. *Evolution*, **36**, 444.

Rienesl, J., & Wagner, G. P. 1992. Constancy and change of basipodial variation patterns: A comparative study of crested and marbled newts *Triturus cristatus, Triturus marmoratus* and their natural hybrids. *J. Evol. Biol.*, **5**, 307.

Shubin, N., & Alberch, P. 1986. A morphogenetic approach to the origin and basic organization of the tetrapod limb. In *Evolutionary Biology*, Hecht, M. K., Wallace, B., & Prance, G. (eds). Plenum Press.

Sokal, R. R., & Rohlf, F. J. 1981. *Biometry. the principles and practice of statistics in biological research*. New York: W. H. Freeman and Company.

Wagner, G. P., Blanco, M. J., & Alberch, P. 1993. Convergent origin of constrained variation patterns in the genus *Triturus*. *Amer. Naturalist*. (Submitted).

3. RETINOIC ACID CANNOT BE THE MORPHOGEN IN REACTION-DIFFUSION MODELS FOR THE FORMATION OF THE CHICK WING BUD

Yannis Almirantis and Spyros Papageorgiou

NRC 'Demokritos'
Aghia Paraskevi
Athens, Greece

3.1 INTRODUCTION

In 1968 Saunders & Gasseling discovered that a group of cells at the posterior margin of the chick wing bud has a remarkable morphogenetic potency: if transplanted to the anterior site of a host limb bud, a mirror image duplication of the normal pattern of digits develops (*cf.* Figure 3.1). This group of cells was labeled the 'zone of polarizing activity' (ZPA) and ever since extensive studies have established many interesting features of the digit pattern formation. A simple model was put forward by Tickle *et al.* (1975) which could describe the experimental facts of pattern duplications. This model was based on the assumption that the ZPA is the source of a morphogen which diffuses along the anteroposterior axis. At the same time the morphogen is supposed to decompose via first order kinetics. Asymptotically, a stable morphogen gradient is established which, by appropriate threshold concentrations, can separate the morphogenetic field into bounded areas for digits 4, 3 and 2 of the wing bud (Tickle *et al.* 1975) (*cf.* Figures 3.2 and 3.3).

All-trans-retinoic acid (RA), if locally applied at the posterior margin, can mimic the action of the ZPA grafting (Tickle *et al.* 1982; Tickle *et al.* 1985): instead of the normal digit pattern 234 a mirror-symmetric pattern 432234 emerges. In the meanwhile it was found that endogenous RA is unevenly distributed in the wing bud, with a higher concentration in the posterior domain than in the anterior (Thaller & Eichele 1987). It is therefore tempting to assume that the putative morphogen released by the ZPA is the retinoic acid itself.

In recent years many efforts were directed towards testing this hypothesis. Wanek *et al.* (1991) have performed the following experiment: beads containing RA were implanted in the anterior margin of the chick wing bud (*cf.* Figure 3.4). After a variable exposure time, anterior cells in the vicinity of the implantation site were transplanted in the anterior margin of some other wing bud. It was found that for an exposure time less than 12 hours no extra structures were formed in the second host wing bud. For more than 15 hours of exposure, supernumerary digits gradually appeared and for 24 hours and longer times the pattern was completely mirror-symmetric: 432234 (Eichele *et al.* 1985). Such a behavior is in obvious conflict with the simple model of a local source (ZPA) producing a diffusible morphogen (RA). If the notion of a static source and passive diffusion fail to describe the observed behavior it is natural to ask whether an autocatalytic mechanism can succeed (Bryant

Experimental and Theorietcal Advances in Biological Pattern Formation,
Edited by H.G. Othmer *et al.*, Plenum Press, New York, 1993

21

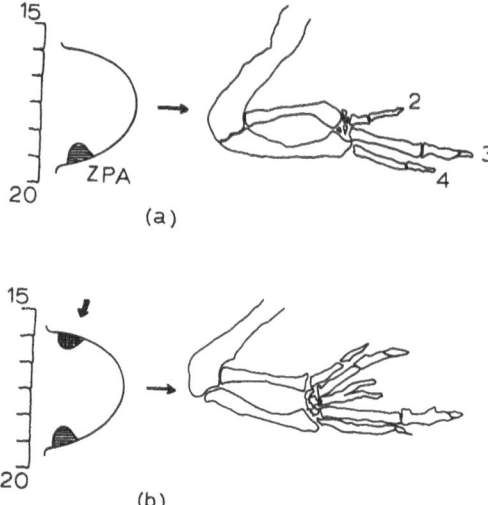

(a)

(b)

Figure 3.1. (a) Left: Dorsal view diagram of a right chick wing bud and adjacent somites. The shaded area stands for the zone of polarized activity. Right: At about 10 days, the skeletal pattern of the wing is shown with the digits 2, 3 and 4. (b) When the ZPA is transplanted at the anterior margin of the bud (left), a supernumerary digital pattern emerges: 432234 (right).

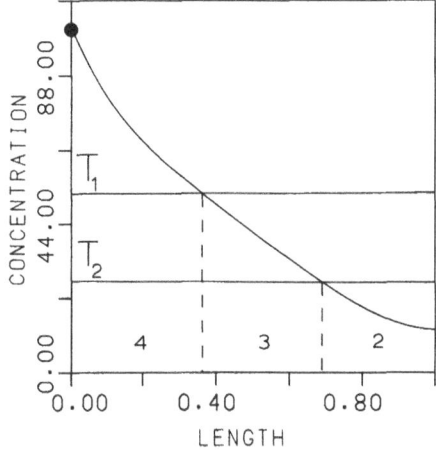

Figure 3.2. Stable morphogen distribution along the anteroposterior axis (length unit of the order of the millimeter). On the left side the morphogen source (ZPA) is fixed arbitrarily at 100. Diffusion constant and rate of decomposition of the morphogen are chosen so that at the anterior end the concentration falls to ten (Tickle *et al.* 1975). The thresholds T_1 and T_2 divide the field in areas of digits (4), (3) and (2).

et al. 1991). More specifically, in a reaction-diffusion framework the morphogen production is not necessarily confined in a specific site - the source. To the contrary, for at least two participating morphogens, their production is self-regulated by feedback responses and

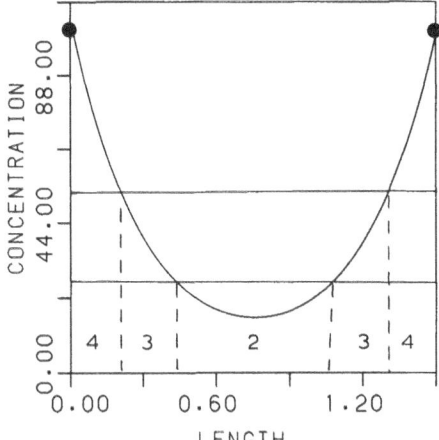

Figure 3.3. A transplanted source at the anterior end of the field produces a pattern duplication (432) $\longrightarrow$ (432234) when the original field is increased to 1.5 length units.

their spatial distribution is autonomously adjusted following the non-linear structure of the particular model in use.

Another set of experiments that we should also compare to the model predictions are the results obtained from RA dose-dependent implantations (Tickle *et al.* 1985): increasing the RA dose released from the microcarrier bead induces gradually the formation of supernumerary digit 2, then digit 3 and finally digit 4 for a complete pattern duplication (432234). If we further increase the doses of RA the following gradual reduction of the digital pattern is observed: 4334, 434, 34, 4, no digits. Notice that no pattern discontinuity is found.

A detailed application of the Gierer-Meinhardt and the Brusselator models has been performed (Papageorgiou & Almirantis 1982) and it was found that these models cannot collectively reproduce the data. However, a negative result for some particular models of the reaction-diffusion theory does not exclude the possibility that some other realization of the theory could accommodate the experimental results. Here we are using the model inspired from the uric acid - oxygen immobilized enzyme - substrate inhibition mechanism. This model was successfully applied by Murray to describe the formation of animal coat patterns (Murray 1982). From the detailed comparison of the different models we can draw some general conclusions concerning the behavior of all autocatalytic systems. It turns out that the experimental data of the chick wing development are incompatible with any simple realization of reaction-diffusion theory where one of the morphogens is the retinoic acid.

3.2 SIMULATIONS BASED ON MURRAY'S MODEL

We apply the following reaction-diffusion scheme put forward by Murray (1982); Murray (1989):

$$\frac{\partial u}{\partial t} = \gamma \, f(u, v) + D_u \frac{\partial^2 u}{\partial r^2}$$
$$\frac{\partial v}{\partial t} = \gamma \, g(u, v) + D_v \frac{\partial^2 v}{\partial r^2} \tag{3.1}$$

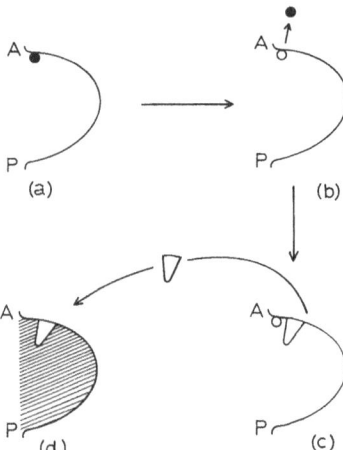

Figure 3.4. Diagram of the Wanek *et al.* (1991) experiment: A bead soaked in a standard solution of RA is implanted in the anterior margin of a chick wing bud (a). The bead is left for 12-24 hours and then it is removed (b). After 90 min, so that the exogenous RA is cleared, a neighboring wedge is excised (c), and grafted anteriorily into an untreated bud (d).

where $f(u,v) = a - u - h(u,v)$, $g(u,v) = \alpha(b-v) - h(u,v)$, and $h(u,v) = \rho uv/(1 + u + Ku^2)$. In the above equations a, b, α, ρ and K are model parameters, while γ is a scale factor characteristic of the geometry and the field domain. The ratio of diffusion constants D_v/D_u is greater than unity. Variables u, v are scaled expressions (Murray 1982) for the chemicals involved in the phenomenological model.

The initial homogeneous state (u_0, v_0) is estimated from the equations $f(u,v) = 0$, $g(u,v) = 0$. For appropriate values of the parameters, a small local disturbance drives the system towards a stable gradient (Figure 3.5). The thresholds T_1 and T_2 separate the field into three domains which correspond to digits 2, 3 and 4. The field length corresponds to the extent of the anteroposterior axis of the bud and it is normalized to unity. We can now simulate the ZPA transplantation to the anterior margin of the bud by changing accordingly the values of u and v on the right side of the gradient of Figure 3.5. The u-distribution that finally is established according to 3.1 is shown in Figure 3.6. Both u and v take up a U-shaped distribution which is interpreted as formation of a mirror symmetric pattern. Growth is an essential element of the developing mechanism in the chick wing. However, it was possible to reproduce the data without taking into account any size increase during pattern formation of the limb. We can therefore ignore the involvement of growth, keeping the simulations intact from any unnecessary complication, since the field growth just facilitates pattern duplication (see next section).

The above results are in agreement with the predictions of the passive diffusion model, where the source was confined in the ZPA region. If an external source (high level of u-concentration) is implanted in the region of low u, the system evolves to a U-shaped pattern similar to the distribution depicted in Figure 3.6. Note that the external source concentration represented by the black spot of Figure 3.6 need not be exactly the high u-value. The autocatalytic nature of 3.1 adjusts the system automatically to the appropriate level. However, if we increase the concentration of the external source, an inversion of the initial gradient occurs (*cf.* Figure 3.7). Furthermore, an excessive increase of the source leads to an inverted

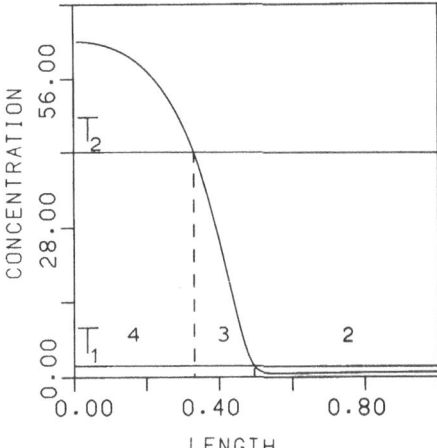

Figure 3.5. The u distribution of 3.1 where the parameters are: $a = 100$, $b = 71.765$, $\alpha = 1.5$, $\rho = 13$, $K = 0.125$, $\gamma = 1$, $D_U = 0.01$, $D_v = 1$. A similar profile is established for v. The thresholds T_1, T_2 divide the field in areas of digits 4, 3 & 2.

profile of a step-like form (Figure 3.8). As will be stressed in the following section, the above model behavior is in disagreement with the experimental observations.

3.3   AUTOCATALYTIC BEHAVIOR VS EXPERIMENTAL RESULTS

It is illuminating to show the 'all-or-none' character of all autocatalytic models as a result of the bifurcating solutions of the nonlinear dynamics. When the external source of Figures 3.6-3.8 is applied for only a limited number of iterations below a critical value N_c the disturbance is absorbed and the system returns to its initial steady state. For a number of iterations $N > N_c$ the distributions are irreversibly changed. The number of iterations directly relates to the exposure time although the time scale itself for the reaction-diffusion dynamics is unknown. This abrupt transition, characteristic of non-linear dynamics, is in gross disagreement with the observed gradual formation of supernumerary structures according to the duration of the exposure time (Wanek *et al.* 1991).

The same step-like behavior of the reaction-diffusion models would have been expected when the concentration C of the external source varies. There are two critical concentrations C_1 and C_2 for which the following transitions take place:

(i) if $C < C_1$ no change of the gradient occurs (Figure 3.5)

(ii) if $C_1 < C < C_2$ the initial gradient goes over to a U-shaped distribution (Figure 3.6)

(iii) if $C_2 < C$ the initial gradient is inverted (Figures 3.7,3.8).

Note that for extremely high values of the source concentration digit 3 shrinks to extinction (Figure 3.4). The dose-dependent results described in the introduction (Tickle *et al.* 1985) are in complete disagreement with the model expectations.

Every model has an inherent degree of rigidity. A 'soft' model retains easily the patterns induced by external influences exhibiting a kind of system memory. The Gierer-Meinhardt is a typical 'soft' model. To the other extreme a 'hard' model has a very pronounced

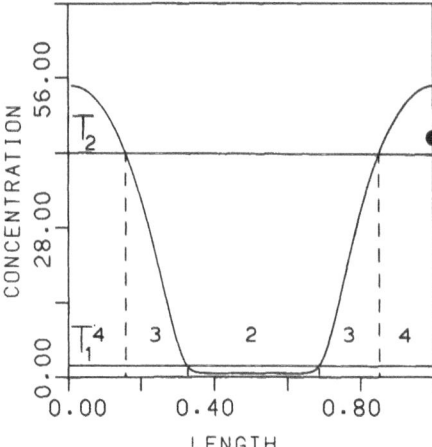

Figure 3.6. Simulation of a ZPA transplantation: the high values of u and v (left side of Figure 3.5) are implanted on the right side. U-shaped distributions are established for both u and v. Here is shown the u profile only. Identical results are obtained if a RA-bead (high value of u only represented by the black spot) is implanted at the anterior margin of the bud.

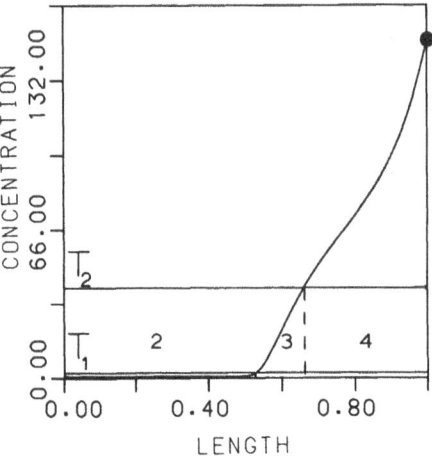

Figure 3.7. For an external source (black spot) higher than the endogenous RA-level, the gradient of Figure 3.5 is inverted.

tendency to relax after removal of external constraints or distortions. The 'Brusselator' shows such a behavior (see comparable observations by Harrison (1987)).

3.4   DISCUSSION

For a realistic treatment of the chick wing development one should consider it as a case of a two-dimensional if not three-dimensional field. However, as numerous simulations have shown, reaction-diffusion systems of high dimensionality respond in a very complicated

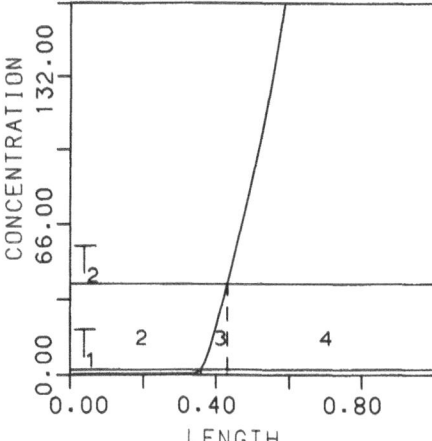

Figure 3.8. A much higher external source ($\times 100$ the endogenous high RA-level) leads to an extremely steep inverted gradient.

way when slightly modified. This is due to the existence in such systems of a multiplicity of modes switching to instability almost simultaneously. This situation prevents the formulation of simple causal scenaria for the mechanism underlying the morphogenetic process. On the other hand, the apparent developmental events occurring in the chick wing bud take place mainly in the AP axis, which justifies the introduction of simple one-dimensional models for their comprehension. Furthermore, it is well known that the 'all-or-none' behavior of the bifurcating solutions is a common feature of the reaction-diffusion systems independent of the field dimensionality.

For the same reasons of minimal complexity we have only considered simple morphogenetic gradients. Such gradients are indispensable for the positional information even in the case of prepatterns where higher wave numbers are introduced in order to account for the normal number of digits (Wolpert 1989). As for the retinoic acid we are dealing with, its distribution along the anteroposterior axis is smoothly monotonic (Thaller & Eichele 1987). (A word of caution: even this simple gradient needs further confirmation).

In our reaction-diffusion version the model parameters are kept constant in the whole extent of the morphogenetic field. We think that even in more complicated cases where the diffusion constants are allowed to vary in space, the bifurcating solutions would behave in agreement with the present analysis. Our conclusion is that not only plain diffusion but also reaction-diffusion cannot admit retinoic acid as a morphogen. We cannot exclude the possibility that RA plays a morphogenetic role. Eventually RA may convert cells into ZPA but then the morphogen produced by the ZPA cannot be the retinoic acid itself (Papageorgiou & Almirantis 1982). On the other hand, there is strong evidence that RA is part of a complicated mechanism for the expression of retinoic acid receptors which in their turn trigger appropriate target genes (Dolle *et al.* 1990).

REFERENCES

Bryant, S. V., Hayamizu, T., Wanek, N., & Gardiner, D. M. 1991. Position dependent properties of limb cells. In *Developmental Patterning of the Vertebrate Limb*, J. M. Hurle, J. R. Hinchliffe & D. Summerbell (ed). pp. 133–142, New York: Plenum Press.

Dolle, P., Ruberte, E., Leroy, P., Morriss-Kay, G., & Chambon, P. 1990. Retinoic acid receptors and cellular retinoid binding proteins: I. A systematic study of their different pattern of transcription during mouse organogenesis. *Development*, **110**, 1133–1151.

Eichele, G., Tickle, C., & Alberts, B. M. 1985. Studies on the mechanism of retinoid-induced pattern duplications in the early chick limb bud: Temporal and spatial aspects. *J. Cell. Biol.*, **101**, 1913–1920.

Harrison, L. G. 1987. What is the status of reaction-diffusion theory thirty-four years after Turing? *J. Theor. Biol.*, **125**, 369–384.

Murray, J. D. 1982. Parameter space for Turing instability in reaction-diffusion mechanisms: a comparison of models. *J. Theor. Biol.*, **98**, 143–163.

Murray, J. D. 1989. *Mathematical Biology*. New York: Springer-Verlag.

Papageorgiou, S., & Almirantis, Y. 1982. Diffusion or autocatalysis of retinoic acid cannot explain pattern formation in the chick wing bud. *Dev. Dynam*, **194**, 282–288.

Saunders, Jr., J. W., & Gasseling, M. T. 1968. Ectodermal-mesenchymal interactions in the origin of limb symmetry. In *Epithelial-Mesenchymal Interactions*, Billingham, R. Fleischmajer & R. E. (ed). pp. 78–97, Baltimore: Williams & Wilkins Co.

Thaller, C., & Eichele, G. 1987. Identification and spatial distribution of retinoids in the developing chick bud. *Nature*, **327**, 625–628.

Tickle, C., Summerbell, D., & Wolpert, L. 1975. Positional signaling and specification of digits in limb morphogenesis. *Nature*, **254**, 199–202.

Tickle, C., Alberts, B., Wolpert, L., & Lee, J. 1982. Local application of retinoic acid to the limb bud mimics the action of the polarizing region. *Nature*, **296**, 564–565.

Tickle, C., Lee, J., & Eichele, G. 1985. A quantitative analysis of the effect of all-trans-retinoic acid on the pattern of chick wing development. *Dev. Biol.*, **109**, 82–95.

Wanek, N., Gardiner, D. M., Muneoka, K., & Bryant, S. V. 1991. Conversion by retinoic acid of anterior cells into ZPA cells in the chick wing bud. *Nature*, **350**, 81–83.

Wolpert, L. 1989. Positional information revisited. *Dev. Biol.*, **107**, 3–12. (Supplement).

4. PATTERN FORMATION IN HETEROGENEOUS DOMAINS

Debbie L. Benson, Philip K. Maini, and Jonathan A. Sherratt

Centre for Mathematical Biology
Mathematical Institute
24–29 St. Giles'
Oxford OX1 3LB, UK

Development of spatial pattern in the early embryo results from the interaction of several processes in a complex hierarchy of mechanisms. Most models for morphogenesis to date have, however, focussed on a particular mechanism. Although such models are capable of capturing some aspects of development they are inconsistent with key experimental observations. Here we consider a two-step hierarchy of patterning mechanisms in which the spatial pattern of a control chemical regulates morphogen diffusivity in an overlying reaction diffusion system. Specifically, we consider the one-dimensional system

$$
\begin{aligned}
u_t &= \gamma f(u, v) + u_{xx} \\
v_t &= \gamma g(u, v) + (D(c)v_x)_x \\
c_t &= \nu c_{xx} - \theta c,
\end{aligned}
\tag{4.1}
$$

where $u(x, t)$ and $v(x, t)$ are morphogen concentrations at position x and time t, $c(x, t)$ is the concentration of control chemical, and f and g describe the chemical reaction kinetics. The system (4.1) has been nondimensionalised such that u, v and c have diffusion coefficients $1, D(c)$ and ν respectively, and we assume that $D = \alpha c$ where α is a positive constant; biologically, this could, for example, reflect an increase in gap junction permeability for v due to the presence of c (Othmer & Pate 1980); c is assumed to degrade with rate θ, and γ is a scale factor proportional to the length of the domain. The equations hold on the space domain $x \in [0, 1]$ with zero flux boundary conditions on u and v. The boundary conditions for c are $c_x(0, t) = 0, c(1, t) = c_0$. If we assume that the equation for c reaches a stable equilibrium on a fast time scale during which insignificant changes in morphogen concentration take place, then $D = \alpha c_0 \cosh(\delta x) / \cosh \delta$ where $\delta = \sqrt{(\theta/\nu)}$. This represents a smoothly increasing diffusion coefficient for v (*cf.* Figure 4.1(a)).

This system is mathematically intractable. However, approximating the diffusion coefficient of v by the step function illustrated in Figure 4.1(b) does not change the qualitative form of the solutions of the original system, but enables us to carry out a linear analysis (Benson *et al.* 1993a) to delimit regions in parameter space wherein different types of pattern occur (*cf.* Figure 4.2). The crucial difference between the patterns illustrated in Figure 4.2 and those of reaction-diffusion models solved in a homogeneous environment is that, in the former case, pattern can be isolated in specific parts of the domain, in marked contrast to the case of homogeneous diffusivity. In the above analysis, the diffusion coefficient of u was assumed constant across the domain. However, if we allow the diffusion coefficients of *both* morphogens to vary in a stepwise manner, the model exhibits patterns in which the wavelength can vary markedly across the domain (see Figure 4.3). We illustrate the application of the two

Experimental and Theorietcal Advances in Biological Pattern Formation,
Edited by H.G. Othmer *et al.*, Plenum Press, New York, 1993

29

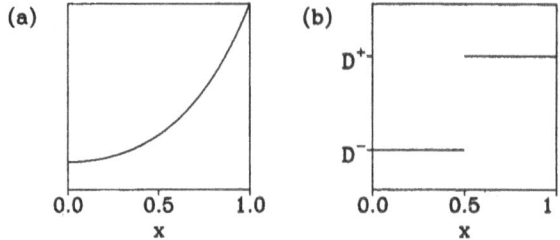

Figure 4.1. Different forms of diffusion coefficient D used for system (4.1): (a) smoothly varying, (b) step function.

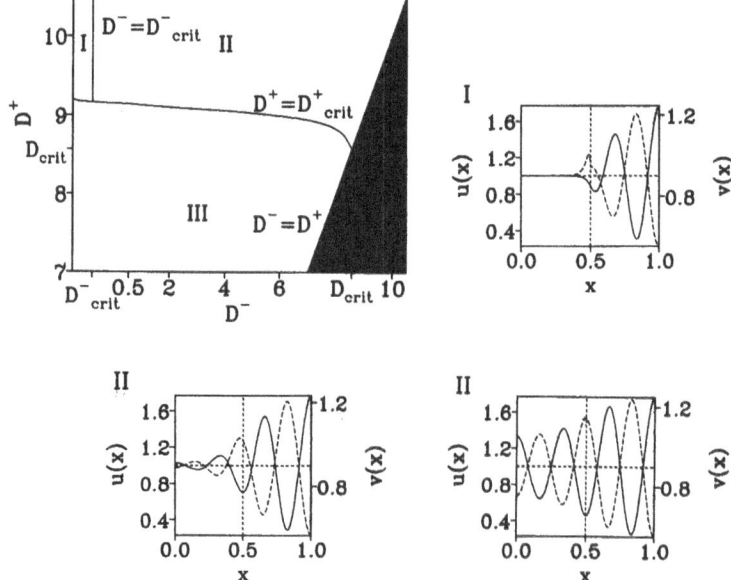

Figure 4.2. Diagram showing the three different regions (I,II,III) in parameter space wherein the three qualitatively different solution types arise. (Computations on the Schnackenberg (1979), reaction diffusion system: solution profiles u ——; v - - - -). See Benson *et al.* (1993b) for parameter values.

step model by considering two experimental observations that contradict the predictions of the standard reaction-diffusion model, namely; the anterior-posterior asymmetry of skeletal elements in the limb (see, for example, Wolpert 1981); and the results of the experiments on double anterior composite limbs (Wolpert & Hornbruch 1990) which show that one can produce limbs with two humeral elements while keeping the domain size constant.

The simulations illustrated by Figure 4.2 show that reaction diffusion systems with spatial variation in diffusion coefficients may produce isolated patterns and asymmetric oscillatory patterns. In our composite model, the underlying spatial pattern in control chemical influences the position and amplitude of peaks in morphogen concentration. If we apply this model to limb morphogenesis, the positional information (Wolpert 1981) supplied by these morphogen profiles could lead, via cell differentiation, to the specification of asymmetrically patterned elements, whose position within the domain can be controlled by the spatial distri-

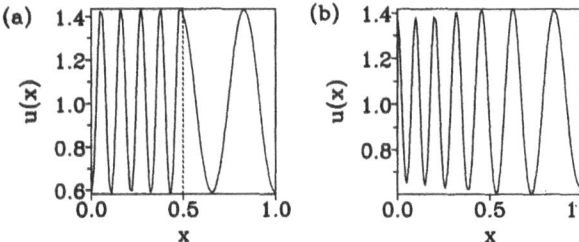

Figure 4.3. Solutions to (4.1) when the diffusion coefficients of **both** morphogens vary (a) stepwise, (b) continuously. Note that in (a) the wavelength varies discontinuously across the domain, but in (b) this variation is continuous. For parameter values, see Benson *et al.* (1993a). (u solution profile only shown – v solution profile is approximately out of phase with that of u.)

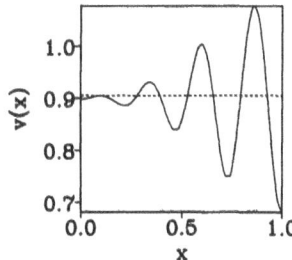

Figure 4.4. Steady state solution exhibited by (4.1) for a certain set of parameter values (see Maini *et al.* (1992)). For a suitably chosen threshold concentration, this prepattern specifies three skeletal elements which are intrinsically distinct because of the varying concentrations of morphogen to which they are each exposed. (u profile only shown).

bution of c. This is in contrast to the standard reaction-diffusion system with spatially uniform parameters, which produces identical and equally-spaced elements. The patterns produced by the model (4.1) are therefore more consistent with those observed in certain developmental processes, for example digit patterning in the vertebrate limb, than those exhibited by the standard model. This is illustrated in Figure 4.4 where, for a suitably chosen parameter set, the composite model exhibits a steady state solution with three peaks in concentration which, in contrast to the standard model, have different amplitudes of oscillation. A further consequence of the isolated solution profiles exhibited by the composite model is that duplication of skeletal elements can be predicted without necessarily increasing the length of the domain. For example, Figure 4.5 shows that imposing symmetric boundary conditions on the control chemical, c, but keeping the domain fixed in size, can lead to duplication. In this case, therefore, one can increase pattern complexity without it being necessary to increase the domain size. Hence, this is consistent with the results of (Wolpert & Hornbruch 1990). Note that in the standard Turing model, one would predict, at least for low mode numbers, that the pattern complexity would remain unchanged for a fixed domain size (Dillon *et al.* 1992).

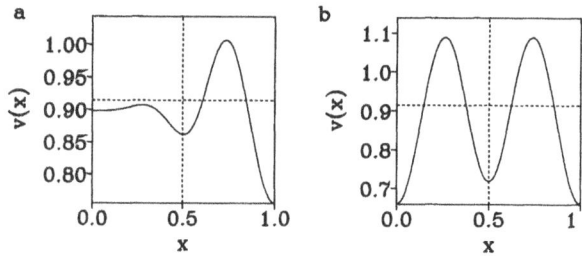

Figure 4.5. For the model (4.1), imposing symmetric boundary conditions on c increases pattern complexity. (a) Solution for boundary conditions $c_x(0,t) = o, c(1,t) = c_o$, (b) solution for symmetric boundary conditions $c(0,t) = c(1,t) = c_o$. (u profiles only shown – see Maini *et al.* (1992) for parameter values).

ACKNOWLEDGMENTS DLB acknowledges the Wellcome Trust for a Prize Studentship in Mathematical Biology. JAS was supported by a Junior Research Fellowship at Merton College, Oxford.

REFERENCES

Benson, D. L., Maini, P. K., & Sherratt, J. A. 1993a. Analysis of pattern formation in reaction diffusion models with spatially inhomogeneous diffusion coefficients. *Math. & Comp. Modelling*. (To appear).

Benson, D. L., Sherratt, J. A., & Maini, P. K. 1993b. Diffusion driven instability in an inhomogeneous domain. *Bull. Math. Biol.*, **55**, 365–384.

Dillon, R., Maini, P. K., & Othmer, H. G. 1992. Pattern formation in generalized Turing systems. I. Steady-state patterns in systems with mixed boundary conditions. (To appear in J. Math. Biol.).

Maini, P. K., Benson, D. L., & Sherratt, J. A. 1992. Diffusion models with spatially inhomogeneous diffusion coefficients. *IMA J. Math. Appl. in Medic.& Biol.*, **9**, 197–213.

Othmer, H. G., & Pate, E. 1980. Scale-invariance in reaction-diffusion models of spatial pattern formation. *Proc. Natl. Acad. Sci. USA*, **77**, 4180–4184.

Schnackenberg, J. 1979. Simple chemical reaction systems with limit cycle behaviour. *J. Theor. Biol.*, **81**, 389–400.

Wolpert, L. 1981. Positional information and spatial patterning. *Phil. Trans. R. Soc. Lond.*, **B259**, 441–450.

Wolpert, L., & Hornbruch, A. 1990. Double anterior chick limb buds and models for cartilage rudiment specification. *Development*, **109**, 961–966.

5. A FIELD MODEL OF SYMMETRY REVERSALS IN THE PATTERN REGULATION OF A CELL

Wendy A. M. Brandts

Centre for Mathematical Biology
Mathematical Institute
24–29 St. Giles'
Oxford OX1 3LB, UK

5.1 INTRODUCTION–SYMMETRIES AND FIELDS IN MORPHOGENESIS

Symmetry reversal in patterns, such as in the formation of mirror image patterns, are frequently observed during regulation in both multicellular and unicellular organisms (*e. g.* amphibians (Totafurno & Trainor 1987); insects (Russel 1985); unicellular systems (Frankel 1988; Shi *et al.* 1991; Suhama 1990; Kumazawa 1979)). A unifying explanation of some reversals, triplications for example, where the symmetry of the middle member is the reverse of that of the other two, has been advanced in terms of a set of phenomenological rules (French *et al.* 1976; Bryant *et al.* 1981). In the present paper, we discuss triplication and mirror image pattern reversals arising in the single celled organism *Tetrahymena*, which show topologically similar patterns –and thus provide a unicellular analogy– to reversals in multicellular systems. We have developed a *quantitative* phenomenological model which both has descriptive and predictive power, as well as forms a foundation for bridging biological understanding to explicit physical and chemical models. By describing unicellular organisms within a conceptual framework that could also apply to multicellular organisms, we encourage a unified interpretation of pattern formation in morphogenesis.

We assume that our morphogenetic field plays an active role in controlling processes that lead to the formation of pattern elements, but we do not need to specify whether the field values represent positional information (Wolpert 1989) which may be associated with the formation of characteristic structures, or alternatively parameter values which themselves directly induce the formation of particular structures (*i. e.* prepattern; Maini *et al.* 1992; Murray 1981b).

5.2 EXPERIMENTAL OBSERVATIONS IN *TETRAHYMENA*

The cell surface structures which serve for the purposes of our model as pattern elements that mark distinct positions around the cell circumference will be briefly described (Frankel 1988). A typical cell (see Figure 5.1) has 18 to 21 longitudinal rows of cilia (CR). A normal cell has one oral apparatus (OA), which defines it as a *singlet* (*cf.* Figure 5.1). It is possible to assign a 'handedness' to the OA, which for the normal configuration is defined to be right–handed. To the cell's right of the OA, located near the posterior end

Experimental and Theorietcal Advances in Biological Pattern Formation,
Edited by H.G. Othmer *et al.,* Plenum Press, New York, 1993

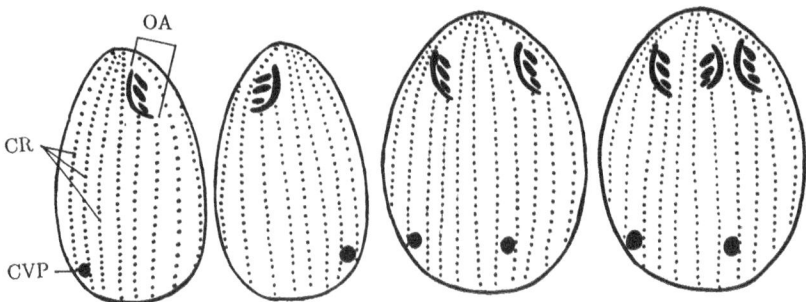

Figure 5.1. *Tetrahymena* configurations: singlet, left–handed singlet, doublet, and triplet.

of one, two, or three adjacent ciliary rows are the contractile vacuole pores (CVPs), or the CVP set. *Doublets* are two cells fused with longitudinal anterior–posterior axes parallel, and a double complement of all cell features. Doublets are not stable but tend to down regulate in size, involving a loss of ciliary rows, the disappearance of one of the OAs, and of one of the CVP sets. In the conversion of a doublet cell bearing *two* oral structures to a singlet cell with *one*, a surprising triplication state with *three* oral structures, the middle one of inverted symmetry, sometimes appears (see Figure 5.1 'triplet'). In addition to the triplication forms described here, throughout development various other configurations of doublets and singlets appear. These have been successfully modelled in detail (Brandts & Trainor 1990a).

Left–handed cells are 'wound' in the opposite direction to right–handed cells: the global positions of all structures that are asymmetrically arranged with respect to one another around the circumference are reversed (see Figure 5.1 'left-handed singlet'). The regulation strategies that cause symmetry reversals can be further probed by examining the behaviors of left–handed doublet cells, experimental counterparts to regulation in right–handed doublet cells. It is particularly interesting to find that many of the features of the cellular organization satisfy mirror–reversal invariance and to understand how in some limited but significant respects only this symmetry is violated (Brandts 1992). For example, in spite of the apparent mirror symmetry of left and right cells, left–handed cells usually regulate through intermediate mirror–reversed configurations, whereas right–handed cells predominantly regulate through triplet configurations (Nelsen & Frankel 1989).

5.3   FIELD MODEL

Our morphogenetic field model (Brandts & Trainor 1990b; Goodwin 1980) is formu- lated for a pattern organized about a *circumference* (*e. g.* a 1–dimensional ring), such as the cell surface features in *Tetrahymena*. The morphogenetic field is expressed as a vector field. It is continuous and periodic around the circumference, and thus every point on the cell has a unique circumferential designation (field value). Different values of the field angle θ relative to the (arbitrary) morphogenetic field axis designate (or induce the formation of) different component structures of the cell pattern.

The expected biological patterns are posited to correspond to the minimum energy configurations of this field. The form of the energy reflects in part phenomenological rules suggested by qualitative models such as the 'polar coordinate model' of French *et al.* (1976) and the 'cylindrical co–ordinate model' of Nelsen & Frankel (1986). The first term reflects the idea that there is an optimal spacing of positional values or pattern features for the cell:

an optimal value of $\partial\theta/\partial x$, where x is the circumferential position. Changes in direction or handedness of the field are discouraged by a smoothing term.

The simplest form for an energy functional which meets the above criteria is a quartic in the field gradient, combined with a second derivative smoothing term with an even power. Explicitly, our original model employed the form:

$$E = \int_0^L \left\{ \left[\left(\frac{\partial\theta}{\partial x} \right)^2 - \left(\frac{2\pi}{L_0} \right)^2 \right]^2 + \beta^2 \left(\frac{\partial^2\theta}{\partial x^2} \right)^2 \right\} dx. \tag{5.1}$$

The weighting of the smoothing term relative to the optimal gradient term is given by the parameter β^2. L_0 is the normal circumference (size) of the cell. The actual cell circumference is L.

Figure 5.2 illustrates the three symmetry classes of solutions $\theta(x)$ obtained in the model: the symmetric (SYM) corresponding to singlets and doublets; reverse intercalated (RI) corresponding to triplets or singlets; and the mirror reversed (MR) solutions corresponding to mirror symmetric states. Which solution has the lowest energy depends on the value of L,

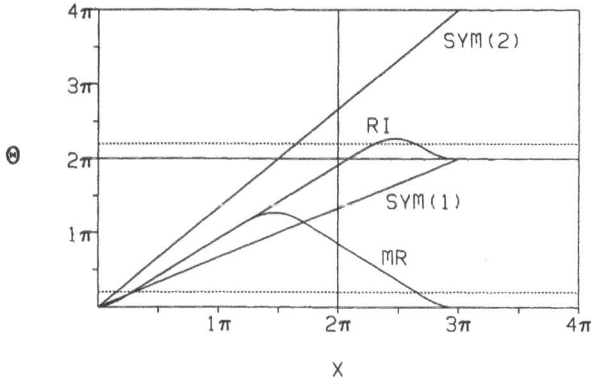

Figure 5.2. The model cell configurations. For example, if the field value $\theta = 0.2\pi (= 2.2\pi)$ corresponds to OA formation, solutions can give singlet (SYM(1)), doublet (SYM(2), MR), or triplet (RI) states.

and as L is reduced from doublet to singlet size, a sequence of transitions between different solutions occurs. This sequence corresponds to the regulation pathways through triplication or mirror reversed configurations observed in *Tetrahymena*. The quantitative details of the patterns correspond closely to experimental results (Brandts & Trainor 1990a).

In this version of the model, right– and left– handed global patterning of the cell are treated as equivalent since the first term in the energy is a *symmetric* double well. We subsequently introduced a small asymmetry into the energy (Brandts 1992), motivated by experiments on left–handed cells, which suggested that left–handed and right–handed winding are locally intrinsically different (Nelsen & Frankel 1989). The small asymmetry did not appreciably affect the right–handed solutions, but it changed the energy of the left–handed solutions sufficiently to induce a different pathway to be taken by left–handed doublets, exactly in agreement with experiment.

5.4   CONCLUSIONS

In this paper we have outlined a model for the formation of regions of reversed patterning in *Tetrahymena*; in particular, pattern triplications, where the symmetry of the middle region is opposite to that of the other two, and mirror image patterns, which contain equal regions of opposite symmetry. The model hypothesizes that the morphogenetic field which specifies the pattern information is controlled by energy minimization. Our model can account in detail for the many different cell configurations seen in doublet regulation experiments, and the different pathways taken by cells of opposite symmetry, and thus strengthens and broadens the experimental support for our field approach to pattern formation.

ACKNOWLEDGMENTS Thanks to Joseph Frankel, Marlo Nelson and Lynn Trainor for discussions on this work. Warm thanks to all the folks at the Centre for Mathematical Biology, especially Philip, Debbie, Gerhard and Jonathan, who helped make my stay enjoyable and productive. The financial support of the Natural Sciences and Engineering Research Council of Canada is gratefully acknowledged.

REFERENCES

Brandts, W. A. M. 1992. *IMA J. Math. Appl. in Medic.& Biol. (In press)*.

Brandts, W. A. M., & Trainor, L. E. H. 1990a. A non-linear field model of pattern formation: Application to intracellular pattern reversal in *Tetrahymena*. *J. Theor. Biol.*, **146**, 57–86.

Brandts, W. A. M., & Trainor, L. E. H. 1990b. A non-linear field model of pattern formation: Intercalation in morphallactic regulation. *J. Theor. Biol.*, **146**, 37–86.

Bryant, S. V., French, V., & Bryant, P. J. 1981. Distal regeneration and symmetry. *Science*, **212**, 993–1002.

Frankel, J. 1988. *Pattern formation: Ciliate studies and models*. Oxford University Press.

French, V., Bryant, P. J., & Bryant, S. V. 1976. Pattern regulation in epimorphic fields. *Science*, **193**, 969–981.

Goodwin, B. C. 1980. Pattern formation and its regeneration in the protozoa. In *The Eukaryotic Microbial Cell: Thirtieth Symposium of the Society for General Microbiology held at the University of Cambridge, March 1980*, Lloyd, D., Gooday, G. W., & Trinci, A. P. J. (eds). Cambridge University Press.

Kumazawa, H. 1979. Homopolar grafting in *Blepharisma japonicum*. *J. Exp. Zool.*, **207**, 1–16.

Maini, P. K., Benson, D. L., & Sherratt, J. A. 1992. Diffusion models with spatially inhomogeneous diffusion coefficients. *IMA J. Math. Appl. in Medic.& Biol.*, **9**, 197–213.

Murray, J. D. 1981. A pre–pattern formation mechanism for animal coat markings. *J. Theor. Biol.*, **88**, 161–199.

Nelsen, E. M., & Frankel, J. 1986. Intracellular pattern reversal in *Tetrahymena thermophila*. I. evidence for reverse intercalation in unbalanced doublets. *Dev. Biol.*, **114**, 53–71.

Nelsen, E. M., & Frankel, J. 1989. Maintenance and regulation of cellular handedness in *Tetrahymena*. *Development*, **105**, 457–471.

Russel, M. A. 1985. Positional information in insect segments. *Dev. Biol.*, **108**, 269–283.

Shi, X., Lu, L., Qui, Z., He, W., & Frankel, J. 1991. Microsurgically generated discontinuities provoke heritable changes in cellular handedness of a ciliate, *Stylonychia mytilus*. *Development*, **111**, 337–356.

Suhama, M. 1990. The regulation of homopolar doublets to singlets in *Glaucoma*. *J. Protozool*, **37**(5), 344–352.

Totafurno, J., & Trainor, L. E. H. 1987. A non-linear vector field model of supernumerary limb production in salamanders. *J. Theor. Biol.*, **124**, 415–454.

Wolpert, L. 1989. Positional information revisited. *Dev. Biol.*, **107**, 3–12. (Supplement).

6. PATTERNING IN LIMBS: THE RESOLUTION OF POSITIONAL CONFRONTATIONS

Susan V. Bryant, Terry F. Hayamizu and David M. Gardiner

Developmental Biology Center
University of California, Irvine
Irvine, CA 92717, USA

6.1   THE NATURE OF POSITIONAL CONFRONTATIONS

Positional confrontations are characteristic of the development of all multicellular organisms. In amphibian embryos, for example, formation of the mesoderm is dependent on contact between vegetal and animal pole cells (Symes *et al.* 1988). Later, mesodermal cells specified for dorsal and ventral fates interact to generate cells with intermediate fates (Dale & Slack 1987). In some insects, the trunk region develops from interactions between head and 'tail' structures (Cohen & Jürgens 1991). Finally, confrontations between differently-specified cells are required for limb outgrowth during development and regeneration (Bryant & Gardiner 1992).

It has been possible to gain considerable knowledge about positional confrontations from the study of limbs because they are regulative systems in which both naturally-occurring and experimentally-created positional confrontations are resolved. For example, we have learned that the basic features of pattern formation are conserved from insects to vertebrates; that positional identity is stable over long periods of time – years in the case of amphibians – and through different states of differentiation, and yet is labile in response to positional confrontations; and that growth is causally linked to the generation of pattern.

6.2   MECHANISMS TO RESOLVE CONFRONTATIONS

Whereas the developmental significance of positional confrontations in limbs and other systems is widely acknowledged, there is less consensus about how they are resolved. A recurrent theme is that the resolution of positional confrontations is fundamentally unequal. This inequality is implicit in the language used to describe the interactions: *e. g.* dominant, controlling, or organizer cells vs. subordinate or responsive cells. Perhaps the most prevalent idea about pattern formation is based on the view that otherwise naive cells acquire positional identity by reference to spatially graded levels of a diffusible morphogen emanating from special cells (Wolpert 1969) (*cf.* Figure 6.1). There is almost no area of developmental biology that has not been interpreted according to this view, but it is in the developing limb that this view has been the most extensively explored.

An alternative view explains pattern formation as a result of cooperative, cell-cell interactions in which none of the participants are naive (French *et al.* 1976; Bryant *et al.*

Experimental and Theorietcal Advances in Biological Pattern Formation,
Edited by H.G. Othmer *et al.,* Plenum Press, New York, 1993

37

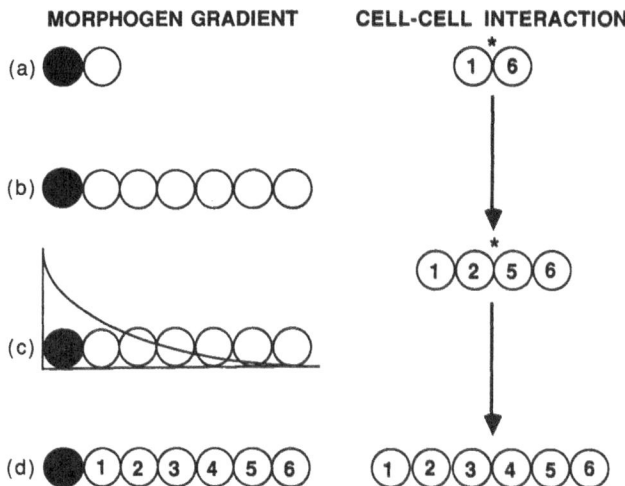

Figure 6.1. *The 'morphogen gradient' view:* (a) There are two initial states; cells specified to produce morphogen (black circle), and developmentally naive cells (empty circles). (b) Prior to specification, the naive cells grow to form the responsive population. (c) Diffusion of morphogen from the source establishes a concentration gradient. (d) Cells adopt different fates (#1 through #6) in response to different morphogen concentrations. *The 'cell-cell interaction' view:* Initially there are two differently-specified cells (in this example #1 and #6) that interact at a positional confrontation (*). The interaction stimulates growth (arrow); new cells acquire new positional values (#2 and #5) that are intermediate between the values at the original confrontation. Newly specified cells generate a new positional confrontation (*), and more growth (arrow) leading to the complete sequence of positional values (#1 through #6).

1981) (Figure 6.1). When differently-specified cells are in contact, growth is stimulated and daughter cells acquire intermediate positional values. This view (formalized in the polar coordinate model) has been applied primarily to appendage development and regeneration in vertebrates and invertebrates. It has also found application in more diverse systems, such as ciliates (Frankel 1989).

In general, much of the available data concerning limb pattern formation can be interpreted in terms of either long-range diffusible morphogens or short-range local cell-cell interactions (Bryant & Muneoka 1986). The difficulty in distinguishing between such different interpretations challenges us to probe more deeply into the as yet hidden machinery of pattern formation.

6.3   DOMINANCE AND MORPHOGENS

The idea that pattern can be specified with reference to a diffusible morphogen gradient, has recently been elegantly validated within the specialized context of early *Drosophila* development. In this system, a graded distribution of a homeobox protein, *bicoid (bcd)*, within the cytoplasm of the single-celled (but multinucleated) early embryo leads to concentration-dependent transcriptional regulation of several genes that divide the embryo into discrete spatial domains (St. Johnston & Nusslein-Volhard 1992). The best studied of the genes controlled by a particular concentration of *bcd* is *hunchback (hb)*. Further subdivision along the anterior-posterior (AP) and dorsal-ventral (DV) axes occurs in response to threshold concentrations of *hb* and other transcription factors (TFs) within the cytoplasm of the syncytial embryo (Ip & Levine 1992; Struhl *et al.* 1992).

Although diffusible morphogens are involved in pattern formation in the early *Drosophila* embryo, evidence for such a role in multicellular systems is less clear cut. In fact, the best candidate for a diffusible morphogen in vertebrates, retinoic acid (RA), has not fared well as an explanation for pattern formation in limbs (Bryant & Gardiner 1992). Given the widespread appeal of diffusible morphogens, we consider it prudent to consider how RA could look so much like a diffusible limb morphogen and yet not be one after all.

Long before RA was cast in the role of the limb morphogen, the AP axis of developing chick limbs was believed to be specified by a morphogen produced by cells of a 'zone of polarizing activity' (ZPA) (Tickle *et al.* 1975). However, evidence for this view was only applicable to chick limbs, with data from other vertebrate limbs being inconsistent with the existence of separate signalling and responding regions (Bryant & Muneoka 1986). Further, RA could not be such a morphogen in amphibian limbs because it alters pattern in all three limb axes (Stocum 1991). Nevertheless, RA was considered to be the long sought-after AP morphogen because it mimics the effect of a ZPA graft (Tickle *et al.* 1982) and is more abundant in the posterior of the wing bud (Thaller & Eichele 1987).

Recent studies have challenged the *RA-as-morphogen* view. We have reported that RA acts not by establishing a graded AP pattern in the cells next to the source of RA, but rather it converts those cells into posterior edge cells (Wanek *et al.* 1991). How this leads to changes in limb pattern is discussed below. Also, data on the expression of retinoic acid receptor beta (RARβ), which contains an RA response element, suggest that there is little or no RA emanating from the ZPA (Noji *et al.* 1991). RARβ is expressed at the base of the limb but not distally; expression is not elevated in the ZPA, nor is it stimulated by proximity to a grafted ZPA; however, expression is stimulated in regions adjacent to an implanted RA-containing bead. Studies of transgenic mice containing reporter gene constructs with an RA response element also fail to indicate the presence of RA in limb buds (Mendelsohn *et al.* 1991; Rossant *et al.* 1991). It is possible that RA and other active retinoids measured in limbs (Thaller & Eichele 1987) are present at the base of the limb, rather than at the tip where pattern formation is occurring. It is our interpretation of the limb data that RA is not the limb morphogen, and that the ZPA does not exert its patterning effect by producing a morphogen of any type (Bryant & Gardiner 1992).

6.4 COOPERATION AND CELL-CELL INTERACTIONS

In contrast to diffusible morphogens, there are numerous examples in which local cell-cell interactions are involved in pattern formation (Greenwald & Rubin 1992). We focus here on the control of limb patterning as detailed in the polar coordinate model (French *et al.* 1976; Bryant *et al.* 1981). Briefly, positional confrontations between A and P (and/or D and V) mesodermal cells arise in the distal region of the limb rudiment. Growth is stimulated and cells with intermediate positional values are created (intercalation). As new positional values arise, further confrontations are generated, leading to successive rounds of growth and intercalation. During growth in response to circumferential positional confrontations, cells progressively adopt more distal positional identities. In theory, limb outgrowth could continue indefinitely; however, growth is not sustainable within symmetrical structures which progressively lose positional diversity (Bryant *et al.* 1981). As the distal part of the limb pattern is generated, the limb circumference is fragmented into digit bases that terminate growth at the ends of the symmetrical digits (Stock & Bryant 1981).

We have recently proposed that RA effects on limb pattern formation are secondary consequences of local cell-cell interactions (Bryant & Gardiner 1992). The primary response of pattern formation-competent cells to RA is to be converted to a single positional value that is posterior, ventral and proximal (P-V-Pr) with respect to the limb. These RA-converted cells then participate in local cell-cell interactions resulting in diverse patterning responses that include teratogenesis after global application, supernumerary limbs after local application, and proximal-distal (Pr-Di) pattern duplication during regeneration. The ability of local cell-cell interactions to account for this diverse array of RA effects further supports this view of limb pattern formation.

An issue that inevitably arises in discussions of RA effects is that of dose response. The observation that RA responses are dose-dependent is often presented as evidence that cells respond directly to a gradient of RA by forming different levels of the AP pattern. However, lineage studies have shown that anterior cells respond to RA by acquiring only posterior identity (Wanek *et al.* 1991), rather than intermediate positional values. These observations are reconciled if the dose response is a result of the *number* of cells that are converted to posterior. At low RA doses (or after shorter exposures), only a few cells are converted to posterior identity; as they are stimulated to divide, they adopt intermediate positional values and the most posterior parts of the pattern are lost. At higher doses (or after longer exposures), more cells are converted; some of these will be able to maintain their most posterior identity. Parallel results have been obtained following grafting of different numbers of posterior (ZPA) cells (Tickle 1981). Consistent with this interpretation is direct evidence that dose-dependent responses occur at the level of cell populations, not at the level of individual cells. In response to glucocorticoids, individual cells either express a glucocorticoid target gene or they do not, and the dose response at the population level results from different numbers of cells responding at different doses (Ko *et al.* 1990). Preliminary results indicate that similar conclusions can be drawn about the mode of action of retinoids (Ko 1991).

6.5 THE RELEVANCE OF *DROSOPHILA* TRANSCRIPTION FACTOR GRADIENTS TO LIMBS

A limitation of the local cell-cell interaction view is the absence of an explicit molecular mechanism for the generation of new positional information through growth (Wilkins & Gubb 1991). At the present time, the only proven mechanism for establishing a series of territories of differently-specified nuclei involves the graded distributions of TFs described for *Drosophila* embryos. In an effort to draw conclusions applicable to multicellular systems from this rather specialized example of pattern formation, it has been assumed that the generalizable feature is the diffusible morphogen. We propose that the feature of significance for patterning in multicellular systems is the establishment of gradients of TFs across fields of nuclei. If TF gradients could be generated across fields of separate cells, concentration-dependent transcriptional regulation of different genes could divide the field into discrete spatial domains; *i. e.* TF gradients could control pattern formation in multicellular systems, as they have been shown to do in syncytial embryos.

It is clear that the mechanism for generating TF gradients in *Drosophila* embryos (diffusion through the cytoplasm) is not feasible for fields of cells. In addition, since transcriptional units appear to be either turned on or off in cells (Ko *et al.* 1990), it seems unlikely that differential rates of transcription will provide an explanation for such results. However, TF gradients could be generated during growth as a result of what we refer to as 'dilution.' By this mechanism, when cells expressing a particular TF are stimulated to divide, their progeny will have less of that TF than those of cells that divide less frequently. When growth is stimulated at a positional confrontation, a field of cells with a graded distribution of the TF would be generated between the originally confronted cells. As in *Drosophila* embryos, this gradient could specify a set of new territories, possibly leading directly to acquisition of a final cell fate. Alternatively, further patterning could occur if cells at the borders of the newly established territories were either stimulated to divide, or were involved in local inductive or inhibitory interactions across the borders.

There are several possible mechanisms to effect the 'dilution' described above. TFs (or their mRNA) could be equally partitioned among daughters, leading to reductions of TF levels in dividing versus non-dividing cells. Alternatively, the distribution of TF at cell division could be unequal. The latter possibility has been suggested as a way for daughter cells to acquire different fates (Stanjevic *et al.* 1991). While dilution is the most direct way to visualize how TF gradients could arise, stability of the transcripts and/or the protein would be a major requirement. A different and more intriguing possibility that we favor is that growth directly interferes with the accumulation of TFs. During *Drosophila* early development, the cell cycle is short relative to the time needed to produce complete Ubx transcripts. Active transcripts abort at mitosis, and are reinitiated *de novo* after mitosis (Shermoen & O'Farrell 1991). Thus, for genes with large transcriptional units, short cell cycle times in embryos could regulate the onset of gene expression. We extend this speculation to the possibility that transcript abortion during growth could establish gradients of TFs with large transcriptional units. The result would resemble dilution; cells that divide less frequently accumulate more TFs than cells that divide more often. Hence, dividing cells at a positional confrontation would accumulate lower levels of TFs; cells further from the disparity will be dividing less often, permitting transcript accumulation.

6.6 CONSEQUENCES AND CONCLUSIONS

A prediction of this view is that TFs that form gradients will arise from large transcriptional units; TFs with short transcriptional units should not show a graded distribution. Many homeobox-containing genes in *Drosophila* have large transcriptional units (up to 100kb), and some have been shown to have graded patterns of protein expression (Celniker *et al.* 1990). For vertebrate homeobox genes, information on the maximal lengths of transcriptional units is as yet incomplete, although it might be anticipated that they will be similar to those of *Drosophila* homologues. In addition, although there are few studies of vertebrate homeobox proteins, some graded distributions have been reported. In early limb buds, *Hox C6* protein is most strongly expressed in proximal and anterior cells. As outgrowth proceeds, strong expression is maintained in these regions, grading to undetectable levels at the distal and posterior (Oliver *et al.* 1988). A similar *Hox C6* expression pattern develops during outgrowth of feather buds (Chuong *et al.* 1990). In limbs, a complementary pattern for *Hox D9* protein (high posterior), has been reported (Oliver *et al.* 1989). These expression patterns are predictable given the starting location of the strongly expressing cells and the growth characteristics of the limb.

The proposed TF gradients would be transient, due to the nature of their origin, and thus the question arises as to how they could be reactivated, as for example during regeneration from different amputation levels in the limb. One possibility is that positional information at different TF levels is encoded in cell cycle times. Hence, when a limb is amputated, cells stimulated to divide at the wound surface do so at a rate characteristic of their positional value. Subsequent TF accumulation would be appropriate for the level of the pattern at which regeneration is initiated. Although data directly addressing this possibility are not available, there are examples linking specification of cell fate to a change in cell cycle time (Foe 1989).

A further consequence of growth-generated TF gradients is that the cells expressing a particular TF will be related by cell lineage, *i. e.* expression domains of TF gradients will also be lineage compartments. This relationship need not be strict; cells not related by lineage could be induced by nearby cells to express a particular TF, and TFs not expressed in a graded fashion may be expressed in cells without lineage relationships. Nevertheless, a relationship between a homeobox expression domain and a lineage compartment has recently been identified in developing rhombomeres (Fraser *et al.* 1990).

The potential relevance of this view to aspects of development other than limbs is extensive. For example, we note that homeobox genes are expressed during hematopoiesis in vertebrates (Shen *et al.* 1992), opening the possibility that decisions about cell fate are regulated by the number and frequency of cell divisions controlling the level of homeobox protein accumulation.

Future challenges are abundant. They include identification of the TFs involved in the specification of positional identity and their relationship to positional molecules expressed at the cell surface. For example, in the limb, *Hox D* and *Hox A* genes appear to be involved specification of AP and PrDi positional information (Izpisúa-Belmonte *et al.* 1991; Yokouchi *et al.* 1991). If homeobox proteins activate expression of cell surface genes that also have large transcriptional units, then gradients of homeobox protein and cell surface molecules could be generated simultaneously during growth. Adjacent cells expressing different amounts of a cell surface molecule would be stimulated to divide, leading to further dilution. Cell division would continue until adjacent cells are generated with similar amounts of this molecule. In this context, we note that N-CAM expression can be controlled by homeobox genes (Jones *et al.* 1992), and that the *Hox C6* protein gradient in the chick feather bud is paralleled by a gradient of N-CAM (Chuong *et al.* 1990; Jiang & Chuong 1992).

In summary, we propose a general theme for pattern formation: TF gradients control the expression of different genes at different threshold levels. In single cells with many nuclei, such gradients can be generated by diffusion from a localized source. In multicellular systems, such gradients can be generated by 'dilution' resulting from the growth stimulated at positional disparities. Dilution could be direct, or more intriguingly, it could arise as a consequence of transcript abortion at mitosis.

REFERENCES

Bryant, S. V., & Gardiner, D. M. 1992. Retinoic acid, local cell-cell interactions, and pattern formation in vertebrate limbs. *Devel. Biol.*, **152**, 1–25.

Bryant, S. V., & Muneoka, K. 1986. Views of limb development and regeneration. *Trends in Genetics*, **2**, 153–159.

Bryant, S. V., French, V., & Bryant, P. J. 1981. Distal regeneration and symmetry. *Science*, **212**, 993–1002.

Celniker, S. E., Sharma, S., Keelan, D. J., & Lewis, E. B. 1990. The molecular genetics of the bithorax complex of *Drosophila: cis*-regulation in the *Abdominal-B* domain. *EMBO J.*, **9**, 4277–4286.

Chuong, C.-M., Oliver, G., Ting, S. A., Chen, B. G. Jegalian H. M., & DeRobertis, E. M. 1990. Gradients of homeoproteins in developing feather buds. *Development*, **110**, 1021–1030.

Cohen, S., & Jürgens, G. 1991. *Drosophila* headlines. *TIG*, **7**, 267–272.

Dale, L., & Slack, J. M. W. 1987. Regional specification within the mesoderm of early embryos of *Xenopus laevis*. *Development*, **100**, 279–295.

Foe, V. E. 1989. Mitotic domains reveal early commitment of cells in *Drosophila* embryos. *Development*, **107**, 1–22.

Frankel, J. 1989. *Pattern formation. ciliate studies and models*. New York: Oxford University Press.

Fraser, S., Keynes, R., & Lumsden, A. 1990. Segmentation in the chick embryo hindbrain is defined by cell lineage restrictions. *Nature*, **344**, 431–435.

French, V., Bryant, P. J., & Bryant, S. V. 1976. Pattern regulation in epimorphic fields. *Science*, **193**, 969–981.

Greenwald, I., & Rubin, G. M. 1992. Making a difference: The role of cell-cell interactions in establishing separate identities for equivalent cells. *Cell*, **68**, 271–281.

Ip, Y. T., & Levine, M. 1992. The role of the *dorsal* morphogen gradient in *Drosophila* embryogenesis. *Sem. Dev. Biol.*, **3**, 15–23.

Izpisúa-Belmonte, J.-C., Tickle, C., Dollé, P., Wolpert, L., & Duboule, D. 1991. Expression of the homeobox *Hox-4* genes and the specification of position in chick wing development. *Nature*, **350**, 585–589.

Jiang, T.-X., & Chuong, C.-M. 1992. Mechanism of skin morphogenesis. I. analyses with antibodies to adhesion molecules tenascin, N-CAM, and integrin. *Dev. Biol.*, **150**, 82–98.

Jones, F. S., Prediger, E. A., Bittner, D. A., DeRobertis, E. M., & Edelman, G. M. 1992. Cell adhesion molecules as targets for hox genes-neural cell adhesion molecule promoter activity is modulated by cotransfection with hox-2.5 and hox-2.4. *Proc. Nat. Acad. Sci.*, **89**, 2086–2090.

Ko, M. S. H. 1991. A stochastic model for gene induction. *J. Theor. Biol*, **153**, 181–194.

Ko, M. S. H., Nakauchi, H., & Takahashi, N. 1990. The dose dependence of glucocorticoid-inducible gene expression results from changes in the number of transcriptionally active templates. *EMBO J.*, **9**, 2835–2842.

Mendelsohn, C., Ruberte, E., LeMeur, M., Morriss-Kay, G., & Chambon, P. 1991. Developmental analysis of the retinoic acid-inducible RAR-B2 promoter in transgenic animals. *Development*, **113**, 723–734.

Noji, S., Nohno, T., Koyama, E., Muto, K., Ohyama, K., Aoki, Y., Tamura, K., Ohsugi, K., Ide, H., Taniguchi, S., & Saito, T. 1991. Retinoic acid induces polarizing activity but is unlikely to be a morphogen in the chick limb bud. *Nature*, **350**, 83–86.

Oliver, G., Wright, C. V. E., Hardwicke, J., & Robertis, E. M. De. 1988. A gradient of homeodomain protein in developing forelimbs of *Xenopus* and mouse embryos. *Cell*, **55**, 1017–1024.

Oliver, G., Sidell, N., Fiske, W., Heinzmann, C., Mohandas, T., Sparkes, R. S., & DeRobertis, E. M. 1989. Complementary homeo protein gradients in developing limb buds. *Genes & Dev.*, **3**, 641–650.

Rossant, J., Zirngibl, R., Cado, D., Shagoand, M., & Giguffre, V. 1991. Expression of a retinoic acid response element-hsplacZ transgene defines specific domains of transcriptional activity during mouse embryogenesis. *Genes & Dev.*, **5**, 1333–1344.

Shen, W.-F., Detmer, K., Mathews, C. H. E., Hack, F. M., Morgan, D. A., Largman, C., & Lawrence, H. J. 1992. Modulation of homeobox gene expression alters the phenotype of human hematopoietic cell lines. *EMBO J.*, **11**, 983–989.

Shermoen, A. W., & O'Farrell, P. H. 1991. Progression of the cell cycle through mitosis leads to abortion of nascent transcripts. *Cell*, **67**, 303–310.

St. Johnston, D., & Nüsslein-Volhard, C. 1992. The origin of pattern and polarity in the *Drosophila* embryo. *Cell*, **68**, 201–219.

Stanojevic, D., Small, S., & Levine, M. 1991. Regulation of a segmentation stripe by overlapping activators and repressors in the *Drosophila* embryo. *Science*, **254**, 1385–1387.

Stock, G. B., & Bryant, S. V. 1981. Studies of digit regeneration and their implications for theories of development and evolution of vertebrate limbs. *J. Exp. Zool.*, **216**, 423–433.

Stocum, D. L. 1991. Retinoic acid and limb regeneration. *Sem. Dev. Bio.*, **2**, 199–210.

Struhl, G., Johnston, P., & Lawrence, P. A. 1992. Control of *Drosophila* body pattern by the *hunchback* morphogen gradient. *Cell*, **69**, 237–249.

Symes, K., Yaqoob, M., & Smith, J. C. 1988. Mesoderm induction in *Xenopus laevis*: responding cells must be in contact for mesoderm formation but suppression of epidermal differentiation can occur in single cells. *Development*, **104**, 609–618.

Thaller, C., & Eichele, G. 1987. Identification and spatial distribution of retinoids in the developing chick limb bud. *Nature*, **327**(625–628).

Tickle, C. 1981. The number of polarizing region cells required to specify additional digits in the developing chick wing. *Nature*, **289**, 295–298.

Susan V. Bryant et al.

Tickle, C., Summerbell, D., & Wolpert, L. 1975. Positional signalling and specification of digits in chick limb morphogenesis. *Nature*, **254**, 199–202.

Tickle, C., Alberts, B., Wolpert, L., & Lee, J. 1982. Local application of retinoic acid to the limb bond [*sic*] mimics the action of the polarizing region. *Nature*, **296**, 564–566.

Wanek, N., Gardiner, D. M., Muneoka, K., & Bryant, S. V. 1991. Conversion by retinoic acid of anterior cells into ZPA cells in the chick wing bud. *Nature*, **350**, 81–83.

Wilkins, A. S., & Gubb, D. 1991. Pattern formation in the embryo and imaginal discs of *Drosophila*: What are the links? *Dev. Biol.*, **145**, 1–12.

Wolpert, L. 1969. Positional information and the spatial pattern of cellular differentiation. *J. Theor. Biol.*, **25**, 1–47.

Yokouchi, Y., Sasaki, H., & Kuroiwa, A. 1991. Homeobox gene expression correlated with the bifurcation process of limb cartilage development. *Nature*, **353**, 443–445.

7. THE DEVELOPMENT OF A SPATIAL PATTERN IN A MODEL FOR CANCER GROWTH

Mark A. J. Chaplain

School of Mathematical Sciences
University of Bath
Claverton Down
Bath BA2 7AY, UK

7.1 INTRODUCTION

Solid tumour growth is a very complicated phenomenon which presents the mathematical modeller with a correspondingly complex set of problems to solve. Deciding which simplifying assumptions to make is a non-trivial task. Experimentally it is possible to grow small avascular nodules (multicell spheroids) whose growth kinetics approximate *in vivo* tumours such as carcinoma. These spherical colonies of cells receive nutrients and dispose of waste products via diffusion processes alone and as such reach a diffusion-limited steady state size a few millimetres in diameter. As the tumour grows and increases in size, cells towards the centre of the tumour are starved of vital nutrients and die. A central necrotic core is thus formed which is surrounded by a thin outer layer of live, proliferating cells.

Previous mathematical models (Burton 1966; Greenspan 1972; Greenspan 1974; Thomlinson & Gray 1955) provide informative descriptions of *in vitro* growth of a tumour that is cylindrical or spherical (or one-dimensional). In these ideal geometries, the net resultant of all intercellular forces within the colony is quite simple in that it is always radially directed; no more precise examination of local cell dynamics is needed. However, this is not the case for growth under asymmetric conditions or for the displacements produced by non-uniform agglomeration. Analysis of these more realistic circumstances requires the specification of the macroscopic laws which govern the jostling movement of crowded cells within an expanding or contracting culture. There is a paucity of relevant data on this subject, but a very reasonable assumption is that gross internal forces may be characterised by a pressure distribution, the non-uniformities of which effect cell motion. In addition, cell adhesion, like molecular attraction, is assumed to produce a surface tension at the outer boundary of the colony that maintains compactness and counteracts internal expansive pressures (it is well known that multicell spheroids are held together by a variety of cell–cell junctions including desmosomes, tight junctions, junctional complexes and gap junctions). The quasi-balance of these forces causes what may be termed regular growth or demise of the population. However, a serious imbalance can occur which affects the basic stability and organisation of the culture. If surface tension is overcome at some stage of development, further colony growth is asymmetric, more spheroidal and subdivision is a possibility when distortions are truly large.

Experimental and Theorietcal Advances in Biological Pattern Formation,
Edited by H.G. Othmer *et al.*, Plenum Press, New York, 1993

This behaviour may relate to the results of similar experiments in which colonies of one type attain equilibrium as spheres or spheroids (Folkman & Hochberg 1973), while another class disintegrates at some stage of growth (Sutherland *et al.* 1971).

7.2   THE MATHEMATICAL MODEL

In this section we present a mathematical model for the growth of a solid tumour which develops and extends an original model due to Greenspan (1976). As we have seen in the introduction, the growth of a tumour is an extremely complex process involving many different variables, and so in order to focus on the main qualitative features of shape and structure, a number of simplifying but realistic approximations are invoked and set out below. These are based on observations from experimental studies on multicell spheroids (Sutherland 1988).

• Although the growth of solid tumours in animals always involves some vascularisation (*i. e.* the existence of a network of blood vessels within the tumour), the earliest stages of development are apparently regulated by the direct diffusion of nutrients and wastes from and to the surrounding tissues. The tumour (or cell culture) and the surrounding tissues (or ambient medium) then, are assumed to be essentially in a state of diffusive equilibrium at all times. The time for diffusion of oxygen (thought to be an indispensable nutrient) over the sizes which characterise the tumour's size and structure is very small compared with the total period of growth. For example, it takes about 10 seconds (1000 seconds) for oxygen to diffuse across a distance of 100μ (1000μ). Both of these time intervals are certainly very short compared to a growth period measured in days, and they may even be regarded as small relative to the twelve- or eighteen-hour period for mitosis (in epithelial cells of a Chinese hamster). The approximation of instantaneous diffusive equilibrium is based on these disparate time periods.

When the tumour is small, every cell receives adequate nourishment by simple diffusion. However, the consumption of a nutrient means that its concentration must decrease towards the centre of the tumour. Eventually the concentration there of a vital nutrient falls below the critical level to sustain life and a central necrotic core develops. We thus restrict ourselves to a discussion of tumours in a stage of growth that has a two-layer structure; an outer shell of viable or proliferating cells that envelops a larger core of necrotic debris. Only the proliferating cells consume the nutrient.

• A tumour cell dies when the available concentration of a vital nutrient denoted by $\sigma(x, y, z, t)$ falls below a critical level σ_1. The local thickness h of the thin layer of living cells depends only on σ_1 and the value of σ at the outer surface of the tumour. This relationship is approximated by

$$h = \begin{cases} \nu\sqrt{\sigma - \sigma_1}, & \text{if } \sigma > \sigma_1 \\ \\ 0, & \text{if } \sigma < \sigma_1 \end{cases} \tag{7.1}$$

a formula known to be exact for the ideal geometries mentioned in the introduction (*cf.* Greenspan 1974).

• All living tumour cells are identical and each is considered to be an incompressible structure of constant volume. Cell division occurs 'instantaneously' relative to the growth time of the tumour and each daughter cell occupies the same volume as any other cell of the population.

• The mitotic index is constant. If dA is an element of surface area of the tumour then the incremental volume of live cells, $hdA = dV$ creates a new cell volume at the rate βhdA. Nutrient is consumed by this volume at the rate γhdA.

• There is an adhesion or surface-tension among the living tumour cells, modelling the cell–cell junctional contacts, which, in a spherical geometry, produces an inward pressure that maintains the tumour as a compact, solid and continuous mass. This surface-tension force T is proportional to the mean curvature of the boundary.

• Necrotic cellular debris continually disintegrates into simpler chemical compounds that are freely permeable through cell membranes. The mass, or equivalently, the cell volume lost in this way in necrosis is replaced by cells pushed inward by the forces of adhesion and surface-tension, and also by stroma and blood vessels. This rate of loss of necrotic mass (or cell volume) per unit volume of debris is taken to be a constant.

• The birth or death of cells produces internal pressure differentials which cause the motion of cellular material. This slow drift is assumed to be governed by

$$\vec{q} = -\nabla p \tag{7.2}$$

where $\vec{q}(x, y, z, t)$ is the particle velocity and $p(x, y, z, t)$ is proportional to the internal pressure.

These last three assumptions constitute an explanation of why cell proliferation can and indeed must continue even when the tumour is in a quiescent steady state. In equilibrium, cells produced in the growth layer flow inward, exactly compensating the loss of cell volume in the necrotic centre. Moreover, a tumour grows when the rate of cell production exceeds the mass and volume loss due to necrotic disintegration. In the reverse situation, the tumour contracts. The loose analogy is to view the tumour as a fluid held together by adhesive forces. Mitosis or cell proliferation acts like a source of incompressible fluid, while necrosis has the role of a fluid sink.

The first objective is to replace the chemical and dynamical processes within the thin layer of viable cells by a set of equivalent conditions that relate variable changes across this region of the tumour. To this end let the outer surface be represented by

$$\Gamma(x, y, z, t) = 0 \ .$$

Conservation of mass applied to a small volume element on the surface of the spheroid implies that since h is small, the mass/volume flow out of the ends of the element $(\vec{q}_+ \cdot \hat{n} - \vec{q}_- \cdot \hat{n})dA$, is approximately equal to the rate of mass/volume production within this small volume, $\beta h dA$. Therefore,

$$\vec{q}_+ \cdot \hat{n} = \vec{q}_- \cdot \hat{n} + \beta h \ . \tag{7.3}$$

Similarly, the rate of nutrient diffusion into dV through the outer surface, $k\hat{n} \cdot \nabla \sigma dA$, is equal to the rate at which nutrient is consumed in this small volume, $\gamma h(\sigma)dA$, that is

$$k\hat{n} \cdot \nabla \sigma = \gamma h(\sigma) \ . \tag{7.4}$$

Since $\sigma = \sigma_1$ in the necrotic core, there is no diffusive transport from the interior. Here the convection of the nutrient across the surface is assumed to be negligible compared to that transported by diffusion in an equal time.

Furthermore, let the proliferation rate be large so that its product with the very small thickness results in an order one quantity, *i. e.*

$$\beta h = \beta \nu \sqrt{\sigma - \sigma_1} = \lambda \sqrt{\sigma - \sigma_1} \ , \quad \text{with } \lambda = \mathcal{O}(1) \ .$$

Similarly,

$$\gamma h = \gamma \nu \sqrt{\sigma - \sigma_1} = \mu k \sqrt{\sigma - \sigma_1} \ , \quad \text{with } \mu = \mathcal{O}(1) \ .$$

To this order of accuracy, the pressure and the tangential velocity components are continuous across the stratum of live cells:

$$p_+ = p_- \; ,$$

$$\vec{q}_+ \times \hat{n} = \vec{q}_- \times \hat{n} \; .$$

The pressure on the surface of the tumour must equal the surface tension force, and if κ_1 and κ_2 are the principal curvatures of the boundary, then

$$p = \frac{\alpha}{2}(\kappa_1 + \kappa_2) = \alpha\kappa \; , \qquad (7.5)$$

where α is a constant and κ is the mean curvature of the boundary.

The preceding conditions apply at the moving interface $\Gamma(x, y, z, t) = 0$, and the equation of motion of a point on this surface is

$$\frac{d\vec{r}}{dt} = \vec{q}_+ \; . \qquad (7.6)$$

The initial geometry of the tumour, of course, must be prescribed.

In the core where the rate of volume loss per unit volume is a constant, S_i, the expression of mass conservation is

$$\nabla \cdot \vec{q} = -S_i \; . \qquad (7.7)$$

The equation for the nutrient concentration σ, which is assumed to be in diffusive equilibrium, is, outside the colony

$$\nabla^2 \sigma = 0 \; , \qquad (7.8)$$

whereas in the necrotic interior

$$\sigma = \sigma_1 \; . \qquad (7.9)$$

A source of nutrient in the finite domain adds an inhomogeneous term to the right-hand side of (7.8). The formulation is completed by specifying the manner in which the tumour is supplied nourishment.

Two interesting problems on the time dependent response of a tumour that can be examined with the theory as formulated are:
• growth with a constant supply of nutrient at infinity $i.\,e.$ $\sigma \to \sigma_\infty$ as $|\vec{r}| \to \infty$;
• the effect of a nearby source of nutrient.
Phenomena with moving boundaries are very difficult to treat analytically and recourse to numerical procedures is eventually necessary in order to obtain complete solutions to any of these.

By using (7.2) in conjunction with (7.7), and hence eliminating the velocity vector from the problem, the mathematical formulation can be given entirely in terms of the scalar functions p and σ. A convenient form for the first boundary value problem to be studied retains interior pressure and the velocity of the tumour surface:

$$\nabla^2 p = S_i \; , \quad \text{inside } \Gamma = 0 \; , \qquad (7.10)$$

$$\nabla^2 \sigma = 0 \; , \quad \text{outside } \Gamma = 0 \; , \qquad (7.11)$$

where on $\Gamma(x, y, z, t) = 0$

$$
\begin{cases}
p = \dfrac{\alpha}{2}(\kappa_1 + \kappa_2) = \alpha\kappa\,, \\[2mm]
\vec{q}_+ \cdot \hat{n} = -\hat{n} \cdot \nabla p + \lambda\sqrt{\sigma - \sigma_1}\,, \\[2mm]
\vec{q}_+ \times \hat{n} = -\nabla p \times \hat{n}\,, \\[2mm]
\hat{n} \cdot \nabla\sigma = \mu\sqrt{\sigma - \sigma_1}\,.
\end{cases}
\tag{7.12}
$$

The bounding surface is defined by

$$
\frac{d\vec{r}}{dt} = \vec{q}_+\,,
\tag{7.13}
$$

and the initial parametric prescription

$$
\vec{r} = \vec{a}(\xi,\ \zeta) \quad \text{at} \ \ t = 0\,.
\tag{7.14}
$$

7.3   THE STABILITY OF GROWING TUMOURS

Under conditions of complete spherical symmetry all cell variables depend only on r and t, and the tumour grows as a sphere of radius $R(t)$. The boundary problem summarised in $(7.10) - (7.13)$ then reduces to

$$
\frac{1}{r^2}\frac{\partial}{\partial r}\left(r^2\frac{\partial p}{\partial r}\right) = S_i\,, \quad r \leq R(t)
\tag{7.15}
$$

$$
\frac{1}{r^2}\frac{\partial}{\partial r}\left(r^2\frac{\partial \sigma}{\partial r}\right) = 0\,, \quad r \geq R(t)
\tag{7.16}
$$

with boundary conditions

$$
\sigma \to \sigma_\infty \quad \text{as} \ \ r \to \infty\,,
\tag{7.17}
$$

and

$$
\begin{cases}
p = \dfrac{\alpha}{R(t)}\,, \\[2mm]
R'(t) = -\dfrac{\partial p}{\partial r} + \lambda\sqrt{\sigma - \sigma_1}\,, \\[2mm]
\dfrac{\partial \sigma}{\partial r} = \mu\sqrt{\sigma - \sigma_1}\,,
\end{cases}
\tag{7.18}
$$

on $r = R(t)$ and $R(0) = a$. The third equation in the boundary conditions (7.12) *i.e.* $\vec{q}_+ \times \hat{n} = -\nabla p \times \hat{n}$, is trivially satisfied since $\vec{q}_+ = (R'(t),\ 0,\ 0)$, $\nabla p = \left(\dfrac{\partial p}{\partial r},\ 0,\ 0\right)$, $\hat{n} = (1,\ 0,\ 0)$ and both vector products give $\mathbf{0}$.

It is convenient to use the relative nutrient concentration $\sigma - \sigma_1$ as a basic variable and this can be accomplished either through a variable change or by setting $\sigma_1 = 0$. The latter convention will be used from now on.

Since the time variable enters the system only in determination of the tumour radius, the preceding system of equations can be solved explicitly for p and σ:

$$
p = \frac{S_i}{6}\left(r^2 - R^2\right) + \frac{\alpha}{R}\,,
\tag{7.19}
$$

$$\sigma = \sigma_\infty - \left(\frac{2\mu\sigma_\infty R^2}{\mu R + \sqrt{\mu^2 R^2 + 4\sigma_\infty}} \right) \frac{1}{r} \,. \tag{7.20}$$

The substitution of these formulae into (7.18) yields a single ordinary differential equation for $R(t)$:

$$R'(t) = -\frac{1}{3} S_i R + \frac{2\lambda\sigma_\infty}{\mu R + \sqrt{\mu^2 R^2 + 4\sigma_\infty}} \,, \tag{7.21}$$

with $R(0) = a$. The last equation can then be integrated to yield an unwieldy closed form solution which will not be quoted here since only the qualitative character of $R(t)$ is of interest.

The tumour (cell colony) grows initially at a linear rate and asymptotes to the steady state radius

$$R_\infty = \frac{3\lambda}{S_i} \left(\frac{\sigma_\infty}{1 + 3\lambda\mu/S_i} \right)^{1/2} \,, \tag{7.22}$$

in a time scale regulated mainly by the rate of volume loss of necrotic debris (this can be obtained by setting the right hand side of (7.21) to zero).

The surface tension which maintains the compactness of the tumour has an otherwise implicit role in the development of the growth of the spherical tumour (cell colony). However, it is of crucial importance in determining whether such symmetrical growth is indeed stable and hence attainable.

Growth is judged unstable to infinitesimal perturbations if any such disturbance amplifies at an exponential rate (exceeding that of the radius $R(t)$). In this circumstance, the instabilities radically alter the shape of the colony and can even lead to fracturing into two or more pieces. The tumour becomes unstable if and when it reaches a critical size beyond which surface tension is overcome by pressure forces. The precise criterion for this is obtained from the analysis that follows.

We now consider again the growth of a spherical tumour (cell colony) which is now subjected to small deviations that are always inevitably present in a real environment. Originally, for simplicity, Greenspan (1976) took the perturbations from complete sphericity to be axially symmetric i. e. independent of the azimuthal angle ϕ in spherical polar coordinates (r, θ, ϕ). Here, however, for completeness, we will include an azimuthal dependence in our perturbations.

To this end, we let $\bar{p}(r,t)$, $\bar{\sigma}(r,t)$ and $R(t)$ denote the basic state of motion given by equations (7.19) – (7.21) and $\epsilon\tilde{p}(r,\theta,\phi,t)$, $\epsilon\tilde{\sigma}(r,\theta,\phi,t)$ and $\epsilon\xi(\theta,\phi,t)$ the perturbations therefrom. The total pressure distribution and nutrient concentration are then represented by

$$\begin{cases} p(r,\theta,\phi,t) = \bar{p}(r,t) + \epsilon\tilde{p}(r,\theta,\phi,t) \\[2mm] \sigma(r,\theta,\phi,t) = \bar{\sigma}(r,t) + \epsilon\tilde{\sigma}(r,\theta,\phi,t) \end{cases} \tag{7.23}$$

and the equation of the moving surface is now given by

$$\Gamma(r,\theta,\phi,t) = r - R(t) - \epsilon\xi(\theta,\phi,t) = 0 \,. \tag{7.24}$$

These expressions are substituted into (7.10)–(7.23) and all terms are developed as power series in ϵ, including the mean curvature and the unit normal vector of the distorted surface. This application of classical perturbation theory immediately yields (7.15)–(7.18), as well as the equations governing the deviations from the basic evolutionary state. It follows that

$$\nabla^2 \tilde{p} = 0 \,, \quad r \le R(t) \,, \tag{7.25}$$

$$\nabla^2 \tilde{\sigma} = 0 \ , \qquad r \geq R(t) \ , \tag{7.26}$$

with $\tilde{\sigma} \to 0$ as $r \to \infty$, while on $r = R(t)$ we have

$$\frac{\partial \xi}{\partial t} = -\left(\frac{\partial^2 \bar{p}}{\partial r^2}\xi + \frac{\partial \tilde{p}}{\partial r}\right) + \frac{\lambda}{2\sqrt{\bar{\sigma}}}\left(\xi \frac{\partial \bar{\sigma}}{\partial r} + \tilde{\sigma}\right) \ , \tag{7.27}$$

$$\frac{\partial^2 \bar{\sigma}}{\partial r^2}\xi + \frac{\partial \tilde{\sigma}}{\partial r} = \frac{\mu}{2\sqrt{\bar{\sigma}}}\left(\xi \frac{\partial \bar{\sigma}}{\partial r} + \tilde{\sigma}\right) \ , \tag{7.28}$$

$$\frac{\partial \bar{p}}{\partial r}\xi + \tilde{p} = -\frac{\alpha}{2R^2}\left[\frac{\partial}{\partial \eta}\left((1-\eta^2)\frac{\partial \xi}{\partial \eta}\right) + 2\xi + \frac{1}{\sin^2 \theta}\frac{\partial^2 \xi}{\partial \phi^2}\right] \ , \tag{7.29}$$

where

$$\eta = \cos\theta \ .$$

Modal solutions of this homogeneous system are given by

$$\tilde{p} = A_n(t)r^n P_n^m(\cos\theta)\cos m\phi \ , \tag{7.30}$$

$$\tilde{\sigma} = B_n(t)P_n^m(\cos\theta)\cos m\phi / r^{n+1} \ , \tag{7.31}$$

$$\xi = C_n(t)P_n^m(\cos\theta)\cos m\phi \ , \tag{7.32}$$

where $n = 0, 1, 2, 3, \ldots$, $0 \leq m \leq n$ and $P_n^m(\cos\theta)$ are the associated Legendre polynomials defined by

$$P_n^m(\cos\theta) = \sin^m\theta\frac{d^m}{d(\cos\theta)^m}[P_n(\cos\theta)]$$

(in the case $m = 0$ these reduce to the Legendre polynomials $P_n(\cos\theta)$ and the whole analysis corresponds to Greenspan's, where there is no azimuthal dependence).

Also, we have from the boundary equations, the following relations:

$$A_n(t) = -\frac{C_n(t)}{R^n}\left[\bar{p}_r - \frac{\alpha}{2R^2}(n+2)(n+1)\right]_{r=R} \ , \tag{7.33}$$

$$B_n(t) = R^{n+2}C_n(t)\left[\frac{\bar{\sigma}_{rr}\sqrt{\bar{\sigma}} - \frac{\mu}{2}\bar{\sigma}_r}{(n+1)\sqrt{\bar{\sigma}} + \frac{\mu}{2R}}\right]_{r=R} \ , \tag{7.34}$$

$$\frac{dC_n(t)}{dt} = C_n(t)\left[-\bar{p}_{rr} + \frac{n}{R}\bar{p}_r - \frac{\alpha n}{2R^3}(n-1)(n+2) + \frac{\lambda}{2}\left(\frac{(n+1)\bar{\sigma}_r + R\bar{\sigma}_{rr}}{(n+1)\sqrt{\bar{\sigma}} + \mu R/2}\right)\right]_{r=R} \tag{7.35}$$

The result of replacing $\bar{p}$, $\bar{\sigma}$ and their derivatives by explicit expressions obtained from equations (7.19) and (7.20) is the fundamental equation governing the growth of infinitesimal instabilities:

$$C_n'(t) = (n-1)f(n, R)C_n(t) \ , \tag{7.36}$$

where

$$f(n, R) = \frac{S_i}{3} - \frac{\alpha n(n+2)}{2R^3} + \frac{(\lambda\mu/2)}{(n+1) + \frac{\mu R}{4\sigma_\infty}\left[\mu R + (\mu^2 R^2 + 4\sigma_\infty)\right]^{1/2}} \ . \tag{7.37}$$

The disturbance amplifies or decays according to whether the product $(n-1)f(n,R)$ is positive or negative. The tumour development is unstable if small perturbations can amplify - otherwise it is stable. For $n = 0$, the time dependent coefficient in $-f(0,R)$ in equation (7.36) is always negative so that this mode is definitely stable. For $n = 1$, $C_1(t)$ is a constant, which is compatible with the fact that this mode represents a translation of axes.

The modes $n \geq 2$ are genuinely unstable when $f(n,R) > 0$. Moreover, since $f(n,R)$ is a decreasing function of the index n for $n \geq 1$, the mode $n = 2$ is the first that can become unstable, and it must always be a component of any growing disturbance. The minimum critical radius R_c at which the developing tumour can become unstable is then given by

$$f(2, R_c) = 0 \ . \tag{7.38}$$

If $R_c > R_\infty$, the steady state radius of the tumour (see (7.22)), then the tumour is stable during its entire growth and attains its symmetric equilibrium configuration. If however $R_c < R_\infty$, then the tumour becomes unstable at some definite time in its growth when small disturbances amplify and change both the structure and the shape of the tumour. In this case, the changing pressure distribution overcomes the surface tension before spherical equilibrium is reached.

The onset of stability in the mode $n = 2$ is manifested as a pinch in the outer surface around the equatorial region. As the tumour grows and increases in size, other modes become unstable and more radical changes in configuration may occur, which can be used to model the tumour invading the surrounding tissues. This process is described in the following section.

7.4   THE LOCAL INVASION OF THE SURROUNDING TISSUES

Clinically, tumours can be classified into two main types - benign and malignant - although these two terms really represent the extremes of a spectrum of tumour growth. We now wish to show that the predictions of, and the parameters involved in, the fundamental equations governing the growth of infinitesimal instabilities - (7.36), (7.37) - and also the modes of instability arising from these equations, describe very well the characteristics of, and the situations which arise in, malignant, benign and those tumours of an intermediate nature. We turn first of all to malignant tumours (*i.e.* cancers) and in particular to the type which makes up the great majority (90%) of malignant tumours, carcinomas [Greek: *Karkinos*, a crab]

At the margin of a carcinoma, single tumour cells and irregular columns or projections of cells infiltrate into the surrounding tissues, extending from the central mass like the legs of a crab. As they proliferate, the invading cells compress the tissue cells and interfere with their blood supply, and highly specialised cells such as the parenchymal cells of internal organs are destroyed. Fibrovascular tissue is more resistant and its growth is stimulated by the tumour cells, which thus secure a supporting stroma and blood supply.

Growth occurs most readily along lines of least resistance such as planes of loose connective tissue. Denser tissue, for example fibrous fasciae, the walls of arteries, cartilage and compact bone, are relatively resistant to invasion and while this is largely explained by their structure, there is evidence that constituent pieces of cartilage can both interfere with the growth of cancer cells and inhibit angiogenesis.

Local invasion is very important, not only because it increases the difficulty of excising a carcinoma completely, but also for its effects. In particular, invasion in and around the walls of hollow viscera or gland ducts very commonly causes their obstruction, while involvement of nerves may cause pain. The sooner a carcinoma gives rise to local effects of this sort, the greater the chance of diagnosis and complete removal.

This invasive power of cancers may be attributed to three main factors:

• the abnormal motility and amoeboid movement of some cancer cells.

• the secretion of lytic enzymes *e. g.* hyaluronidase which dissolve the ground substance and digest formed tissue elements and damage cell membranes.

• the reduced adhesiveness of cancer cells to one another.

When considered alongside two other important features of malignant tumours, *i. e.* a large rate of cell production and extensive necrosis, we see from (7.37), that high modes of instability are possible.

We recall that

$$f(n, R) = \frac{S_i}{3} - \frac{\alpha n(n+2)}{2R^3} + \frac{(\lambda\mu/2)}{(n+1) + \frac{\mu R}{4\sigma_\infty}[\mu R + (\mu^2 R^2 + 4\sigma_\infty)]^{1/2}} \ .$$

Thus the combination of a high value for S_i (extensive necrosis), a high value for λ (fast proliferation rate) and a small value for the surface tension parameter α (reduced cell-to-cell adhesiveness) leads to the possibility of an instability being dominated by a high mode number n. The mode n that actually dominates the instability is selected by the function $R(t)$ which sets the amplification rate and the time available for each perturbation to amplify, while the selection of the m-mode [*cf.* $P_n^m(\cos\theta)$] will be determined by external influences such as the denseness of the surrounding tissue and the external distribution of the enzymes secreted by the tumour. Surface energy considerations may also play a part, and these will be discussed in the next section, along with cell mobility. Final domination by a mode such as $P_6^3(\cos\theta)$ leads to the typical crab-like appearance assumed by carcinomas, as illustrated in Figures 7.1(a) - (f). These show dominant modes $P_2^0 + P_5^2, P_4^2, P_5^2, P_5^2, P_4^2, P_5^3$, respectively.

At the other end of the spectrum, we have benign tumours. Generally these show little or no necrosis, have a slower rate of cell proliferation than malignant tumours and are very often surrounded by a fibrous capsule. Even when no such capsule exists, the margin between the tumour and the surrounding tissues remains sharp, without evidence of local invasion. Again, upon consideration of the function $f(n,R)$, this time with a low value for S_i (little sign of necrosis), a low value for λ (slower cell proliferation rate) and a high value for α the surface-tension parameter (modelling the effect of the fibrous capsule), we see that the possibility of domination by a high mode number n is very small. Indeed, the presence of the capsule, which keeps the tumour a compact spherical mass, effectively rules out the growth of any small instability except perhaps the mode $n = 0$, which is simply associated with radial growth. This models the non-invasive nature of benign tumours.

For tumours of an intermediate nature, *i.e.* combining some of the properties of both malignant and benign tumours, the choice of mode number n will not be as clear cut, depending upon a more complicated interplay between the parameters involved and the external influences. This will vary upon the degree of malignancy from tumour to tumour.

7.5   THE SPREAD OF TUMOUR CELLS BY LYMPHATICS AND THE BLOOD

As well as invading surrounding tissues, tumour cells also invade the walls of lymphatics and blood vessels in and around the tumour. This gives rise to the possible spread of the tumour to other parts of the body, where secondary tumours or metastases then form. We examine this process of metastasis in the light of the abnormal motility of some cancer cells and the reduced cell adhesiveness using the third equation of (7.12).

Smaller lymphatics, whose walls consist of little more than an endothelium with loose intercellular junctions and an incomplete basement membrane, are readily invaded by

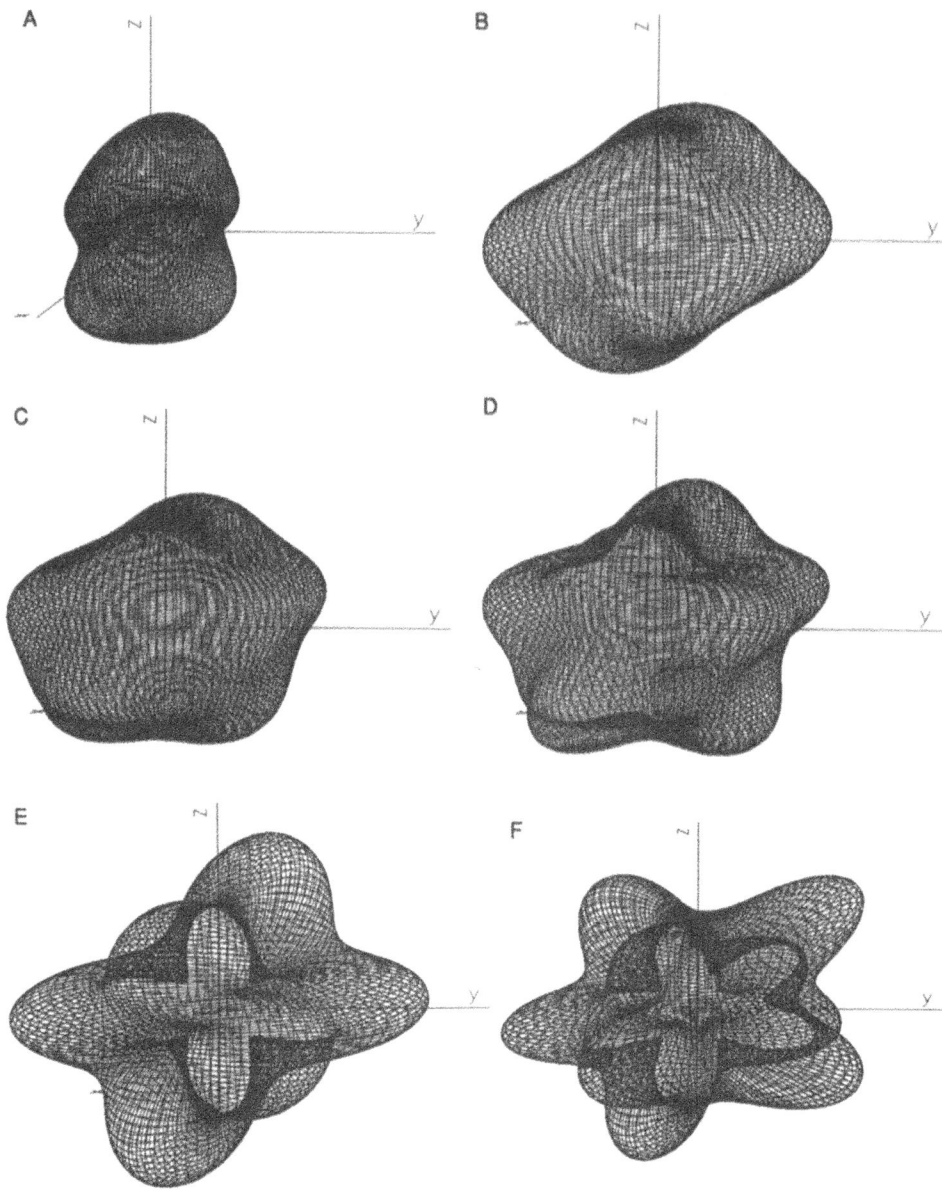

Figure 7.1.

carcinoma cells which may either become detached and carried in the lymph to the next draining lymph node, or they may remain attached at the site of invasion and proliferate within the lymphatic to form a continuous column of tumour which distends and obstructs the lumen.

Carcinoma cells also invade the walls of small vessels, particularly venules, in and around the tumour, and single cancer cells or small groups of cells frequently become detached and are carried away in the venous blood where they are likely to become impacted as emboli in the next microvascular network.

There is clinical and experimental evidence that rough handling of a malignant tumour promotes the escape of tumour cells into the blood and increases the risk of metastasis. Like the primary tumour, metastases are locally invasive and their cells also enter the lymphatics and blood vessels and cause further dissemination by these routes.

Using the third equation of (7.12), *i. e.*

$$\vec{q}_+ \times \hat{n} = -\nabla p \times \hat{n}$$

enables us to follow the locus of a particular cell as it moves about on the surface of the tumour. Initially, since all variables depend only upon r and t, the velocity of a particular cell has a radial component only *i. e.* $\vec{q}_+ = (R'(t), 0, 0)$. Upon consideration of the introduction of small perturbations, we assume that each cell now has tangential components of velocity, describing its possible motion in both the θ- and ϕ-directions. Assuming that the new angular velocity components are of first order in ϵ, we thus have

$$\vec{q}_+ = (R'(t) + \epsilon\xi_t, \ \epsilon R\dot{\theta}, \ \epsilon R\dot{\phi}\sin\theta) \tag{7.39}$$

where the dots denote differentiation w.r.t. time.

Calculation of the vector products to order ϵ, then gives us the following equations for the angular components of velocity:

$$\begin{cases} R\dot{\phi}\sin^2\theta = -\left[\dfrac{2S_i}{3} + \dfrac{R'}{R}\right]\xi_\phi - \dfrac{\tilde{p}_\phi}{R} \\[4mm] R\dot{\theta} = -\left[\dfrac{2S_i}{3} + \dfrac{R'}{R}\right]\xi_\theta - \dfrac{\tilde{p}_\theta}{R}. \end{cases} \tag{7.40}$$

For highly malignant tumours, we saw that it was possible for an instability to be dominated by a high mode number n. It is also known that certain malignant tumour cells display an abnormal degree of cell motility. Thus we can see from (7.40) that if we have a dominant high mode number then this allows for the cells on the surface of the tumour to have fairly substantial movement in both the θ- and the ϕ-directions and so to move about quite freely over the tumour surface. When this is viewed in the light of the fact that malignant tumours have a greatly reduced cell to cell adherence and also that the surface tension distribution over the tumour surface is given by

$$\frac{\alpha}{R} + \frac{\epsilon\alpha C_n(t)}{2R^2}(n-1)(n+2)P_n^m(\cos\theta)\cos m\phi \ , \tag{7.41}$$

we see that a cell or groups of cells may come across areas of the tumour surface where the surface tension is reduced, and so there is the possibility that single cells or groups of cells may become detached and carried away in the bloodstream, as previously described, leading to the formation of secondary tumours elsewhere in the body. Again, the more malignant the tumour is, the more mobile cells will be on the surface and so the more likely metastasis is to occur.

7.6 SELECTION OF THE MODE OF INSTABILITY

Since the onset of instability in a malignant tumour and its subsequent invasion of the surrounding tissues is characterised by a process involving a certain furrowing of the tumour surface and the projection of crab-like legs of tumour cells from the central mass outwards into the tissue, we attempt to describe this invasiveness partly as the result of a minimisation procedure on a function which in some way represents a generalised surface energy. The invasion of the surrounding tissue by a malignant tumour can then be viewed partly as a series of transformations to successively higher energy levels. To this end, we assume that there exists some field function u defined over the surface of the tumour such that its nodal lines represent the lines of least resistance to a furrowing or "pinching" process manifested in the onset of instability of the tumour. In our case, it is assumed that u is some sort of order-disorder parameter relating to the organisation of the live, proliferating cells on the tumour surface. It is reasonable to suppose (Goodwin & Trainor 1980; Landau & Lifschitz 1959) in problems of this nature that the surface energy density is given by an expression of the form

$$\mathcal{E}(\theta, \phi) = D\left[\left(\frac{\partial u}{\partial \theta}\right)^2 + \frac{1}{\sin^2 \theta}\left(\frac{\partial u}{\partial \phi}\right)^2 + \beta u^2\right], \tag{7.42}$$

where D and β are constants incorporating the physiological properties of the tumour surface (e. g. cell to cell adhesion, cell mobility etc.). The total surface energy E is then given by:

$$E = \int_{dA} \mathcal{E}(\theta, \phi) dA. \tag{7.43}$$

We now suppose that the onset of instability corresponds to a minimum of E i. e.

$$\delta E = \delta \int_0^{2\pi} \int_0^{\pi} \mathcal{E}(\theta, \phi) \sin \theta d\theta d\phi = 0, \tag{7.44}$$

subject to a conservation law on u^2

$$\int_0^{2\pi} \int_0^{\pi} u^2(\theta, \phi) \sin \theta d\theta d\phi = 1. \tag{7.45}$$

Equations (7.44), (7.45) are then required to satisfy the Euler-Lagrange equation

$$\frac{1}{\sin \theta}\frac{\partial}{\partial \theta}\left(\sin \theta \frac{\partial u}{\partial \theta}\right) + \frac{1}{\sin^2 \theta}\frac{\partial^2 u}{\partial \phi^2} - \gamma u = 0, \tag{7.46}$$

where γ incorporates the parameter β and an undetermined multiplier of (7.45).

With the conditions that u must be finite, single-valued and continuous over the tumour, this leads to the solution

$$u(\theta, \phi) = \sqrt{2} N_{mn} P_n^m(\cos \theta) \cos m\phi, \tag{7.47}$$

where N_{mn} are normalisation constants and the parameter γ takes the value $n(n + 1)$. From (7.43) we can now easily calculate the surface energy corresponding to each mode number n. Thus we have

$$E_n = D[n(n + 1) + \beta]. \tag{7.48}$$

However, not every characteristic state of (7.48) is realised in the growth and invasion process, since there are many external influences which impose restrictions on the choice of mode number n. The degree of malignancy of the tumour will also play a part. For highly malignant tumours whose cells have reduced adhesion to one another and greater mobility, coupled with the fact that secretion of enzymes by the tumour will affect the surrounding tissues, we envisage the possibility of dominance by a high mode number n which gives the tumour a sufficiently large surface energy E to invade the surrounding tissues extensively.

7.7 MODIFICATION OF THE NUTRIENT SUPPLY

We now consider the behaviour of a tumour which has reached its steady state radius R_∞ (and as such is in a stable state of equilibrium) to a slight modification in its supply of nutrient. We can model this by placing a weak source or sink of nutrient somewhere in the surrounding tissue. The reaction of a tumour to a nearby sink models the interaction of two neighbouring tumours in competition for the same nutrient supply, while the reaction of a tumour to a nearby source could be taken to model the effect of extra nutrient supply derived from neighbouring blood vessels after the secretion of Tumour Angiogenesis Factor. In order to account for a weak source placed at, say, $x = 0$, $y = 0$, $z = l$ we write (7.8) as

$$\nabla^2 \sigma = \epsilon \sigma_\infty R_\infty \delta(x)\delta(y)\delta(z - l)H(t) , \qquad (7.49)$$

where $H(t)$ is the Heaviside Function defined by

$$H(t) = \begin{cases} 1, & t > 0 \\ 0, & t < 0. \end{cases}$$

All the remaining equations and boundary conditions are the same, with the exception that we now assume that the tumour is initially in a stable state of equilibrium *i. e.* a sphere of radius R_∞. We again use perturbation theory in our analysis, replacing $R(t)$ by R_∞ and equation (7.26) by

$$\nabla^2 \tilde{\sigma} = \sigma_\infty R_\infty \delta(x)\delta(y)\delta(z - l)H(t) . \qquad (7.50)$$

Now the inhomogeneous system (7.25), (7.50), (7.27)–(7.29) is solved with initial conditions $\tilde{p} = 0$, $\tilde{\sigma} = 0$, $xi = 0$. Assuming axial symmetry, the solution to the above system is given by

$$\tilde{p} = \sum_0^\infty A_n(t)P_n(\cos\theta)r^n , \qquad (7.51)$$

$$\tilde{\sigma} = \frac{\sigma_\infty R_\infty H(t)}{4\pi \left(r^2 - 2rl\cos\theta + l^2\right)^{1/2}} + \sum_0^\infty \frac{B_n(t)P_n(\cos\theta)}{r^{n+1}} , \qquad (7.52)$$

$$\xi = \sum_0^\infty C_n(t)P_n(\cos\theta) . \qquad (7.53)$$

Using the well known expansion

$$(r^2 - 2rl\cos\theta + l^2)^{-1/2} = \sum_0^\infty \frac{r^n}{l^{n+1}}P_n(\cos\theta) \quad \text{for } r \le l, \qquad (7.54)$$

and then substituting equations (7.51)–(7.53) into (7.27)–(7.29) leads to the inhomogeneous counterparts of (7.33)–(7.35), for $n = 0, 1, 2, \ldots$. Now for the mode $n = 1$, representing a linear translation, we have

$$C_1(t) = \frac{R_\infty^2 \lambda}{16\pi l^2} \left[\mu R_\infty + \left(\mu^2 R_\infty^2 + 4\sigma_\infty\right)^{1/2}\right] t + C_1(0) . \qquad (7.55)$$

Thus we can see that the tumour begins to move towards the nutrient source with a velocity that is inversely proportional to the square of the distance l. Correspondingly, if we have a weak sink at $x = 0$, $y = 0$, $z = l$ then the tumour will move away from it.

7.8   DISCUSSION AND CONCLUSIONS

The study of the asymmetric development of a compact, spherical tumour growing initially by expansion, is based on several simplifying assumptions about the macroscopic nature of the intercellular forces. A surface-tension force proportional to the mean curvature of the tumour surface, is assumed to maintain the tumour as a compact and continuous mass, consisting of a thin layer of live cells undergoing mitosis surrounding a necrotic core. As new cells form in the outer mitotic layer and expand to their proper size, they must push aside neighbouring cells of the tumour. These forces of displacement are transmitted with attenuation, cell to cell, throughout the crowded population. The total pressure in this way causes the internal migration of cells and a drift of the entire tumour as it builds in the direction of a richer nutrient supply. A "seepage" law of motion is adopted in which internal motion is proportional to the negative gradient of pressure. In essence, the model constructed may be compared to the motion of an incompressible fluid within a closed, extensible membrane which has a distribution of sources and sinks that depend on time and other factors.

It is shown that under certain conditions, the tendency of the tumour to distort by either growth or the elimination of material in the necrotic core overcomes the stabilising effect of surface tension. The tumour development is then unstable, and it may achieve a non-linear equilibrium, pinch apart or invade extensively the surrounding tissues, depending upon the dominant modes of instability and their development in time. These possibilities are linked to the degree of malignancy of the tumour, with more malignant tumours being more likely to show asymmetrical development because of certain factors inherent in their make-up. Most notably, a greatly reduced cell to cell adhesion, which diminishes the surface-tension force holding the tumour together as a compact mass, an increased cell mobility, allowing individual cells and groups of cells to move freely over the tumour surface with the possibility that upon coming across an area of the tumour surface of reduced surface-tension, the cells may become detached and carried away in the bloodstream to form metastases in distant parts of the body. The effect of the secondary perturbations introduced in the model, as well as accounting for the deviations in pressure and nutrient supply, may also be taken to simulate the rough handling of tumours which is known to cause the detachment of cells.

By attributing a surface energy E to the live cells, it is shown that the process of invasion by a malignant tumour may be thought of in part as a transition from a low energy level (that of the initial spherical mass) to a high energy level (that of the invading malignancy), with the final dominant mode number n being sufficiently high to allow the tumour to overcome any repressive external influence, such as dense surrounding tissue, and to assume the typical crab-like shape.

Finally, we demonstrated that a tumour which has developed to its stable, steady state equilibrium can react both to the presence of another tumour nearby and to a change in its nutrient supply. In the former situation the two tumours tend to move apart, while in the latter, the tumour moves towards the extra nutrient supply.

REFERENCES

Burton, A. C. 1966. Rate of growth of solid tumours as a problem of diffusion. *Growth*, **30**, 157–176.

Folkman, J., & Hochberg, M. 1973. Self-regulation of growth in three dimensions. *J. exp. Med.*, **138**, 745–753.

Goodwin, B. C., & Trainor, L. E. H. 1980. A field description of the cleavage process in embryogenesis. *J. Theor. Biol.*, **86**, 757–770.

Greenspan, H. P. 1972. Models for the growth of a solid tumor by diffusion. *Stud. ApplMath.*, **51**, 317–340.

Greenspan, H. P. 1974. On the self-inhibited growth of cell cultures. *Growth*, **38**, 81–95.

Greenspan, H. P. 1976. On the growth and stability of cell cultures and solid tumors. *J. Theor. Biol.*, **56**, 229–242.

Landau, L. D., & Lifschitz, E. M. 1959. *Theory of Elasticity*. London: Pergamon Press.

Sutherland, R. M. 1988. Cell and environment interactions in tumour microregions: the multicell spheroid model. *Science*, **240**, 177–184.

Sutherland, R. M., McCredie, J. A., & Inch, W. R. 1971. Growth of multicell spheroids in tissue culture as a model of nodular carcinomas. *J. Natn. Cancer Inst.*, **46**, 113–117.

Thomlinson, R. H., & Gray, L. H. 1955. The histological structure of some human lung cancers and the possible implications for radiotherapy. *Br. J. Cancer*, **9**, 539–549.

8. SEQUENTIAL AND SYNCHRONOUS SKIN PATTERN FORMATION

Gerhard C. Cruywagen[1], P. K. Maini[2] and J. D. Murray[1]

[1]Department of Applied Mathematics FS-20
University of Washington
Seattle, WA 98195, USA

[2]Centre for Mathematical Biology
Mathematical Institute
24–29 St. Giles'
Oxford OX1 3LB, UK

8.1 INTRODUCTION

Mathematical modelling has become a widely accepted method for examining how and why vertebrate skin structures are laid down in an orderly and organized fashion. Although various theoretical models have been proposed for examining the morphogenetic processes responsible for the large variety of patterns observed on animal skin, these processes are still not well understood. By examining a mechanochemical tissue interaction model based on recent experimental evidence we therefore hope to contribute towards the understanding of skin morphogenesis.

Vertebrate skin is composed of two layers: the epidermis, made up of sheets of columnar cells, overlies the dermis, consisting of motile cells which move about on the extracellular matrix (ECM). These two layers are separated by the fibrous basal lamina. The first stages of appendage formation is characterized by dermal cell aggregation centres, the papillae, which form directly underneath and simultaneously with epidermal thickenings, the placodes. Experimental evidence (for example, Gallin *et al.* 1986; Chuong & Edelman 1985) indicates that interaction between the epithelial and dermal layers plays an important role during skin appendage formation.

Apart from the model of Cruywagen & Murray (1992), all the theoretical tissue interaction mechanisms proposed so far involve reaction-diffusion systems (see for example, Nagorcka 1986; Shaw & Murray 1990). For a brief review on the modelling of tissue interaction refer to Murray *et al.* (1993). There is, however, still little experimental evidence for the existence of such reacting and diffusing morphogens in skin pattern formation. Therefore we examine here the two-dimensional synchronous and sequential pattern formation capabilities of Cruywagen & Murray's (1992) mechanochemical model.

Experimental and Theoretical Advances in Biological Pattern Formation,
Edited by H.G. Othmer *et al.*, Plenum Press, New York, 1993

8.2   THE TISSUE INTERACTION MODEL

The model of Cruywagen & Murray (1992) is a continuum tissue interaction model and consists of two parts: an equation to describe dermal cell movement and two equations for describing epithelial sheet deformation. These two sub-models are coupled by introducing tissue interaction. We briefly describe this model here and refer the reader to the original paper for full details.

The epithelial sheet is modelled as a two-dimensional, visco-elastic continuum (see for example, Murray & Oster 1984; Murray 1989). As the system is in a low Reynold's number regime, we assume that the visco-elastic and cell traction stresses within the epidermis are balanced by the external body forces. The force balance equation takes the form

$$\nabla \cdot \left\{ \overbrace{\frac{E}{1+v} \left[\varepsilon - \beta_1 \nabla^2 \varepsilon + \frac{v}{1-2v}(\theta - \beta_2 \nabla^2 \theta)I \right]}^{\text{elastic stress}} \right.$$

$$\left. + \overbrace{\mu_1 \frac{\partial \varepsilon}{\partial t} + \mu_2 \frac{\partial \theta}{\partial t}I}^{\text{viscous stress}} + \overbrace{\tau(s)I}^{\text{traction}} \right\} = \overbrace{\rho u}^{\text{body forces}}, \tag{8.1}$$

where the variable $u(x, t)$ is the displacement at time t of a material point in the epithelial layer which was initially at position x, $\varepsilon = (\nabla u + \nabla u^T)/2$ is the strain tensor, $\theta = \nabla \cdot u$ the dilation, T denotes the transpose, and s is the concentration of a signalling chemical. The parameter E is Young's modulus, v is Poisson's ratio, μ_1 and μ_2 are the shear and bulk viscosities, respectively (Landau & Lifshitz 1970), and I is the unit tensor. The parameters β_1 and β_2 measure the long range elastic stresses (see Murray 1989 for a discussion). The epidermis is attached to the basal lamina with adhesion tethers; ρ reflects the strength of these attachments. The epithelial sheet exerts active traction which we assume depends on the signal chemical s, which diffuses from the dermis into the epidermis, thus introducing tissue interaction. We model this traction as the switch, $\tau(s) = \tau s^2(n)/(1 + cs^2(n))$, where τ and c are positive constants (see Murray & Oster 1984).

An epithelial cell conservation equation relates the epidermal cell density $N(x, t)$ to the displacement u. Since the only contribution to cell flux is convection, the equation is simply

$$\frac{\partial N}{\partial t} = -\overbrace{\nabla \cdot N \frac{\partial u}{\partial t}}^{\text{convection}}. \tag{8.2}$$

To model dermal morphogenesis we consider a chemotaxis equation, related to the cell-chemotaxis model of (Oster & Murray 1989) and based on the *Morphoregulator Hypothesis* of Edelman (see for example, Edelman 1986). According to this hypothesis skin organ morphogenesis is controlled by cell-cell adhesion mechanisms mediated by cell adhesion molecules (CAMs). Because chemical modulation can have a marked effect on the binding rates and binding strengths of CAMs (Grumet & Edelman 1988), we assume that a chemical signal concentration e, diffusing from the epidermis into the dermis, is responsible for CAM expression. The conservation equation for dermal cell density, $n(x, t)$, then takes the form

$$\frac{\partial n}{\partial t} = \overbrace{\nabla \cdot D\nabla n}^{\text{diffusion}} - \overbrace{\nabla \cdot n\nabla\alpha(e)}^{\text{chemotaxis}} + \overbrace{rn(n_0 - n)}^{\text{mitosis}}, \tag{8.3}$$

where D is the coefficient of random diffusion and where we have assumed that cell growth obeys the logistic law, with r and n_0 positive constants. It is assumed that the function $\alpha(e)$ models the chemo-attraction and that it has the linear form $\alpha(e) = \alpha e(N)$, where α is a non-negative constant and e is a function of N (see Cruywagen & Murray 1992).

The system (8.1), (8.2) and (8.3) constitutes the field equations of our tissue interaction model. The full system is extremely complex, but by making a few reasonable biological assumptions, (Cruywagen & Murray 1992) reduced it to two coupled nonlinear equations, thus making it more amenable to analysis while still retaining the essential biological features of the full model.

They obtained this simpler version by taking the divergence of the epidermal tensor equation (8.1), thus reducing it to a scalar dilation equation in θ. They also applied the small strain assumption to equation (8.2) to obtain a linear relationship between epidermal cell density and dilation. The reduced model involves only the epithelial dilation θ and the dermal cell density, n.

8.3   SYNCHRONOUS AND SEQUENTIAL PATTERN FORMATION

Initially we examined the synchronous pattern formation capabilities of Cruywagen & Murray's (1992) reduced model. We solved the reduced system numerically on rectangular domains using zero-flux boundary conditions in cell density and dilation. As initial conditions small random perturbations about the homogeneous steady state were specified. Parameter values were chosen so that the linearized version of the problem had only one unstable eigenvalue satisfying the boundary conditions. This eigenvalue could, for example, be a simple or a multiple eigenvalue (see Cruywagen *et al.* 1993 for an explanation). The model can exhibit several different types of patterned solutions such as squares, rhombi, rolls, hexagonal and mixed mode solutions.

However, in many developmental situations spatial pattern formation occurs sequentially. Examples of such sequential patterning include chick feather germ initiation (see for example, Chuong & Edelman 1985) and alligator skin pigmentation (Murray *et al.* 1990). (Nagorcka 1986) examined the sequential pattern formation capabilities of reaction-diffusion systems on two-dimensional domains. Here we apply the tissue interaction model described in Section 2 to sequential pattern formation.

Numerical simulations of the model revealed that the spatial pattern which propagates across the domain depends crucially on the pattern that forms initially at one end of the domain. Figure 8.1 shows that if spots are initially specified at one end the resultant propagating pattern is rhombic. On the other hand, if the initial pattern is stripes, then the resultant two-dimensional pattern would be stripes. A detailed discussion of the sequential aspects of pattern formation is presented in (Cruywagen *et al.* 1993).

ACKNOWLEDGMENTS GCC would like to thank The Rhodes Trust, Oxford, and the South African Foundation for Research Development for their financial support. This work (JDM) was in part supported by a Grant DMS-9106848 from the U. S. National Science Foundation.

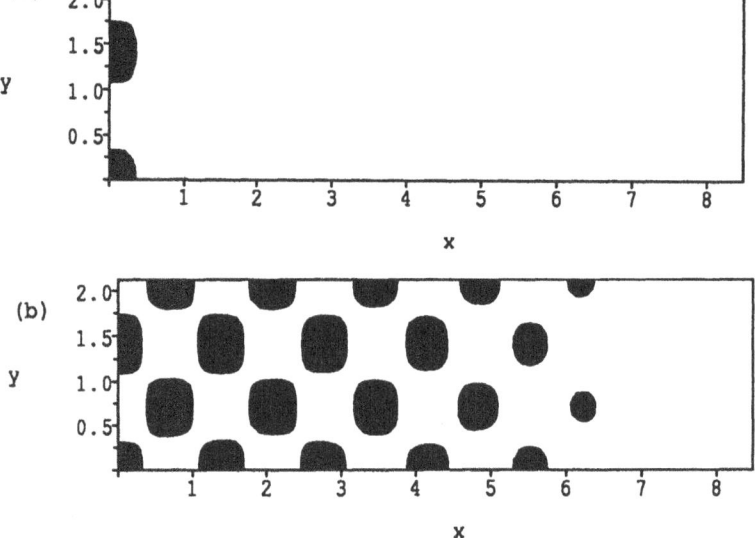

Figure 8.1. Numerical solutions of the caricature tissue interaction model. An initial row of spots (a) propagates across the domain to form a rhombic pattern (b).

REFERENCES

Chuong, C.-M., & Edelman, G. M. 1985. Expression of cell adhesion molecules in embryonic induction. I. Morphogenesis of nestling feathers. *J. Cell Biol.*, **101**, 1009–1026.

Cruywagen, G. C., & Murray, J. D. 1992. On a tissue interaction model for skin pattern formation. *J. Nonlinear Sci.*, **2**, 217–240.

Cruywagen, G. C., Maini, P. K., & Murray, J. D. 1993. Sequential pattern formation in a model for skin morphogenesis. *IMA J. Maths. Appl. Med. & Biol.* (In press).

Edelman, G. M. 1986. Cell adhesion molecules in the regulation of animal form and tissue pattern. *Annu. Rev. Cell Biol.*, **2**, 81–116.

Gallin, W. J., Chuong, C.-M., Finkel, L. H., & Edelman, G. M. 1986. Antibodies to liver cell adhesion molecules perturb inductive interactions and alter feather pattern and structure. *Proc. Natl. Acad. Sci. USA*, **83**, 8235–8239.

Grumet, M., & Edelman, G. M. 1988. Neuron-glia cell adhesion molecules interact with neurons and astroglia via different binding mechanisms. *J. Cell Biol.*, **106**, 487–503.

Landau, L. D., & Lifshitz, E. M. 1970. *Theory of Elasticity*. 2nd edn. New York: Pergamon.

Murray, J. D. 1989. *Mathematical Biology*. New York: Springer-Verlag.

Murray, J. D., & Oster, G. F. 1984. Generation of biological pattern and form. *IMA J. Maths Appl. Med. & Biol*, **1**, 51–75.

Murray, J. D., Deeming, D. C., & Ferguson, M. W. J. 1990. Size dependent pigmentation pattern formation in embryos of *Alligator Mississipiensis*: time of initiation of pattern generation mechanism. *Proc. Roy. Soc.*, **B239**, 279–293.

Murray, J. D., Cruywagen, G. C., & Maini, P. K. 1993. *Pattern formation in tissue interaction systems*. Heidelberg: Springer-Verlag. (In press) *Lect. Notes in Biomathematics* **100**.

Nagorcka, B. N. 1986. The role of a reaction-diffusion system in the initiation of skin organ primordia. I. The first wave of initiation. *J. Theor. Biol.*, **121**, 449–475.

Oster, G. F., & Murray, J. D. 1989. Pattern formation models and developmental constraints. *J. exp. Zool.*, **251**, 186–202.

Shaw, L. J., & Murray, J. D. 1990. Analysis of a model for complex skin patterns. *SIAM J. Appl. Math.*, **50**(2), 628–648.

9. CONTROL OF GAP JUNCTION PERMEABILITY CAN CONTROL PATTERN FORMATION IN LIMB DEVELOPMENT

R. Dillon and H. G. Othmer

Department of Mathematics
University of Utah
Salt Lake City, UT 84112, USA

9.1 INTRODUCTION

It is known that in many developing systems the developmental pathway taken by a cell depends not only on its genome and history, but also on where it is in relation to other cells (Wolpert 1969). Consequently, in order to understand how a relatively unstructured egg leads to an adult organism with many different cell types arranged in the appropriate spatial pattern, it is necessary to understand how the spatial pattern of cellular differentiation is controlled. Spatially-dependent differentiation, or pattern formation for short, is often thought to result from the response of individual cells to an underlying spatial pattern of one or more substances called morphogens, which directly or indirectly initiate cellular differentiation (Wolpert 1969; Meinhardt 1982). Turing (1952), who coined the term morphogen, suggested a mechanism by which an initially-homogeneous distribution of morphogens could give rise to a spatial pattern through the interaction of reaction and diffusion.

Avian limb development is a model system for the study of pattern formation, and there is a substantial body of experimental information on which to base mathematical models (reviewed in Maini & Solursh 1991). In chick, the wing site in the flank of the embryo is determined by Hamburger-Hamilton stage 8, limb outgrowth is visible by stage 17, and the first skeletal element is visible at stage 24. Pattern formation is described relative to three axes, the proximal-distal (PD) axis, which extends from somites to wing tip, the anterior-posterior (AP) axis, and the dorsal-ventral (DV) axis. The skeletal elements (humerus, radius and ulna, wrist, and digits) form in a proximal-distal and posterior-anterior sequence. Fate maps show that the anterior half of the limb bud gives rise to part of the humerus, the radius, and digit 2, while the posterior half gives rise to part of the humerus, the ulna, and digits 3 and 4 (Stark & Searls 1973).

The ectoderm at the distal tip of the limb bud forms the apical ectodermal ridge (AER), which appears at stage 18. Removal of the AER between stages 18 and 28 stops outgrowth of the limb and leads to a truncated limb with deficiencies in the distal skeletal elements (Summerbell 1974b). The rapidly dividing mesodermal cells adjacent to the AER comprise the so-called progress zone. A homeobox gene that is expressed in the ridge and one that is expressed in the subridge region have recently been identified (Coelho & Kosher 1991a; Coelho *et al.* 1992). The AER is maintained by a factor produced either in the progress zone or in a specialized group of cells, the zone of polarizing activity (ZPA), that lies at the posterior

Experimental and Theorietcal Advances in Biological Pattern Formation,
Edited by H.G. Othmer *et al.,* Plenum Press, New York, 1993

65

margin of the progress zone (Zwilling 1961). Transplants of the ZPA to the anterior margin of the limb usually lead to duplication of skeletal elements, the pattern of which depends on the location of the transplant relative to the ZPA and the AER (Wolpert & Hornbruch 1987; Tickle *et al.* 1975). Moreover, the effect of the ZPA is dependent on the number of ZPA cells in the transplant (Tickle 1981), and only cells in the progress zone can respond to the polarizing action of the ZPA (Summerbell 1974a). Finally, functional gap junctions seem to be required for communication between ZPA cells and anterior mesenchyme, since blocking antibodies to gap junctional proteins prevent ZPA-induced limb duplications (Allen *et al.* 1990). Thus direct cell-to-cell communication seems to play an important role in ZPA transplants, and it is reasonable to suppose that it also does in normal development.

In the past, pattern formation along the PD axis and along the AP axis have generally been treated as separate, uncoupled processes. Wolpert (1969) postulated that the ZPA produces a morphogen which diffuses throughout the tissue and is degraded in it. This would establish a gradient in the AP direction which could provide positional information and could lead to a spatial pattern of differentiation. Tickle *et al.* (1975) estimated that such gradients have to be established within 10 hours across a distance of 500–1000 μm, and concluded that transport by diffusion is sufficiently rapid. If an impermeable barrier is placed along the PD axis, then skeletal development occurs only on the posterior side of the barrier, which suggests that a diffusible morphogen is produced at the ZPA (Summerbell 1979). The gradient model predicts that transplants of the ZPA at positions along the AP axis of a stage 16 wing bud should result in either the elimination of the humerus, or its duplication, or the formation of a mirror-image duplicate of a single humerus, depending on the position of the graft and on the threshold concentration. However, in no cases is the humerus eliminated and rarely is it duplicated. Usually either a normal or a mirror-image duplicate humerus forms (Wolpert & Hornbruch 1987). The theory also predicts that multiple ZPA grafts should lead to fused or abnormally thick digits, which again is contrary to observation. Wolpert and Hornbruch (1987) conclude from this that there is another mechanism at work which controls the thickness of the digits.

The progress zone model, in which differentiation is controlled by the number of divisions a cell undergoes while in the progress zone, was proposed to explain pattern formation and differentiation in the PD direction (Wolpert *et al.* 1975). It predicts that removal of the AER will lead to distal truncation, as is observed, and it makes other predictions on the outcome of grafting a donor wing tip onto a host stump which agree closely with observation (Summerbell & Lewis 1975). However, it does not have any regulative properties and thus cannot account for the regulative capability of the early limb bud. For instance, removal of slices of the early limb bud perpendicular to the PD axis can lead to normal limbs (Summerbell 1977), but according to the model, this should produce deletions along the PD axis of the final pattern.

While one-dimensional models have given some useful insights, there are experimental results which suggest that to describe spatial patterning in limbs requires at least a two-dimensional model, and ultimately a three-dimensional model that incorporates both growth of the limb and cell movement. For example, transplant results show a dependence on the distance between the ZPA and the AER (Wolpert & Hornbruch 1987), and on the position along the PD axis at which the graft is implanted in the host (Javois *et al.* 1981). Furthermore, an analysis of the movement of marked cells in a growing limb during stages in which pattern formation is believed to occur shows significant preaxial movement of tissue. This produces a significant change in the geometry of the limb, and hence in the relative spatial relationships between cells (Bowen *et al.* 1989). Formal models which incorporate some of these aspects have been proposed (Wilby & Ede 1975; French *et al.* 1977), but to date these do not have a mechanistic basis.

Our objective is to develop a model that is sufficiently general to enable us to test various hypotheses concerning the nature of the interaction between the ZPA, the AER, and the underlying mesoderm, both in normal development and in transplants, and to explore the role of the ectoderm in the inhibition of chondrogenesis. This will require an understanding of the role of cell-cell communication in growth control and pattern formation, and of the control of cell movement and shape changes. In this paper we report a number of results from a preliminary version of the more general model. The model is a generalized Turing model, but it incorporates several significant extensions of Turing's original ideas. Firstly, we do not require that all morphogens satisfy the same boundary conditions, and as we have shown elsewhere (Dillon *et al.* 1993), this leads to more reliable pattern formation in a sense made precise there. Secondly, we incorporate the possibility that gap junction permeability can be modulated by the concentration of a control species whose concentration may vary throughout the limb. In essence, the self-organizing pattern formation capability inherent in a Turing system is coupled with spatial gradients of a control species that modulates the patterning process. In a sense this model combines the two dominant models of pattern formation, but invokes the components of these processes in a novel way. Earlier a related scheme was proposed to achieve scale invariance in reaction-diffusion systems (Othmer & Pate 1980), but in that model all cells produced the control species, whereas here the production of the control species is localized at the boundary. Since the modulation of gap junction permeability is an integral part of the model, we discuss some of the experimental evidence for such control in the following section, and we describe how it translates into a continuum description of transport between cells. In Section 3 we describe the complete mathematical model, and in Section 4 we present some numerical results obtained from the model.

9.2 GAP JUNCTION MODULATION

9.2.1 Experimental observations

Gap junctions are found in the AER (Fallon & Kelley 1977) and throughout the mesoderm of the developing limb bud (Coelho & Kosher 1991b). Furthermore, their permeability can vary spatially in a graded manner. For instance, Coelho & Kosher (1991b) found an anterior-posterior gradient of gap junctional permeability in stage 20-21 limb buds. They found that the permeability to Lucifer yellow dye was high in the posterior mesenchymal tissue, considerably lower in the center of the limb bud, and very low in the anterior mesenchymal tissue.

Recent work suggests that retinoic acid (RA) is produced in the ZPA (Smith *et al.* 1989), and that it may have a role in the control of growth, cell-to-cell communication, and pattern formation. Thaller and Eichele found that RA is graded in the AP direction *in vivo*: levels of RA were found to be 2.5 times greater in the posterior 1/4 of the limb bud than in the anterior portion (Thaller & Eichele 1987). An implanted bead which releases RA at the anterior margin of the limb can mimic ZPA transplants in that the digit pattern is dose dependent (Tickle *et al.* 1985). Low concentrations lead to a normal digit pattern, intermediate concentrations produce supernumerary digits, and high concentrations lead to wings in which only the humerus and a knob of cartilage are formed. ZPA tissue is also thought to produce a mitogen (McLachlan 1991), but whether this mitogen is RA or some other substance is not known.

Mehta *et al.* (1989) have shown that retinoic acid has a biphasic effect on gap junctional communication at noncytotoxic concentrations in several cell lines. Communication in 10T1/2 cells is enhanced in the 10^{-9} - 10^{-7}M range but inhibited in the 10^{-10} - 10^{-9}M range. In addition, *in vitro* saturation density is found to be inversely correlated with gap junctional permeability. Thus mitotic rates may be inversely related to gap junctional permeability. As we remarked earlier, Allen *et al.* (1990) found that in chick limb, functional gap junctions seem to be required to produce pattern duplication following ZPA transplants.

These results suggest that RA may have several distinct effects on limb development. Mehta *et al.* (1989) postulate that RA reduces gap junction permeability on a fast time scale, but enhances it on a longer time scale. Given the inverse relation between growth and junctional communication observed in other cells, one scenario for the effect of RA is that it stimulates growth on a fast time scale, either directly or indirectly, by reducing cell-cell communication, while enhancing communication and suppressing mitosis on a longer time scale. This is consistent with the observation that there is large-scale, transient mitotic enhancement 17 hours after a ZPA transplant, which is correlated with a decrease in cell packing density (Cooke & Summerbell 1980). However, this scenario may be too simple, for both the mitotic rate and gap junction communication is normally highest in the posterior portion of the limb.

9.2.2 Modulation of diffusion by a control species

In the following section we develop a model of pattern formation in which morphogens are produced within cells and diffuse freely within cells. Intercellular transport of the morphogens is via diffusion through gap junctions that couple the cells, and the permeability of these junctions is controlled by the local concentration of a control species, such as RA, that is produced in the ZPA (*cf.* Figure 9.1). A complete mathematical description of reaction and transport in such an array requires a system of reaction-diffusion equations for morphogen concentrations in the intracellular space, coupled to diffusion equations for transport through the gap junctions (Othmer 1983). However, an asymptotic analysis of these equations shows that for the types of solutions of interest here, namely those for which a characteristic spatial scale is large compared to the cell length, a simpler description is applicable (Othmer 1983). That description entails reaction-diffusion equations in which, as a first approximation, the diffusion coefficients of the morphogens are equal to their value in the gap junctions, suitably weighted, and the reaction term is that in the intracellular space, again suitably weighted. In the model, the diffusion coefficient of the ith morphogen is a function of the local concentration of the control species w. In this paper we assume that this functional dependence is

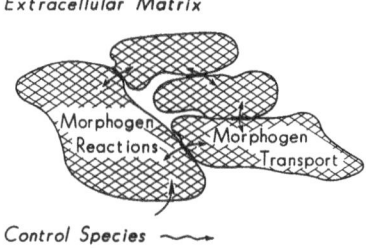

Figure 9.1. A schematic of the model. The morphogens are produced within cells, and intercellular transport of the morphogens is via gap junctions. A control species that diffuses within the extracellular space modulates the permeability of gap junctions by an unspecified mechanism.

linear, and so write

$$D_i(w)(\mathbf{x}, t) = \mathcal{D}_i(w(\mathbf{x}, t) + \alpha_i) \qquad (9.1)$$

where $w(\mathbf{x}, t)$ is the concentration of the control species at the point $(\mathbf{x}, t)$, and $\mathcal{D}_i$ and α_i, are constants. The motivation for assuming linear dependence arises from two facts: (a) the concentration of retinoic acid in chick limb is of the order of 10^{-8}M (Thaller & Eichele 1987), and (b) the experimentally-obtained relationship for some other cell types is approximately linear when the retinoic acid concentration is in the range 10^{-9} to 10^{-7}M (*cf.* Figure 9.2). However, non-linear diffusion modulation functions might may be applicable in some situations, and their effect will be discussed in the conclusions section.

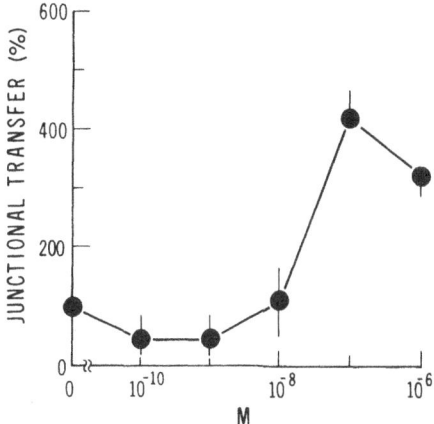

Figure 9.2. The dependence of the gap junctional permeability on the concentration of all-*trans* retinoic acid in methylcholanthrene-transformed 10T1/2 cells (modified from Mehta *et al.* 1989).

9.3 THE MATHEMATICAL MODEL

We treat the limb as a two-dimensional rectangular region of constant width, which can be thought of as a section through the limb taken at the centerline in the DV direction (*cf.* Figure 9.3). Although the two-dimensionality precludes analyzing transplants in which the DV polarity is altered, it is an essential first step, given the complexity of the system. The limb also changes shape in later stages, but it is reasonable to treat it as a rectangle up to about stage 24 or 25. The length increases from ~ 0.25 mm to ~ 1.75 mm in the 36 hours between stage 18 and stage 25, and it is known that patterning occurs in this time interval, but we do not include growth as a dynamic variable. Similarly, the differential or localized growth that occurs after a ZPA transplant will only be incorporated as a parametric change. The shape changes that occur in the transplants and in later stages of normal limb development will be incorporated in succeeding models. The model involves two diffusible morphogens that are produced by all cells in the interior of the region, and which diffuse throughout it. Since cells in the AER are tightly coupled (Kelley & Fallon 1983), we will model this region as a uniform (with respect to variations along the boundary) source of a substance which maintains cells in the undifferentiated state. In this paper the detailed spatial distribution of the AER substance will not be considered; we simply postulate that when this substance drops below a threshold the source terms in the morphogen kinetics are turned off, perhaps by turning off the enzymes involved. We postulate that the active pattern-forming region extends

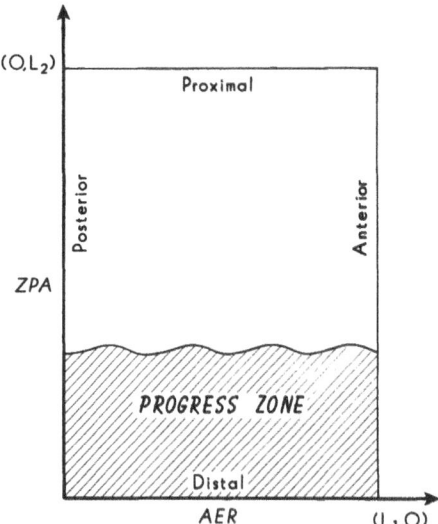

Figure 9.3. A schematic of the idealized limb bud used in the model. The boundary along the x-axis is the AER, and the boundary on the y-axis is the ZPA.

$200 - 400$ μm from the AER, which corresponds approximately to the progress zone. The hypothesized role of the substance produced by the AER is based on several observations: (a) the AER can delay the differentiation of mesenchyme and stimulate mesenchymal outgrowth *in vitro* (Reiter & Solursh 1982), and (b) terminal wing parts fail to differentiate *in vivo* following removal of the AER and the level of truncation depends upon the stage of the bud at the time the AER is removed (Saunders, Jr 1948; Summerbell 1974b). Furthermore, the AER determines the size of the limb bud and it seems to maintain its shape (Hornbruch & Wolpert 1970; Lee & Tickle 1985). In essence, we assume that the AER produces a substance which modulates the morphogen reactions in a concentration-dependent manner. A similar mechanism may be responsible for the pattern of expression of the homeobox gene, Hox-7.1, which correlates with the progress zone and is thought to be regulated by a signal produced by the AER (Muneoka & Sassoon 1992).

Maps of the potency of grafted tissue from a donor limb to produce mirror-image duplication in the host show that it varies spatially and temporally, and that it differs somewhat between the leg and the wing bud (Honig & Summerbell 1985; Hinchliffe & Sansom 1985). In chick wing this potency is highest from stage 19 - 25, and at stage 21 is approximately uniformly distributed along the posterior edge between the flanks and the proximal edge of the AER. The strength falls off rapidly where the limb bud joins the flank and also distally to the posterior edge of the AER. By stage 25 the activity is skewed distally but still falls off rapidly in the distal tip beyond the edge of the AER. Our hypothesis is that the ZPA is the source of the control species, and in a more complete model it should mirror the spatial and temporal distribution of the polarizing activity. However, to simplify the numerical computations, we assume here that the control species is held fixed at one value along the entire posterior boundary and at a second (lower) value along the anterior boundary. This implies that the control species varies only in the AP direction, but this should not drastically affect the results given later for the simple reason that the morphogen kinetics are turned off at the proximal end of the domain, which is where the major PD variation of the control species would occur when its source is strongest at the distal end. The effect of this assumption and the assumption that the ZPA can be adequately modeled as a source on the boundary, rather than extending into the interior of the limb, will be tested in a future model.

In the absence of any effect by the AER substance, the governing equations for the morphogen concentrations in the interior of the limb are

$$\frac{\partial u}{\partial t} = \nabla \cdot (D_1(w)\nabla u) + f(u,v)$$

$$\frac{\partial v}{\partial t} = \nabla \cdot (D_2(w)\nabla v) + g(u,v). \tag{9.2}$$

Throughout this paper we use the following kinetic functions, which arise from a simplified version of a model for glycolysis (Othmer & Aldridge 1978; Ashkenazi & Othmer 1978):

$$\begin{aligned}
f(u,v) &= \omega(\beta - \kappa u - uv^2) \\
g(u,v) &= \omega(\kappa u + uv^2 - v).
\end{aligned} \tag{9.3}$$

where ω^{-1} is a characteristic time scale.

The control species w satisfies the equation

$$\frac{\partial w}{\partial t} = D_3 \nabla^2 w - k^2 w$$

$$w(0,y,t) = w_0 \qquad\qquad w(L_1,y,t) = w_1$$

$$w_y(x,0,t) = 0 \qquad\qquad w_y(x,L_2,t) = 0 \tag{9.4}$$

$$w(x,y,0) = W_0(x).$$

We suppose that the spatial distribution of w relaxes to its steady state value rapidly on the time scale of changes in the morphogen concentrations, and therefore we ignore transients in w. As a result, $w(x,y,t)$ is a function of x only, and satisfies the equation

$$\frac{d^2w}{d\zeta^2} - \gamma^2 w = 0 \quad \text{for} \quad \zeta \in (0,1)$$

$$w(0,t) = w_0 \qquad w(1,t) = w_1, \tag{9.5}$$

where $\gamma = kL_1/\sqrt{D_3}$ is a non-dimensional degradation parameter and $\zeta \equiv x/L_1$. The solution is

$$w(\zeta) = (w_0 - A)e^{\gamma\zeta} + Ae^{-\gamma\zeta}, \tag{9.6}$$

where

$$A = \frac{w_1 - w_0 e^{\gamma}}{e^{-\gamma} - e^{\gamma}}. \tag{9.7}$$

As was discussed in the previous section, we assume that the diffusion coefficients of the morphogens depend linearly on the concentration of the control species, and we assume that $\alpha_1 = \alpha_2 \equiv \alpha$. Then in dimensionless spatial variables we have

$$D_i(\zeta,\eta) = \mathcal{D}_i(w(\zeta) + \alpha) \tag{9.8}$$

for $i = 1,2$, where $w(\zeta)$ is given by (9.6) and $\eta \equiv y/L_2$. Incorporating these expressions for the diffusion coefficients, as well as the hypothesis concerning the effect of the AER substance on the morphogen kinetics, into (9.2), we obtain the system

$$
\begin{aligned}
\frac{\partial u}{\partial \tau} &= \nu_1 \frac{\partial}{\partial \zeta} \left((w(\zeta) + \alpha) \frac{\partial u}{\partial \zeta} \right) + \nu_2 \frac{\partial}{\partial \eta} \left((w(\zeta) + \alpha) \frac{\partial u}{\partial \eta} \right) + F(u, v, \zeta, \eta) \\
\frac{\partial v}{\partial \tau} &= \delta \nu_1 \frac{\partial}{\partial \zeta} \left((w(\zeta) + \alpha) \frac{\partial v}{\partial \zeta} \right) + \delta \nu_2 \frac{\partial}{\partial \eta} \left((w(\zeta) + \alpha) \frac{\partial v}{\partial \eta} \right) + G(u, v, \zeta, \eta).
\end{aligned}
\tag{9.9}
$$

Here $\nu_1 \equiv \mathcal{D}_1 / \omega L_1^2$, $\nu_2 \equiv \mathcal{D}_1 / \omega L_2^2$, $\delta \equiv \mathcal{D}_2 / \mathcal{D}_1$, and $\tau \equiv \omega t$. The functions F and G are related to f and g as follows.

$$
\begin{aligned}
F(u, v, \zeta, \eta) &= H(\eta_0 - \eta)(\beta - uv^2) - \kappa u \\
G(u, v, \zeta, \eta) &= H(\eta_0 - \eta)uv^2 + \kappa u - v
\end{aligned}
\tag{9.10}
$$

where H is the Heaviside function

$$
H(\eta_0 - \eta) = \begin{cases} 0 & \text{if} \quad \eta_0 - \eta < 0 \\ 1 & \text{if} \quad \eta_0 - \eta > 0. \end{cases}
\tag{9.11}
$$

Thus F and G reduce to $\omega^{-1} f$ and $\omega^{-1} g$, respectively, whenever $\eta < \eta_0$. Boundary conditions will be specified later, but we remark here that the standard Turing model consists of (9.9) with $\alpha = 0$ and $w(\zeta) \equiv 1$, together with boundary conditions that are the same for both u and v.

Except where otherwise noted, the parameters β, κ, δ, and $\mathcal{D}_1 / \omega$ are fixed at $1.0, 0.001$, 0.14, and 0.001 in the remainder of the paper. A typical value for the diffusion coefficients is $\mathcal{D}_1 \sim 1 \times 10^{-5}$ cm^2/sec, and thus the time scale implicit in the foregoing parameters is $\omega^{-1} \sim 100$ sec.

9.4   NUMERICAL RESULTS

9.4.1   One-dimensional systems

In order to gain some insight into the effect of spatially-varying diffusion coefficients on pattern formation in a generalized Turing mechanism, we first compare the solutions for a standard Turing system in one space dimension with those for a generalized Turing system in one space dimension. The one-dimensional system can be regarded as a cross section of the limb near the AER. The governing equations for the generalized Turing system are

$$
\begin{aligned}
\frac{\partial u}{\partial \tau} &= \nu_1 \frac{\partial}{\partial \zeta} \left((w(\zeta) + \alpha) \frac{\partial u}{\partial \zeta} \right) + f(u, v) \\
\frac{\partial v}{\partial \tau} &= \delta \nu_1 \frac{\partial}{\partial \zeta} \left((w(\zeta) + \alpha) \frac{\partial v}{\partial \zeta} \right) + g(u, v),
\end{aligned}
\tag{9.12}
$$

together with boundary conditions that will be specified shortly. Figure 9.4 shows the steady state spatial profiles of the second morphogen v for the standard Turing system under homogeneous Neumann boundary conditions on both components. As is well known (Dillon *et al.* 1993), solutions of the standard Turing model have a characteristic wavelength set by the balance between reaction and diffusion, and as the length L of the system increases, more

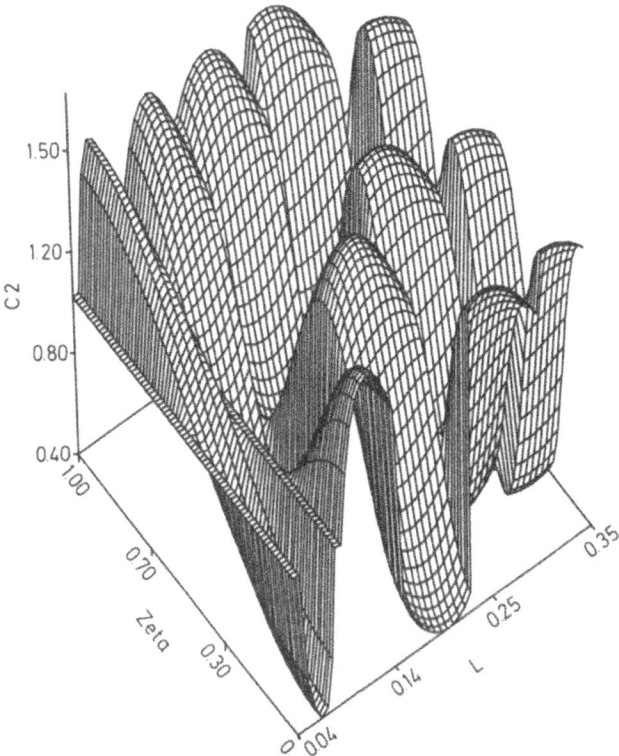

Figure 9.4. The steady-state spatial distribution of the morphogen v, as a function of L (in cm) and ζ, in the case of homogeneous Neumann boundary conditions on both u and v. A monotonic profile exists for $L \in (0.05, 0.08)$, and all other non-constant solutions have at least one internal maximum or minimum. The maximum and minimum of v for all solutions shown are 1.5 and 0.4, respectively. In this figure $\mathcal{D}_1/\omega = 2.67 \times 10^{-3}$. (From Pate and Othmer (1984), with permission.)

repetitions of this basic pattern appear, as can be seen in Figure 9.4. Since the spatial scale and the diffusion coefficient only appear in the dimensionless group ν_1, the results in Figure 9.4 can also be interpreted in terms of a variable diffusion coefficient on a domain of fixed length. In particular, this interpretation shows that as the diffusion coefficient is reduced, the spatial frequency of the non-constant solutions increases. This conclusion is strictly applicable only if the diffusion coefficient is uniform in space, but in a sufficiently large system, the local variation of the solution will depend primarily on the local diffusion coefficients, and one will observe high-frequency (resp., low-frequency) variation in regions where the diffusion coefficients are low (resp., high). We illustrate this using a generalized Turing system as follows. We consider (9.12) with the mixed boundary conditions[1]

$$\frac{\partial u}{\partial \zeta} = 0 \quad \text{at} \quad \zeta = 0, 1$$
$$v = 0 \qquad \text{at} \quad \zeta = 0, 1. \tag{9.13}$$

---

[1]This combination of boundary conditions, as we showed elsewhere (Dillon *et al.* 1993), produces the most robust pattern formation in one space dimension for these kinetics. Benson *et al.* analyze the effect of variable diffusion coefficients in a standard Turing system elsewhere in this volume.

In addition, we suppose that in the diffusion modulation function $w(\zeta) + \alpha$, $\alpha \equiv 0$ and $w(\zeta)$ is the step function

$$w(\zeta) = \begin{cases} 1.0 & 0 < \zeta < \frac{1}{2} \\ \\ 0.1 & \frac{1}{2} < \zeta < 1. \end{cases} \tag{9.14}$$

The numerical results are shown in Figure 9.5, and one can see there that the spatial frequency on the right half of the domain is two to three times the frequency on the left half.

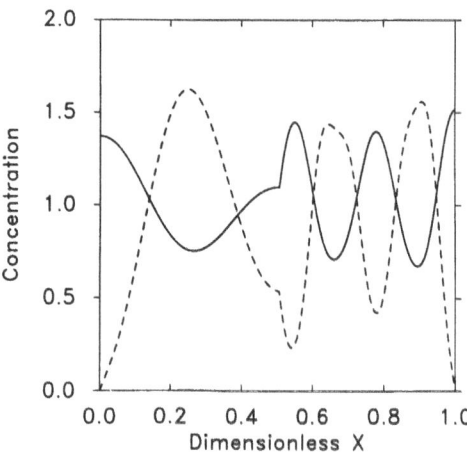

Figure 9.5. Steady state spatial profiles of the morphogens u and v for a discontinuous, piecewise constant diffusion modulation function. Here —— denotes u, - - - - denotes v, and $L = 2$ mm. Solutions of (9.12) and (9.13) were obtained using a finite difference approximation to the derivatives and a continuous approximation to $w(\zeta)$ with $w = 0.55$ at the midpoint of the interval.

Of course a step function is unrealistic if the modulation is generated by a spatially-distributed control species, and in Figure 9.6 we show the solutions of (9.12), with the diffusion modulation function as in (9.6), for several values of the dimensionless decay rate γ. We may classify these solutions by the number of local maxima in the second morphogen v, and according to this criterion, the solutions shown in Figure 9.6 demonstrate that the sequence I, II, III, IV can be achieved by increasing the control parameter γ. At $\gamma = 10$ the solution is qualitatively similar to that shown in Figure 9.5. Similar results (not shown) are obtained using w_0 as the control parameter while holding γ fixed. The results of numerous computations for various combinations of w_0 and γ are summarized in Figure 9.7. One sees there that the number of local maxima in v increases as the source strength w_0 decreases and/or the dimensionless degradation rate γ increases. In Figure 9.8 we show the results of a numerical "ZPA transplant" experiment. In Figure 9.8(a) the length L is 1 mm, which is approximately the width of the limb bud at stage 19. The concentration of the control species, which is shown in Figure 9.9(a), is fixed at one on the left hand boundary and zero on the right, and the degradation rate γ is chosen to give a solution of type III. We mimic the transplant experiment by increasing the domain size to $L = 1.6$ mm, and imposing identical boundary conditions for the control species on the left and right. The degradation rate is kept the same as in the Figure 9.8(a), and the resulting profile of the control species is shown in Figure 9.9(b). The morphogen distribution, which is shown in Figure 9.8(b), is a type VI distribution. Simulations with slightly smaller L produced a type V or a solution in the transition region from a type V to a type VI. If one reduces the concentration of the control

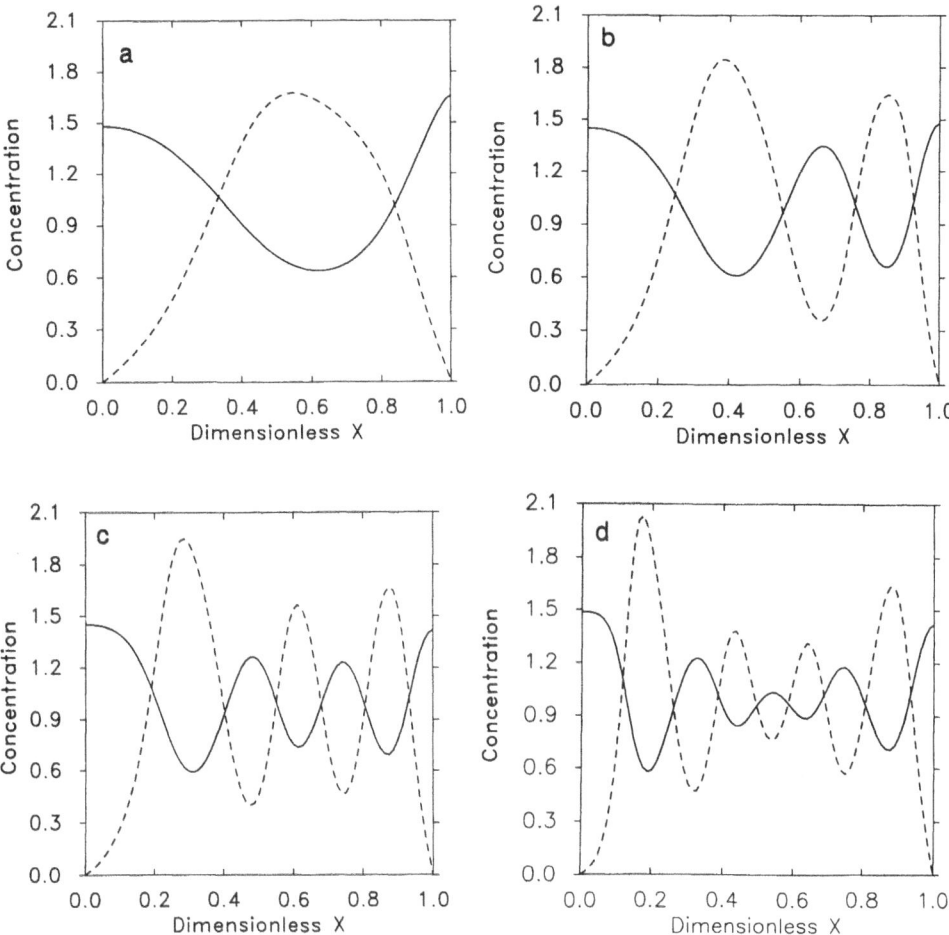

Figure 9.6. The steady-state concentration profiles obtained from (9.12) and (9.13) for fixed $w_0 = 3$. (a) $\gamma = 1$, (b) $\gamma = 5$, (c) $\gamma = 10$, (d) $\gamma = 30$.

species on the right hand boundary in the transplant experiment, it is possible to obtain a type VI solution at smaller values of L.

9.4.2 Two-dimensional systems

As we remarked earlier, a minimal description of pattern formation in the limb requires a two-dimensional model, and in this section we present some preliminary results obtained from the model described in the previous section. Throughout we fix L_2, the length in the PD direction at 1 mm, and set the parameter η_0, which determines where the source terms are cut off, at 0.2 mm or 0.3mm. Thus the autocatalytic and constant production components of the reaction kinetics are absent sufficiently far from the distal boundary, yet the morphogen degradation terms are unaffected by the AER substance. Localizing the pattern near the AER by cutting off the production terms reduces the importance of the choice of boundary

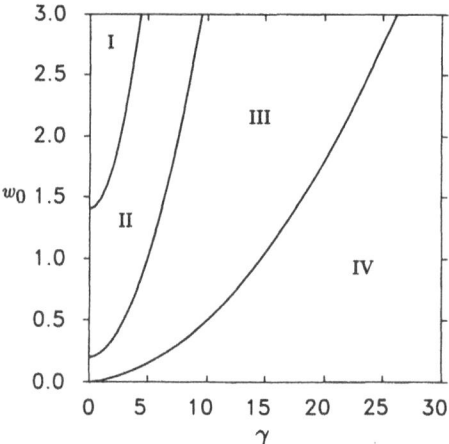

Figure 9.7. The type of pattern, as characterized by the number of local maxima in v, as a function of γ and w_0. Roman numerals denote the number of maxima in the regions outlined. From a mathematical standpoint the boundaries of the regions are sharp, because new maxima in v arise via the splitting of an existing interior maximum. However, if a threshold interpretation function is used the boundaries between the regions may be fuzzy.

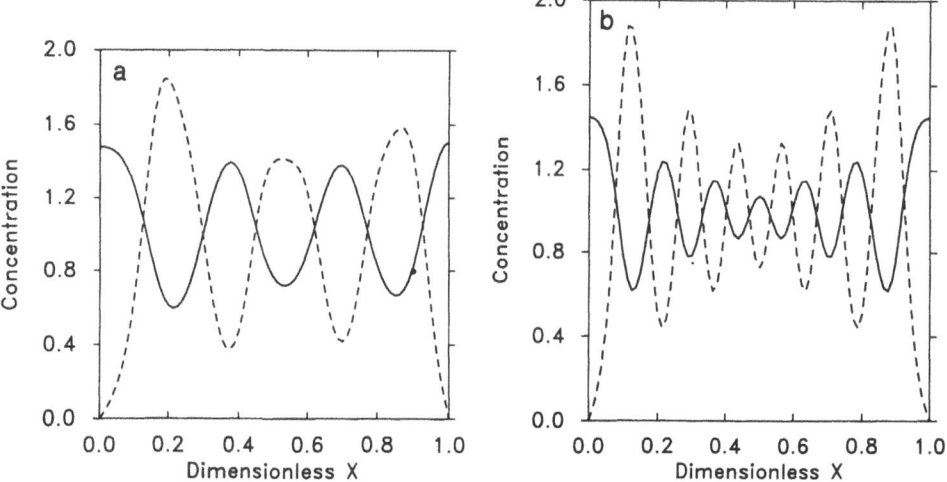

Figure 9.8. Steady state solution profiles before and after the numerical ZPA transplant. (a) A solution profile of type III at $L = 1.0$ with a single source of unit strength on the posterior boundary. (b) A solution of type VI at $L = 1.6$ mm with a source of unit strength on the posterior and the anterior boundaries. All other parameters are the same in both solutions. Here $k/\sqrt{D_3} = 150$ cm^{-1}.

conditions at the proximal border. The results described shortly are obtained using mixed boundary conditions at the ends of the AP axis, as in the one-dimensional system described earlier, and homogeneous Neumann boundary conditions at the ends of the PD axis. In

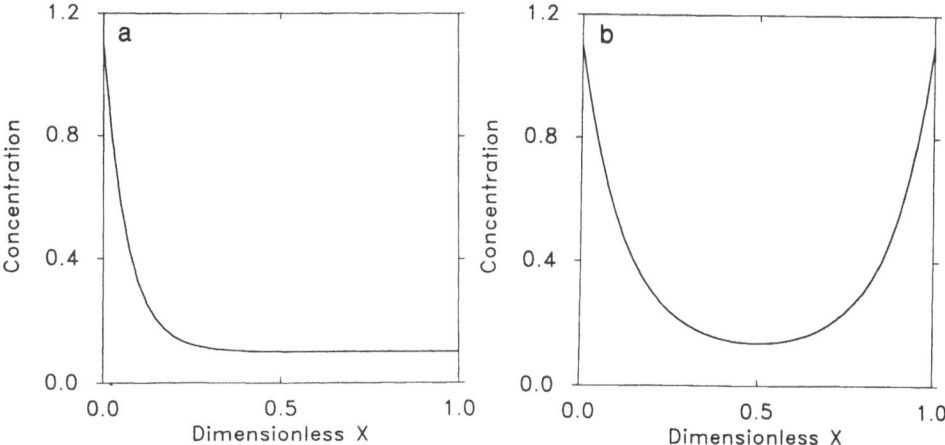

Figure 9.9. Spatial profiles of the control species w at fixed $\gamma = 15$. (a) $w_0 = 1$, $w_1 = 0$, $L_1 = 1$ mm. (b) $w_0 = 1$, $w_1 = 1$, $L_1 = 1.6$ mm.

mathematical terms the boundary conditions are

$$\frac{\partial u}{\partial \zeta} = 0 \quad v = 0 \quad \text{at} \quad \zeta = 0, 1$$
$$\frac{\partial u}{\partial \eta} = 0 \quad \frac{\partial v}{\partial \eta} = 0 \quad \text{at} \quad \eta = 0, 1 \tag{9.15}$$

The boundary conditions on the control species will be specified shortly.

As in the one-dimensional study we consider the ability of a control species to produce the normal sequence of cartilage pattern on a limb bud of constant width. In Figure 9.10 we show a sequence of solutions for several values of the parameters (w_0, γ) with control strength held at zero on the posterior border. The contour plots show regions in v above a fixed threshold level in black. We use the same classification scheme for these solutions as in the one-dimensional case by counting the number of local maxima along a section in the AP direction at $\eta = \eta_0/2$. As in Figure 9.7 above, the solutions obtained here evolve from type I to type IV as w_0 decreases and/or γ increases.

Figure 9.11 shows the results of the numerical "ZPA transplant" experiment carried out in 2D. Figure 9.11a is the "normal" case with a single source of the control species on the posterior boundary. Figure 9.11b shows a numerical "ZPA transplant" where length in the x direction is increased by 65%. In addition, a second source of the control species equal in strength to that at the posterior boundary is specified on the anterior boundary. In the "normal" case the pattern is of Type III while in the "ZPA transplant' case it is of Type VI. As in the 1D simulations, the critical parameter is the increase in width. If the increase is somewhat less, the numerical experiment produces a Type V pattern.

9.5 CONCLUSIONS

In this model the morphogen distribution depends upon many factors: the concentration distribution of the control species, the local interpretation of this concentration with

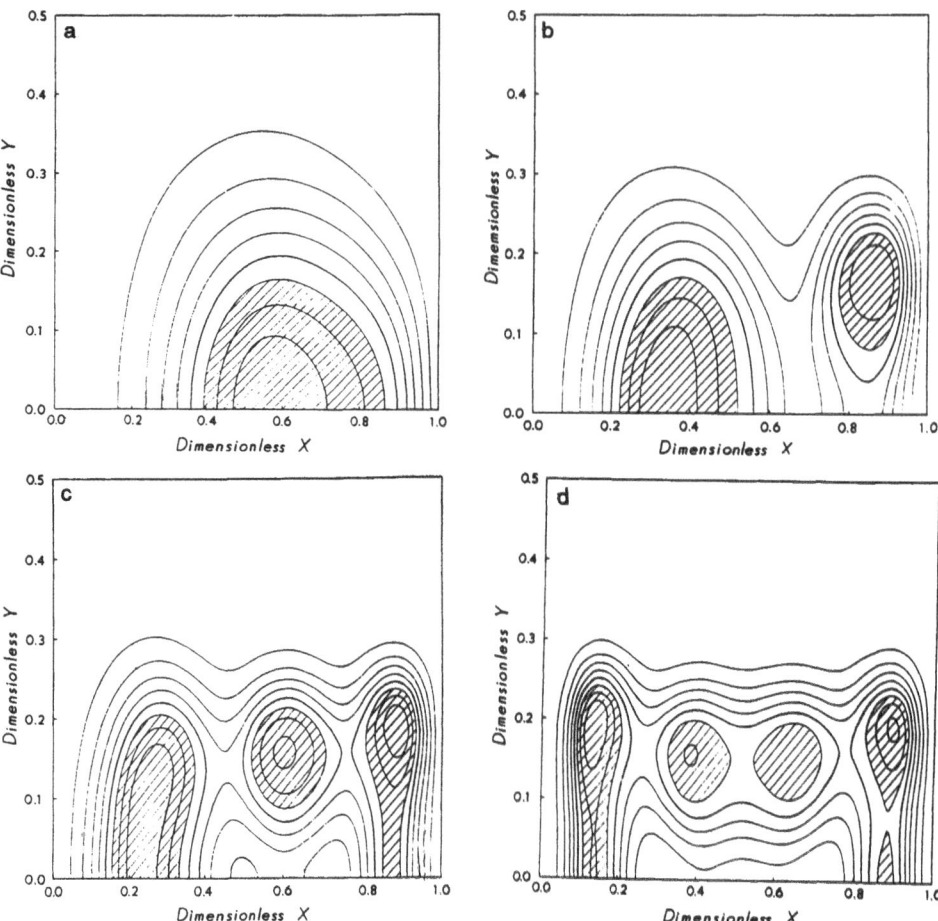

Figure 9.10. Contours of the morphogen v at steady state for several choices control parameters. (a) $w_0 = 3.0$, $\gamma = 1$; (b) $w_0 = 2.0$, $\gamma = 4$; (c) $w_0 = 1.0$, $\gamma = 6$; (d) $w_0 = 0.5$, $\gamma = 30$; In each figure $L_2 = 1$ mm, $\eta_0 = 0.3$ mm, $\mathcal{D}_1/\omega = 0.0005$. In both the 1D and 2D simulations a system of algebraic equations was obtained using finite difference schemes of Crank Nicholson type. The 2D solver (based on the Peaceman-Rachford scheme) is an alternating direction implicit (ADI) method (Mittelmann & Weber 1980). The resulting non-linear equations were solved using the software package *nksolve* (Brown & Saad 1987).

respect to gap junctional permeability, the interaction of a diffusible factor produced in the AER on the reaction kinetics as well as the reaction kinetics of the morphogens, limb bud geometry, and boundary conditions. In this preliminary study we have shown that control species modulation of gap junctional permeability can reliably produce a morphogen prepattern, which by a simple threshold mechanism, could lead to the I,II,III,... pattern required for limb development. Moreover, our numerical "ZPA transplant" experiments in both one and two spatial dimensions give results consistent with experimental results in that width increases of 60-70% lead to complete mirror-image duplication.

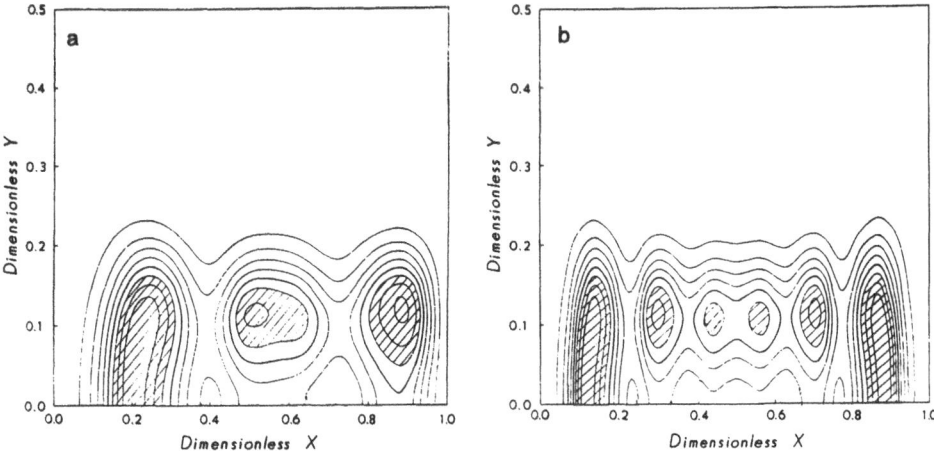

Figure 9.11. Steady state contours of v for the "Numerical ZPA transplant experiment". (a) $L_1 = 1$ mm, $w_0 = 2$, $w_1 = 0$ (b) $L_1 = 1.65$ mm, $w_0 = 2$, $w_1 = 2$ In both Figures $L_2 = 1$ mm, $\eta_0 = 0.2$ mm, $\alpha/\sqrt{D_3} = 150$.

The variety of pattern possible in two and three spatial dimensions is far richer than in the one dimensional case. In our model the morphogen reactions are localized near the AER, thereby constraining the morphogen prepattern to the subdistal region. In normal limb development in chick, there is significant elongation of the proximo-distal axis during stages 18-24 while the anterior-posterior axis remains relatively constant (Stark & Searls 1973). With autocatalysis localized in the distal subridge, spatial pattern is also localized near the AER and is independent of PD length (as long as this length is not too small). Intuitively, one would expect that different localization mechanisms would lead to variation in pattern. Indeed, preliminary results (not shown) suggest that proximal-distal segmentation of the pattern elements can be controlled by variation in the AER localization mechanism. It may be desirable to localize pattern away from the immediate neighborhood of the AER. This could be accomplished by using a square-wave function in place of (9.11) or by a assuming that cells adjacent to the AER do not respond to the morphogen prepattern.

ACKNOWLEDGMENTS This work was supported in part by NIH Grant # GM 29123.

REFERENCES

Allen, F., Tickle, C., & Warner, A. 1990. The role of gap junctions in patterning of the chick limb bud. *Development*, **108**, 623–634.

Ashkenazi, M., & Othmer, H. G. 1978. Spatial patterns in coupled biochemical oscillators. *Jour. Math. Biol.*, **5**, 305–350.

Bowen, J., Hinchliffe, J. R., Horder, T. J., & Reeve, A. M. F. 1989. The fate map of the chick forelimb-bud and its bearing on hypothesized developmental control mechanisms. *Anat. Embryol.*, **179**, 269–283.

Brown, P. N., & Saad, Y. 1987. Hybrid Krylov methods for nonlinear systems of equations. *LLNL Report*, **UCRL-97645**(Nov.).

Coelho, C. N. D., & Kosher, R. A. 1991a. Gap junctional communication during limb cartilage differentiation. *Dev. Biol.*, **144**, 47–53.

Coelho, C. N. D., & Kosher, R. A. 1991b. A gradient of gap junctional communication along the anterior-posterior axis of the developing chick limb bud. *Dev. Biol.*, **144**, 529–535.

Coelho, C. N. D., Sumoy, L., Kosher, R. A., & Upholt, W. B. 1992. Ghox-7: a chicken homeobox-containing gene expressed in a fashion consistent with a role in patterning events during embryonic chick limb development. *Differen.* (In press).

Cooke, J., & Summerbell, D. 1980. Cell cycle and experimental pattern duplication in the chick wing during embryonic development. *Nature (Lond)*, **287**, 697–701.

Dillon, R., Maini, P. K., & Othmer, H. G. 1993. Pattern formation in generalized Turing systems. *J. Math. Biol.* To appear.

Fallon, J. F., & Kelley, R. O. 1977. Ultrastructural analysis of the apical ectodermal ridge during vertebrate limb morphogenesis II. Gap junctions as distinctive ridge structures common to birds and mammals. *J. Embryol. Exp. Morphol.*, **41**, 223–232.

French, V., Bryant, P. J., & Bryant, S. V. 1977. Pattern regulation in epimorphic fields. *Science*, **193**, 969–981.

Hinchliffe, J. R., & Sansom, A. 1985. The distribution of the polarizing zone in the leg bud of the chick embryo. *J. Embryol. Exp. Morphol.*, **86**, 169–176.

Honig, L. S., & Summerbell, D. 1985. Maps of strength of positional signaling activity in the developing chick wing bud. *J. Embryol. Exp. Morphol.*, **87**, 163–174.

Hornbruch, A., & Wolpert, L. 1970. Cell division in the early growth and morphogenesis of the chick limb. *Nature*, **226**, 764–766.

Javois, L. C., Iten, L. E., & Murphy, D. J. 1981. Formation of supernumerary structures by the embryonic chick wing depends on the position and orientation of a graft in a host limb bud. *Dev. Biol.*, **82**, 343–349.

Kelley, R. O., & Fallon, J. F. 1983. A freeze-fracture and morphometric analysis of gap junctions of limb bud cells: initital studies on a possible mechanism for morphogenetic signalling during development. In *Limb Development And Regeneration, part A*, Fallon, J. F., & Caplan, A. I. (eds). pp. 119–130, New York: Alan. R. Liss, Inc.

Lee, J., & Tickle, C. 1985. Retinoic acid and pattern formation in the developing chick wing: SEM and quantitative studies of early effects on the apical ectodermal ridge and bud outgrowth. *J. Embryol. Exp. Morphol.*, **90**, 139–169.

Maini, P. K., & Solursh, M. 1991. Cellular mechanisms of pattern formation in the developing limb. *Int. Rev. of Cytology*, **129**, 91–133.

McLachlan, J. C. 1991. Growth factors produced by the polarising zone - A complement to the retinoic acid system. In *Developmental Patterning of the Vertebrate Limb*, Hinchliffe, J. R. (ed). pp. 115–122, New York: Plenum Press.

Mehta, P., Bertram, J. S., & Loewenstein, W. R. 1989. The actions of retinoids on cellular growth correlate with their actions on gap junctional communication. *J. Cell Biol.*, **108**, 1053–1065.

Meinhardt, H. 1982. Models of Biological Pattern Formation. London: Academic Press.

Mittelmann, H. D., & Weber, H. 1980. Numerical methods for bifurcation problems - A survey and classification. In *Bifurcation Problems and Their Numerical Solution*, ISNM 54. pp. 1–45, Birkhäuser Verlag.

Muneoka, K., & Sassoon, D. 1992. Molecular aspects of regeneration in developing vertebrate limbs. *Dev. Biol.*, **152**, 37–49.

Othmer, H. G. 1983. A continuum model for coupled cells. *J. Math. Biol.*, **17**, 351–369.

Othmer, H. G., & Aldridge, J. 1978. The effects of cell density and metabolite flux on cellular dynamics. *J. Math. Biol.*, **5**, 169–200.

Othmer, H. G., & Pate, E. F. 1980. Scale invariance in reaction-diffusion models of spatial pattern formation. *Proc. Nat. Acad. Sciences*, **77**, 4180–4184.

Pate, E., & Othmer, H. G. 1984. Applications of a model for scale-invariant pattern formation in developing systems. *Differentiation*, **28**, 1–8.

Reiter, R. S., & Solursh, M. 1982. Mitogenic property of the apical ectodermal ridge. *Dev. Biol.*, **93**, 28–35.

Saunders, Jr., J. W. 1948. The proximo-distal sequence of origin of the parts of the chick wing and the role of the ectoderm. *J. Exp. Zool.*, **108**, 363–403.

Smith, S. M., Pang, K., Sundin, O., Wedden, S. E., & Thaller, C. 1989. Molecular approaches to vertebrate limb morphogenesis. *Development (Suppl.)*, pp. 121–131.

Stark, R. J., & Searls, R. L. 1973. A description of chick wing bud development and a model of limb morphogenesis. *Dev. Biol.*, **33**, 138–153.

Summerbell, D. 1974a. Interaction between the proximo-distal and antero-posterior co-ordinates of positional value during the specification of positional information in the early development of the chick limb-bud. *J. Embryol. Exp. Morphol.*, **32**, 227–237.

Summerbell, D. 1974b. A quantitative analysis of the effect of excision of the AER from the chick limb-bud. *J. Embryol. Exp. Morphol.*, **32**, 651–660.

Summerbell, D. 1977. Regulation of deficiencies along the proximal distal axis of the chick wing-bud: a quantitative analysis. *J. Embryol. Exp. Morphol.*, **41**, 137–159.

Summerbell, D. 1979. The zone of polarising activity: evidence for a role in normal chick morphogenesis. *J. Embryol.Exp.Morph*, **50**, 217–233.

Summerbell, D., & Lewis, J. H. 1975. Time, place and positional value in the chick limb-bud. *J. Embryol. Exp. Morph.*, **33**, 621–643.

Thaller, C., & Eichele, G. 1987. Identification and spatial distribution of retinoids in the developing chick limb bud. *Nature*, **327**, 625–628.

Tickle, C. 1981. The number of polarising region cells required to specify additional digits in the developing chick wing. *Nature*, **289**, 295–298.

Tickle, C., Summerbell, D., & Wolpert, L. 1975. Positional signalling and specification of digits in chick limb morphogenesis. *Nature*, **254**, 199–202.

Tickle, C., Lee, J., & Eichele, G. 1985. A quantitative analysis of the effect of all-trans-retinoic acid on the pattern of chick wing development. *Dev. Biol.*, **109**, 82–95.

Turing, A. M. 1952. The chemical basis of morphogenesis. *Phil. Trans. R. Soc. Lond. B*, **237**, 37–72.

Wilby, O. K., & Ede, D. A. 1975. A model for generating the pattern of cartilage skeletal elements in the embryonic chick limb. *J. Theor. Biol.*, **52**, 199–217.

Wolpert, L. 1969. Positional information and the spatial pattern of cellular differentiation. *J. Theor. Biol.*, **25**, 1–47.

Wolpert, L., & Hornbruch, A. 1987. Positional signalling and the development of the humerus in the chick limb bud. *Development*, **100**, 333–338.

Wolpert, L., Lewis, J., & Summerbell, D. 1975. Morphogenesis of the vertebrate limb. In *Cell Patterning*, pp. 95–119, Amsterdam, 1975. Ciba Fdn. Symposium 29, new series, Associated Scientific Publishers.

Zwilling, E. 1961. Limb morphogenesis. *Adv. in Morph.*, **1**, 301–330.

10. WAVE PATTERNS IN ONE-DIMENSIONAL NONLINEAR DEGENERATE DIFFUSION EQUATIONS

Faustino Sánchez-Garduño[1,2] and Philip K. Maini[1]

[1]Centre for Mathematical Biology
Mathematical Institute
24–29 St. Giles'
Oxford OX1 3LB, UK

[2]Departamento de Matemáticas
Facultad de Ciencias, UNAM
Circuito Exterior
C. U. México 04510, D.F.,Mexico

10.1 INTRODUCTION

Several different types of wave patterns occur in physiology, chemistry and biology. In many cases such phenomena are modelled by reactive-diffusive parabolic systems (see, for example, Fisher 1937; Kolmogorov *et al.* 1937; Winfree 1988; Murray 1989; Swinney & Krinsky 1992). In many biological and physical situations, dispersal is modelled by a density-dependent diffusion coefficient, for example, the bacterium *Rhizobium* diffuses through the roots of some *leguminosae* plants according to a nonlinear diffusive law (Lara-Ochoa & Bustos 1990); nonlinear diffusion has been observed in the dispersion of some insects (Okubo 1980) and small rodents (Meyers & Krebs 1974).

Here, we restrict ourselves to analyzing the problem of the existence of travelling wave solutions (TWS) in an one-dimensional domain for the special case of a nonlinear diffusion coefficient which is degenerate at $u = 0$. That is, we look for a solution $u(x,t) = \phi(x - ct)$ for equation

$$\frac{\partial u}{\partial t} = \frac{\partial}{\partial x}\left[D(u)\frac{\partial u}{\partial x} \right] + g(u); \quad (x,t) \in \mathbf{R} \times \mathbf{R}^{+}, \tag{10.1}$$

where D and g are defined on $[0, 1]$ and:

1. $g(0) = g(1) = 0$, $g(u) > 0 \ \forall \ u \in (0,1)$

2. $g \in C^2[0,1]$ with $g'(0) > 0$ and $g'(1) < 0$

3. $D(0) = 0$ with $D(u) > 0 \ \forall \ u \in (0,1]$

4. $D \in C^2[0,1]$.

with $u(x,0) = u^0(x)$, $0 \leq u^0(x) \leq 1 \ \forall \ x \in \mathbf{R}$.

Different types of one-dimensional TWS, for example, fronts, pulses, sharp and oscillatory, have been reported in the literature (Fife 1979; Sánchez-Garduño & Maini 1992).

Experimental and Theorietcal Advances in Biological Pattern Formation,
Edited by H.G. Othmer *et al.*, Plenum Press, New York, 1993

The problem of the existence of TWS of (10.1) has been studied for a few particular cases by a number of authors, for example, Aronson (1980); Newman (1980); Murray (1989) and Lara-Ochoa & Bustos (1990). In this communication we generalize previous analysis. The proof of our results uses a qualitative approach in which to look for TWS for (10.1) is equivalent to finding the appropriate parameter space (which includes the speed c) for which there exists a heteroclinic trajectory of a certain system of ordinary differential equations. The boundary conditions for the TWS are given by the coordinates of the equilibrium states connected by the heteroclinic trajectory.

   In the following section we only state the results and present some illustrative examples. Proofs, full details and other examples can be found in Sánchez-Garduño & Maini (1992).

10.2   RESULTS AND EXAMPLES

Theorem 1 If the functions D and g in (10.1) satisfy the above conditions and $D'(u) > 0 \ \forall \ u \in [0, 1]$ with $D''(0) \neq 0$, then the reaction-diffusion equation (10.1) possesses at most a TWS $u(x, t) = \phi(x - c^*t)$ of sharp type such that for $c^* > 0$: $\phi(-\infty) = 1$, $\phi(\xi) = 0$ for $\xi \geq \xi^*$; $\phi'(-\infty) = 0$, $\phi'(\xi^{*-}) = -\frac{c^*}{D'(0)}$, and $\phi'(\xi^{*+}) = 0$. Moreover

   1. For $0 < c < c^*$ there are no TWS

   2. For each $c > c^*$, (10.1) has a TWS of front type satisfying the boundary conditions: $\phi(-\infty) = 1$ and $\phi(+\infty) = 0$.

An application of this theorem is illustrated in the following example.

Example 1. Here we consider the equation

$$\frac{\partial u}{\partial t} = \frac{\partial}{\partial x}\left[(\beta u + u^2)\frac{\partial u}{\partial x}\right] + u(1 - u)[1 - u(1 - u)], \tag{10.2}$$

where $\beta > 0$. In this case the qualitative behaviour of the solution does not depend on the value of β. We illustrate in Figure 10.1 the phase portrait for the case $\beta = 2.0$, for which $c^* \approx 0.98$.

Theorem 2 If the functions D and g in equation (10.1) satisfy the conditions 1-4 in the Introduction, with the further conditions on D, that, $D'(0) = 0$, $D'(u) > 0 \ \forall \ u \in (0, 1]$ and $D''(0) > 0$, then for each c such that

$$c \geq \sup\left\{\frac{d}{d\phi}\left[\rho(\phi)D(\phi)\right] + \frac{g(\phi)}{\rho(\phi)}\right\}, \tag{10.3}$$

where ρ is a continuously differentiable non-negative function in the interval $[0, 1]$ with $\rho(0) = 0$ and $\rho'(0) > 0$, there exists a TWS of front type for the equation (10.1) satisfying the boundary conditions, $\phi(-\infty) = 1$ and $\phi(+\infty) = 0$. An application of this theorem is illustrated in example 2.

Example 2. Here we consider the equation

$$\frac{\partial u}{\partial t} = \frac{\partial}{\partial x}\left[u^2\frac{\partial u}{\partial x}\right] + u(1 - u). \tag{10.4}$$

Typical phase portraits are illustrated in Figure 10.2.

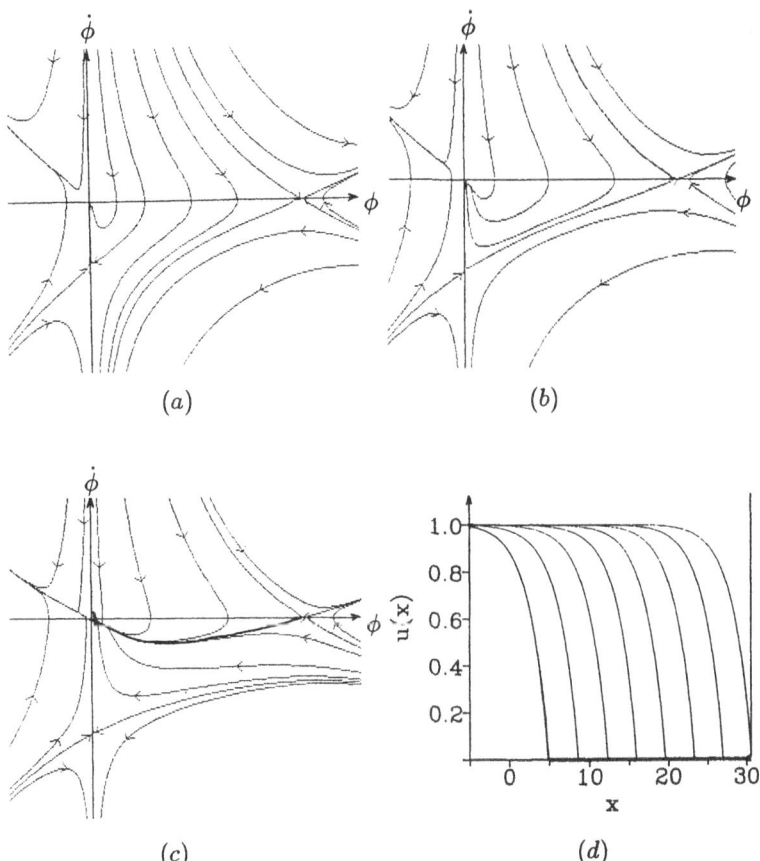

Figure 10.1. Phase portraits for example 1 with $\beta = 2.0$ for different values of the speed c. (In this case $c^* \approx 0.98$.) (a) $c = 0.7$, (b) $c = 0.98$, (c) $c = 1.5$. For $c < c^*$ there is no connection between the equilibrium points and hence no travelling wave. For $c = c^*$, the saddle-saddle connection implies a travelling wave of sharp type, while for each $c > c^*$ we have a travelling wave of front type. The solution for the partial differential equation is shown in (d). The numerically calculated wavespeed is $c^* = 0.9797$.

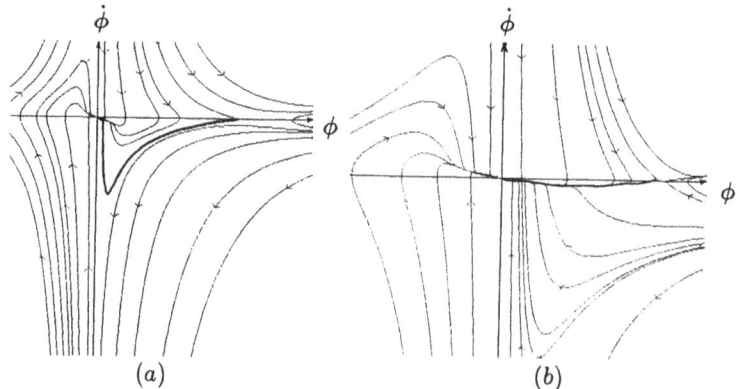

Figure 10.2. Phase portraits for example 2 for different values of the speed c. (a) $c = 0.5$ and (b) $c = 2.0$. In both cases there exists a connection between the two equilibrium points suggesting the existence of a travelling wave of front type.

REFERENCES

Aronson, D. G. 1980. Density-dependent interaction-diffusion systems. In *Dynamics and Modelling of Reactive Systems*, Stewart, Warren E. (ed). Academic Press.

Fife, P. 1979. *Mathematical aspects of reaction diffusing systems*. Vol. 28. Springer-Verlag. Lecture Notes in Biomathematics.

Fisher, R. A. 1937. The wave of advance of advantageous genes. *Ann. Eugenics.*, **7**, 353–369.

Kolmogorov, A., Petrovsy, I., & Piskounov, N. 1937. Study of the diffusion equation with growth of the quantity of matter and its applications to a biological problem. In *Applicable Mathematics of Non-Physical Phenomena*, Oliveira-Pinto, F., & Conolly, B. W. (eds). John Wiley and Sons, 1982 edn.

Lara-Ochoa, F., & Bustos, V. P. 1990. A model for aggregation-dispersion dynamics of a population. *BioSystems*, **24**, 215–222.

Meyers, M. R., & Krebs, C. 1974. Population cycles in rodents. *Sci. Am.*, **230**, 38–46.

Murray, J. D. 1989. *Mathematical biology*. New York: Springer-Verlag, Berlin.

Newman, W. I. 1980. Some exact solutions to a non-linear diffusion problem in population genetics and combustion. *J. Theor. Biol.*, **85**, 325–334.

Okubo, A. 1980. Diffusion and ecological problems: mathematical models. In *Biomathematics*, volume 10. Springer-Verlag.

Sánchez-Garduño, F., & Maini, P. K. 1992. Travelling wave phenomena in some degenerate reaction-diffusion equations. (Submitted for publication).

Swinney, H. L., & Krinsky, V. I. 1992. Waves and Patterns in Chemical and Biological Media. Special Issue of Physica D.

Winfree, A. 1988. *When time breaks down*. Princeton University Press.

11. MAPPING GENE ACTIVITIES INTO MORPHOLOGICAL PATTERNS IN *DROSOPHILA*

Brian Goodwin[1] and Stuart Kauffman[2]

[1]Development Dynamics Research Group
The Open University
Walton Hall
Milton Keynes, MK7 6AA, UK

[2]Santa Fe Institute
1660 Old Pecos Trail, Suite C
Santa Fe, NM 87501, USA

11.1 INTRODUCTION

In *Drosophila*, one of the veils that obscures our understanding of has been drawn and we can now see the choreography of gene activity that is involved in generating the changing shapes of the developing embryo. The immediate question that accompanies this revelation is how the various combinations of gene products in different parts of the embryo relate to overt patterns of differentiated cell state. It tends to be assumed that there must be a simple and direct combinatorial code, but this is not yet obvious. In this paper, we make use of the spatial patterns of segmentation gene products and characteristic properties of mutant phenotypes to construct a mapping that suggests how the code may work, and how to test it experimentally. This may begin to draw a second veil. But there is yet another before we really begin to understand morphogenesis, and that is the direct causal relations between epigenetic processes (of which gene products are but one category of variable) and the actual physical shapes and properties of differentiated cells. This may not be nearly as difficult a job as it appears to be, because the multitudinous variables of the epigenetic system all funnel through some fairly simple cellular properties, which they modulate in various ways. But we restrict our attention here to the problem of the combinatorial epigenetic code that maps gene product concentrations onto differentiated patterns. The particular model presented here has been previously described with somewhat different emphases in Kauffman & Goodwin (1990) and in Goodwin & Kauffman (1992).

11.2 DELETIONS AND MIRROR SYMMETRIES

The mutant morphologies that we examine in this paper are illustrated by the series of phenotypes generated by alleles of the *bicaudal* gene (Nüsslein-Volhard 1977). Figure 11.1 shows the normal wild-type first instar larva and the extreme mirror-symmetric phenotype of the mutant, while Figure 11.2 illustrates the range of mutant morphologies generated by

Experimental and Theorietcal Advances in Biological Pattern Formation,
Edited by H.G. Othmer *et al.*, Plenum Press, New York, 1993

87

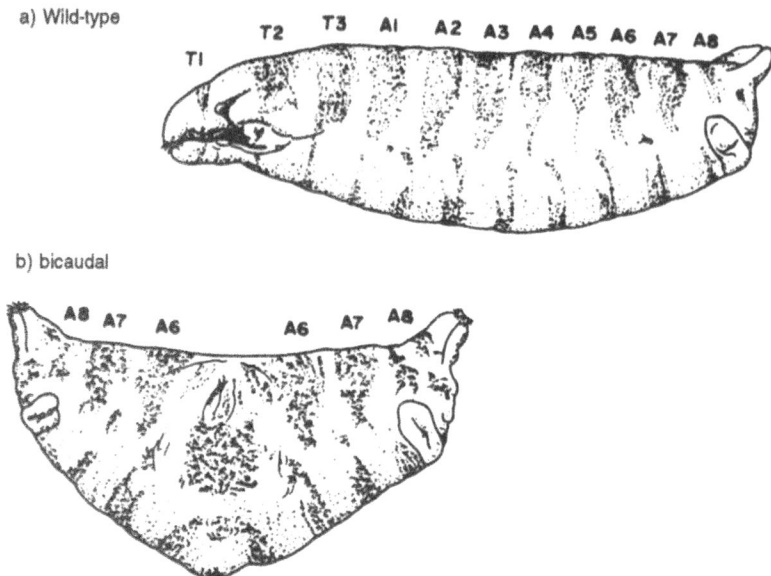

Fig. 11.1. (a) Normal cuticular pattern in first instar *Drosophila* larvae (b) mirror symmetric *bicaudal* phenotype.

different alleles at this locus. Evidently the plane of mirror reflection can occur anywhere within a segment, and the resulting form can have either equal or unequal numbers of segments on either side of the reflection plane. Figure 1 shows that there are fewer dorsal than ventral pattern elements reflected, and yet there is complete coherence of pattern from ventral to dorsal, without discontinuity. Weaker forms of the allele result in deletions without mirror reflection, as in Figure 11.2(f) which shows a headless embryo with complete thoracic and abdominal segments. There are also cases where one side of the embryo shows mirror reflection of the denticle band pattern while the other has the deletion phenotype (Figure 11.2e). In such a case the character of a segment changes in the mid-line from abdominal 2 on one side to a mirror-reflected abdominal 3 on the other, again without any discontinuity. Even more remarkable are cases such as those observed in the beetle, *Collosobruchus maculatus* (Van der Meer 1984), in which induced bicaudal phenotypes include embryos that are fully normal except for a longitudinal stripe in which a mirror duplication has occurred. How can such striking transformations of pattern arise? Clearly there are no compartments or sub-divisions of the embryo that specify the positions of mirror-reflection; and in some sense the spatially adjacent, but radically transformed overt patterns indicate that the different generating states, are neighbours in some space, despite the discontinuities of pattern that they generate.

Bicaudal is in no sense an isolated instance of a mirror symmetric mutant pattern in *Drosophila*. The complementary double-headed phenotype arises as the mutant *dicephalic* (Lohs-Schardin 1982). Furthermore, the three categories of segmentation gene (gap, pair-rule, and segment polarity) all have mutants that reveal the same series of deletion to mirror symmetry phenotypes as those described for *bicaudal*, but on different spatial wavelengths characteristic of these categories. Weak mutant alleles of the gap gene *Krüppel*, for instance, result in deletions of thoracic and anterior abdominal segments, posterior segments being relatively normal; while in stronger alleles the absent thoracic and abdominal segments are replaced by a partial mirror reflection of the remaining posterior abdomen, as shown in Figure 11.3 (Jäckle *et al.* 1986). Mutants of *hunchback*, another gap gene, have effects

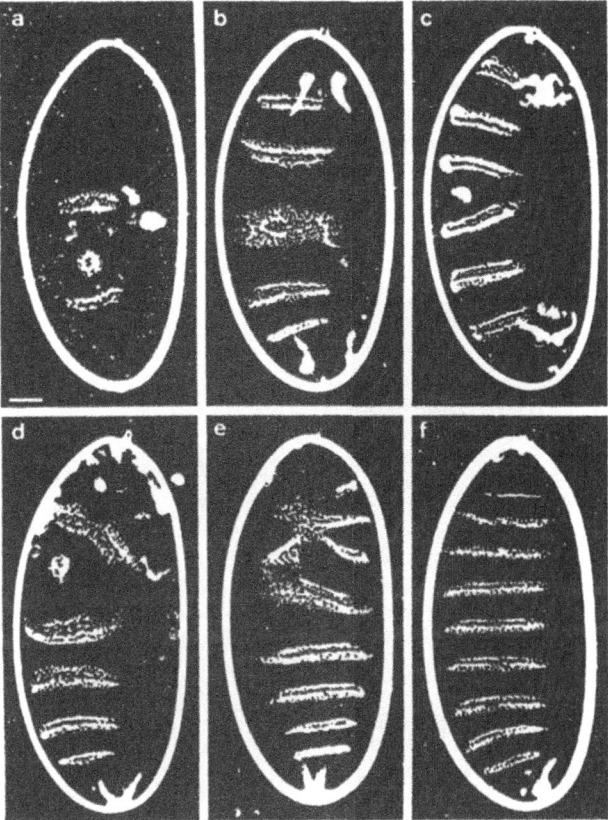

Figure 11.2. The range of phenotypes generated by different alleles of the *bicaudal* gene.

centered on head and thoracic segments, these being deleted by weaker alleles or replaced by mirror-image duplications of the anterior abdominal segments in cases of stronger alleles. *hunchback* mutants also have a posterior domain of action, abdominal segments 7 and 8 being deleted (Lehmann & Nüsslein-Volhard 1987).

Pair-rule mutants show similar deletion and mirror-symmetric patterns, but now the spatial domain has a characteristic wavelength of 2 segments. Weak mutants of *hairy* are characterized by loss of the anterior part of each even-numbered abdominal segment and the posterior part of each odd-numbered segment, with corresponding deletions to the thoracic segments. Strong *hairy* alleles result in a mirror-reflection of the denticle bands of odd-numbered abdominal segments and T_2, together with a loss of all naked cuticle, so that the ventral abdomen is a continuous lawn of setae in 2-segment mirrored arrays (Ingham *et al.* 1985). Similar mutant morphologies arise with *runt* (Nüsslein-Volhard & Wieschaus 1986), but the deletions are phase-shifted by about 100° relative to hairy so that they are centered on T_2 and the odd-numbered abdominal denticle bands. In stronger *runt* alleles, the deleted domains are replaced by mirror-image duplications of the remaining pattern elements (Gergen & Wieschaus 1985).

Segment polarity mutants characteristically involve a replacement of one half of each denticle band by a mirror-symmetric duplication (Nüsslein-Volhard & Wieschaus 1986). Each different mutant is phase-shifted relative to the others so that different segmental regions are mirror-imaged in *gooseberry, hedgehog, patched*, and so on. There are also alleles that show deletions without mirror-symmetry, so these conform to the same properties as mutations in genes of the other segmentation categories. Evidently we are dealing with a typical or generic property of the mapping between gene activities and phenotypes. In order to discover what this may be, it is necessary to examine the characteristics of gene product distribution patterns in embryos and to define an appropriate mapping between these and morphogenesis.

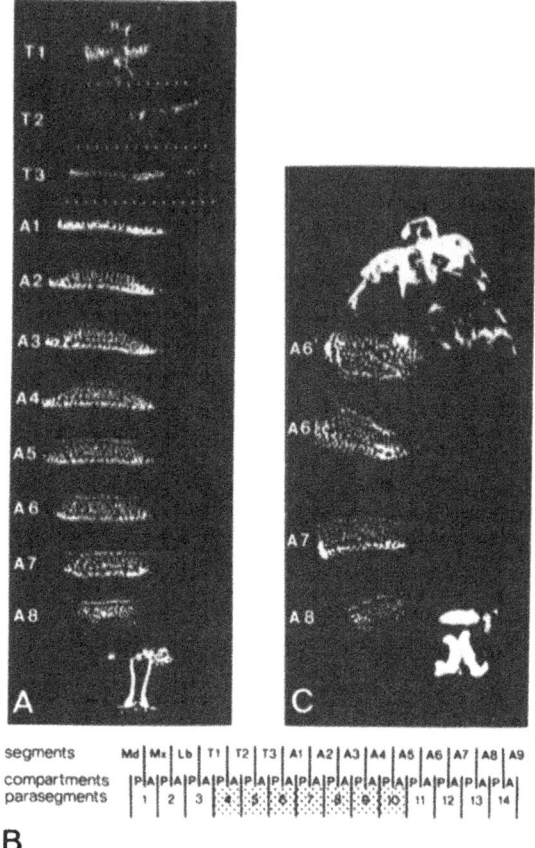

Figure 11.3. The phenotypes of a strong *Krüppel* mutant (C) compared with the normal cuticular pattern of the larva (A), with the range of affected segments shown in B.

11.3 SPATIAL PERIODICITIES IN GENE PRODUCT DISTRIBUTIONS

One of the major revelations of molecular techniques applied to *Drosophila* embryos is the hierarchical sequence of spatially periodic patterns of segmentation gene products (transcripts and protein) that reflect the domains of action of these genes in mutant alleles. Segment polarity gene products are distributed in stripes within each segmental domain, each gene phase-shifted relative to others in close correlation with mutant defects; pair-rule gene products have a two-segment periodic pattern and relative phase-shifts; gap genes are expressed in bands extending over several segments, each with a characteristic domain in the embryo; while the maternals like *bicoid* have protein distributions that extend over roughly half the embryo. But whereas segment polarity and pair-rule gene products have spatially periodic patterns that conform to expectations from the mutant data, the gap genes show an interesting inconsistency. *Krüppel* mutants, for example, typically have a single deletion domain in the center of the embryo (thoracic and anterior abdominal segments deleted). But the product of this gene is found not only in this central domain, but also at both poles, where there is no sign of their presence from the mutant phenotypes, which have normal terminalia. Similarly *knirps*, another gap gene with a single domain of defect in mutants in the abdomen, has *two* domains of gene expression: a broad band where the defects occur and a narrower one anteriorly. The third major gap gene, *hunchback*, has two domains of the embryo affected by mutations, one anterior and one posterior, and corresponding gene product distributions. So these three genes actually have spatially periodic product distributions, each phase-shifted relative to one another, despite the fact that for two of them only one domain is expressed in segmental structure. The periodicities may be telling us something about the underlying dynamics of pattern formation in terms of globally periodic pattern production, as discussed elsewhere (Goodwin & Kauffman 1990; Hunding *et al.* 1990).

11.4   HOW DO GENE PRODUCTS MAP INTO EXPRESSED PATTERNS?

It tends to be assumed that genes influence patterns of cell differentiation by a combinatorial code that is effectively determined by the simple presence or absence of gene products in different combinations. This can be generalized to the notion of a threshold value of each gene product above which it has one influence, designated by a +, and below which its influence is described by a –. A set of 4 genes, A, B, C, and D, for example, with spatially periodic distributions, each phase-shifted with respect to one another as shown in Figure 11.4(a), can then be assigned combinatorial code words on different regions of the domain over which they cycle according to whether they are above or below threshold, taken to be the mean of the concentration range. Plotting a single spatial cycle of the 4 gene products as a circle divided into 8 sectors, each sector can be assigned a distinct code-word determined by the combination of 4 binary variables describing whether each of the gene products is above or below threshold, as shown in Figure 11.4(b).

Suppose now that gene A mutates so that its product is absent - a strong mutation. Will the coding pattern have mirror symmetry? The resulting code-words are shown in Figure 11.4(c). The two sectors with ringed code-words are now meaningless or illegal, and there is no mirror-symmetry to the pattern generated. Therefore a simple binary code for the mapping from gene product space to expressed pattern is not consistent with the evidence, and it is necessary to consider a different relationship.

An important clue to an alternative mapping is provided by an experimental observation by Coulter & Wieschaus (1988). They produced flies with two pair-rule mutations, one in *even-skipped*, the other in *odd-skipped*. If gene effects are additive, this double mutant should have an absence of both even- and odd-numbered denticle bands, and so have nothing but naked cuticle. In fact it has eight partial denticle bands, each a small mirror duplication of the normal patterns of abdominal segments 1 - 8. Another result was equally revealing. The pair-rule mutant, *runt* has an odd-skipped phenotype for weak deficiency alleles. But in a mutant that overproduces *runt* product, a complementary phenotype arises in which deletions

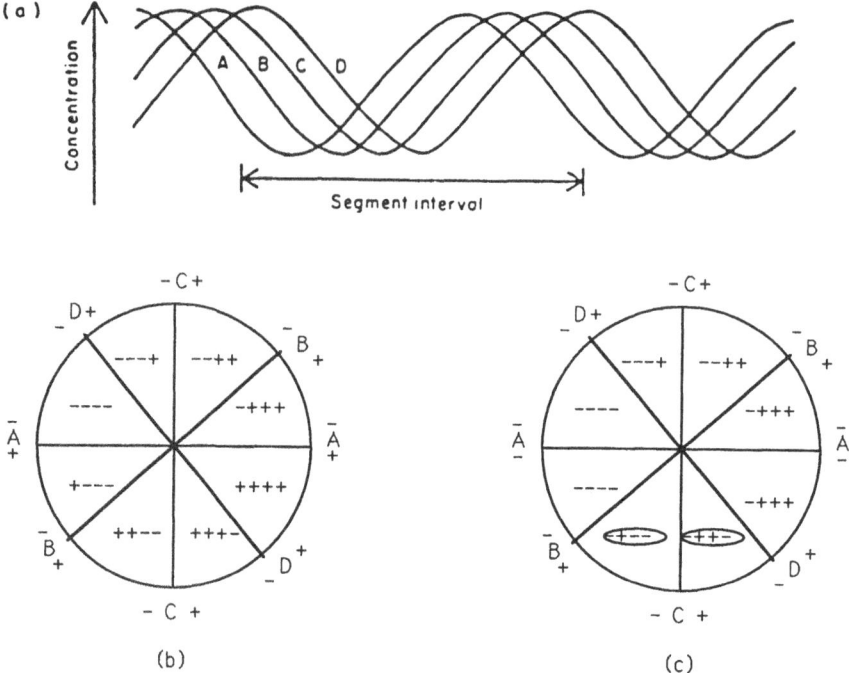

Figure 11.4. Combinatorial code words for a binary code of four genes, A,B,C and D, phase-shifted over a segment interval as shown in (a). On a map in which space is represented by position on a circle read in counter-clockwise direction, the code words specifying eight distinct domains over a single sement are as shown in (b). If there is a deletion mutant of A, the code-words are as in (c), resulting in the two illegitimate combinations shown encircled.

occur in the even-numbered denticle bands: too much *runt* is like too little *eve* (Gergen & Wieschaus 1985; Gergen *et al.* 1986). These observations led the authors to propose that gene products exert their influence as a function of their ratios. So let's see what results this gives.

We start with a description of the phase relations of the three major pair-rule gene products, *hairy, eve*, and *runt*, and their positions relative to denticle bands, as shown in Figure 11.5. Only a part of the total pattern is shown, but it simply repeats over the segmentation domain. Of course the detailed character of segments such as A1, A2, and A3 is specified by the combined influence of all the segmentation genes, and also the homeotics, as they act in these domains. But we restrict our attention to the pair-rules for the moment.

To examine a ratio mapping, plot eve against *hairy* over one complete spatial cycle, in Figure 11.6. Spatial position is now represented by points on this wheel, and the positions of the denticle bands in this 2-segment cycle can be designated as shown, the cycle repeating over every double segment domain. We now make some specific assumptions about the role of ratios of gene products in specifying pattern.

Assume that gene product concentrations as they affect pattern are measured from the center of the circle (which is the mid-range of the concentration amplitudes in Figure 11.5). This is similar to the previous assumption about a combinatorial code that is measured above and below a threshold level, but now we assume that concentrations are read as continuous variables measured from the center of the wheel. Call these concentrations x for *hairy* and y for *eve* gene product.

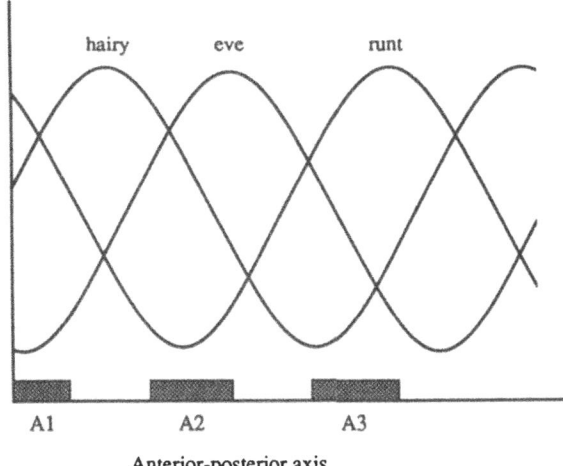

Figure 11.5. The spatial pattern of the three major pair-rule genes, *hairy, eve,* and *runt,* showing their approximate phase relations.

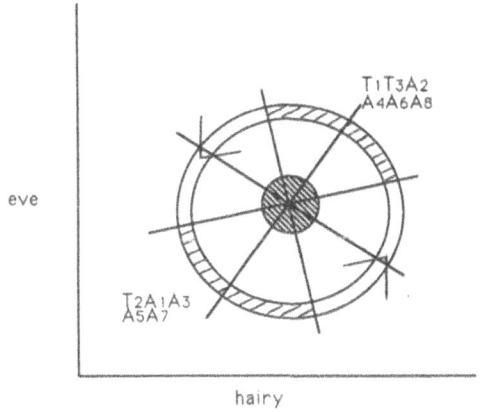

Figure 11.6. A phase plot of *eve* against *hairy* over one pair-rule cycle, showing the positions of the denticle bands in this colour wheel.

Now assume that the way a particular pair of values of *hairy* and *eve* gene product concentrations, x and y, affect pattern is by specifying a line of equivalent points that makes an angle θ with a reference line through the center of the circle and parallel to the x- axis, this angle being defined by $\tan \theta = y/x$, the ratio of concentrations.

Then each radius vector from the center of the wheel at a specific angle θ to the reference line consists of points that have the same effect on pattern formation. Therefore the exact position of the wheel in Figure 11.6 is not important, so long as it circles the origin and covers all angles $0 - 2\pi$. This means that the spatially periodic concentration of the gene products in Figure 11.5 can vary substantially without affecting the specification of pattern. In fact, the important properties of the mapping relate primarily to topological properties, which gives the model robustness, as we shall see. This is why we can use circles for our phase maps, which actually occur only if the phase differences between the waves in Figure 11.5 are 90°.

The origin of the radius vectors at the center of the wheel is a point with all angles, θ, and so it does not specify any identifiable pattern. Since the sensitivity of the pattern-specification process is limited to some tolerance range around any point, this indeterminate region is actually a disc - the null disc shown as the shaded circle at the center of the wheel. Within this disc, no pattern is specified. Similarly there will be discrete sectors defined by ranges of θ or the ratio, y/x, within which no discriminations can be made, all cells within such a range following the same pathway of differentiation in terms of the contribution of these two genes. So the whole space, called tissue specificity space (TSS) by Winfree (1980), is quantized. These quantized sectors we identified metaphorically with the colour spectrum, and the sectored cycle is described as a colour wheel (Kauffman & Goodwin 1990). The language reflects a topologically similar analysis carried out by Winfree (1980) on the periodic temporal organization of organisms, particularly biological clocks, in which the concept of the isochron was introduced to describe states that map into points of equal time in the dynamic space of biological oscillators. There are some deep qualitative similarities between these treatments of time and of space, but these will not be pursued in this paper.

11.5   DELETIONS AND MIRROR SYMMETRIES EXPLAINED

The colour wheel model can now be used to give an explanation of the mutant phenomena, and to produce some experimentally testable predictions. Figure 11.7 shows the

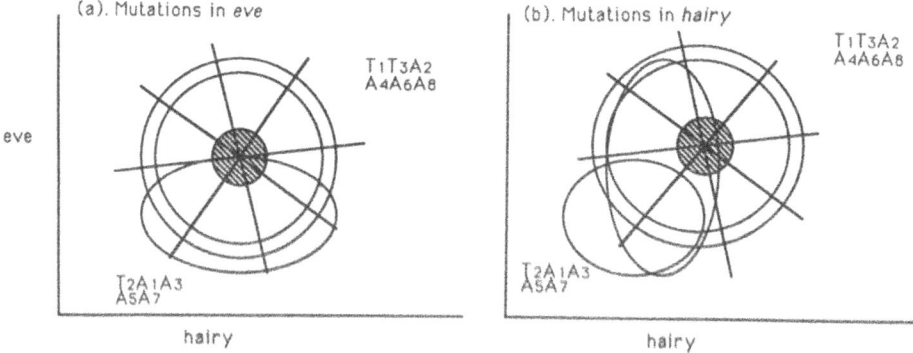

Figure 11.7. (a) The effect of a mutation in *eve* on the colour wheel, showing a deletion pattern of loss of even-numbered segments. (b) Effects of mutations in *hairy* showing both deletions and a mirror-symmetric pattern.

expected effects of mutations in *eve* and *hairy*. Considering *eve* first, Figure 11.7(a) shows the effect of a mutation in which there is an under-production of *eve* product, so that the colour wheel flattens towards the *hairy* axis as a result of reductions in *eve*. There is still a spatially periodic pattern of *eve* product, but the amplitude and the mean are decreased, whereas *hairy* has its normal amplitude. The pattern that results is read by following the ellipse around in antero-posterior direction (counter-clockwise). All the even-numbered abdominal and the odd-numbered thoracic segments disappear, replaced by indeterminate states, while the odd-numbered abdominal segments and T2 remain. This is the even-skipped phenotype.

Now consider mutations in *hairy* (Figure 11.7b). The phenotype of a typical weak mutant is characterized by loss of the anterior part of each even-numbered abdominal segment and the posterior part of each odd-numbered abdominal segment, with corresponding deletions to the thoracic segments. In the model, the effect of a *hairy* mutation is a displacement of the colour wheel towards the *eve* axis, as the amplitude of the *hairy* spatial periodicity decreases. This is shown as the ellipse that intersects the null disc in Figure 11.7(b). The posterior (naked) parts of the odd-numbered segments are deleted, as are the anterior (denticle band) regions of the even-numbered segments, conforming to observation. Stronger *hairy* alleles show the mirror-symmetric phenotype: each of the odd-numbered denticle bands is mirror-reflected, and there is no naked cuticle. This is shown by the ellipse in the lower-left of the colour wheel, which lies entirely to one side of the null disc. Following around this curve in counterclockwise direction gives an odd-numbered denticle band first in normal (antero-posterior) polarity and then in reverse polarity, the curve intersecting the same sectors twice in opposite directions and so resulting in a continuous lawn of setae in mirror-symmetric array. This cycle repeats over every 2-segment domain, resulting in the strong *hairy* phenotype. But why is the cycle displaced towards the lower left of the colour wheel rather than parallel to the *hairy* axis? The reason is that *hairy* represses *runt* which represses *eve* (Carroll & Vavra 1989). Therefore decreased levels of *hairy* product will result in elevated *runt*, hence in reduced *eve*. So the observed strong *hairy* phenotype is what is expected from the model. This shows that we have to take account of interactions in a higher-dimensional space than simply the two described in Figures 11.6 and 11.7, which are 2-dimensional projections of a much more complex space which must include all the segmentation and homeotic genes to provide a full description of the mapping from gene products to patterns of differentiated cell state. More of this later.

Let's now see how this procedure gives an explanation of the initially rather puzzling results of *runt* mutants in which under-production and over-production give complementary phenotypes. To analyze this, plot *runt* against *hairy* and locate the segmental pattern elements on the colour wheel as shown in Figure 11.8. As previously mentioned, *runt* and *hairy* interact by mutual repression, so that if one decreases the other increases, and vice-versa.

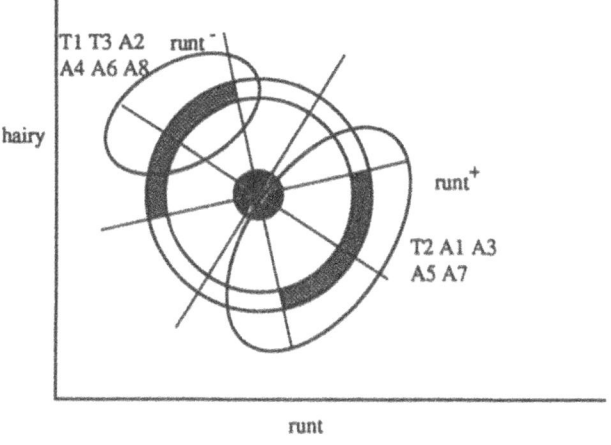

Mutations in runt.

Figure 11.8. Distortions of the colour wheel for too much and too little *runt* product, showing deletions and mirror-symmetry.

A *runt* [+] mutant therefore shifts the curve towards the right and down, as shown by the ellipse that cuts the null disc. This gives the even-skipped phenotype described for these mutations. A strong *runt* [-] mutant will have a spatial pattern described by an ellipse shifted up and to the left, lying entirely to one side of the null disc. Following this curve around in antero-posterior (counter-clockwise) direction and reading off the pattern elements from the colour wheel, we get a mirror-symmetric pattern of even-numbered abdominal and odd-numbered thoracic denticle bands, as observed (Gergen & Wieschaus 1985). Weaker *runt* $^-$ alleles give the odd-skipped phenotype complementary to *runt* $^+$, in which the cycle is displaced upwards towards the left and intersects the null disc to give the deletion phenotype. So we get a full correspondence between the observed phenotypes and the expectations from the model.

However, the model gives some predictions, as models should. A stronger *runt* $^+$ mutation should result in a closed curve lying entirely to the lower right of the colour wheel in Figure 11.8. This is a mirror-symmetric pattern complementary to strong *runt* $^-$. So far as we are aware, this has not been reported. It could be generated either by a mutation or by injecting *runt* message into an early embryo, using the techniques that have been employed in mutant rescue experiments by microinjection. Or genetic constructs that over-produce *runt* could be used. The model in fact predicts that any of the pair-rule genes should give the full range of phenotypes from mirror-symmetric patterns due to severe deficiency of the product through deletions to the complementary deletion and then the complementary mirror-symmetric phenotypes due to over-production of gene product. Testing these predictions are now well within the scope of basic molecular genetic techniques.

11.6 THE GAP MUTANT PHENOTYPES

The three major gap genes, *hunchback*, *Krüppel*, and *knirps*, have phase-shifted distributions of gene products that cover the segmentation domain as shown in Figure 11.9, which describes amplitude variation in a strictly qualitative manner. The three head segments, mandibular (Ma), maxillary (Mx), and labial (La) are included. There are other gap genes such as *giant, tailless, unpaired*, and *hopscotch*, whose influence is significant in determining segment character, but the same principles apply to the analysis of their effects as will be demonstrated now for the three shown in Figure 11.9.

Plotting *hunchback* (hb) against *Krüppel (Kr)* and locating the segments on the closed curve that now represents part of the anteroposterior axis of the embryo, we get the circle

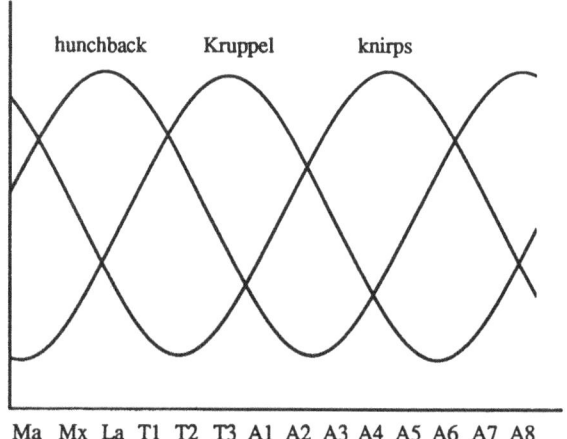

Figure 11.9. The spatial pattern of the three major gap genes, showing their phase relationships relative to segmental positions.

of Figure 11.10(a). Consider the effect of mutations in *hunchback*, resulting in the ellipse that passes through the null disc. Deletions occur in the head and thoracic segments as well as posterior abdomen (A_7 and A_8) where *hunchback* activity increases in a second cycle (Figure 11.9). Stronger mutations result in further displacement of the closed curve

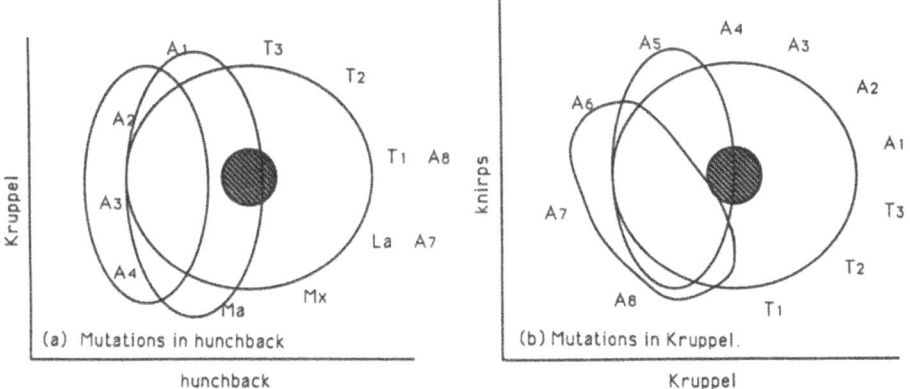

Figure 11.10. (a) The gap colour wheel and *hunchback* mutations giving deletions and mirror symmetry. (b) *Krüppel* mutations resulting in deletions and mirror symmetry.

towards the *Krüppel* axis, giving ellipses that lie entirely to one side of the null disc as shown. These describe mirror-symmetrical patterns involving anterior abdominal segments, as observed. The equivalent plot of *knirps* and *Krüppel* is shown in Figure 11.10(b), with deletions of thoracic and abdominal segments resulting from weak *Kr* mutants while strong mutants produce the mirror-symmetrical pattern depicted in the closed curve that intersects the A_6 domain twice in opposite directions. Once again, as for the pair-rule genes, we predict that complementary patterns of deletions and mirror-symmetry to those observed should be obtained by increasing *hunchback* and *Krüppel* activity either as a result of over-production mutants or by injection of message.

The gap and pair-rule genes do not function as independent systems. Not only are there interactions between different gap gene products, such as a repression of *hunchback* by *Krüppel*, but the gap genes also influence pair-rule gene expression (Carroll & Vavra 1989; Gaul & Jäckle 1989). Furthermore, all the mappings that we have been considering are 2-dimensional projections of higher-order mappings. If the three variables of Figure 11.9 were plotted in a 3-dimensional phase space, they would define a closed curve that surrounds a null disc from which radiate solid 'cones', 3-D analogues of the quantized sectors between radii of the colour wheel, these 'cones' containing the closed curve of the colour wheel itself. The two dimensional maps are obtained by projecting the 3-D colour wheel onto a 2-D subspace. Extending this picture to the seven gap genes raises the dimension to 7, but the topology remains the same.

When the pair-rule genes are included in the description of the mapping from segmentation gene products to pattern, the picture gets more complex, with different spatial periods wrapped together. The gap colour wheel in 2D can be described as a circle around which the pair-rule cycle is wrapped, as a spiral on a torus. Reading off cell states requires simultaneous reading of gap and pair-rule gene products, the periodic repeats of the latter being combined with gap gene values to provide distinct states over the gap cycle. However, since the gap genes have a spatial cycle that is less than the overall length of the embryo, as shown in Figures 11.9 and 11.10, unique specification of state requires variables with a cycle longer than that of the gaps. This is provided by the maternal genes, whose products are distributed in gradients (a fraction of a full spatial cycle), as described for the case of *bicoid* (Driever & Nüsslein-Volhard 1988). Since maternal gene products define less than a full cycle of values, they specify a colour fan rather than a wheel, as described in Goodwin & Kauffman (1992), but otherwise the analysis of deletions and mirror-symmetries resulting from mutations is topologically the same as that for the other segmentation genes. And at the other end of the frequency spectrum, the segment polarity genes, with spatial wavelengths of one segment, provide detailed pattern information within single segments. The whole set of genes thus provide a complex high-dimensional and multiply-periodic mapping from gene product to cell state space.

What has been identified in this analysis is a generic topological property of this mapping, occurring on four spatial wavelengths, that is a robust consequence of a particular way of reading gene products as ratios from an origin that is located in the middle of the amplitude range of the variables. Neither of these assumptions is an obvious one to make, but they provide explanations of the observations and make testable predictions. The topological discontinuity that arises from the existence of a null domain at the center of the colour wheels explains how it is that very different patterns, such as deletions and mirror symmetries, can coexist in neighbouring domains of a mutant embryo, as in Figure 11.2(e). These are neighbouring states across the topological discontinuity resulting from a null disc and a ratio map, so they are readily accessible as alternative expressions of a mutational disturbance within a single embryo. The multiple colour wheel model described here addresses only one of the levels of epigenetic analysis between the molecular and the morphological domains. What is still conspicuously missing is a model of the way in which the complex, multidimensional molecular level is mapped into the pattern of cell properties that define overt patterns, such as the array of setae in denticle bands. However the mapping discussed in this paper, which is essentially a vector field model of segment patterning, is consistent with an attractive possibility at the level of physical state, which is a liquid crystal description of cell state (Bouligand 1985; 1986). The phase transitions that such a system undergoes in response to a variety of disturbances are suggestive of those observed in *Drosophila*.

REFERENCES

Bouligand, Y. 1985. Brisures de symétrie et morphogénèse biologique. *La Vie de Sciences, Comptes rendus, série générale*, 2(2), 121–140.

Bouligand, Y. 1986. Theory of microtomy artefacts in anthropod cuticle. *Tissue and Cell*, 18, 621–643.

Carroll, S. B., & Vavra, S. H. 1989. The zygotic control of *Drosophila* pair-rule gene expression. II, spatial repression by gap and pair-rule gene products. *Development*, 107, 673–683.

Coulter, D. E. R., & Wieschaus, E. 1988. Gene activities and segmental patterning in *Drosophila*: analysis of odd-skipped and pair rule double mutants. *Genes and Dev.*, 2, 1812–1823.

Driever, W., & Nüsslein-Volhard, C. 1988. A gradient of *bicoid* protein in *Drosophila* embryos. *Cell*, 54, 83–93.

Gaul, U., & Jäckle, H. 1989. Analysis of maternal effect combinations elucidates regulation and function of the overlap of *hunchback* and *krüppel* gene expression in the *Drosophila* blastoderm embryo. *Development*, 107, 651–662.

Gergen, J. P., & Wieschaus, E. F. 1985. The localized requirements for a gene affecting segmentation in *Drosophila*: Analysis of larvae mosaic for *runt*. *Dev. Biol.*, 109, 321–335.

Gergen, J. P., Coulter, D., & Wieschaus, E. 1986. Segmental pattern & blastoderm cell identities. In *Gametogenesis and the early Embryo*, pp. 195–220. Alan R. Liss, Inc.

Goodwin, B. C., & Kauffman, S. A. 1990. Spatial harmonics and pattern specification in early *Drosophila* development. part I, bifurcation sequences and gene expression. *J. theoret. Biol.*, 144, 303–319.

Goodwin, B. C., & Kauffman, S. A. 1992. Deletions and mirror symmetries in *Drosophila* segmentation mutants reveal generic properties of epigenetic mappings. In *Principles of Organization of Organisms*, Mittenthal, J., & Baskin, A. (eds). Addison-Wesley.

Hunding, A., Kauffman, S. A., & Goodwin, B. C. 1990. *Drosophila* segmentation: supercomputer simulation of prepattern hierarchy. *J. theoret. Biol.*, **145**, 369–394.

Ingham, P. W., Pinchin, S. M., Howard, K. R., & Ish-Horowitz, D. 1985. Genetic analysis of the *hairy* locus in *Drosophila melanogaster*. *Genetics*, **111**, 463–486.

Jäckle, H., Tautz, D., Schuk, R., Seifert, E., & Lehmann, R. 1986. Cross-regulatory interactions among the gap genes of *Drosophila*. *Nature*, **324**, 668–670.

Kauffman, S. A., & Goodwin, B. C. 1990. Spatial harmonics and pattern specification in early *Drosophila* developments. part II, the four colour wheel model. *J. theoret. Biol.*, **144**, 321–345.

Lehmann, R., & Nüsslein-Volhard, C. 1987. *Hunchback*, a gene required for segmentation of an anterior and posterior region of the *Drosophila* embryo. *Dev. Biol.*, **119**, 402–417.

Lohs-Schardin, M. 1982. *Dicephalic* - A *Drosophila* mutant affecting polarity in follicle organization and embryonic patterning. *Wilhelm Roux's Archives*, **191**, 28–36.

Nüsslein-Volhard, C. 1977. Genetic analysis of pattern formation in the embryo of *Drosophila melanogaster*. characterisation of the maternal-effect mutant bicaudal. *Roux's Arch*, **183**, 244–268.

Nüsslein-Volhard, C., & Wieschaus, E. 1986. Mutations affecting segment member and polarity in *Drosophila*. *Nature*, **287**, 795–801.

Van der Meer, J. M. 1984. Parameters influencing reversal of segment sequences in posterior egg fragments of *Callosobruchus (Coleoptera)*. *Wilhelm Roux Arch. Dev. Biol.*, **193**, 339–356.

Winfree, A. T. 1980. *The geometry of biological time*. N.Y.: Springer-Verlag.

12. TRAJECTORIES OF SWIMMING MICRO-ORGANISMS AND CONTINUUM MODELS OF BIOCONVECTION

N. A. Hill

Department of Applied Mathematical Studies
University of Leeds
Leeds LS2 9JT, UK

12.1 INTRODUCTION

A random walk model has been developed to describe the motion of individual swimming cells, such as *Chlamydomonas* and *Peridinium*. These types of motile cells swim along helical paths and change their direction continuously and smoothly. They do not exhibit the run-and-tumble behaviour of bacteria like *E. coli*. The length scale of interest in this study is that on which the cells' mean direction of motion changes significantly. This length is much greater than the radius of the helical trajectories so that we regard the trajectories as smooth lines. Consequently, it is the continuous limit of the random walk as the time step $\tau \to 0$ which is used in the modelling. This model has been successfully tested and is being used to provide the data for calculating the macroscopic parameters needed in continuum models of spontaneous pattern formation in suspensions of swimming micro-organisms, which is known as bioconvection. Details of the continuum models are given in the review by Pedley & Kessler (1992).

12.2 THE RANDOM WALK MODEL

We begin by assuming that each cell swims with a constant speed, which is supported by the experimental data, and for simplicity here we suppose that the cell's trajectory lies in a two dimensional plane. The model can be extended to three dimensions, and full details are given by Hill & Häder (1993). The direction in which the cell is swimming at time t is given by the unit vector $\hat{\mathbf{p}}(t)$, which moves on the unit circle and makes an angle $\theta(t) \in (-180°, 180°)$ to the preferred direction of motion along the Oz axis. At each time step τ, $\theta(t)$ is either unchanged or changed by a small angle $\pm\Delta(\theta)$, which itself depends upon θ. We define the probability density function $f(\theta_0, \theta, t)$ such that

$$f(\theta_0, \theta, t)\,\delta\theta = \text{Prob}\left[\theta \le \theta(t) < \theta + \delta\theta \mid \theta(0) = \theta_0\right].$$

In other words, $f(\theta_0, \theta, t)$ is the probability that at time t the cell is swimming in the direction θ, given that it was swimming in the direction θ_0 initially. By expanding $f(\theta_0, \theta, t)$ in a Taylor series and considering the limit in which $\tau \to 0$, we find that $f(\theta_0, \theta, t)$ satisfies a

Experimental and Theorietcal Advances in Biological Pattern Formation,
Edited by H.G. Othmer *et al.*, Plenum Press, New York, 1993

Fokker-Planck partial differential equation,

$$\frac{\partial f}{\partial t} = -\frac{\partial}{\partial \theta}\left[\mu(\theta)f\right] + \frac{1}{2}\frac{\partial^2}{\partial \theta^2}\left[\sigma^2(\theta)f\right],$$

where

$$\mu(\theta) = \lim_{\tau \to 0}\frac{\mathrm{E}\left[\theta(t+\tau) - \theta(t) \mid \theta(t)\right]}{\tau}$$

and

$$\sigma^2(\theta) = \lim_{\tau \to 0}\frac{\mathrm{var}\left[\theta(t+\tau) - \theta(t) \mid \theta(t)\right]}{\tau}$$

are the mean and variance of the rate of change of direction of the continuous limit of the random walk, respectively. The macroscopic parameters that are needed in the continuum models are found by solving the Fokker-Planck equation. For example, the steady, or equilibrium, distribution of swimming directions is the long-time limit of the probability distribution function *i. e.*

$$\lim_{t \to \infty} f(\theta_0, \theta, t).$$

A particular case in which the cell orientation mechanism is understood is that of upswimming (negative geotaxis). Cells such as those of *Chlamydomonas nivalis* are bottom heavy because of the position of the organelles within them and thus they tend to right themselves in quiescent water, so that on average they are pointing upwards. There is always some inherent randomness in their motion, however, so that at any one instant some cells will be pointing in other directions. The gravitational torque on a cell is readily shown to be proportional to $\sin\theta$ and consequently we would predict that $\mu(\theta) \propto \sin\theta$ in this case.

12.3 DATA ANALYSIS

The data on cell trajectories consists of the positions of the centroids (x_i, y_i) of swimming cells measured at successive timesteps $t_i = i\tau, (i = 1, 2, \ldots)$. From this, we calculate the changes in direction $\Delta_i(\theta)$ between successive data points and plot these in a histogram. By fitting a suitable probability distribution, such as the von Mises distribution, which can be thought of as the circular analogue of the Normal distribution, we derive the mean, $\overline{\Delta}(\theta)$, and the variance, $\overline{\sigma^2}(\theta)$, for the changes in direction. Finally, provided that the timesteps in the data are sufficiently small, the coefficients in the Fokker-Planck equation are given by

$$\mu(\theta) = \overline{\Delta}(\theta)/\tau \quad \text{and} \quad \sigma^2(\theta) = \overline{\sigma^2}(\theta)/\tau.$$

12.4 RESULTS AND CONCLUSIONS

For negatively geotactic cells, it is indeed found that $\mu(\theta) \propto \sin\theta$, which gives us confidence in the model and the analysis and, for the first time, we have a reliable measure of the coefficient of proportionality. We also find that $\sigma^2(\theta)$ is constant, which was not previously known at all.

In experiments on phototaxis, in which the cells swim towards a light source, we have been able to show that $\mu(\theta) \propto \theta$ and that $\sigma^2(\theta)$ has a minimum at $\theta = 0°$ and a maximum at $\theta = 180°$. Thus positive phototaxis is not directly analogous to negative geotaxis, as has been suggested (Pedley & Kessler 1992), and cannot be described as an 'effective torque'.

In conclusion, it has been demonstrated that for micro-organisms, which do not make *sudden* changes in direction, it is possible to model their trajectories as the continuous limit of a random walk model that is based on a Wiener process. We have found that this approach leads to a reasonably robust and reliable method of data analysis that can be used to calculate the coefficients in the Fokker-Planck equation for the probability distribution function for the direction of motion, from which macroscopic parameters such as the mean cell velocity and diffusivity can be derived. Because this approach provides modellers with detailed information about the movement of the cells, it is possible to make detailed predictions about and comparisons with the fluid mechanics of the full suspension. Finally, we emphasize that this method is not restricted to swimming cells and may prove to be a very useful way of describing other biological systems in which the cells are motile.

REFERENCES

Hill, N. A., & Häder, D.-P. 1993. A random walk model for swimming micro-organisms. (In preparation).

Pedley, T. J., & Kessler, J. O. 1992. Hydrodynamic phenomena in suspensions of swimming micro-organisms. *Ann. Rev. Fluid Mech.*, **24**, 313–358.

13. TESTING THE THEORETICAL MODELS FOR LIMB PATTERNING

J. R. Hinchliffe[1] and T. J. Horder[2]

[1]Dept of Biological Sciences
University of Wales
Aberystwyth, Dyfed
SY23 3DA, Wales, UK

[2]Dept of Human Anatomy
South Parks Road
Oxford OX1 3QX
England, UK

The present paper aims to examine the current models of limb development in the light of the experimental production of "extradigits" and other experimental data, which provide the means to subject the models to critical testing.

13.1 PARADIGM MODELS FOR LIMB DEVELOPMENT

Essentially, recent research on limb development begins with the Saunders/Zwilling hypothesis of ectoderm-mesenchyme interaction (Saunders, Jr 1977, reviewed by Hinchliffe & Johnson 1980). According to this, in the chick limb bud, the apical ridge (AER) induces the outgrowth of the underlying mesenchyme while being itself dependent on the mesenchyme, through the AEMF, or apical ectodermal maintenance factor. This hypothesis was soundly based on experimental grafting work, for example with non-limb ectoderm or mesenchyme being substituted for limb tissue within the bud. While accounting for outgrowth, the model did not directly explain skeletal patterning other than by attributing limb type control (wing or leg) to the mesoderm source.

As part of these experiments, Saunders & Gasseling (1968) discovered that turning the limb bud tip through 180° relative to its stump provoked a duplication of the wing skeleton. Refinement of this experiment led Saunders & Gasseling (1968) to the discovery that the posterior marginal cells of the limb bud had special properties. When these cells, which became known as the zone of polarising activity (ZPA) were transplanted preaxially to the limb bud, the anterior host tissue responded by forming supernumerary digits, mirror imaging the normal skeleton (Figure 13.1). Always, the most posterior of the digits was the one closest to the graft, to which responsibility for the control of the antero-posterior axis of differentiation was attributed. While a large number of workers confirmed the polarising action of the ZPA in preaxial grafts, there was considerable controversy over whether the ZPA had such a role in normal development. On the basis of the lack of specificity in the ZPA (several other tissues had the same effect following preaxial grafts), and because normal

Experimental and Theorietcal Advances in Biological Pattern Formation,
Edited by H.G. Othmer *et al.*, Plenum Press, New York, 1993

105

limbs could be obtained following ZPA extirpation from the wing bud, Saunders, Jr (1977) came to doubt that the ZPA had a polarising effect during normal development.

However, most workers continued to support the idea of a polarising role for the ZPA, partly because in the ZPA extirpation experiments it was difficult to prove that all the ZPA had been removed (see Summerbell 1979; Hinchliffe & Gumpel-Pinot 1981). The hypothesis was transformed into a more precise form, that the ZPA specified the position of the digits and their type. Lewis Wolpert, Dennis Summerbell and Cheryll Tickle were particularly prominent in designing experiments to test this hypothesis, and Tickle (1981) was able to demonstrate a quantitative relation between the number of ZPA cells transplanted preaxially and the number and type of supernumerary digits formed. Wing digit 4, the most posterior, required about 100 cells, while digit 2, the most anterior required only about 35 cells. These quantitative studies led to the hypothesis that the ZPA was a source of "positional information" for the remainder of the limb field, in which cells were sensitive to the level of a ZPA-based signal and on this basis differentiated into the structure appropriate for the signal level experienced (Figure 13.1). The ZPA is considered to cease to have a role in control of pattern at stage 22 or 24 (Hinchliffe *et al.* 1984) or even as early as stage 17 (Fallon & Crosby 1975). According to the hypothesis the signal was molecular in nature and took the form of a morphogen profile across the limb field, with the highest level in the ZPA source, but declining towards the anterior "sink". At this time, the "morphogen" remained purely hypothetical.

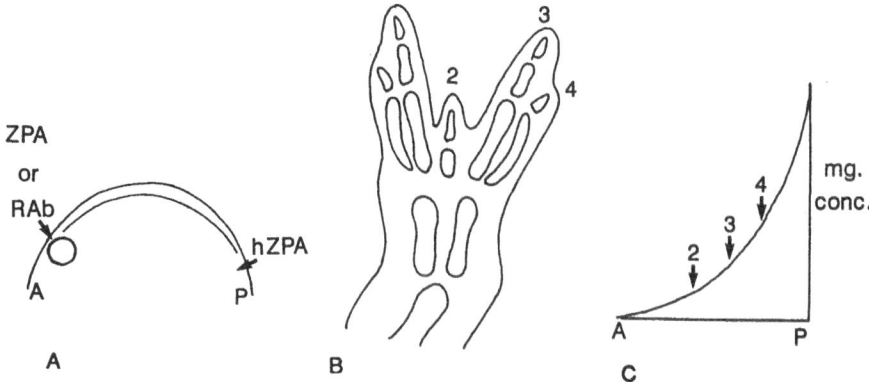

Figure 13.1. Implantation of a grafted ZPA or retinoic acid in a carrier bead (RAb) preaxially into a chick limb bud (A) provokes duplication of the limb skeleton, as illustrated by the wing (B). According to the morphogen profile hypothesis (C) the graded distribution of a morphogen, possibly retinoic acid, acts as a positional signal, governing the differentiation of the distal mesenchyme into specific digits (2,3,4). (A, anterior; h, host; mg. conc., morphogen concentration; P, posterior).

Parallel with this work, control of patterning along the proximo-distal (P-D) axis was analysed. On the basis of experiments showing a lack of regulation of heterochronic grafts of limb bud tips onto stumps by Summerbell (1977) and Wolpert *et al.* (1975), the progress zone hypothesis was proposed. According to this, differentiation along the P-D axis is determined by the number of divisions undergone by the mesenchymal cells underlying the AER (the "progress zone" cells) at the time they leave the progress zone (Summerbell *et al.* 1973). Cells which leave early, with few cell divisions, form proximal structures, while those leaving late, with more divisions, form more distal structures. Thus cells are specified by different mechanisms along the P-D axis (cell division number), and along the A-P axis (morphogen signalling distance from ZPA). It should be noted, however, that comparable experiments carried out by Kieny, (1964a; 1964b; Sengel 1975) produced quite different results, involving regulative differentiation of cells which were long out of the progress zones, and the conflict in experimental evidence has never been satisfactorily resolved (see Hinchliffe & Johnson 1980).

However, it has been the ZPA control of patterning across the A-P axis which has caught the imagination of the developmental biology world (see Alberts *et al.* 1989; Gilbert 1991; Browder *et al.* 1991; Shostak 1991), and eventually retinoic acid (RA) was identified as a candidate molecule ("We have a morphogen": Slack 1987). The duplicating effects of the ZPA were mimicked by implanting preaxially retinoic acid carriers (first newspaper, later 200μ latex plastic beads) (Tickle *et al.* 1982; Summerbell 1983). Tickle *et al.* (1985) was able to produce quantitative effects with RA, ranging from duplication of a full set of digits, to that of only digit 2, similar to the effects produced by varying the number of ZPA cells implanted. Endogenous RA was identified in the limb bud, more concentrated posteriorly than anteriorly (Thaller & Eichele 1987). The RA hypothesis took the form that the profile might be set up by RA cytoplasmic receptors (more concentrated anteriorly), and that RA in association with nuclear receptors might regulate "pattern control" genes.

Candidate "pattern control" genes have meanwhile been discovered in the form of homeotic or homeobox-containing regulatory genes whose activity is related to pattern formation in *Drosophila* (reviewed Duboule 1992). One complex, Hox 4, is expressed in nested form roughly along the anterior- posterior axis. Along the posterior margin of the Stage 21 limb-bud, all the Hox 4 genes (4.4, 4.5, 4.6, 4.7, 4.8) are expressed, while anteriorly only 4.4 is expressed (Figure 13.2, Dolle *et al.* 1989; Morgan *et al.* 1992).

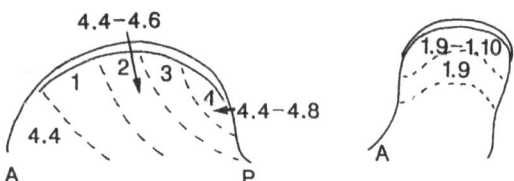

Figure 13.2. In the early limb bud, members of the Hox 4 gene complex are expressed in a nested set of overlapping domains ranging from 4.4 anteriorly to 4.4-4.8 posteriorly, while members of the Hox 1 complex are localised distally in overlapping domains of later stages. According to Morgan *et al.* (1992) and Tabin (1992), the Hox 4 domains specify the positional identity governing differentiation of digits 1-4. (A, anterior; P, posterior).

There is thus a series of 5 domains of gene expression from anterior to posterior in the limb bud. Another homeogene, Hox 1, is expressed in the "progress zone" but not more proximally. There is evidence that the Hox 4 domains can be altered significantly by experimental embryological techniques, such as preaxial implants of ZPA or RA, which provoke limb duplication (Izpisua-Belmonte *et al.* 1991). A ZPA graft will convert adjacent cells from an anterior expression pattern (Hox 4.4) to a posterior one (Hox 4.4-4.8), prior to their forming posterior duplicated digits. In further experiments, the normal pattern of homeogene expression is altered by inserting Hox genes into chick limb bud cells. Hox 4.6 - not normally expressed anteriorly - can be experimentally expressed anteriorly in a domain which, it is claimed, later forms a more posterior digit than usual (digit 2, with three phalangeal elements, rather than digit 1, which has two) (Morgan *et al.* 1992). The central idea here is that Hox gene expression can encode the positional address of the cells of the limb bud.

13.2 "EXTRADIGITS" AND THEIR IMPLICATIONS

The "extradigit" experiments initially described by Hurle and his co-workers (Hurle & Ganan, 1986; 1987; Hurle *et al.* 1989; 1991) present what is possibly a uniquely critical test of positional information models of limb development.

Hurle and his co-workers found that removal of the AER or dorsal ectoderm in the interdigital spaces between digits 3 and 4 in the 8 day chick leg bud resulted in the formation of extra digits. This is a surprising result for a number of reasons. AER removal is normally considered to prevent the development of more distal skeletal elements (Saunders, Jr 1948), while such extradigits are forming long after the ZPA is considered to specify the positioning of digits. The extradigit phenomenon was puzzling enough to warrant further investigation. Our results (Figure 13.3), reported in greater detail elsewhere (Hinchliffe & Horder 1993) are summarised briefly here.

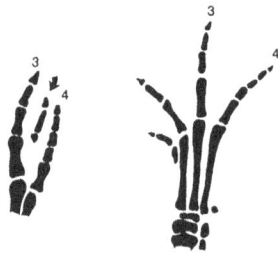

Figure 13.3. An "extradigit" with 3 elements between normal digits 3 and 4 (other digits not shown), produced by surgical intervention in the leg interdigit at stage 28 (Hinchliffe & Horder 1993). Right; normal hind limb (at half-magnification).

We discovered that the most effective way of provoking extradigit formation was to make a T-cut in the leg interdigital tissue at Stage 28. While we generated extradigits in other interdigital spaces (*e. g.* 2-3), the 3-4 interdigital area is the most favourable, probably because it is the largest interdigit, broad enough to prevent the extradigits fusing with adjacent normal digits. Ectodermal healing is very rapid, while the cutting of the mesenchyme appears to result in local mesenchymal contraction (Saunders, Jr 1948), and temporary avascularity. We are inclined to attribute the post-operational pre-chondrogenic condensation directly to such contraction, rather than to the ectodermal lesion, as Hurle *et al.* (1991) believe. Aside from the likelihood that the AER can regenerate, ectoderm removal - whether enzymatic or microsurgical - is likely always to damage the underlying mesoderm.

Rather than attempting to identify the precise triggering mechanisms initiating the pre-chondrogenic condensation, we consider it more relevant to focus on the nature of the events set in train by the intervention. Some condensations may disappear, as remarked by Hurle *et al.* (1989), or if close to an existing digit, may fuse with it. But given a critical size and central position for the condensation in the interdigit, the following events can be directly visualised (Figure 13.4) in the thin semi-transparent distal mesenchyme. First, the condensation becomes a cartilage nodule followed by distal addition of mesenchymal tissue which also chondrifies. The initial cartilage becomes proximo-distally elongated possibly in part through stretching by the adjacent digits. In many cases further elongated cartilage elements are added distally, so that the extradigit may comprise up to 3 elements complete with joints and terminating in a claw (Figure 13.3). For such well-patterned extradigits, it appears essential for the initial condensation to maintain distally an intact AER, which normally regresses in the interdigit, but is maintained over the normal digits to Stage 30/31. Presumably the maintained AER generates mesenchyme utilised in extradigit construction.

The morphology of such extradigits may be classified and identified as any digit type, *i.e.* corresponding to the distal part of any normal leg digit. Extradigits are found which are similar to digit 5 (a vestigial structure) or digit 1, while the longer 3-element extradigits correspond with the distal parts of digit 3 or 4 (see Figure 13.3). The variable morphology of the "extradigits" probably relates to a critical condensation size (Hurle *et al.* 1989), and the length of time the overlying AER continues its activity. The ease of transition between morphological types of digit, whether normal or experimental, suggests that relatively simple quantitative differences in the starting conditions will generate qualitative differences in the morphological out-turn. In the leg, specific digits (especially 1-4) consist of repeated modular metatarsals and phalanges and differ mainly in the number of the latter which are otherwise relatively similar in size and shape. The capacity of the mesenchyme to form a proximal to distal sequence of condensations, together with the extension (or contraction) of the period of AER action in generating mesenchyme distally goes far to explain morphological differences between longer or shorter digits.

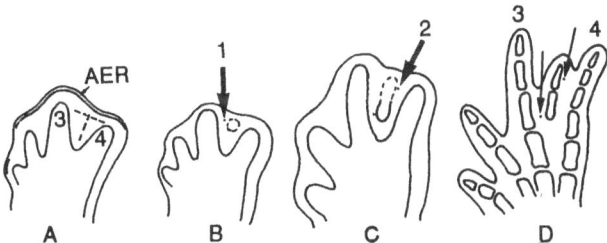

Figure 13.4. Diagram illustrating development of an extradigit. (A) T-cut in leg interdigit, (B) 14 hrs later, condensation visible and carbon 1 mark inserted, (C) 34 hrs post-op, elongated chondrogenic element, carbon 2 mark inserted, (D) 70 hrs post-op, fixed and stained. The carbon marks show the proximo to distal sequence of extradigit development.

Such extradigits, with the exception of the initiating step, appear to share the same developmental processes and final morphology of normal digits. AER survival over the developing digit tip is one common process, while their structural detail - joints, claws, associated tendons (Hurle *et al.* 1990) - appears entirely normal. The process of digit formation thus involves a sequential cascade of developmental events.

Extradigits are difficult to explain according to the "progress zone" and "ZPA control" theories. The way in which the initial experimental condensation organises tissue more distally (provoking chondrogenesis in tissue whose normal fate is death, and AER survival instead of regression) goes against the central idea of the progress zone model that cell fate along the P-D axis is determined once the cells leave it, with the information flow from distal to proximal, and with differentiation unaffected by the developmental state of more proximal tissue. (However, it should be noted that the P-D level of differentiation - whether metatarsals or phalanges - of the extradigit is appropriate for the developmental age of the undifferentiated distal mesenchyme).

Extradigits represent crucial evidence against which the theory of ZPA control of digit position and identity can be tested. The residual ZPA area of stage 28 developing legs is not disturbed in our experiments. The fact that extradigits form at all in the interdigital space, that they closely resemble normal digits, and vary in type, implies that information about position in the developing limb cannot be used to explain control of choices between different morphogenetic pathways. There is no reason that positional values should have been changed by our intervention and there is no reason, on the positional information model, that interdigital tissue should change its development. In turn, this must raise doubts about the ZPA theory of patterning control of normal digits.

13.3 PROBLEMS WITH SINGLE FACTOR POSITIONAL EXPLANATIONS OF PATTERNING

What the models of limb development reviewed earlier have in common is that they explain pattern as being controlled, or caused by the action of a single factor which acts over a relatively long distance as a source of "positional information". (We use the term "single factor" to refer to models which essentially attribute final patterning to the action of single - or a few - initial causes in early limb buds).

There is considerable evidence, in addition to that provided by extradigits, which makes it difficult to assign a clear role to the ZPA in the positioning and identity allocation of digits. We now discuss the various problems with such "single-factor" positional explanations as the ZPA control theory and its subsequent transformations into RA morphogen gradient or homeobox domain theories. A number of such problems have been discussed previously (Horder, 1976; 1978; 1981; 1983; 1989; Bowen *et al.* 1989).

It is convenient to describe adult limbs in terms of the 3 Cartesian axes (antero-posterior, proximo-distal, dorso-ventral), and recent work on limb development has implicitly assumed that separate developmental mechanisms operate along these 3 axes. This assumption stems partly from the early work of Harrison, who sought in amphibians to demonstrate the independent determination of the axes at different times in development. Later, Hamburger (1938) designed very similar experiments for the chick limb bud.

Harrison (1921) claimed that the antero-posterior (AP) axis was invariably already in-built (no pre-AP stage was found) and that the dorso-ventral (DV) axis was independently established later, but it is to be noted that DV axis fixation could only be defined indirectly by reference to two alternative modes of development (namely limbs of left or right handedness, all other parameters having already been defined by the supposedly prior event of AP determination). There are reasons to question Harrison's conclusion that the proximo-distal axis is the last to be laid down; the techniques used (Swett 1927) (it was necessary to remove limb ectoderm) mean that it is difficult to exclude regeneration (with secondary atrophy of the original proximo-distally rotated stump) as an explanation for normal limbs obtained at earlier stages. It seems more likely that the proximo-distal axis is the first formed in that it is a precondition for the outgrowth of the limb-bud in the first place and that, at the earliest stages, the centre of outgrowth may coincide with the ZPA itself (Stocum & Fallon 1982; Stocum & Fallon 1984). Even later it is clear in the chick that the orientation of the AER is a major prior contributor to digit outgrowth and alignment (Horder 1989).

The developmental reality of the three axes is challenged by the evidence of Bowen *et al.* (1989) on the fate map of the chick wing bud, based on chimeric grafting or on carbon-particle marking. Mesenchymal tissue behaviour does not conform to an axial description. The fate map (summarised in Figure 13.5) shows that the cells undergo continuous changes in position, cutting across the Cartesian axes. Thus posterior tissue becomes stretched along the P-D axis relative to anterior tissue, while posterior distal tissue expands much more than corresponding anterior distal tissue, thus moving its position in an anterior direction.

Doubts about axial concepts must also lead to doubts as to whether there are separate causal mechanisms operating along the supposed different axes. Moreover, since the ZPA control hypothesis requires distal mesenchymal cells to have their position specified with respect to the posterior border, a new complication is added if these cells are successively further away from the signal source. A similar complication arises as a result of tissue growth.

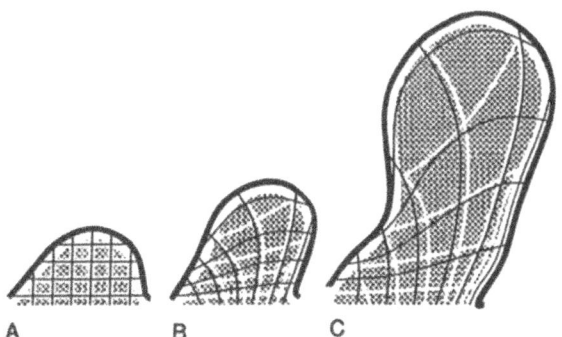

Figure 13.5. Summary fate maps for the forelimb based on carbon marking and quail transplants. Core mesoderm (white grid) and ectoderm (black grid) at stages (A) 21, (B) 24 and (C) 27, from Bowen *et al.* (1989). Note the relatively greater postaxial mesoderm growth.

Removal of the posterior half (including ZPA) from a wing bud results within 24 hours in large-scale distal mesenchymal cell death and poor skeleton development (Hinchliffe & Gumpel-Pinot 1981). Implanting a ZPA distally rescues such an anterior half, inhibiting cell death, permitting vigorous outgrowth and good skeletal development, frequently of all 3 digits (Wilson & Hinchliffe 1987). Frost was able to show that when a ZPA was implanted the cell division rate of the anterior half (as monitored by tritiated thymidine labelling over 22 hours following operation) was nearly 4 times that of the control (Frost & Hinchliffe 1991; 1993). There was a similar difference between high cell division rate following a ZPA implant and a low rate after a retinoic acid bead implant, which in most cases failed to inhibit cell death in the anterior half. This result suggests that RA beads, whether they play a patterning role or not, fail to provide the growth stimulation of a ZPA graft (McLachlan 1991). It further suggests that posterior mesenchyme plays a growth control role, separate from any possible patterning role by the ZPA.

Besides the "extradigit" evidence, other experiments call into question the digit positioning role attributed to the ZPA. Pautou (1973) and MacCabe *et al.* (1973) disaggregated chick limb mesoderm, replacing it in an ectodermal jacket, and obtained branching skeletal structures, terminating in several modular jointed digits which lacked a specific identity. In the disaggregation process the ZPA had been randomised, so that Pautou's modular digits formed in the absence of localised ZPA.

Partly in response to the difficulty in explaining the Pautou result on the positional information model, an isomorphic prepattern has been proposed in addition to positional information (Wolpert, 1989; 1991). As supporting experimental evidence, it was shown that it is difficult to induce a double humerus by widening an early limb bud with a ZPA graft (Wolpert & Hornbruch 1987). This does not appear to us to distinguish between the two mechanisms unless one already knows the distribution of the morphogen in the experimental conditions and the range of values used by cells in interpreting their position as appropriate to the formation of a humerus. This result may follow from the greater difficulty in rapidly and significantly altering the shape of the proximal limb bud (with its rounded cross-section, compared with the digital plate, with its elliptical cross-section: see Hinchliffe 1991). To add prepattern as a mechanism not only complicates a model whose great virtue was simplicity, but the origin of the prepattern itself would still remain to be explained. In their 1990 paper, Wolpert and Hornbruch find that double anterior half limb-buds give rise to double humeri when constructed at a time "before cartilage condensation". They therefore argue that specification of cells had already occurred and that this must have been by a positional information mechanism. The weakness of this argument is that it is difficult to exclude the possibility that chondrogenesis might already have been underway; the earliest stages might be unobservable.

As discussed earlier, the ZPA patterning control theory initiated "positional information" and retinoic acid "morphogen profile" interpretations of limb development. However, more recent work suggests that RA beads may well initiate ZPA-like properties in the adjacent mesenchyme. In that case, RA-induced limb duplication would be due not to the gradient of RA through the limb, but to the effect on the remainder of the wing bud of anterior mesenchyme adjacent to the bead newly converted into cells having ZPA-like properties (Bryant & Gardiner 1992). Frost's work previously mentioned also strongly suggests major differences in the mode of action of transplanted RA beads and ZPA grafts, further questioning the RA "morphogen profile" evidence.

Finally, and most significantly, given that all the models we have so far considered depend ultimately on the ZPA effect, we return to an aspect of the extradigit phenomenon which has a direct bearing on the nature of the ZPA effect itself. We interpret extradigits as falling into a class of phenomena well known to embryologists in general: the non-specific ectopic release of a potential for a self-organising developmental programme. The simplicity of the initial triggering conditions is indicated by the nature of the stimulus and by the nature of the required initial change of state of the responding embryonic tissue. We would put extradigits in the same class as the triggering of extra limbs in the flank (Balinsky 1957) or tail (Mohanty-Hejmadi *et al.* 1992) or the classical mimicking of Spemann's organizer in the initiation of an entire secondary embryonic axis by diverse extrinsic factors. In all such cases the nature of the extrinsic stimulus (and indeed its normal intrinsic equivalent) is in itself not of major significance given its non-specificity and largely releasing action.

In particular we suggest that the ZPA effect itself should be regarded in the same light as a non-specifically initiated simple change of state in anterior limb-bud tissue leading to the release of an ectopic, progressively unfolding developmental cascade (Hinchliffe & Horder 1993). The ZPA effect may differ in its scale of effect from extradigits simply because an earlier stage in the developmental cascade is involved. Certainly all the evidence suggests that the ZPA effect is non-specific and organisationally simple. Saunders, Jr (1977) and Saunders, Jr. & Gasseling (1983) have emphasised that the same limb duplication can be triggered by many non-ZPA releasers. The list includes chorio-allantoic membrane (McLachlan & Phoplonker 1988), somites, flank, mesonephric tissue and tail bud (Saunders, Jr 1977; Saunders, Jr. & Gasseling 1983), colchicine (Gabriel 1946) and retinoic acid (Tickle *et al.* 1982; Summerbell 1983; Hinchliffe *et al.* 1991). Hensen's node also evokes duplication (Hornbruch & Wolpert 1986), possibly due to retinoic acid release (see Bryant & Gardiner 1992). Overall, when taken together with the previous results showing that digits form without a ZPA, these arguments suggest that any involvement of the ZPA in pattern formation is remote and indirect.

13.4 ALTERNATIVE, MULTIFACTORIAL APPROACHES: THE CONCEPT OF A DEVELOPMENTAL CASCADE OF LOCAL TISSUE INTERACTIONS

We come, in this concluding section, to one last set of considerations which we see as arising from the extradigit phenomenon. Extradigits demonstrate the potential of embryonic tissues to generate complexity from simple, purely local starting conditions. Our observations show some of the ways in which the final emergence of a differentiated digit is the result of a cascade of events. Taken together with what we have said above about the ZPA, digit formation can be seen as a late stage of a cascade of developmental processes traceable back to the point at which ZPA grafts have their far-reaching effects. Although extradigits can be described as "self-organising", we would argue that organisation is achieved as the result of multiple, local interactions and that final, integrated organisation is the result of the sequential build up of these interactions in time from an initial organisationally simple starting state.

The current domination of the analysis of limb development by single-factor, one-step mechanisms of patterning has had the effect of diverting attention from other possible approaches and has led to the relative neglect of experiments demonstrating - or designed to demonstrate - sequential, local interactions such as inductions or morphogenetic relative cell movements among tissues. On any theory of limb development attention must be paid to the complexity of the processes of differentiation and patterning occurring between the time of supposed determination of undifferentiated mesenchyme and its arrival at definitive form. Among the many known examples of such processes are the various phases of skeletal differentiation, from condensation, through chondrogenesis to osteogenesis, interactions between different tissues such as muscle and connective tissue (Kieny *et al.* 1986), or digit and tendon (Hurle *et al.* 1990) (see also Alberch 1990) and morphogenetic fusion patterns occurring between cartilage elements (Hinchliffe 1985; Muller 1991). Local epithelial/mesenchymal inductions have been demonstrated in a number of cases. These include AER induction of the blood vessels of the subjacent marginal sinus (Feinberg & Saunders, Jr. 1982), ectoderm inhibition of chondrogenesis in adjacent mesenchyme (Globus & Vethamany-Globus 1976; Solursh 1984), induction of bone by adjacent cartilage (Fell 1956), and the interaction of mesenchyme and ectoderm in forming scales (legs) and feathers (*e. g.* the wing primaries, Sengel 1976). A probable ectodermal effect in tendon development has been shown by Hurle *et al.* (1990). A few simple cellular rules may account for the basic limb pattern (Waddington 1962; Horder 1978, 1983) as in Ede's attempts (reviews 1971; 1982; 1983) to relate patterning to a condensation process and the mechanical effects of condensation centres in creating zones of inhibition, thus relating element number to area of mesenchyme. Mechanical forces have also been invoked (Shubin 1991; Shubin & Alberch 1986), as in a condensation branching model (Oster *et al.* 1983; 1988) as a direct explanation of skeletal patterning. Similarly, Stopak & Harris (1982) have demonstrated the role of stress patterns in the limb in localising injected collagen in developing tendons.

The involvement of multiple intervening causes is disregarded in the recently proposed hypothesis that homeobox gene expression domains encode the positional address of limb bud cells, thus directly governing the final morphology. The five Hox 4 domains in chick and mouse limb buds have been related to the 5 digits of the pentadactyl limb (Morgan *et al.* 1992; Tabin 1992). Implicit in this interpretation is a simple relation between genes and morphology. In view of the past history of attempts to verify and identify positional information signals one is perhaps justified in inserting a note of caution about the homeobox results. An alternative (or parallel) explanation is possible regarding the transfection experiments, where a broadened "digit 2" domain in the chick limb bud is followed by the formation in some cases of two digits 2 (Morgan *et al.* 1992). Since leg digits in both mouse and chick differ mainly in their number of phalangeal elements, it is possible that prolonged AER activity alone would generate sufficient additional distal mesenchyme for a digital ray long enough to be segmented into the additional element. The frequent syndactyly reported may itself secondarily affect the digits involved. The extradigit phenomenon reported here suggests that differences in digital morphology may well be a consequence of relatively minor variation in the dynamics of events. Our results have indicated how easily digit type transformations can be achieved and we have argued that anterior limb bud is particularly open to "posteriorisation" (Hinchliffe & Horder 1993). The ZPA effect suggests that anterior limb-bud tissue can occupy one of only very few (perhaps two) states: the ability of the ZPA to polarise disaggregated mesoderm and to provoke mirror-image duplications may be the result of no more than a simple, nonspecific release of the preconditions for a posterior mode of development. Some indication that the transfection may have involved transformations akin to the variations we see in our extradigits is suggested by the fact that Morgan *et al.* (1992) sometimes obtained extra interdigital cartilages similar to those found in our studies.

In the light of the above considerations it can be seen that the evidence for the ZPA, gradients and positional information is indirect - and much influenced by the remarkable regulative and self-organising, integrative powers of embryonic systems - and is open to alternative explanations. We suggest that these remarkable powers, which so easily seem to require some single, overall supervisory mechanism, can equally well be accounted for on a cascade mechanism based on the spatio-temporal summation of multifactorial interactions. Not only does a cascade concept provide an alternative explanation for limb pattern: the wide range of known local cell interactions as pattern determinants is in itself against the whole notion of positional information.

We have argued against "single factor" models of pattern formation and emphasised problems inherent in the concept of positional information, even though these are approaches which owe much of their current interest specifically to studies of the limb. Whatever the merits of the alternative cascade approach, it may well be asked why the problem of limb pattern formation remains so unresolved despite such intensive study. We would suggest that the explanation lies in a feature of the limb as a model embryonic system, which leads to an emphasis on unified long-distance models and makes the study of local and inductive-type interactions particularly difficult. This we suggest is due to the fact that the limb bud is essentially a single, continuous mesodermal structure. Classical demonstrations of local tissue/tissue interactive phenomena such as induction have concerned epithelia and epithelio-mesodermal interactions (Horder, 1976; 1983; 1989). An example is provided by analysis of the eye as a sequence of tissue inductions between ectoderm, neurectoderm and mesenchyme in the differentiation of lens and cornea (Horder 1983). Mesodermal development itself is characterised by small-scale, self-organising effects which are difficult to study and isolate when compared to classical systems like the eye or neural induction.

Thus we suggest that, despite its obvious and well-known practical advantages, as a model for the analysis of pattern generation, the limb is in some respects a system that has been misleading. What we have said with regard to the relative merits of local versus long-distance positional models can in fact be better substantiated by reference to other parts of the embryo where induction, in association with morphogenesis, is clearly the dominant developmental patterning mechanism. Evaluating the choice between two fundamentally different approaches, we consider that there is considerable evidence to support an explicitly multi-factorial, cascade approach which, in principle, can account fully for the complexity, integration, predictability and accuracy of pattern.

Footnote This review does not address the polar coordinate theory, an alternative positional information interpretation of limb development. Analysis of limb development has mainly involved experiments on chick embryos and explanatory models (ZPA, morphogen profile, RA, homeogene) as described above. The polar coordinate theory emerged from urodele and arthropod limb regeneration experiments (Bryant *et al.* 1981) and there is no reason to assume that regeneration exactly repeats development: indeed it could well be expected to be considerably more complicated. It has been applied to the chick limb by Bryant and her co-workers (Bryant & Gardiner 1992; Javois 1984; Iten 1982). This theory assigns positional value to all parts of the limb bud (or regenerate), with intercalation of missing values following heterotopic grafts (such as a preaxial ZPA graft). While it "explains" the preaxial graft result at least as well as the "ZPA control" hypothesis, it is less successful with the (largely negative) results from anterior tissue grafted postaxially (see Honig 1983; Dvorak & Fallon 1987). Thus its relevance to chick patterning has not been widely accepted: it is, for example, unclear how polar coordinate values are set up in the first place and how the concept of intercalation would be relevant in normal development. Moreover, from the point of view of the present paper, the polar coordinate model presents the same difficulties as ZPA control models in that it does not deal with the developmental processes which intervene between the hypothesized initial allocation of positional values, and the emergence of definitive morphology.

REFERENCES

Alberch, P. 1990. The logic of monsters: evidence for internal constraint in development and evolution. In *Ontogenese et Evolution*, pp. 21–57. Colloque International du CNRS Geobios. Suppl. 12.

Alberts, B., Bray, D., Lewis, J., Raff, M., Roberts, K., & Watson, J. D. 1989. *The molecular biology of the cell*. 2 edn. New York: Garland.

Balinsky, B. I. 1957. New experiments on the mode of action of the limb inductor. *J. Exp. Zool.*, **134**, 239–273.

Bowen, J., Hinchliffe, J. R., Horder, T. J., & Reeve, A. M. F. 1989. The fate map of the chick forelimb-bud and its bearing on hypothesized developmental control mechanisms. *Anat. Embryol.*, **179**, 269–283.

Browder, L. W., Erikson, C. A., & Jeffery, W. R. 1991. *Developmental biology*. Third edn. Saunders.

Bryant, S. V., & Gardiner, D. M. 1992. Retinoic acid, local cell-cell interactions, and pattern formation in vertebrate limbs. *Devel. Biol.*, **152**, 1–25.

Bryant, S. V., French, V., & Bryant, P. J. 1981. Distal regeneration and symmetry. *Science*, **212**, 993–1002.

Dolle, P., Izpisua-Belmonte, J-C., Falkenstein, J., Renucci, A., & Duboule, D. 1989. Coordinate expression of the murine *Hox-5* complex homeobox-containing genes during limb pattern formation. *Nature*, **342**, 767–772.

Duboule, D. 1992. The vertebrate limb: a model system to study hox/hom gene network during development and evolution. *Bioessays*, **14**, 375–384.

Dvorak, L., & Fallon, J.F. 1987. The ability of the chick wing bud to regulate positional disparity along the anterior-posterior axis. *Dev. Biol,*, **120**, 392–398.

Ede, D. A. 1971. Control of form and pattern in the vertebrate limb. In *Control Mechanisms of Growth and Differentiation*, Davis, D. D., & Balls, M. (eds), pp. 235–254. Symp. Soc. Exp. Biol., Vol. 25, Cambridge University Press.

Ede, D. A. 1982. Levels of complexity in limb mesoderm cell culture systems. In *British Society for Cell Biology Symposium, Vol. 4. Differentiation* in Vitro, Yeoman, M. M., & Truman, D. E. S. (eds), pp. 207–230. Cambridge University Press.

Ede, D. A. 1983. Cellular condensation and chondrogenesis. In *Cartilage 2: Development, differentiation and growth*, Hall, B. K. (ed). pp. 143–185, New York: Academic Press.

Fallon, J. F., & Crosby, G. M. 1975. Normal development of the chick wing following removal of the polarizing zone. *J. Exp. Zool.*, **193**, 449–455.

Feinberg, R. N., & Saunders, Jr., J. W. 1982. Effects of excising the apical ectodermal ridge on the development of the marginal vasculature of the wing bud in the chick embryo. *J. Exp. Zool.*, **219**, 345–354.

Fell, H. B. 1956. Skeletal development in tissue culture. In *The Biochemistry and Physiology of Bone*, Bourne, G. H. (ed). pp. 401–441, London: Academic Press.

Frost, S. C., & Hinchliffe, J. R. 1991. Retinoic acid effects on experimental chick wing bud: patterns of cell death and skeletogenesis. In *Developmental Patterning of the Vertebrate Limb*, Hinchliffe, J. R., Hurle, J. M., & Summerbell, D. (eds). pp. 157–160, New York: Plenum Press. NATO ASI Series A, Vol. 205,.

Frost, S. C., & Hinchliffe, J. R. 1993. Retinoic acid effects on cell death and division patterns in experimental chick wing buds. In *Limb Development and Regeneration, et al.*, J. F. Fallon (ed). J. Wiley. (In press).

Gabriel, M. L. 1946. The effect of local applications of colchicine on leghorn and polydactylous chick embryos. *J. Exp. Zool.*, **101**, 339–350.

Gilbert, S. F. 1991. *Developmental biology*. Third edn. Sunderland, Massachusetts: Sinauer.

Globus, M., & Vethamany-Globus, S. 1976. An *in vitro* analogue of early chick limb bud outgrowth. *Differentiation*, **6**, 91–96.

Hamburger, V. 1938. Morphogenetic and axial self-differentiation of transplanted limb primodia of 2-day chick embryos. *J. Exp. Zool.*, **77**, 379–400.

Harrison, R. G. 1921. On relations of symmetry in transplanted limbs. *J. Exp. Zool.*, **32**, 1–136.

Hinchliffe, J. R. 1985. "One, two, three" or "two, three, four": an embryologist's view of the homologies of the digits and carpus of modern birds. In *Beginnings of Birds, et al.*, M. K. Hecht (ed). pp. 141–147, Eichstatt: Jura-Museums.

Hinchliffe, J. R. 1991. Developmental approaches to the problem of transformation of limb structure in evolution. In *Developmental Patterning of the Vertebrate Limb*, Hinchliffe, J. R., Hurle, J. M., & Summerbell, D. (eds). pp. 313–323, New York: Plenum Press.

Hinchliffe, J. R., & Gumpel-Pinot, M. 1981. Control of maintenance and antero-posterior skeletal differentiation of the anterior mesenchyme of the chick wing bud by its posterior margin (the ZPA). *J. Embryol. Exp. Morph.*, **62**, 63–82.

Hinchliffe, J. R., & Horder, T. J. 1993. Lessons from extradigits. In *Limb Development and Regeneration*, Fallon, J. F. (ed). (4th Limb Development International Conference, California), J. Wiley. (In press).

Hinchliffe, J. R., & Johnson, D. R. 1980. *The development of the vertebrate limb*. Oxford: Oxford University Press.

Hinchliffe, J. R., Gumpel-Pinot, M., Wilson, D. J., & Yallup, B. L. 1984. The prospective skeletal areas of the chick wing bud: their location and time of determination in the limb field. In *Matrices and Cell Differentiation*, Kemp, R. B., & Hinchliffe, J. R. (eds). pp. 543–570, New York: A. R. Liss.

Hinchliffe, J. R., Hurle, J. M., & Summerbell, D. 1991. *Developmental patterning of the vertebrate limb*. Series A, vol. 205. New York: Plenum Press. NATO ASI.

Honig, L. 1983. Does anterior (non-polarizing region) tissue signal in the developing chick limb? *Devel. Biol.*, **97**, 424–432.

Horder, T. J. 1976. Pattern formation in animal development. In *The Developmental Biology of Plants and Animals*, Graham, C. F., & Wareing, P. F. (eds). pp. 169–197, Oxford: Blackwell.

Horder, T. J. 1978. Functional adaptability and morphogenetic opportunism. the only rules for limb development? *Zoon*, **6**, 181–192.

Horder, T. J. 1981. On not throwing the baby out with the bath water. In *Evolution Today*, Scudder, G. G. E., & Reveal, J. L. (eds), pp. 163–180, Pittsburgh, 1981. Hunt Inst. for Botanical Documentation, Carnegie-Mellon University.

Horder, T. J. 1983. Embryological bases of evolution. In *Development and Evolution*, B. C. Goodwin, *et al.* (ed). Cambridge University Press.

Horder, T. J. 1989. Syllabus for an embryological synthesis. In *Complex Organismal Functions: Integration and Evolution in Vertebrates*, Wake, D. B., & Roth, G. (eds), pp. 315–348. Dahlem Conference No. 45, Chichester, J. Wiley.

Hornbruch, A., & Wolpert, L. 1986. Positional signalling by Hensen's node when grafted to the chick limb bud. *J. Embryol. Exp. Morph.*, **94**, 257–265.

Hurle, J. M., & Ganan, Y. 1986. Interdigital tissue chondrogenesis induced by surgical removal of the ectoderm in the embryonic leg bud. *J. Embryol. Exp. Morph.*, **94**, 231–244.

Hurle, J. M., & Ganan, Y. 1987. Formation of extra-digits induced by surgical removal of the apical ectodermal ridge of the chick embryo leg bud in the stages previous to the onset of interdigital cell death. *Anat. Embryol.*, **176**, 393–399.

Hurle, J. M., Ganan, Y., & Macias, D. 1989. Experimental analysis of the in vivo chondrogenic potential of the interdigital mesenchyme of the chick leg bud subjected to local ectodermal removal. *Devel. Biol.*, **132**, 368–374.

Hurle, J. M., Ros, M. A., Ganan, Y., Macias, D., Critchlow, M., & Hinchliffe, J. R. 1990. Experimental analysis of the role of ECM in the patterning of the distal tendons of the developing limb bud. *Cell Diff. and Devel.*, **37**, 97–108.

Hurle, J. M., Macias, D., Ganan, Y., Ros, M. A., & Fernandez-Teran, M. A. 1991. The interdigital spaces of the chick leg bud as a model for analysing limb morphogenesis and cell differentiation. In *Developmental Patterning of the Vertebrate Limb*, Hinchliffe, J. R., Hurle, J. M., & Summerbell, D. (eds). pp. 249–259, New York: Plenum Press.

Iten, L. E. 1982. Pattern specification and pattern regulation in the embryonic chick limb bud. *Amer. Zool.*, **22**, 117–129.

Izpisua-Belmonte, J-C., Tickle, C., Dolle, P., Wolpert, L., & Duboule, D. 1991. Expression of the homeobox hox-4 genes and the specification of position in chick wing development. *Nature*, **350**, 588–589.

Javois, L. C. 1984. Pattern specification in the developing chick limb. In *Pattern Formation*, Malacinski, G. M., & Bryant, S. V. (eds). New York: Macmillan.

Kieny, M. 1964a. Etude du mecanisme de la regulation dans le developpement du bourgeon de membre de l'embryon de poulet. *J. Embryol. Exp. Morph.*, **12**, 357–371.

Kieny, M. 1964b. Etude du mecanisme de la regulation dans le developpement du bourgeon de membre de l'embryon de Poulet. I. Regulation des excedents. *Devel. Biol.*, **9**, 197–229.

Kieny, M., Pautou, M. P., Chevallier, A., & Mauger, A. 1986. Spatial organisation of the developing limb musculature in birds and mammals. *Biblthca anat.*, **29**, 65–90.

MacCabe, J. A., Saunders, Jr., J. W., & Pickett, M. 1973. The control of the anteroposterior and dorsoventral axes in embryonic chick limbs constructed of dissociated and reaggregated limb-bud mesoderm. *Devel. Biol.*, **31**, 323–335.

McLachlan, J. C. 1991. Growth factors produced by the polarising zone - a complement to the retinoic acid system. In *Developmental Patterning of the Vertebrate Limb*, Hinchliffe, J. R., Hurle, J. M., & Summerbell, D. (eds). pp. 115–122, New York: Plenum Press.

McLachlan, J. C., & Phoplonker, M. H. 1988. Limb reduplication effects of chorio-allantoic membrane and its components. *J. Anat.*, **158**, 147–155.

Mohanty-Hejmadi, R., Dutta, S. K., & Mahapatra, P. 1992. Limbs generated at the site of tail amputation in marbled balloon frog after vitamin A treatment. *Nature*, **355**, 352–353.

Morgan, B. A., Izpisua-Belmonte, J-C., Duboule, D., & Tabin, C. J. 1992. Targeted misexpression of hox 4.6 in the avian limb bud causes apparent homeotic transformations. *Nature*, **358**, 236–239.

Muller, G. B. 1991. Evolutionary transformation of limb pattern: heterochrony and secondary fusion. In *Developmental Patterning of the Vertebrate Limb*, Hinchliffe, J. R., Hurle, J. M., & Summerbell, D. (eds). pp. 395–405, New York: Plenum Press. NATO ASI Series A, Vol. 205.

Oster, G., Shubin, N. H., Murray, J., & Alberch, P. 1988. Evolution and morphogenetic rules. The shape of the vertebrate limb in ontogeny and phylogeny. *Evolution*, **42**, 862–884.

Oster, G. F., Murray, J. D., & Harris, A. K. 1983. Mechanical aspects of mesenchymal morphogeneses. *J. Embryol. Exp. Morph.*, **78**, 83–125.

Pautou, M. P. 1973. Analyse de la morphogenese du pied des oiseaux a l'aide de melange cellulaires inter-specifiques. i. Etudes morphologiques. *J. Embryol. Exp. Morphol.*, **29**, 174–196.

Saunders, Jr, J. W. 1948. The proximo-distal sequence of the origin of the parts of the chick wing and the role of ectoderm. *J. Exp. Zool.*, **108**, 363–404.

Saunders, Jr, J. W. 1977. The experimental analysis of chick limb bud development. In *Vertebrate Limb and Somite Morphogenesis*, Ede, D. A., Hinchliffe, J. R., & Balls, M. (eds). pp. 1–24, Cambridge University Press.

Saunders, Jr., J. W., & Gasseling, M. T. 1968. Ectodermal-mesenchymal interactions in the origin of limb symmetry. In *Epithelial Mesenchymal Interactions*, Fleischmajer, R., & Billingham, R. F. (eds). pp. 78–97, Baltimore: Williams & Wilkins.

Saunders, Jr., J. W., & Gasseling, M. T. 1983. New insights into the problem of pattern regulation in the limb bud of the chick embryo. In *Limb Development and Regeneration, Part 1*, Fallon, J. F., & Caplan, A. I. (eds). pp. 67–76, New York: Alan R. Liss.

Sengel, P. 1975. Discussion to Wolpert. In *Cell Patterning*, pp. 119–121. Ciba Fdn. Symposium 29, new series, Associated Scientific Publishers.

Sengel, P. 1976. *Morphogenesis of skin*. Cambridge University Press.

Shostak, S. 1991. *Embryology*. New York: Harper Collins.

Shubin, N., & Alberch, P. 1986. A morphogenetic approach to the origin and basic organization of the tetrapod limb. In *Evolutionary Biology*, M. K. Hecht, B. Wallace, & Prance, G. (eds). Plenum Press.

Shubin, N. H. 1991. The implications of the "Bauplan" for development and evolution of the tetrapod limb. In *Developmental Patterning of the Vertebrate Limb*, Hinchliffe, J. R., Hurle, J. M., & Summerbell, D. (eds). pp. 411–421, New York: Plenum Press.

Slack, J. M. W. 1987. We have a morphogen! *Nature*, **327**, 553–554.

Solursh, M. 1984. Cell-matrix interactions during limb chondrogenesis. In *Matrices and Cell Differentiation*, Kemp, R. B., & Hinchliffe, J. R. (eds). pp. 47–60, New York: A. Liss.

Stocum, D. L., & Fallon, J. F. 1982. Control of pattern formation in urodele limb ontogeny: a review and a hypothesis. *J. Embryol. Exp. Morphol.*, **69**, 7–36.

Stocum, D. L., & Fallon, J. F. 1984. Mechanisms of polarization and pattern formation in urodele limb ontogeny: a polarizing zone model. In *Pattern Formation*, Malacinski, G. M., & Bryant, S. V. (eds). pp. 507–520, New York: Macmillan.

Stopak, D., & Harris, A. K. 1982. Connective tissue morphogenesis by fibroblast traction 1) tissue culture observations. *Devl. Biol.*, **90**, 383–398.

Summerbell, D. 1977. Regulation of deficiencies along the proximal distal axis of the chick wing-bud: a quantitative analysis. *J. Embryol. Exp. Morph.*, **41**, 137–159.

Summerbell, D. 1979. The zone of polarizing activity: evidence for a role in normal chick limb morphogenesis. *J. Embryol. Exp. Morph.*, **50**, 217–233.

Summerbell, D., & Harvey, F. 1983. Vitamin A and the control of pattern in developing limbs. In *Limb Development and Regeneration*, Fallon, J. F., & Caplan, A. I. (eds). pp. 109–118, New York: A. R. Liss.

Summerbell, D., Lewis, J. H., & Wolpert, L. 1973. Positional information in chick limb morphogenesis. *Nature, London*, **244**, 482–496.

Swett, F. H. 1927. Differentiation of the amphibian limb. *J. Exp. Zool.*, **47**, 385–440.

Tabin, C. J. 1992. Why we have (only) five fingers per hand: Hox genes and the evolution of paired limbs. *Development*, **116**, 289–296.

Thaller, C., & Eichele, G. 1987. Identification and spatial distribution of retinoids in the developing chick limb bud. *Nature*, **327**, 625–628.

Tickle, C. 1981. The number of polarizing region cells required to specify additional digits in the developing chick wing. *Nature*, **289**, 295–298.

Tickle, C., Summerbell, D., Wolpert, L., & Lee, J. 1982. Local application of retinoic acid to the limb bud mimics the action of the polarizing region. *Nature,* London, **296**, 564–565.

Tickle, C., Lee, J., & Eichele, G. 1985. A quantitative analysis of the effect of all-trans-retinoic acid on the pattern of chick wing development. *Devl. Biol.*, **109**, 92–95.

Waddington, C. H. 1962. *New patterns in genetics and development*. New York: Columbia University Press.

Wilson, D. J., & Hinchliffe, J. R. 1987. The effect of polarizing activity on the anterior half of the chick wing bud. *Development*, **99**, 99–108.

Wolpert, L. 1989. Positional information revisited. *Dev. Biol.*, **107**, 3–12. (Supplement).

Wolpert, L. 1991. Some problems in limb development. In *Developmental Patterning of the Vertebrate Limb*, Hinchliffe, J. R., Hurle, J. M., & Summerbell, D. (eds). pp. 1–7, New York: Plenum.

Wolpert, L., & Hornbruch, A. 1987. Positional signalling and the development of the humerus in the chick limb bud. *Devel.*, **100**, 333–338.

Wolpert, L., & Hornbruch, A. 1990. Double anterior chick limb buds and models for cartilage rudiment specification. *Devel.*, **109**, 961–966.

Wolpert, L., Lewis, J., & Summerbell, D. 1975. Morphogenesis of the vertebrate limb. In *Cell patterning*, pp. 95–119, Amsterdam, 1975. Ciba Fdn. Symposium 29, new series, Associated Scientific Publishers.

14. THE CHICKEN AND THE EGG

T. J. Horder

Department of Human Anatomy
South Parks Road
Oxford, OX1 3QX, UK

"Which came first, the chicken or the egg?" For all its innocence, this is a question which, as this paper will show, obliges one to think deeply about the terms in which one ultimately seeks to explain biological systems. "A hen is only the egg's way of making another egg", in Butler's (1878)[1] apt phrase, sums up the view that organisms can most simply be explained as the results of causes laid out in the egg or, in contemporary terms, as the results of their inherited sets of genes. Today's biologist might well extend the metaphor and say that "organisms are only expressions of a sequence of the four nucleotides of DNA". This eminently successful way of explaining biological systems is examined in the first part of the present paper. Yet there remains a nagging worry about this viewpoint. Can this really be the whole story? In the second section, I attempt to identify the possible missing element and consider whether the other attributes of organisms might be more than mere genetic epiphenomena. In the third section I suggest how, due to the continuing impact of now outdated methodological assumptions, we face a choice of methods which greatly affects our ability to tackle the real problem of biological organization.

14.1   THE EGG

The strategy adopted in this paper is to analyze, in the simplest possible terms, the principal ways in which the problem of the nature of biological organization has been conceptualized. The aim is to identify the underlying rationale - the conceptual categories used, the articulations of concepts and data used to support the explanatory framework - unfamiliar though it might sometimes be, since the full rationale is so rarely spelled out. As a start to the analysis, I begin by reviewing how the structure of thought today has in practice been built up. This can be seen as falling into a series of logical steps, which can be reduced to a few, brief (and approximately historical) characterizations.

14.1.1   Preformation

Among approaches to defining and explaining the nature of organized structure, pre-formation can be considered the first systematically-argued attempt to have an influence on the modern era. According to this view, pattern is directly anticipated, perhaps only invisibly, in a corresponding organization of the contents of the egg (Figures 14.1 and 14.2). Although easily dismissed today as failing to *explain* anything - the concept implies an infinite regress and an unexplained initial setting up of the first pattern -, it was vigorously defended and, when understood in the context of the times, it was an entirely rational position.[2]

Experimental and Theorietcal Advances in Biological Pattern Formation,
Edited by H.G. Othmer *et al.,* Plenum Press, New York, 1993

Figure 14.1. The eighteenth century concept of preformation. This is Hartsoeker's depiction of the structure of a spermatozoon; others inferred a similar homunculus in the egg.

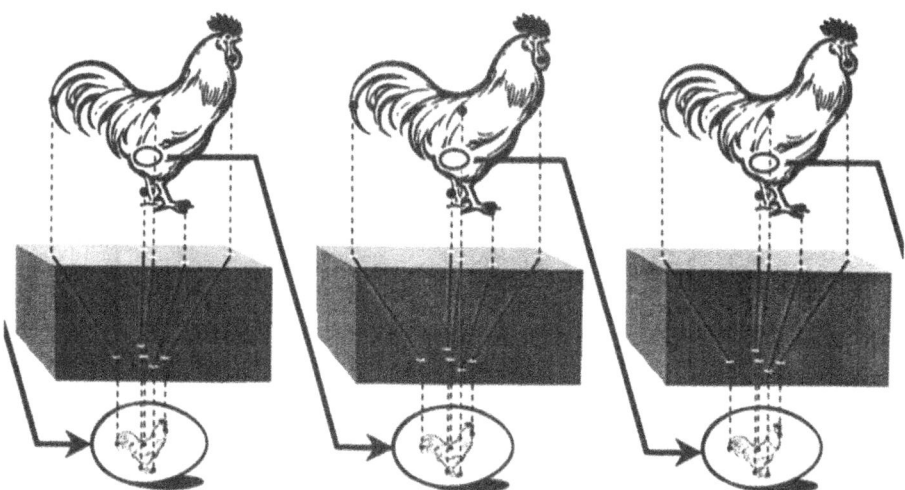

Figure 14.2. Diagram illustrating the logic entailed in the pre-Darwinian concept of preformation. The only role implicitly ascribed to embryogenesis (depicted as a "black box") is visible differentiation of pattern and size increase.

14.1.2   The impact of evolution theory

By finally overturning the fixity of species, Darwin's theory transformed the nature of the problem. What now needed to be explained was morphological flexibility, and in particular progressive change, including its twin aspects, ancestry and novelty. Emphasis was now drawn to the adaptedness of adult structure. Darwin went as far as to outline the mechanisms underlying the transmission through generations of the changes acquired by natural selection[3] (Figure 14.3).

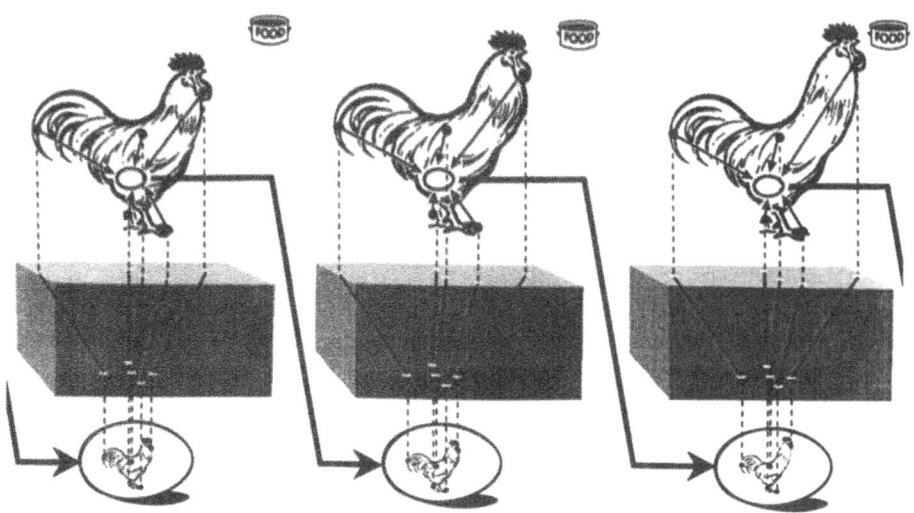

Figure 14.3. Diagram showing the logic of Darwin's theory of heredity ("pangenesis") designed to explain the transmission of evolutionarily acquired changes in morphologies. Through a series of generations, the results of adaptive variations acquired at one stage are transferred directly into the egg, which is the sole vehicle for transmission to the next stage: the mechanism is a Lamarckian inheritance of the effects of the environment (*e. g.* growth of the neck in response to an acquired food source). The onus of explanation of pattern change was on the hereditary mechanism, while the events in the black box of embryogenesis are non-contributory and, implicitly, preformative.[3,4]

14.1.3   The split between genetics and embryology

The next comprehensive recasting of the conceptual framework, due largely to Weismann[5] (Figure 14.4), established the separation of the pathway for transmission of hereditary factors between generations from the actual manifestations of the hereditary factors (*e.g.* adult structure and function, including developmental causes). By focussing on the biparental assortment of hereditary factors and removing the Lamarckian element, this opened the way directly to modern genetics.

With the rapidly ensuing re-discovery of Mendel's laws and introduction of gene theory, the conceptual split between the embryonic and genetic domains triggered by Weismann intensified unavoidably. As explanations of adult morphology the choice now lay between putting the explanatory weight directly on the genetic factors, as Weismann had done, or taking into account the possible role of the invisible processes in the embryo. Events at that time, particularly Driesch's discovery of embryonic regulation (Figure 14.7), established the clear possibility that embryogenesis itself played a crucial role in pattern formation.[6]

14.1.4   The major post-genetic classes of models of pattern formation

Broadly viewed it is possible to see two main trends in the approaches that have been taken among attempts to reconcile genetics and embryology.

(a) The evolutionary approach.

D'Arcy Thompson[7] epitomizes a common starting point in which form and evolu-

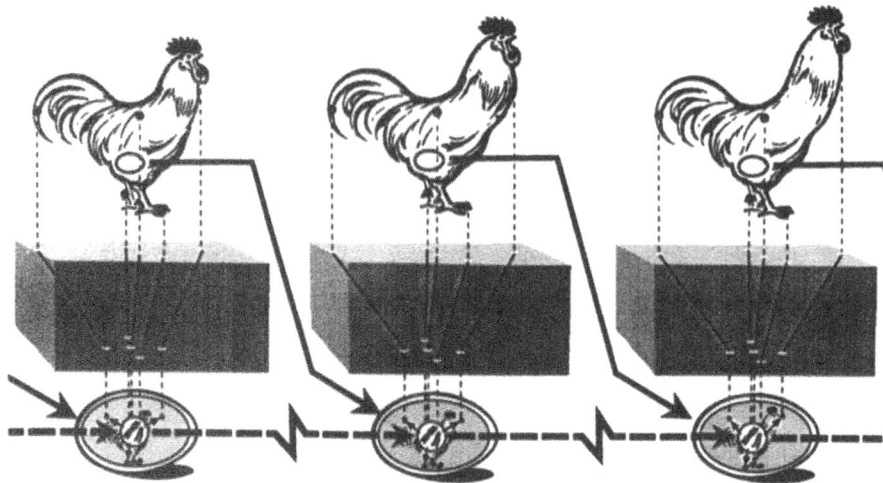

Figure 14.4. Diagram of Weismann's formulation. The hereditary mechanism is now identified as nuclear and chromosomal. New morphologies are seen as the result of new assortments (due to crossing over and biparental mixing: shown as interruption in the broken line) of independent hereditary factors acquired directly from the previous generation. Pattern expression depends entirely on the way the factors are transferred from the chromosomes: Weismann hypothesized patterned localization within the egg cytoplasm. Later embryonic events as such were ignored, implying therefore a preformationist view of embryogenesis.[6]

tionary change in form, is explained as the result of the immediate, primarily physical and mechanical, forces operating on the developing embryo (Figures 14.5 and 14.6). .

This approach has been continued, essentially unchanged, by "Neo-Darwinist" evolutionary biologists in their attempts to integrate evolution theory with genetics. Much the same type of explanation of morphology can be said to underlie the concepts of allometry and heterochrony (as used, for example, by Huxley, Wright, Goldschmidt, Waddington),[8] the genetic bases for which were explained by invoking "rate genes" or "gene balance".[8]

(b) The embryological approach

For those whose focus was primarily embryological, the regulative properties of embryos (Figure 14.7) have become a dominating consideration.

The problem first opened up by Weismann (namely how randomly arranged discrete particles in the nucleus can be selected out in different cells in order to produce the patterned differentiation of adult morphology) has been reinforced by the increasing certainty that, not only does the egg cytoplasm provide minimal preformed organization, but that, throughout development, all cells have the same (complete and often multipotential) array of genes.[9] In order to explain the adjustments that must underlie regulation and the way in which integrated pattern generation might be achieved through embryogenesis, a variety of concepts have been introduced (Figure 14.8).

In fact, both the evolution and embryology-centred approaches, and indeed all current viewpoints (Figures 14.9, 14.10), have come to share the following common features. Accommodating the need to explain regulation, embryogenesis is seen as coordinated by an initially single integrating signal (Figures 14.8, 14.9 and 14.10). This signal is the basis

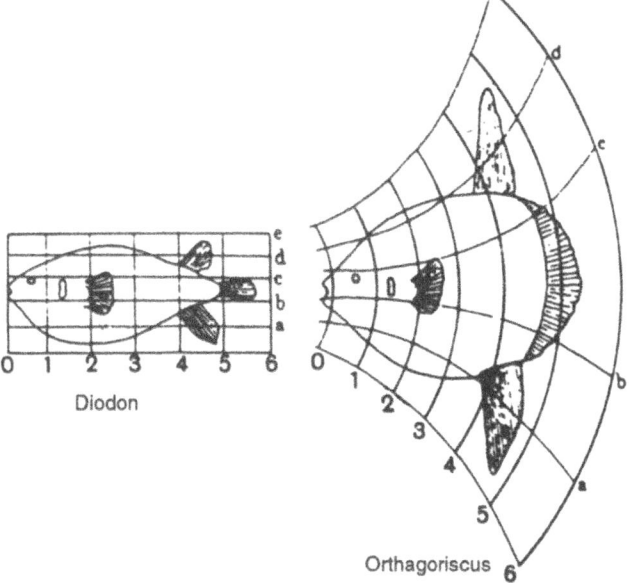

Figure 14.5. An example of a D'Arcy Thompson transformation.[7]

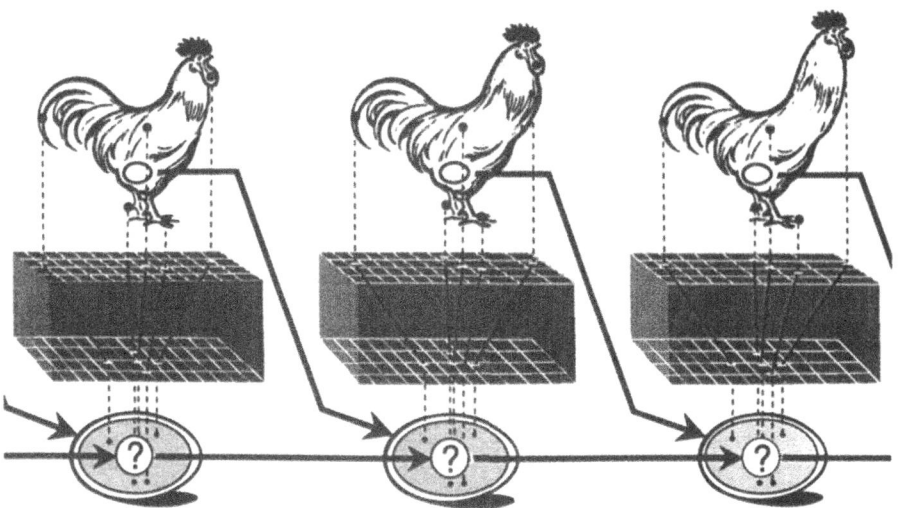

Figure 14.6. The logic of allometric explanations, as exemplified by D'Arcy Thompson's transformations. Changing morphologies are explained solely as the result of coordinated differential growth during development (shown as gradually spreading out (transverse) grid lines in the black box).

(through the resulting switching on of gene expression) for the emergence of patterned cellular differentiation in an overall, integrated array. The positional information model (Figure 14.9) formalises this scheme in the most explicit way: pattern formation is controlled by a direct readout of position, implying therefore a direct, one step, relation between position and final

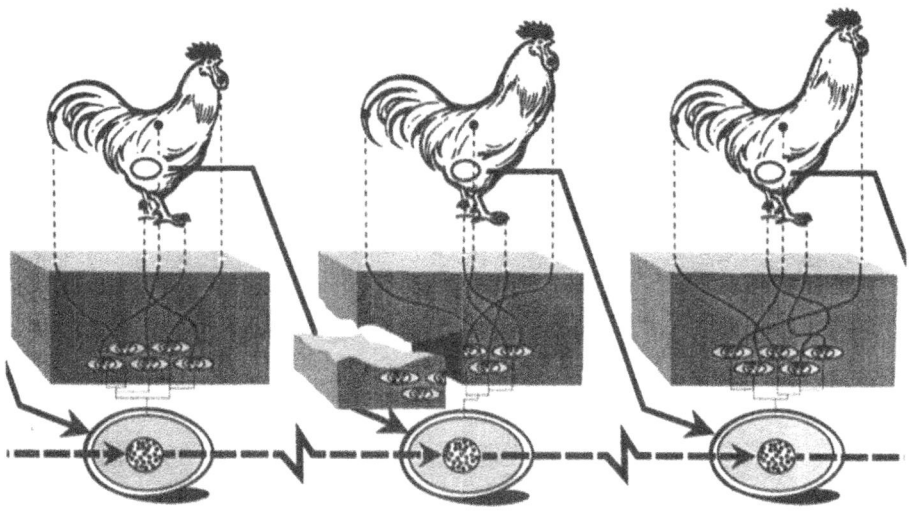

Figure 14.7. The problem of embryonic regulation. As first demonstrated by Driesch, removal of a part of the early embryo does not result in a partial adult ("mosaic development" as predicted by Weismann) but a complete final morphology. It follows that pattern is not already established in the egg - parts of the egg and early embryo are in fact totipotent (shown as array of equivalent cells at the start of development) - and that the later events of embryogenesis (shown as interdependent unscrambling of the information in the initial cells) themselves are responsible for pattern creation.

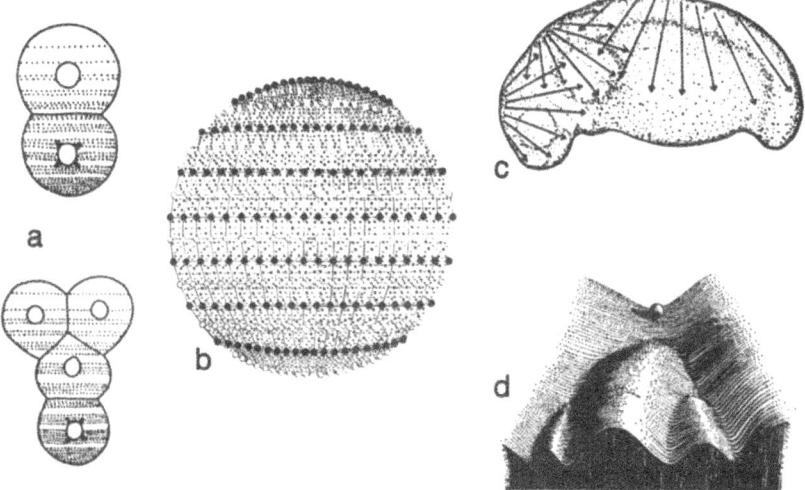

Figure 14.8. Four influential concepts hypothesizing mechanisms of embryogenesis, which explain regulation and the integration of emerging pattern; (a) The concept of a gradient built in to the egg and subdivided in daughter cells of the early embryo, first introduced by Boveri. (b) Harrison's concept of polarisation as applied to the egg, (c) Spemann's field concept, (d) Waddington's depiction of the "epigenetic landscape" which anticipated "standing wave" models of positional information.[10]

pattern. Figure 14.10 summarises evidence for *Drosophila*, where an initial gradient and the putative, required position-sensitive genes have actually been identified.

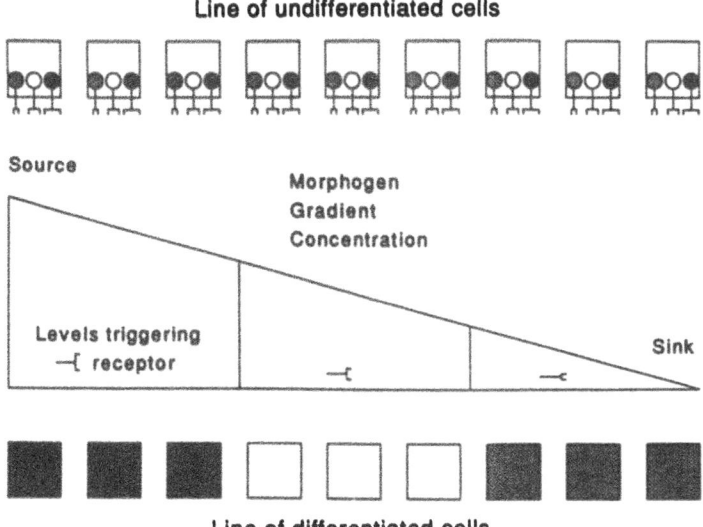

Line of undifferentiated cells

Source

Morphogen Gradient Concentration

Levels triggering receptor

Sink

Line of differentiated cells

Figure 14.9. The underlying logic of the positional information model. Position is defined by an initial gradient of a morphogen. Differentiation (here into three regions) follows as the result of a one-step readout. The model requires an already built-in array of receptors (with ability to discriminate morphogen levels to a degree which must match the grain of the final pattern), which mediate DNA switching and which themselves must be programmed into the pre-existing genome.[10]

14.1.5 Conclusions

Grossly abbreviated though this account of historical events may be, the point I want to bring out is the way in which genetics and embryology have long been conceptually interdependent.

The most consistent trend has been towards an ever more self-confirming understanding of the genetic domain, culminating in its description in terms of DNA. The reduction to the molecular level is particularly powerful because, in one single conceptual framework, a coherent foundation is simultaneously provided for most of the major themes of biology, including the origin of life and cell biology on one extreme, and on the other the mechanisms and reconstruction of evolution, as understood in terms of DNA recombinations and mutation, and species as gene pools.[8] As the above described outline indicates, we have reached a point where it appears that a full and complete account of a chicken can be given by specifying, not just the contents of the egg, but merely the DNA sequences within it, while evolution can be viewed simply as a succession of genomes.

This concentration on the molecular level was, however, only achieved by way of a splitting off, in terms of theory and techniques, of embryological considerations and, moreover, was only possible given a certain class of models of embryogenesis. Increasingly the events of embryogenesis have been treated as if they amounted to a one-step transformation to adult pattern from preconditions in the egg. Accepting that the egg cytoplasm provides a minimal contribution to patterning, it has then become increasingly feasible to see pattern as a simple reflection of the direct translation of genetic information. Any solution to the problem of pattern formation depends on the particular way in which the domains of genetics and embryology have been articulated: to an extent that is not often recognized, there is a trade off between the roles ascribed to them. In so far as the outline given here covers all

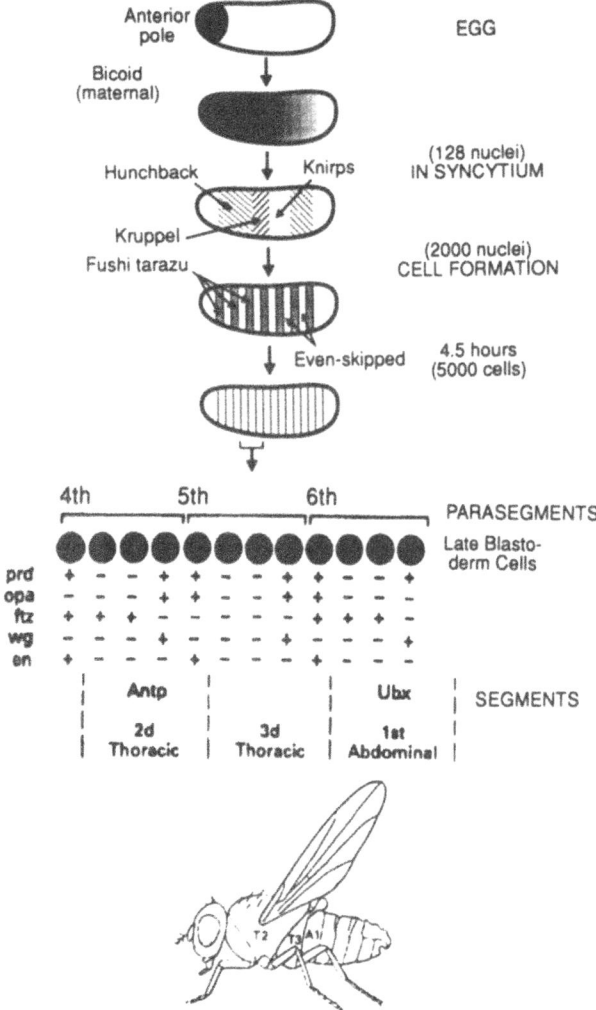

Figure 14.10. Summary of the current model of *Drosophila* development. An initial, maternal gradient in the egg is the basis for the sequential, localized expression of a number of identified genes which, (when combined specifically in particular cells, as shown in the table below), leads to the eventual differentiation of segmental body regions.[10]

the currently accepted "theories of development", it is clear that the explanatory burden has largely been handed over to DNA.

However, this increasingly explicit and direct reference to DNA raises problems. Most obviously, does this scheme not imply the need for an unacceptably large number of genes? In order to explain the required selection of specific genes for expression, from a starting condition of unrestricted totipotentiality in both the genome and the egg cytoplasm, there must be, somewhere and in some form, components in the genetic apparatus capable of discriminating each position at which choices of differentiation have to be made.[11] The notion of locus-responsive control genes ("pattern genes") is implicit in much of today's thinking (*e. g.* Figure 14.10), although its form remains obscure. It has been suggested that the actual sequence of genes on the chromosome mediates this role (*i. e.* that patterned expression is achieved colinearly by sequential read out of ordered DNA sequences),[12,13] but it could equally be achieved by a hierarchy of control genes[14] or a specific position- receptivity of the promoter of each expressed gene.[14] As indicated in Figure 14.9, a specific gene is, potentially, needed for each cellular element in the final array in order to mediate the read out of its differentiation in accordance with its specific position. The problem reduces to the following fundamental difficulty: the selection of expressed genes which determines adult pattern must be controlled by other genes. This problem of how the genome controls itself is not only unresolved: it also seems to demand a vast genetic complexity and imposes consequential constraints on how the genome functions and evolves.[11]

The genetic approach encapsulates, in the most clear-cut possible form, the basic logic of the view that the egg is a sufficient cause and explanation of the chicken. Regardless of the details of the various conceptual mixes included among current theories, the common thread of direct reference to DNA ultimately faces one crucial underlying dilemma. The implied one-to-one causal relation between genes and pattern effectively means that the end result is being explained by presupposing an earlier stage or state which itself has equivalent complexity. This simply redescribes the problem, at least until the imponderables of gene control are solved: the burden of explanation has merely been shifted into different terms. Thus, ironically, direct reduction to DNA leaves us trapped in the logic of preformation: the position we have described has all the characteristics of "genetic preformationism". As is likely to be the case with all models that explain pattern as due to structural preconditions in the egg, the problem with a preformationist way of thinking[15] is that it has little explanatory value: it merely begs the question of how the preconditions themselves are explained.

14.2   BUT WHAT ABOUT THE CHICKEN?

Impeccable though the rationale of reduction to DNA may in theory appear to be, how are we to avoid its untoward consequences? Can we escape the cramping preformationist logic of this position? Is there any equally clearly definable alternative? In what follows, inadequacies in the logic are sought out and identified; four sets of considerations are presented, on the basis of which a second perspective can be arrived at.

14.2.1   Has the central problem been appropriately defined?

We should begin by reminding ourselves of our ultimate target. If one stops to consider the adult chicken, its single most remarkable feature must be its adaptedness. A chicken is adapted to its environment in all aspects of its function and behaviour, in ways that are all in some sense built into its pattern of structural organization. Ability to explain adaptedness is the real test of any theory of pattern formation.

The most obvious explanation for adaptedness, namely that "environmental factors" directly modulate adult structure according to functional requirements, has, since Weismann, been treated warily.[16] Yet the evidence for such mechanisms is considerable. Cases in which truly external factors dictate morphology are perhaps few and only general in their effects; dependence of growth on nutrition is a clear enough example. But something like modelling of bone by mechanical forces is a powerful illustration, reflecting not only the effects of external environmental forces on the growth and behaviour of bone cells (*i. e.* the indirect effects of gravity) but also internal ones, *e. g.* stresses and strains within and between bones. There are few tissues which do not at least respond in size changes to changed physiological demands. The extent of functional modulation to adult morphology is probably considerable[16] and it cannot be dismissed on the false logic that, one suspects, often comes into play; *i. e.* that genetic factors mediating heredity and somatic function cannot interact, since this would imply "Lamarckian" effects. The expression of genes and functioning of cells can readily be conditional on responses to "extrinsic" factors since responsiveness could itself be programmed by the genome, without any implication that such factors affect the hereditary pathway itself.

Evidence such as this suggests that a purely DNA-based view must be incomplete. Much depends on how the DNA is used and "function" is therefore necessarily part of causal explanations. Genes are irrelevant unless they are expressed, directly or indirectly: gene control is equal in importance to the expressed DNA itself. At the most basic and familiar level of the cell, the expression of any gene is the result of factors extrinsic to the gene *per se*. Most obviously, translation itself depends on the availability of the machinery for transcription and protein synthesis, and in turn, more widely, it ultimately depends on metabolism, cell structure, specific regulatory factors, feedback controls and so on.[17] It is of course true that most of these "extrinsic" causes contributing to gene expression are themselves gene-based, but the main point I wish to make is that the interactions and continuities of functional factors internal and external to the cell mean that "environmental" factors extrinsic to, and remote from, the genetic apparatus are always, at least potentially, contributors to causality. The fact that gene expression needs "functional" interactions, due to other genes and "non-genetic" contributions, means that DNA without its context is a biologically meaningless abstraction.

Thus the prospect arises that direct reference to DNA sequences alone may be a biologically unrealistic approach to pattern, and that we must redefine the terms in which an explanation of pattern formation is to be expected. It seems clear that we must adopt a wider view of cause and explanation.[18] Compared to the essentially descriptive, structural and static account which characterizes the DNA view, functional causes are much harder to specify; they are likely to be multifactorial, interactive and "open" in range of possible combinations. Without going further into a problem area which, in its broader form as the nature-nurture dichotomy, is one of the most generally confused issues in biology, it is clear that genes and environment cannot be treated as opposed or even separable. Biological systems are inevitably in part the result of "non-genetic" causes and the very concept of "genetic" is fundamentally flawed.[19] We must consider the possibility that all aspects of the chicken's structure, function and interaction with the external environment should be incorporated as part of any complete explanatory scheme.

If these conceptual difficulties apply on the genetic side, what about the other partner in any explanation of pattern formation, embryology? Embryogenesis is the domain in which the relation between genes and pattern becomes manifest, *i. e.* where control of gene expression is revealed as cell differentiation. It is in this domain that the causality of differentiation belongs and the question of how the genome can control itself must be solved.

14.2.2   Have embryonic events been adequately characterized?

If one looks more closely at the phenomena characteristic of embryogenesis, and looks in particular for a known mechanism most directly involved in the localized initiation of cell differentiation, embryonic induction immediately stands out.[20] This is an intercellular interaction in which the onset and specific type of differentiation of a given tissue is determined by a specific, adjacent tissue. It has been intensively documented and has been demonstrated in virtually every organ system in vertebrates (Figure 14.11).

In order to account for the adjacency and localization of the inducing tissue in relation to the cells responding to it, it is necessary to invoke the phenomenon of morphogenetic cell movements, another equally general feature of embryos. Together these two phenomena can account for the adult pattern of spatially coordinated, differentiated cell types, through their mediation of an integrated, spatio-temporal cascade of events, in which inductors resulting from early interactions lead to morphogenetic rearrangements bringing cells into new adjacencies that in turn result in new types of inductions.

This account of embryogenesis readily accommodates the fundamental fact of the initial totipotentiality of the egg. A feature of this analysis is that responding cells do not need sources of developmental information other than that due to induction. The cascade can originate from a single, initial inductive interaction, and progress, on the basis of initially equivalent cells, all sharing the same totipotentiality due automatically to their direct derivation from the egg by mitotic division. Although the nature of inductive signals is only partially defined,[20,21] the signals correlate with the inductor cell's own state of differentiative commitment and can be thought of as part of normal gene expression leading to cell differentiation. Thus, all that has to be provided to allow the inductive cascade to proceed is pre-programmed receptivity and responsiveness to inductive signals in induced cells and this is inherent in the shared totipotentiality common anyway to all cells.[9] The inductive cascade resolves the dilemma of how one part of the genome can control another: genes in one cell control the identical genome in another cell, by way of their gene products, through the route of inductive intercellular interactions. We can conclude that, given the appropriate characterization of the rules of developmental interactions, it is possible to avoid any requirement for preformed egg organization.

On this analysis, in contrast to the position described in Section 14.1, embryogenesis is indirect, multistage and multicausal. The important developmental signals are local, cell-to-cell inductive interactions rather than position as such or long distance gradients.[20,21,22] Furthermore, few assumptions need be made about gene organization. Few alternative cell states have to be anticipated in the genetic organization (*i. e.* of the order of the number of cell types themselves, 100-200) and the same cell type can contribute within a variety of adult morphological arrays without further genetic control complications.[14,20,22] The true nature of embryogenesis is reflected in the evident indirectness and progressiveness of events: the cascade of morphogenetic and differentiative events is a directly observable fact and induction is fully definable in operational terms by experimental means.

Pattern is the highly indirect, interactive result of readout of the common potential of inductive responsiveness shared among all cells. To this must be added one further precondition: the contribution of the egg cell itself. In so far as all zygotic cells are derived by mitosis from the egg cell - and thereby *directly* acquire much of their organization and organelles - the egg cell as such is in itself a channel for the inheritance of organization independent of zygotic DNA: it is therefore another example of how zygotic DNA is only part of the explanation of the organism. The egg also influences future morphogenesis (*e. g.* by way of its impact on the spatial organization of early mitosis), but how much this affects adult pattern is much less certain.[15] In combination the various, overlapping developmental considerations that have been addressed here (from the morphogenetic integration of the cascade to the programmed functional modelling of adult tissues) can fully account for the final adaptive pattern of the adult, including its integration on all scales of organization. This new view of the nature of embryological events - added to turnover, growth and its latent potential for metaplasia and regeneration, all of which indicate the retention of the properties of embryonic tissues - suggests that the adult state is in some respects an arbitrary and misleading notion. All aspects and phases of the life cycle can be seen as continuously integrated and the adult can be regarded as merely a relatively stationary phase within a unified series of expressions of the potentialities inherent in each cell of the organism.[9]

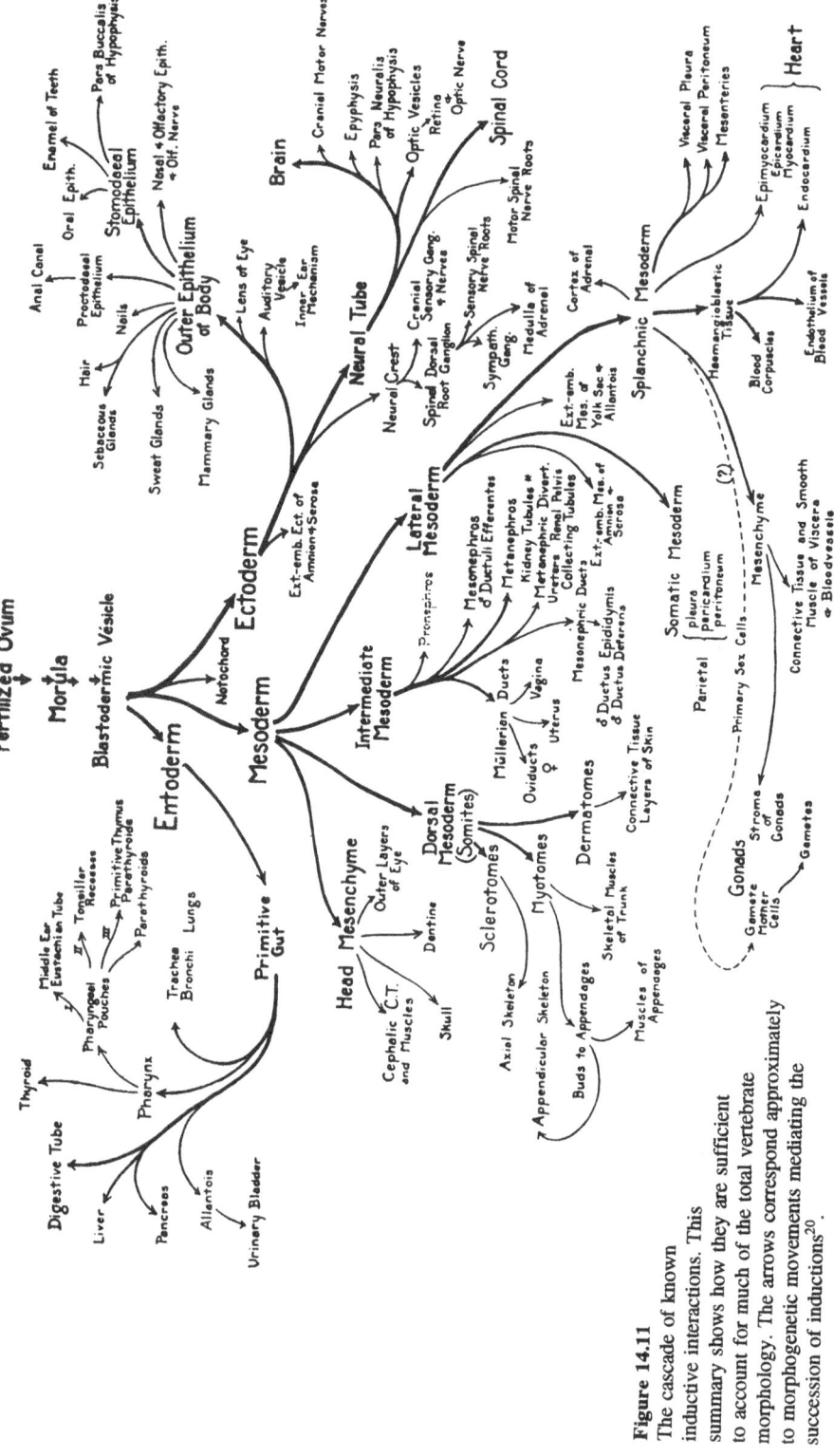

Figure 14.11

The cascade of known inductive interactions. This summary shows how they are sufficient to account for much of the total vertebrate morphology. The arrows correspond approximately to morphogenetic movements mediating the succession of inductions[20].

14.2.3   How is pattern mapped in the genome?

What does classical, rather than molecular, genetic evidence tell us directly about the gene organization mediating pattern formation? As is well known, this source of evidence typically demonstrates the following features;[23] genes or mutations affecting morphology show considerable variation among individuals in their precise effects; patterns of effects fall into relatively few major groupings, so that only certain patterns and aspects of morphology are affected; frequently many genes have the same type of effect. It is therefore clear that morphology is far from being mapped in a one-to-one fashion in the genome. Some of the reasons for the complexities of these genetic phenomena are also well understood; the effect of any one gene is dependent on (often many) other genes; major classes of defect are often mimicked by non-genetic ("phenocopy") interventions, which implies that mutations themselves often act through a multifactorial pathway interacting with, and dependent on, environmental factors. In the light of such considerations, and bearing in mind the indirectness of embryogenesis, it is hardly surprising that the effects of a single mutation very often involve diverse and irregularly arranged adult tissues ("pleiotropy").

Despite the considerable limitations in our knowledge of genes mediating pattern,[14,19,22,23] it seems safe to say that any element of morphological pattern is multi-determined in terms of the number of genes involved and other causal influences. Furthermore, many arguments converge on the conclusion that, in general, the chromosomal order of genes is unimportant; *e. g.* the lack of effect on morphology of the surprisingly rapid evolutionary redistribution of DNA sequences within and between chromosomes, the dispersed mapping of genes involved in closely related functions (such as metabolic pathways or contributions to specific protein molecules or differentiated cell types) and the lack of effect of experimental dissociation of neighbouring genes.[24]

Thus there is every reason to believe that pattern is represented in the genome as a randomly dispersed collective of genes. Such a situation is entirely compatible with the induction-based view of embryogenesis, as just outlined, according to which pattern is built up by purely local interactions between cells with discrete states, such that it is unnecessary to posit genes responsive to position as such or control genes whose specific purpose is to control pattern layout. Control of gene expression can be explained by direct access to specific expressed genes (via their promoters) - or via an intracellular cascade[14] - so that their chromosomal location is immaterial. The developmental evidence suggests that gene control is mediated on a cell by cell basis, requiring as signals gene products expressed independently (and quite possibly as part of other functions) in the inducing cell. One only has to consider how a coordinated, multicellular event like the morphogenetic movements of gastrulation is genetically programmed to recognize how developmental and adult morphological phenomena can only be implicitly programmed in terms of cellular responsiveness to normally expected intercellular forces and signals: the programme, in genetic terms, must be exceedingly remote and complex.

14.2.4   In what terms is integration of parts ultimately to be explained?

The explanatory problem has now become further refined. With the onus for developmental integration now placed on cooperativity between dispersed genes, how are we to account for the coexistence of the particular sets of genes and the all-important implied matching relationships between them?

The answer lies in the cumulative nature of evolution.[25] The problems associated with the reconstruction of the course of evolution are many and well known, but the principles of the process itself are familiar, are extremely well documented and apply at all levels whether molecular or morphological. Evolution is characterized by the fact that all elements of the organism are open to variation (e. g. at the DNA level; mutation, reassortment, gene duplication, etc), but subject to the dual constraints of selection for improved function and non-survival if they have any deleterious effects of any kind. The considerable data we now have on molecular changes in evolution[26] amply confirm how each stage of evolution is a (usually minor) substitutive change in the foregoing state. The effect, as evolution proceeds cumulatively, is that older stages are gradually replaced and that the evolutionary record of successive stages is lost.[27] It can be assumed that at each intermediate stage only favourable changes compatible with the genome as a whole will survive. The important point is that at every stage of evolution the mutually adapted matching of a specific set of genes will be ensured.

Any given combination of genes at any one time is the outcome of an unimaginably long, drawn out, and complex cumulative process of this kind. Even though the intervening steps are lost and can only be reconstructed inferentially,[25,27] the cumulative evolutionary mechanism is fully capable, in principle, of accounting for the particular combinations of specific genes needed to account for pattern formation. Evolutionary constraints will apply to each and every one of the elements of the genome and to every role they may have in the collectivity of elements.[28]

Of course the ultimate involvement of the adult chicken *per se* is in its role in mediating selection. Here, above all, the DNA view of biology breaks down because selective forces entirely outside the organism (e. g. external environmental forces such as competition within and between species, ecological change, etc) impinge as part of the causal chain. It is important to remember that selection does not only operate on adult characters: it applies equally, though less directly, to events during development.[29] Thus all aspects of the life cycle are subject to selection, as, in turn, must be the whole assembly of dispersed genes contributing, however indirectly, to pattern formation.

When one tries to contemplate such a complexity of causes, operating across the vast numbers of organisms generated over the timescale of evolution, one begins to understand why our view of pattern formation is so incoherent. The set of causal factors stored in any given, existing genome gives us only a fragmentary glimpse of the totality of causes actually explaining the organism: most causes (e. g. the chain of intervening ancestors and the selective forces applied to them) no longer exist.

14.2.5 Conclusions

Here we have arrived at a new viewpoint, which is distinct from the one outlined in Section 14.1. The core of this conceptual framework lies in the idea of a developmental cascade and it is this that makes it possible to escape from genetic preformationism.[30] Although it is difficult to envisage the dynamic, interactive and cumulative complexity of the cascade in any simple or complete sense, it is possible, in principle, to follow the causal chain of events through to patterned adult morphology. This conceptual framework cannot be formalised as a single, complete description, but it is consistent with a very wide variety of data on all levels. The essence of the above analysis is that embryological phenomena and morphology have been integrated into an internally consistent explanatory scheme by linking higher level data (morphogenesis, induction, potentialities, receptivity) to lower level data (cell biology, the fixed common genome) through a network of causal relationships which, in spite of being incompletely defined, are plausible in the light of the whole body of knowledge. Perhaps the most important outcome is the demonstration that alternative ways of conceptualizing embryology lead to profound changes in one's understanding of the nature of biological systems in general and to entirely new ways of fitting data together in many other specific biological domains, including genetics and evolution.

On three grounds, it has been argued that DNA alone is not a sufficient explanation for biological systems, namely the integral contributory role of "environmental" factors, the independently derived starting conditions provided in the egg cell itself and the causal elements lost during evolutionary progression. The new view is in no way independent of the molecular basis of biological systems, but DNA is now put in context.

Now the domains of genetics and embryology are conceptually bridged in a way that relates realistically to our knowledge about DNA and cells. There is no need to think in terms of genetic anticipation of pattern in any congruent, one-to-one fashion, because pattern is a property of the collectivity of genes, represented only implicitly as dispersed, multiple genetic elements and in the form of actual and potential genetic and cellular interactions. It is unnecessary to hypothesize a special class of "pattern" genes (or any other such specific principle of genetic organization), nor are exorbitant numbers of genes required. There is no mystery about the complexity, indirectness of effect and matching of the various genetic and developmental mechanisms required on this scheme: the past complexity of the evolutionary process is ample explanation. Receptivity (whether at the level of molecules, whole cells or the responsiveness of adult tissues to functional demands) is programmed to be highly tuned to the appropriate stimuli because evolution targets precisely that: natural selection is a tuning to the environment and adaptedness is the result.

Within this new framework concepts such as function, environment, adaptedness and integration find a clearer and undeniable place in explanations of biological systems. Integration becomes a very real thing at all levels, *e. g.* at the level of morphology (as in the morphogenetic cascade and the functional, mechanical, growth and endocrine modulations of adult structure), in evolutionary changes across the genome and in the unity of functions within the cell. The capacity to generate a cumulative developmental process depends on initial potentialities and starting conditions, but its realization depends on the actual circumstances of each stage and how succeeding stages build on their predecessors through multicausal interactions and summations. This cumulative property emerges as a most characteristic feature of living systems - in evolutionary terms as well as in embryogenesis - and it provides a conceptual alternative to the logic of preformation. Without taking away from genetic factors many of their traditional roles in biological explanation, the above considerations show that it is possible and necessary to think in a cumulative way in all biological domains; developmental, genetic and evolutionary.

The missing element in the purely DNA-based account can, then, be identified as a failure to cover the totality of causes. A complete understanding requires explanations which include the concept of a "whole chicken" in all its complex aspects, of which the genome is only a part. Moreover, when seen in evolutionary perspective, phenotypes and genotypes are causally inseparable and the evolution of each must advance step by step with the other; the chicken as such is as important as its egg or its DNA. Unlike the position described in Section 14.1, explanation is here seen as being multicausal and potentially infinite in its ramifications and complexity.

Having thus hopefully established the case for a second perspective, it remains for us to define the nature and origin of the difference in the two approaches as precisely as possible. This we turn to in the final section.

14.3 THINKING BIOLOGICALLY

Given the successes of the approach described in Section 14.1, it could well be asked why there can be any possibility of any true alternative and why, if there is any such alternative, its nature, and role in relation to the first, is not already well defined and established.

14.3.1 Traditions of "scientific method"

The potted history outlined in Section 14.1 records how, by way of a long and consistent trend, physico-chemical reduction came to dominate as a method in biology.[31,32] This trend was no doubt inevitable, but it was inspired for much of the time by one overriding concern, namely the question of whether biological systems are intrinsically different from physical ones. Opposed initially by vitalism, the steady progress of biochemistry, cell theory and genetics led to an increasingly certain "No". Molecular genetics supplied a comprehensive and seemingly almost definitive "explanation" for virtually the whole range of biological phenomena: not only did it eventually provide the single most persuasive argument against any special status of living systems, but its astonishing successes in themselves further validated the method.

The parallel history of the conceptualization of embryology presents some extraordinary contrasts. Embryonic development was prominent among phenomena originally invoked as evidence for vitalism. But, given the existing limitations of knowledge, such arguments could have been based on little more than mere assertion or an appeal to deliberately selected properties of biological systems on grounds of supposed incompatibility with, or inexplicability by, physico-chemical reduction. The later, equally influential, concept of organicism was difficult to disentangle from vitalism and the reasoning behind it. It is hard to discern any clear, sustained line of advance in the discipline of embryology over the course of this century. As illustrated in Section 14.1, embryogenesis was overwhelmingly given over to explanations in terms of models.[32] To an extent that is not perhaps widely realized today, the direct translation of biological phenomena into the phenomenology of physics, often fully expressible mathematically, was regarded not only as feasible, but also as offering realistic solutions to the actual biological situation.[33,34] For much of this century, embryogenesis has, in one guise or another, been reduced directly to physical forces, a chemical gradient or pattern genes.

Thus we can trace the origins of all major elements of the currently predominant view of pattern formation to their treatment according to a specific tradition of scientific methodology, modelled particularly on the physical sciences. Physical sciences have long been the prime exemplars of the ideals of "scientific method", such as empiricism, materialism, mechanistic determinism and proof, and it can be argued that they have largely shaped our approaches to biological systems; not just directly in terms of laboratory techniques, but in the way they have bequeathed a particular type of scientific reasoning.

14.3.2 Two ways of thinking

We have seen how today's consensus view of pattern formation was gradually arrived at by means of a particular way of fitting data together. Beginning with a splitting off of specific categories of data (*e. g.* embryology from genetics),[35] these categories later came to be linked conceptually in a direct, one-to-one fashion, with consequences for the way each domain itself was ultimately considered (*e. g.* the notion of pattern genes or embryogenesis as controlled by a gradient).[36] Faced with higher order phenomena like embryogenesis, it is tempting to treat them like a black box, by reducing them to the simplest possible one-step transform to accommodate the known or assumed starting conditions and endpoints. But in such phenomena, the underlying complexity is hidden: we notice above all their integrated, orderly and predictable aspects and for this reason simple causal concepts and explanatory models often seem sufficient. But any such models or abstractions are bound to misrepresent complex biological phenomena (if only because, in the explication of one specific mechanism, others are, implicitly, ignored; and because too many alternative intervening mechanisms would be equally compatible with the model). The objective of abstract simplicity may well be appropriate to the physical sciences, but biology is, by its nature, specific and complex: diversity, variation and specialization are pre- conditions in evolution for the very existence of organisms.

Molecular reduction is the logical culmination of the single-minded application to biology of the above described traditions. In the most overt and literal sense, it is an approach which immediately satisfies the ideals of traditional scientific methodology. It is a method which caters perfectly to the perennial, but vain, quest for explanations of biological phenomena in terms of specifically identifiable, and preferably materially demonstrable, causes.

Here we have come to the central purpose of this paper, which has been to consider the effect on our understanding of biological systems of the terms in which we try to explain them. The account briefly sketched out in Section 14.2 highlights some of the problems arising from reduction. The result of this type of thinking is a distinctly narrow view of causality, which ignores much of the data available: the extent of the misrepresentation of the actual biological phenomena can be judged by comparing the two accounts of pattern formation. Virtually all aspects of the position presented in Section 14.1 need revision in the light of further considerations. The problem is not just reduction as such, but the ways of thinking that go with it. Very gradually, and almost imperceptibly, our criteria for a satisfactory explanation for biological phenomena may be shifting. Because molecular explanations are seen as ultimate and entirely explicit, they are easily taken to be the ideal, and it sometimes seems as though, against this background, higher order features of biological systems have become suspect and in some way illegitimate as evidence. Yet, even if reduction to molecular levels is the primary explanatory strategy and, by implication, alone provides a sufficient explanation, it is ultimately only valid and useful as an explanation if higher order phenomena can be linked to it: as we have seen, reduction pure and simple inevitably implies a one-to-one relationship and with it the logic of preformationism. The need for a different way of thinking in biology is not just theoretical. I have tried to show how it matters in practice; reductionism can lead to biologically quite unrealistic conclusions.

A prime feature of the approach outlined in Section 14.2 is the insistence that all the data must be covered and that higher order phenomena must be regarded as equal to molecular data in their validity as objective and demonstrable facts admissible as evidence. Morphogenesis, induction, the developmental cascade and the integration of morphology, for example, are not just abstractions. They are just as real as phenomena as DNA and they are essential contributors to the full explanatory network. Higher order data not only reveal otherwise unsuspected properties of the complex biological system, but they set limits on, and put into context, the fine detail of molecular data.

A second feature is the demonstration of the unavoidable complexity of any "explanation" of any aspect of such a fundamentally multicausal situation as is represented in biological systems. Once this is recognized, it is possible to cross between levels[37] by moving through an explanatory path which parallels the causality of the complex, cumulative and fundamentally continuous[38] properties of the system itself. The way in which embryological phenomena were used in Section 14.2 to establish linkages with adult form, cell biology, genetics and evolution illustrates this methodology. In the notion of the cumulative properties of biological systems, as exemplified by the developmental cascade, a constructional approach to the data - to contrast with reduction - can be taken.[39] This exercise is not equal and opposite to reduction: reduction to DNA sequences entails a totally explicit specification of explanatory data, but a symmetrical and equally complete reconstitution of reduced data is out of the question, given the complexity of the causal system. Nonetheless, the cascade of embryogenesis can be traced out "in principle", supported by ample illustrative examples and the internal consistency of all the data as a whole. At all stages a dual methodology is required; the use of bridging concepts which are generalizations from relevant data, alongside reference to available specific illustrative examples in the raw data.

These requirements for a way of thinking appropriate to biology inevitably follow from the nature of evolution. In approaching a body of data which is the result of a process as infinitely variable and complex as evolution, abstractions and generalizations soon become detached from reality; the complexity, specifics and actualities of biology are all- important and the key to explanation is to keep as close to the data as possible, which means embracing all the data available non-selectively. In the linking up of data by explanatory - *i. e.* predominantly causal - concepts one must be wary of one-to-one assumptions or one-step rules of translation (*i. e.* "theory reduction" between levels),[37] and indeed of all simplifications and abstractions (as built in to biological "laws", rigid categorizations of data or the remarkably common all-or-none or dichotomous uses of biological terms and concepts);[35,40] all of these conceptual devices are liable to misrepresent the multicausal complexity which evolution actually entails. Linking concepts must strike a careful balance between specificity and generality and - far from the ideals of theory reduction[37] - explanatory schemes must potentially mirror the actual complex causal network.

Given the potentially limitless nature of biological explanations, a biological way of thinking must always involve approximations, limitations and uncertainties. Features like receptivity, potentiality, induction, morphogenesis or functional modulation of morphology are unlikely to be precisely definable in cellular and molecular terms. But, given the increasing comprehensiveness (and unity)[38] of biological knowledge, it is possible to outline, in principle, the appropriate terms in which they become perfectly explicable, and indeed it is becoming harder to understand embryology in any other terms. Just because an explanatory scheme is complex or imprecise, this does not mean that it does not explain; one could ill-afford to throw out the idea of evolution just because one cannot prove, observe or fully define it. In the end the wealth, diversity and specificity of the biological data is itself the basis for confidence about the prospects of improved methods. Such data can only be explained by an explanatory network which is itself unique and specific: given the appropriate ways of linkage, all types and levels of data can be accounted for in an increasingly mutually supporting manner. Increasingly then, as the properties of intermediate levels (*e. g.* the integrated genome or the whole cell) are better understood and consolidated, these can be assumed and generalized, and, in turn, used as components in the explanation of yet higher levels.

14.3.3   Conclusions

What of the innocent question with which we began this paper: what finally can we conclude about the fundamental matter of how best to "explain" biological systems? It is now possible to see how the choice between egg and chicken exemplifies the many simplistic splits and false dichotomies which beset biological thinking. The egg is best regarded as just an integral part of a much wider concept of a chicken. Indeed, when seen in evolutionary context, neither egg nor chicken has any fixed or independent meaning. With respect to the wider issue of the alternative ways of thinking by which to explain biology, it is not a question of either/or. Important as they are, nucleotide sequences are not enough. But neither is there any outright alternative. Within the broader way of thinking presented above, molecular reduction plays a crucial, though only limited, part. The most basic methodological point that this paper attempts to address is the way in which these alternative approaches can be fitted together as contributions to the most comprehensive "explanations" in biology.

With the pre-eminence of ideals of scientific method traditionally modelled on the physical sciences, alternatives applicable to biology, though often suggested,[32,33,39] have uniformly failed to achieve any accepted status. And, it must be said, the case made for alternatives has often been woefully inadequate.[38] But it is possible that arguments, approaches and aims that were entirely appropriate in the past have, for reasons of historical and philosophical inertia, persisted well after the original concerns have disappeared. For philosophers of biology, as primary guardians of "scientific methodology", the long running theme of whether biology is distinct (in terms of reducibility, teleology or lawfulness, etc) from physical sciences is still a leading preoccupation.[37] But the issue of physico-chemical reducibility as such has long since been won and the problems for both biologists (and philosophers of biology)[42] have moved on, above all perhaps to the intrinsically difficult matter of methods of analysis for tackling high level biological organization and of how to meet the challenge of the single, most characteristic feature of biology, complexity!

The possibility must be faced that the dominance of the reductionist method might actually handicap our full understanding of biology. My argument has been that there are very significant limitations to a purely reductionist methodology. There are certainly major limits in the extent to which the human observer can comprehend or envisage the true complexity of events at the DNA level (*e. g.* cooperation across the whole genome, its control and the interactions of all the functions of its products: the impossibility of linking DNA to higher level phenomena (*i. e.* Mendelian genetics) has been particularly well demonstrated),[37] let alone the extent to which the information can be reconstituted in order to account for all the phenomena of embryogenesis. As an explanatory logic, molecular reduction is inevitably narrowing; in its fundamental tendency towards becoming merely a description based solely on structure (*i. e.* DNA sequences), it cuts out functional and multicausal considerations. Even if it provides full explanations in material terms, this is still only a partial explanation; the claim that an organism is explained by DNA expresses only a formal truth. Molecular biology as such will never fully explain embryology, behaviour or phylogeny.

Using pattern formation as an illustrative example of a characteristically complex biological phenomenon, this paper has argued for the necessity of a way of thinking very different from reduction. Although the issues apply throughout biology (*e. g.* to neuroscience, phylogenesis, behaviour),[43] pattern formation is an especially relevant case in point: it occupies a particularly crucial middle ground between many other biological disciplines and much depends on the conceptual terms in which this key area of uncertainty is bridged. Most importantly, it is also amenable to solution. By comparison with molecular reduction, the alternative method must inevitably remain imprecise. But today we have one immense advantage over earlier biologists: we have reached a point, in terms of the relative comprehensiveness of our knowledge of biological systems on all levels, where it is now possible to establish a reasoned case for the legitimacy of higher level phenomena as integral components of explanatory networks and of conceptual frameworks which cover all this data, because these claims are so well supported by the massive internal consistency of the evidence.

Reductionist thinking only becomes a problem when it is used as if it were alone adequate to provide full biological explanations. Even though we may take it that biological systems are entirely accounted for by their physical and chemical constituents, in practice - simply because of the complexity of biology - analysis in these terms comes up against insuperable limits. Even though it is clear that in terms of scientific reasoning or procedures there can in theory be no difference between biology and other scientific disciplines, in practice the way towards full explanations in biology involves relinquishing of the safety of ideal methods and accepting the relative uncertainties of a more appropriate methodology, which must embrace both ways of thinking described here. If "explanation" (and therefore "understanding") is the ability to fit data together coherently, according particularly to causal relationships,[44] then the case of pattern formation offers an unrivalled demonstration of the requirements. As I have tried to show, an enlarged methodology, additional to reduction, is imperative given the nature of biology.

NOTES AND REFERENCES

1. Butler, S. 1878, *Life and Habit*. London: Trubner.

2. Preformation dominated eighteenth century thinking about the origins of biological systems (Hartsoeker's drawing dates from 1694), chiefly due to the overturning of the notion of spontaneous generation, on which the earlier idea of "epigenesis" had depended (see; Needham, J., 1934, *A History of Embryology*, Cambridge: Cambridge University Press). The idea of an infinite succession of preformed germs implied an original ("special") creation and later invariance of species; the word "evolution" indeed originally meant expression of the preformed germ: Bowler, P.J., 1975, The changing meaning of "evolution", *J. Hist. Ideas* 36: 95-114.

3. The "provisional theory of pangenesis" in Darwin, C., 1868, *The Variation of Animals and Plants under Domestication*, London: Murray.

4. It is difficult for us now to put ourselves back in Darwin's conceptual framework, in which heredity and embryonic development were in effect indistinguishable. The embryo as such was poorly understood, even descriptively. His contemporary, Haeckel, considered embryonic stages (seen as recapitulating phylogenetic stages) to be the actual manifestations and vehicles of heredity. Against this view, His argued that embryogenesis must be explained in terms of physical causes at the time of development.

5. Weismann, A., 1893, *The Germ-Plasm. A Theory of Heredity*, London: Walter Scott. His theory was originally proposed in 1885.

6. Weismann hypothesized a process of assignment of specific nuclear formative factors to an organized pattern in the egg cytoplasm. Hertwig and Wilson, whose writings vividly illustrate the struggle to clarify the domains of genetics and embryology at the time, recognized clearly that Weismann's theory implied preformed egg organization and that it did not explain embryonic regulation (first shown by Driesch in 1891) or regeneration. Hertwig, O., 1896, *The Biological Problem of To-day*, London: Heinemann; Wilson, E.B., 1896, *The Cell in Development and Inheritance*, London: MacMillan.

7. Thompson, D'A. W., 1917, *On Growth and Form*, Cambridge: Cambridge University Press. D'Arcy Thompson's influence is notoriously difficult to characterize, but is probably considerable (see; Medawar, P.B. in Thompson, R.D'A., 1958, *D'Arcy Wentworth Thompson. The Scholar-Naturalist, 1860-1948*, London: Oxford University Press). Like His, his aim was "The study of organic Form by methods which are the common-place of physical science. It is the skilled and learned mathematician who must ultimately deal with such problems" (Preface). He was unconcerned with genetics or with embryology as such. As theories of development or evolution, modern versions of his approach (*e. g.* approaches based on the concepts of allometry and heterochrony) are very limited: they fail, for example, to explain changes in number or differentiation of body parts (see; Horder, T.J. 1989. Syllabus for an embryological synthesis. In *Complex Organismal Functions; Integration and Evolution in Vertebrates.* Wake, D.B. and Roth, G., eds, pp 315-348. Chichester: Wiley).

8. "Neo-Darwinism" or "the modern evolutionary synthesis" of the 1930s and 1940s was primarily aimed at bridging the divide between genetics and evolution. Gradualistic evolution and adaptation were shown to be compatible (through population genetics) with the discrete effects of genes and mutations. The aim was achieved at the cost of considerable abstraction and detachment from specific data, particularly concerning phylogenesis, the specific forms of organisms and the actual phenomenology of embryogenesis. Changes in adult morphologies were seen as the results of gradualistic "growth" changes, using concepts like allometry, in turn linked to genetics in concepts such as "rate genes" and statistical gene combinations (Wright's "shifting balance"). For analysis of methodological limitations see; Mayr, E., 1959, Where are we? *Cold Spring Harbor Symp. Quant. Biol.* 24: 1-14; Horder (1989) (op cit).

9. Regeneration, metaplasia and embryonic regulation all demonstrate the retention (and potential expressibility) of the entire genome in most cells throughout the life cycle, later confirmed by nuclear transplantation.

10. Figure 14.8 (a) Boveri, T., 1910, Die Potenzen der *Ascaris*-Blastomeren bei abgeanderter Furchung: Zugleich ein Beitrag zur Frage qualitativ ungleicher Chromosomenteilung, Festschr. R. Hertwig, Jena. III; 133-214. (b) Harrison, R.G., 1945, Relations of symmetry in the developing embryo, *Trans. Conn. Acad. Arts Sci.* 36: 277-330. (c) Spemann, H., 1938, *Embryonic Development and Induction*, New Haven: Yale University Press. (d) Waddington, C.H., 1940, *Organisers and Genes*, Cambridge: Cambridge University Press. Waddington, C.H., 1957, *The Strategy of the Genes*, London: Allen and Unwin.

    Figure 14.9; Derived from Wolpert, L., 1974, "The Development of Pattern and Form in Animals", *Oxford Biology Reader No 51*. Oxford: Oxford University Press.

    Figure 14.10; For details see; Lawrence, P.A., 1992, *The Making of a Fly. The Genetics of Animal Design*, Oxford: Blackwell Scientific Publications.

11. On the positional information model, cells must have morphogen receptors capable of discriminating levels to a degree equal to the number of morphological positions at which cells undergo alternative forms of differentiation. A particular difficulty for locus-specific control mechanisms concerns the differentiation of identical structures at multiple, different locations: on the positional information model, specificity would still be needed at each locus even if the same cell type differentiates (as acknowledged in the associated concept of "non-equivalence"). This independence of locus receptivity and differentiation makes it

difficult to see how evolution could change pattern, since two, matching mutations serving the two functions would always be needed simultaneously.

12. Lewis, E.B., 1978, A gene complex controlling segmentation in *Drosophila. Nature* 276: 565-570.

13. "It has become almost a truism to assume that the position of a gene, either on a chromosome or else within a nucleus, is of critical importance to its functional capabilities". John, B. and Miklos, G., 1988, *The Eukaryote Genome in Development and Evolution*, London: Allen and Unwin, p193.

14. The concept of master genes - hypothesized as functioning to control and integrate groups of genes mediating pattern or mediating the differentiation of specific cell types (see; Davidson, E.H. and Britten, R.J., 1979, Regulation of gene expression: possible role of repetitive sequences. *Science* 204: 1052-1059; MacIntyre, R.J., 1982, Regulatory genes and adaptation: past, present, and future, *Evol. Biol.*, 15: 247-285) - carries preformationist implications. Such overall control elements may be unnecessary given that control might be explained as the result of a cascade, whereby the products of one structural gene are used as signals for another, and so on in a chain reaction.

15. Although maternally imposed patterning in the egg cytoplasm is well established, this mechanism has occupied a diminishing place in developmental theories. Adequate testing increasingly reveals other mechanisms (*e. g.* see; Lambie, E.J. and Kimble, J., 1991, Genetic control of cell interactions in nematode development. *Ann. Rev. Genet.* 25: 411-436). Most maternal factors are involved in forming the egg cell as such and their role in pattern control must be restricted if they are not to limit zygotic variation and rates of possible future evolution. Compared to other patterning mechanisms, their role is always very limited and is minimal in vertebrates.

Hidden forms of preformationist logic are often to be found in developmental biology: in many concepts (*e. g.* prepattern, instructive induction, models of development based on the clonal history of cells), and also in many attempts to explain pattern in terms of specific molecular factors (*e. g.* growth factors or extracellular matrix) patterning is already assumed in the explanatory mechanism and has not itself been explained.

16. Examples of functional modulation of morphology are reviewed in Goss, R.J., 1978, *The Physiology of Growth*, New York: Academic Press; Horder, T.J., 1983, Embryological bases of evolution. In *Development and Evolution*, Goodwin, B.C., Holder, N. and Wylie, C.C., eds, pp 315-352. Cambridge: Cambridge University Press; Holmes, S.J., 1948, *Organic Form and related Biological Problems*, Berkeley: University of California Press. Why such evidence has not generally featured in theories of development is unclear; apart from the Lamarckian overtones and the difficulties of defining the underlying genetic and cellular mechanisms (in common with many developmental phenomena),[20] there is perhaps a tendency to think of "environmental factors" as external to the organism: *i. e.* to disregard "internal environmental" factors.

17. Darnell, J., Lodish, H. and Baltimore, D., 1990, *Molecular Cell Biology*, New York: Scientific American Books.

18. Three types of cause, commonly combined in biological phenomena, can loosely be distinguished; the source of the specificity of the factor or function under consideration (an "instructive" cause); the conditions needed to trigger or regulate events ("elective"); and

other preconditions necessary for events, but not directly contributing to them ("faculta-tive", "permissive").

19. There are many contending definitions of the term "genetic" (Kitcher, P., 1982, Genes *Brit. J. Phil. Sci.* 33; 337-359). Even the nearest to an ultimate definition - the description of a specific DNA sequence - involves ambiguities (*e. g.* whether to include variations between individuals, introns, initiating and control regions). The concept of "genetic" factors is not just difficult to define: it is ultimately meaningless, because it cannot be separated from "environmental" factors. The two types of consideration will always interact, because no genetic effects can occur unless they are expressed and expression depends on factors extrinsic to the DNA sequence itself. What is more, there is no complete way of determining the extent of involvement of either consideration, given the complexity of their interactions and the practical impossibility of testing all possible environmental variants. Despite the importance and frequency of the appeals to a supposed separability (as in "nature/nurture" discussions), these points are rarely made and the literature on the subject is extraordinarily restricted. For the best available attempts at clarification see: Oyama, S., 1985, *The Ontogeny of Information. Developmental Systems and Evolution*, Cambridge: Cambridge University Press; Rose, S., Kamin, L.J. and Lewontin, R.C., 1984, *Not in Our Genes*, Harmondsworth: Penguin Books. For an important critique of geneticism see; Tauber, A.I. and Sarkar, S., 1992, The human genome project: has blind reductionism gone too far? *Perspect. Biol. Med.* 35: 220-235.

The gene concept has always referred primarily to a structural entity. In practice "identify-ing the gene for a function" means localizing an involved chromosomal site, or sequencing it or its products. These operations are usually merely handles for further practical pro-cedures: the specificity of the products and their functions are usually in themselves of secondary importance. Compared to their structure, the functions of genes, via their prod-ucts, are difficult to investigate and can never be delimited with certainty. Despite the problems, the search for genetic explanations continues apace; Owen, M. and McGuffin, P., 1992, The molecular genetics of schizophrenia, *Brit. Med. J.* 305: 664-665; Alper, J.S. and Natowicz, M.R., 1992, The allure of genetic explanations, *Brit. Med. J.* 305: 666.

20. Figure 14.11; adapted from Patten, B.M., 1946, *Human Embryology*, London: Churchill. The phenomenon of induction is massively documented and Figure 14.11 summarises many of the known cases. See also; Nakamura, O. and Toivonen, S., 1978, *Organizer - a milestone of a half-century from Spemann*, Elsevier: Amsterdam. Regulation, coordination and precision (potentially to the level of individual cells) of adult pattern (the main reasons for hypothesizing gradients) are explicable by the developmental cascade, as are varieties of distributions and combinations of differentiated cell types. Induction has been a focus for much dispute, largely due to methodological problems (*e. g.* multi-causality due to double assurance, involvement of competence, reciprocal interactions between inductor and induced tissues) inherent in the analysis of any higher level embryological phenomena. For reviews of these problems and the role of morphogenetic movements and receptivity to inductors see; Horder (1983) (op cit); Horder, T.J., 1976, Pattern formation in animal development. In *The Developmental Biology of Plants and Animals*, Graham, C.F. and Wareing, P.F., eds, pp 169-197. Oxford: Blackwell Scientific Publications.

21. Induction and positional information are opposing concepts. In induction the control of one cell's fate is locally determined; only "position" in relation to neighbouring cells is relevant and it is unnecessary that a cell has information about its position in the embryo as a whole. For further discussion see; Hinchliffe and Horder (this volume).

22. The genetic analysis of *Drosophila* development (Figure 14.10) illustrates some of the limitations of genetic methodology in general. It is essentially descriptive. Increasingly, it is becoming clear that a full explanation of development requires more than just the identification of genes or description of their products and their spatio-temporal patterns of expression; it also requires explanation of the distributions and interactions of these products, since it is these considerations which determine succeeding gene switches. It is likely that any one switching factor (*e. g.* the *bicoid* gradient) is only responsible for a limited number (around three) of spatially distinct patterns as expressed in the next round of gene switches. Pattern is therefore actually built up gradually (*i. e.* involving multiple interactive, dependent stages - the gene product of one stage having free, direct access to the promoters of other genes which are then selectively switched[14] - so that final adult differentiation is not the result of direct one-to- one read out of initial position, *e. g.* as defined by the *bicoid* gradient; interactions include induction and morphogenesis (Horder, 1983, op cit)). Early embryogenesis and mutations affecting segmentation in *Drosophila* may be poor models for other organisms; *e. g. Drosophila* embryos do not regulate (see; Sander in Goodwin et al. (1983) (op cit)) and equivalent ("homeotic") mutations are unknown in vertebrates (Horder, 1976;1983, op cit).

23. See Horder (1989) (op cit), which includes arguments against the notion of pattern (*e. g.* limb pattern) genes. There is a wide spectrum of possible relationships between genes and their ranges of effect morphologically; *e. g.* genes may be locus or organ-specific, cell-type specific (with mutations often leading to patchy effects) or may affect metabolism (with mutations having diffuse effects). In all cases pleiotropy may be due to knock-on effects, secondary to the primary gene action. There are major limits to our ability to analyse the genetic programming of pattern; *e. g.* due to incomplete identification of genes or mutations and difficulties in separating component genes in situations of polygenetic control.

24. On the evolutionary flexibility of chromosomal organization see; Berry, R.J., 1977, *Inheritance and Natural History*, London: Collins; Wasserman, M., and Wasserman, F., 1992, Inversion polymorphism in island species of *Drosophila*, *Evol. Biol.* 26: 351-381; O'Brien, S.J., and Seuanez, H.N., 1988, Mammalian genome organization: an evolutionary view, *Ann. Rev. Genetics* 22: 323-351; John and Miklos (1988) (op cit). As an example of the dispersed mapping of protein molecules originating by duplication from one ancestral form, see; Wilkie, T.M., Gilbert, D.J., Olsen. S.A., Chen, X-N., Amatruda, T.T., Korenberg, J.R., Trask., B.J., de Jong, P., Reed, R.R., Simon, M.I., Jenkins, N.A. and Copeland, N.G., 1992, Evolution of the mammalian G protein alpha-subunit multigene family, *Nature Genetics*, 1: 85-91. Haemoglobin sub-units illustrate dispersed mapping of a single final molecule: see also; Wissinger, B., Schuster, W., and Brennicke, A., 1991, Trans splicing in *Oenotheran* mitochondria: *nad1* mRNAs are edited in exon and *trans*-splicing Group II intron sequences, *Cell* 65: 473-482. Bacteria may not be good models here; they typically show integrated polycistronic genome control and very stable gene sequences in evolution; Riley, M. and Krawiec, S., 1987, Genome organization. In *Escherichia coli and Salmonella typhimurium: Cellular and Molecular Biology*, Neidhardt, F.C., Ingraham, J.L., Low, K.B., Magasanik, B., Schaechter, M. and Umbarger, H.E. eds. pp 967- 981. Washington: American Society of Microbiology.

In general gene organization, in higher organisms, tends towards random order. Given appropriate control mechanisms[14,22] this is entirely compatible with integrated function. Many special factors may explain instances where particular gene sequences show close relationships; *e. g.* recent evolution by gene duplication, "supergenes", "position effects", linkage disequilibrium, "gene complexes". On the lack of effect of dissociating neighbouring genes see: Struhl, G., 1984, Splitting the bithorax complex in *Drosophila, Nature* 308: 454-7; John and Miklos (1988) (op cit).

25. See; Horder, T.J., 1991, Molecular biology and evolution: two perspectives. In *Developmental Patterning of the Vertebrate Limb*, Hinchliffe, J.R., Hurle, J.M. and Summerbell, D. eds, pp 423-438. New York: Plenum; Horder, T.J., 1993, Three glimpses of evolution. In *Formation and Regeneration of Nerve Connections*, Sharma, S.C. and Fawcett, J.W., eds. pp 222- 238. Boston: Birkhauser.

26. Nei, M. and Koehn, R.K., 1983, *Evolution of Genes and Proteins*, Sunderland, MA: Sinauer; Li, W-H. and Graur, D., 1991, *Fundamentals of Molecular Evolution*, Sunderland, MA: Sinauer.

27. "Causality" in the organism may sometimes come to differ widely from the causal chain of events in evolution; *e. g.* in the case of DNA as the "cause" of RNA and in turn proteins, it seems likely that the evolution of RNA preceded that of DNA (see; Darnell et al. (1990) (op cit), Chapter 26). However, other characters (such as the DNA genetic code itself) are so constant and universal among organisms that we can confidently infer their early origin and subsequent evolutionary inflexibility. Sequences of developmental events in extant organisms often provide direct evidence on which evolutionary inferences can be based and sometimes "recapitulate" or retain the record of the evolutionary sequences of events (Horder, 1993, op cit).

28. Recombination is evidently almost as basic a property of DNA sequences as the DNA code itself. Maintaining the "integration" of a genome tending towards random order becomes a matter, not only of the selective advantages, matching and compatibility of structural gene products, but also of the regulation of the units of DNA rearrangement (Plasterk, R.H.A., 1992, Genetic switches: mechanisms and function, *Trends in Genetics*, 8: 403-406) and expression control,[11,14] by other, nonexpressed DNA components.

29. It is easy to think of natural selection as only operating directly on the adult; however, despite its protected circumstances, the embryo is indirectly under even greater selective pressures because multiple adult characters depend on each developmental step; hence the relative evolutionary stability of developmental processes (Horder, 1993, op cit).

30. The available terminology is inadequate;[39] the term "epigenesis" comes closest to covering the concept we have attempted to characterize, but is unsatisfactory because of its connotations as the antithesis of preformation and of genetics.

31. Roll-Hansen, N., 1978, *Drosophila* genetics: a reductionist research program, *J. Hist. Biol.* 11: 159-210.

32. On vitalism, physico-chemical reduction and the origin and use of "models", see; Hall, T.S., 1969, *History of General Physiology*, Chicago: University of Chicago Press; Oppenheimer, J.M., 1967, *Essays in the History of Embryology and Biology*, Cambridge: M.I.T. Press:; Haraway, D.J., 1976, *Crystals, Fabrics, and Fields. Metaphors of Organicism in Twentieth-Century Developmental Biology*, New Haven: Yale University Press; Oyama (1985) (op cit).

33. On the origin of embryological concepts, see; Needham (1934) (op cit); Holmes (1948) (op cit); Oppenheimer (1967) (op cit); Haraway (1976) (op cit); Horder, T.J., Witkowski, J.A. and Wylie, C.C., 1985, *A History of Embryology*, Cambridge: Cambridge University Press; Gilbert, S.F., 1991 *A Conceptual History of Modern Embryology*, New York: Plenum. In this area of biology - notable for its importation of ex-physical scientists, *e. g.* Turing, Schrodinger, Whitehead, Weiss, Wolpert - many concepts have been borrowed from the physical sciences; *e. g.* field, polarity, system, double assurance. On the coextensivity of thinking in physical and biological sciences see, for example; Stebbing, L.S., 1937, *Philosophy and the Physicists*, Harmondsworth: Penguin; Hull, D.L., 1974, *Philosophy of Biological Science*, Englewood Cliffs: Prentice-Hall; Yoxen, E.J., 1979, Where does Schroedinger's "What is Life?" belong in the history of molecular biology? *Hist. Science*, 17: 17-52. As case histories, see; Whyte, L.L., 1963, *Focus and Diversions*, London: Cresset Press; Fischer, E.P. and Lipson, C., 1988, *Thinking about Science. Max Delbruck and the Origin of Molecular Biology*, New York: Norton.

34. "The purpose of this paper is to discuss a possible mechanism by which the genes of a zygote may determine the anatomical structure of the resulting organism. The theory does not make any new hypotheses; it merely suggests that certain well-known physical laws are sufficient to account for many of the facts". Turing A.M., 1952, The chemical basis of morphogenesis, *Phil. Trans Roy. Soc.* 237B: 37-72.

35. Problems of biological explanation often revolve around a conceptual dichotomy or antithesis, both elements of which very often turn out to be interrelated and equally relevant in the end (*e. g.* nature/nurture)[19] or the bridging of the domains of evolution and genetics (Horder, 1989, op cit). Such disjunctions often originate in the way in which data are initially categorized and classified; *e. g.* the distinguishing of genetics and embryology (as described above), pre-functional and functional phases of development (Holmes, 1948, op cit) or morphogenesis from pattern formation and differentiation (see, Waddington, 1957, op cit; Wolpert, 1974, op cit). Many biological concepts are used in a disjunctive or all-or-none manner (*e. g.* homology, species, phases of life cycle, genotype/phenotype, nature/nurture) leading to typological and saltationist thinking, which can often be seen to create artefacts which misrepresent the underlying continuities of biological processes.[38]

36. It could be argued that in practice what has happened is that, caught between the far more coherent and well-developed fields of genetics and evolution theory, embryology has been conceptualized in conformity with assumptions and expectations derived from other fields.

37. On the theory of "reducibility of theories" and transitions between levels, see; Nagel, E., 1961, *The Structure of Science*, London: Routledge and Kegan Paul. It has become increasingly recognized that reduction cannot be fully realized in practice, even in the paradigm case of translating between Mendel's laws and molecular genetics: the rules of translation cannot themselves be closely enough specified (see; Rosenberg, A., 1985, *The Structure of Biological Science*, Cambridge: Cambridge University Press; Hull (1974) (op cit); Hull, D.L., 1976, Informal aspects of theory reduction, *Boston Studies in the Phil. of Science*, 32: 653-670). Similar problems apply to axiomatization; see, Ruse, M., 1975, Woodger on genetics. A critical evaluation, *Acta Biotheoret.* 24: 1-13.

38. Given a large enough scale of view, biological functions and processes (with very few exceptions and in marked contrast to biological structures) can be seen as continuous; *e. g.* continuity of genome and cell organization across generations, morphogenesis, phases

of the life cycle, growth, morphological integration. The discrete manifestations of evolution immediately available to us (*e. g.* the individuality of organisms, gaps between known species, the quantal nature of mutations) encourage disjunctive thinking.[35] However, there must be limits to the size of unit steps of evolutionary change compatible with the survival of intermediate stages (particularly given the requirements for genome integration)[11,28] and the concept of the species as a gene pool containing a reservoir of possible variants implies a diffuse, statistical basis for change: the nature of embryogenesis provides strong grounds for inferring a close continuity of morphologies during phylogenesis (Horder, 1993, op cit).

The essential consistency of molecular genetic mechanisms across the vast evolutionary scale represented by known organisms is strong evidence for the universality of these fundamental features of the evolutionary process and for the underlying unity of all biological phenomena. All biological explanations ultimately go back to causes in unprovable evolutionary scenarios and the possibility of quite arbitrary hypotheses; however, given our mounting awareness of the unity and internal consistency of biological mechanisms, the gaps in our knowledge of the evolutionary record can increasingly be bridged (or at least explained).

39. The term "emergence" refers to the appearance of new (and, by implication, unpredictable) properties as the result of the combination of elements in a complex system (Nagel, 1961, op cit). As Nagel argues, whether new properties are "unpredicted" all depends on how well understood the contributing elements are in the first place: if the complexity of the system is adequately recognized the emergent properties become less surprising. A single complex molecule is often a sufficient and fully understandable explanation for high level physiological functions (*e. g.* haemoglobin). A type of thinking based on emergence underwrites many anti-reductionist claims about biological systems (Simpson, G.G., 1963, Biology and the nature of science, *Science* 139: 81-88; Polanyi, M. 1968, Life's irreducible structure, *Science* 160: 1308-1312; Weiss, P.A., 1971, *Hierarchically Organized Systems in Theory and Practice*, New York: Hafner).

40. On laws in biology, see Rosenberg (1985) (op cit). Due to the opportunistic nature of evolution, it is inevitable that there will be few broad generalizations in biology and that most will eventually have exceptions. Only at the level of chemistry (*i. e.* molecular biology) will laws (in the usual sense of absolute rules) apply. High level concepts such as evolution are more like generalizations than laws.

41. Fundamental to what I have been saying is the importance of an awareness of the procedures involved in the use of data and of the concepts needed to link them. Embryology has been dogged by the inadequacies of its heritage of concepts and terminology (*e. g.* epigenesis, organicism, holism, emergence, etc): suspicions regarding their imprecision, abstractness or merely metaphorical character, and of the theories built on them, were often justified.

42. That the priorities of philosophers of biology should lag behind those of biologists is perhaps inevitable given the philosophers' dependence on the state of already established scientific knowledge and their traditional affinities with the physical sciences; Hull, D.L., 1969, What philosophy of biology is not, *J. Hist. Biol.* 2: 241-268; Hull (1974) (op cit).

43. For a penetrating analysis of the requirements of scientific method in biology (as applied to psychology), see Kaplan, A., 1964, *The Conduct of Inquiry*, San Francisco: Chandler.

44. Nagel (1961, op cit) defines "explanation" as "systematized knowledge"; Kaplan (1964, op cit, p329) as "concatenated description... each element of what is being described shines, as it were, with light reflected from all the others".

15. SUPERCOMPUTER SIMULATION OF TURING STRUCTURES IN *DROSOPHILA* MORPHOGENESIS

Axel Hunding

Chemistry Department C116
H. C. Orsted Institute
University of Copenhagen
Universitetsparken 5
DK 2100 Copenhagen 0, Denmark

15.1 INTRODUCTION

Biological pattern formation is a process which so far has gone largely unexplained. Experimentally working biologists favour mechanisms based on morphogenetic gradients, in which the activation of a certain gene, in a well defined spatial region is seen as a response to a concentration gradient set up by another gene thus giving rise to positional information. Among theoreticians the pattern forming processes are believed to some extent to be dependent upon truly symmetry breaking processes known to be possible in nonlinear (biochemical) control systems. This latter approach is explored here to discuss the early morphogenesis in *Drosophila*. This system has emerged as one of the currently most important species for which detailed experimental data have accumulated to such an extent that a beginning is made in understanding the processes which govern early embryogenesis. A hierarchy of genes seems to control the initial transition from the egg to a segmented embryo. Regions of activity of specific genes and their proteins are now available, and this has caused some revision in previous models. In present models maternal genes *bicoid* and *nanos* are believed to control the activation of the gap gene *hunchback*, and *hb* then controls the formation of the remaining gap genes *Krüppel, knirps* and *giant*. In the next level of the hierarchy the primary pair-rule genes appear. These genes are each expressed in a series of 7 stripes. The mechanism for the formation of these 'zebra' stripes is unknown. Activation by a combination of maternal and gap genes seems to be involved in the expression of particular stripes. It is thus believed that a mechanism similar to the one which activates the gap gene level could also operate on the primary pair-rule gene level. This requires a sufficient number of gradients provided by the gap level to define at least 7 distinct stripes and it is not clear that the gap and maternal genes could provide sufficient information for this. How a number of independent particular stripe generators (cues) could cooperate to form the observed equally spaced stripes is a much more sinister problem. Theoreticians have pointed out that the 'zebra' stripes alternatively may be generated by a truly symmetry breaking mechanism such as Turing's mechanism, that is, by an autocatalytic reaction-diffusion system which is known to be capable of producing such stripes autonomously. The particular pair-rule stripes could then be activated by a combination of maternal, gap and Turing pattern interactions.

Experimental and Theorietcal Advances in Biological Pattern Formation,
Edited by H.G. Othmer *et al.,* Plenum Press, New York, 1993

It is believed that most if not all the genes involved in the pattern formation processes in the gap and pair-rule levels are known. The interactions yielding the observed patterns are still a matter of much discussion. Biological experiments yield crucial information on these mechanisms but it seems that quite a number of seemingly direct effects of one gene upon another are indirect, that is, mediated through the combined effect of several gene interactions. In such a situation the control system should be treated as a whole which is possible in computer simulations of the combined interactions.

Recent reviews have appeared by Ingham (1988); Hülskamp & Tautz (1991); Pankratz & Jäckle (1990) and Nüsslein-Volhard (1991). References to Turing type models may be found in Hunding *et al.* (1990). For information on Turing structures in general see Turing's original work (1952), Nicolis & Prigogine (1977) and Murray (1989).

15.2   CURRENT MODELS FOR GENE CONTROL IN *DROSOPHILA*

An overview of the genes under discussion here will be given shortly. The egg contains *maternal genes*, and thus an asymmetry from the very start, as the anterior part of the egg contains the gene *bicoid* and consequently a gradient of *bcd* protein. Similarly the posterior part contains the gene *nanos* and presumably a similar gradient from this end in *nos* gene products. It is believed that *nos* represses *bcd*.

Maternal *hunchback* is present as well, and the function of *nos* is to suppress expression of *hb* in the posterior part. The gene *hb* is however activated in the eggs own DNA, and thus this zygotic *hb* places this gene among the first level in the hierarchy activated by the maternal genes: the gap genes. Activation of zygotic *hb* is due to *bcd*, and *hb* in turn activates the gap gene *Krüppel*. The result is an expression of *Kr* in the middle of the embryo. Finally gap gene *knirps* may be activated through *Kr*. Gene *hb* represses *kni* so the result is expression of *kni* on the posterior side of the *Kr* region with some overlap.

This is the current model for the gap gene expression. Although this model accounts for a number of experimental facts, it is known not to be a complete model. Several aspects of the gap gene level go unexplained with this model in its basic form. Thus zygotic *hb* is not only expressed anteriorly, where *bcd* suffices to be the activator, but it eventually becomes expressed in the posterior part as well. The reason for this activation is so far unknown. The gene *Kr* is expressed not only in a band in the middle of the embryo, but eventually also at both the anterior and posterior pole. Finally *kni* is not only expressed posterior to *Kr* but in the anterior part as well.

Recently it has been argued that the gene *giant* is a genuine gap gene and that its presence is essential to the expression of the other three. The gene is expressed in two broad bands on both sides of the central peak of *Kr*, but in mutants deficient in *bcd* the anterior band disappears. Similarly the posterior band is absent in *nos*-mutants (Capovilla *et al.* 1992). They argue that an alternative explanation of the activation of *kni* by *Kr* would be inhibition of *kni* by *gt* and inhibition of *giant* by *Kr*. At present the activation mechanism for *kni* is unknown.

Recently yet another aspect of the control mechanism at the gap level has been proposed by Struhl *et al.* (1992), who argue that *hb* has a key role in organizing gap gene expression in the posterior half. The expression of *gt* and *kni* is believed to be due to global activation (by a so far unknown factor), followed by repression of *gt* and *kni* by *hb*. The repression of *gt* is overruled by *bcd* anteriorly. The wild type maternal *hb* is sufficient to depress *gt* and activate *Kr*. Thus this work supports the view that, say, *hb* protein may be crucial even in regions of the embryo where current techniques barely are able to detect it, a

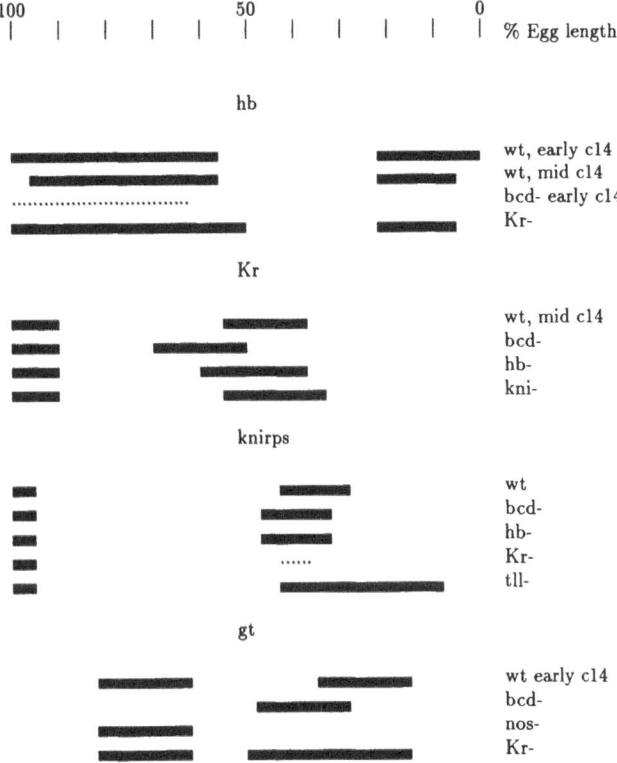

Figure 15.1. Experimentally established gap gene expression in the wildtype (wt) and mutant embryos. Expressions of *hb, Kr, kni* and *gt*. Anterior to the left (100 % egg length), posterior to the right. A weak expression only is indicated with ···. Gene *hb* occurs as maternal *hb* only in *bcd*- mutants. Mutants indicated with *hb*- means devoid of zygotic *hb*. The actual expressions are dynamic and thus change over nuclear cycle 14. Thus all data are approximate. Mutants not included usually means that little if anything changes. Anterior *gt* splits up in three regions at the time of cellularization (mid c14). Most patterns are three dimensional with smaller regions dorsally (up) than ventrally. *kni* has a broader anterior expression ventrally than shown.

feature which further complicates the use of available expression patterns for model building.

A collection of experimentally-recorded gene expressions in the wildtype (wt) and mutant embryos is reproduced in Figures 15.1 and 15.2.

The next level in the gene hierarchy is the primary pair-rule genes *hairy, runt* and *even-skipped*. After the gap genes start to be expressed, but before they reach quasistationary spatial positions, the primary pair-rule genes generate 7 zebra stripes in the middle of the embryo. Gene *hairy* is expressed also at the anterior tip. Gene *runt* is expressed between the stripes of *hairy*, whereas *eve* is slightly phase shifted from *hairy* to the posterior site. The zebra stripes were first seen in the gene *fushi tarazu* (or *ftz* for short) (Hafen *et al.* 1984). Subsequently it was inferred that the seven stripes were controlled as a single unit (Hiromi *et al.* 1985). This prompted speculations that the underlying mechanism was connected to Turing structures. However the *ftz* gene activation is now believed to be preceded by genes *hairy, runt* and *eve* for which reason these genes are known as the primary pair rule genes. Turing's mechanism for these genes is at present not the prevailing model after the discovery

Figure 15.2. Experimentally established pair-rule gene expressions at mid-cycle 14, shortly before cellularization. Weak expression indicated by .... In the wild-type (wt), *eve* and *hairy* are almost on top of each other. *ftz* and *runt* are roughly between these stripes, shifted posteriorly. Actual expressions are three-dimensional and more spread out ventrally, and the stripes do not emerge completely simultaneously. In *runt-* or *hairy-* the *eve* stripes are almost normal before cellularization. Afterwards the *eve* stripes start to merge somewhat in *runt-* indicating repression by *runt* in the wild-type.

that partial deletions in the *hairy* DNA caused particular *hairy* stripes to vanish. The discovery of these region specific *hairy* alleles (Howard *et al.* 1988) and the subsequent discovery that similar fragments of the *eve* gene expressed particular stripes (Goto *et al.* 1989) have given rise to the model most videly used at present. It is believed that gradients formed above in the gene hierarchy activate particular stripes. This idea has recently been exploited to give a detailed model for control of *eve* stripe 2. Interactions from *bcd, hb, gt, Kr* are shown to play an important role in activation of this stripe (Stanjevic *et al.* 1991). However if these kinds of interactions were the only ones necessary for generating specific stripes, this interpretation raises the very serious question how 7 individual stripes could be generated with *equal spacing*. It is inferred that *hairy* and *runt* depress each other and *eve* depresses *ftz*. Also *hairy* depresses *ftz* and *runt* depresses *eve*. Such interactions and evolutionary selection are held responsible for the eventual appearance of the final regular pattern.

This picture of specific cues for each stripe is not without its problems though. Such a mechanism would require a sufficient number of gradients from the gap gene level and certainly the three 'canonical' genes *hb, Kr* and *kni* are not enough. Addition of other genes from the gap class like *giant, hücklebein,*and *tailless* are needed. Moreover, the formation of stripes begins before the gap genes reach their final positions. One would expect to see the initial stripes move around until they found their final position, but this is not observed. Models of the refinement required to yield an equally spaced pattern by mutual interactions between primary pair-rule genes must meet the experimental observation that *hairy* yields some stripes in the absence of all the other primary pair-rule genes. The same holds for *eve*.

The pair-rule genes may not just respond to other pair-rule genes, but to some general, redundant striping mechanism which implies that the cues read by all pair-rule genes should be the same. Carroll & Vavra (1989) point out however, that the very different effects of *hb, kni* double mutants on *eve* and *hairy* indicate that the mechanisms for *eve* and *hairy* control from the gap level are widely different. This indicates that the two genes do not read the same cues from the gap level, but then it becomes a sinister question how they manage to resolve into 7 stripes which very nearly *overlap* in the wild type.

It is difficult however to see what common striping mechanism could be present if it does not have its origin in cues set up by the gap (and maternal) genes. A common striping mechanism for the pair-rule genes could be of Turing type, but then it should be sought in low molecular weight components which could diffuse rapidly enough. It has been pointed out by Lacalli & Harrison (1991) that many such factors have multiple essential functions such that any deficiency would be lethal before an effect on pattern formation was detectable, so they would not necessarily turn up in conventional screens for pattern genes.

15.3 SUPERCOMPUTER SIMULATION OF PATTERN FORMATION

The method of lines was used and thus the system of nonlinear partial differential equations was converted to a large system of ordinary differential equations by discretization of the Laplacian in three curvilinear coordinates. The resulting system is stiff and solved accordingly (modified Gear code). Indeed the use of 3 D elliptic cylinder coordinates yields some regions with comparatively very small mesh sizes. As the condition number of the problem is inversely proportional to $(\Delta x)^2$ and thus directly proportional to n^2, where n is the number of mesh points, this results in a problem which is substantially more stiff than simulations in a 3 D cube with equal mesh size over the region. Thus simulations of pattern formation in 3 D shapes like *Drosophila* embryos are much more time consuming with non-stiff codes than simulations with such codes on 2 D or 3 D regions with equal mesh size. A factor of 500 speed up is achieved with the stiff code relative to conventional non-stiff codes.

The Jacobian used in the corrector step is a sparse banded matrix which may be rearranged (chessboard numbering of mesh points) to yield large blocks within which the solution vector elements may be iterated in parallel (RBSOR method). This makes such algorithms suitable for modern high speed computers. Implementation on vector computers results in a huge speed up: the parallel RBSOR code runs efficiently and close to the top speed of machines like the CRAY X-MP, and on the Fujitsu/Amdahl VP1200 it yields 490 MFLOPS on this code, thus reducing the CPU time to a mere 6 minutes for 12 interacting substances simulated in 3 curvilinear coordinates. This makes the numerical study of three dimensional pattern formation possible and thus direct comparison to biological experiments feasible.

The actual rates of the chemical interactions involved in *Drosophila* are unknown. From the proposed interactions for a particular model it is possible however to write up matching rates. For example it is believed that *Kr* is activated by *hb*. A rate law comprising this feature could be

$$\frac{d(Kr)}{dt} = \frac{kz^n}{K + z^m} - k_1(Kr) \tag{15.1}$$

with z = *hb*. This is a Hill type rate law which would yield cooperative activation for small values of *hb* for $n > 1$. If one choses $n = 1$, no cooperative, S-shaped activation occurs and thus no threshold of the activation is imposed. Finally, for $m > n$, inhibition occurs for large values of z. It is important to have cooperative gene regulation to obtain activation or repression in sufficiently narrow regions. This is a feature of any model, based on cues or Turing structures.

In the recent paper by Struhl *et al.* (1992) they point out that the activation of the gap genes is very much an all or none situation. The gradient of *bcd* is demonstrably able to turn *hb* on in regions where *bcd* is only a factor of two greater than in regions in which no detectable activation occurs. The same holds for the activation of other gap genes by *hb*. This means that the activation is highly cooperative. A factor of two with a Hill constant of only 3 means only a factor $2^3 = 8$ difference in the rates. To account for the observed on/off properties seen experimentally it seems reasonable to assume rate differences of at least a factor 100 but this implies effective Hill constants in excess of 6. The implication is that gene activation has kinetics which is much more cooperative than is usually found in enzyme kinetics, where Hill constants rarely exceed 3.

At first sight it may seem that there are so many parameters in the combined kinetics that one may get anything out, but this is not so. The known experimental properties of say activation of *hb* by *bcd* is essentially a *spatial* feature and since the gradient of *bcd* protein has been measured the experimentally-recorded features of *hb* expression impose severe restrictions upon the available constants in the rate law for *hb*. With these settled one proceeds to *Kr* and again there are only limited freedom to choose rate constants if the simulated expression of *Kr* should come close in space to the experimentally recorded expression region. Thus the rate laws of the gap genes *hb, Kr, kni, gt* may be explored in succession and almost independently, which greatly facilitates the search for a sufficiently realistical model of their simultaneous expression in space and time.

Rate laws for the pair-rule genes following this model may be illustrated by *eve*:

$$\frac{d(eve)}{dt} = \frac{(A)^n}{1 + K_2(A)^n} \times \frac{1}{1 + K_3(\text{runt})^m} \times f(\text{gap,mat}). \tag{15.2}$$

The substance A is one of the components of the stripe defining Turing system, which is activating *eve* expression. Both *A* and *runt* repression are here taken to be cooperative with Hill constants n and m greater than two. The last term f(gap,mat) may be included to define activation or repression by cues defined by combinations of gap (or maternal) genes. For example one may insert a sum of such terms, one for each cue:

$$f(gap, mat) = \sum_i f_i(gap, mat). \tag{15.3}$$

The rate for *ftz* is analogous to that of *eve*, with *hairy* replacing *runt*, and an additional term for *eve* repression. The interaction with the Turing system is now through the second component B of this Turing system. As B is high where A is low and *vice versa*, this yields activation of *ftz* in stripes positioned between those of *eve*.

With respect to the choice of kinetics for the Turing system it is well known among physical chemists that many different models yield the *same* qualitative patterns. A Turing system of the first kind is defined as

$$\partial \mathbf{c}/\partial t = \mathbf{F}(\mathbf{c}) + \mathbf{D}\Delta\mathbf{c} \tag{15.4}$$

The Jacobian is defined as $\mathbf{J} = \partial\mathbf{F}/\partial\mathbf{c}$. There are only two classes of such Turing systems defined by the following two Jacobians

$$\mathbf{J} = \begin{pmatrix} + & - \\ + & - \end{pmatrix} (a) \qquad \mathbf{J} = \begin{pmatrix} - & - \\ + & + \end{pmatrix} (b) \tag{15.5}$$

The first class (a) is known as an activator-inhibitor system. The second class (b) has no obvious classification of activators or inhibitors. In class (a) the two substances A and B are coincident with respect to maxima: A is high where B is high and *vice versa*, contrary to some conventional wisdom. The second class (b) has high A where B is low etc. We have used a member of this class to simulate the low molecular weight components A and B to give alternating stripes. A particular such model is the one previously used by us, originating from Sel'kov:

$$\frac{\partial c_1}{\partial t} = \nu - \frac{k_1 c_1 c_2^\gamma}{1 + K c_2^\gamma} + D_1 \Delta c_1 \tag{15.6}$$

$$\frac{\partial c_2}{\partial t} = \frac{k_1 c_1 c_2^\gamma}{1 + K c_2^\gamma} - k_2 c_2 + D_2 \Delta c_2. \tag{15.7}$$

The influence from the maternal and gap systems on this system may be taken through a rate constant, say k_1. If the enzyme activity controlling this step is modified by the gene products from the maternal and gap levels, k_1 varies with position, and a Turing system of the second kind arises. In general the effective diffusion constants may change with position as well, and in general a system of the following form emerges

$$\partial \mathbf{c}/\partial t = \mathbf{F}(\mathbf{k}(\mathbf{r}), \mathbf{c}) + \nabla \cdot \mathbf{D}(\mathbf{r})\nabla\mathbf{c}. \tag{15.8}$$

The patterns which arise in Turing systems of the first and second kind are independent of the chemistry details, but not of the spatial dependencies. Also the patterns depend upon the boundary conditions imposed *i. e.*the shape of the embryo. Thus stripes may bend close to the anterior or posterior ends because they tend to be perpendicular to the local boundary. The independence of the geometry of the emerging patterns of the details of the chemistry is a result from bifurcation theory valid for small amplitudes of the arising pattern. (If the pattern builds up to large amplitudes in some regions and virtually zero in other regions, peculiar chemistry effects may arise though.) Essentially the present model is then a model of spatial coupling in a self-organizing system of the global field type.

15.4 DISCUSSION

The idea that gap genes may influence rate constants in a Turing system which generates stripes is common for recent models (Lacalli 1990; Hunding *et al.* 1990). The present model is an extension of previous Turing-based models as it takes into account actual gap gene expressions. Also, the current gap model(s) are substantially different from those used in previous models. Indeed the emergence of the new gap model(s) have made most previous zebra stripe models outdated.

The first actual implementation of a model which takes into account both maternal and gap gene control of zebra stripe formation, as presented here, may seem to confront a rather formidable number of possible interactions. Even though the model is confined to a Turing type mechanism only, the creation and stabilization of the underlying stripe pattern may at first sight involve a huge amount of possible interactions. If we assume that the 6 genes *bcd, nos, hb, Kr, gt, kni* may enter a rate constant in the Turing system in essentially 5 different manners (strong activation, moderate activation, absent, moderate inhibition, strong inhibition) we already have 5^6 or some 15000 different models. Even with a fast supercomputer code which could run each such model in 3 spatial dimensions to check for stable stripes, this would exhaust anyone's annual budget. The search for possible interactions is thus governed by a successive approach where the best model using maternal genes only is found first and this is then supplemented with the addition of gap genes. This reduces the search for possible combinations to less than a few hundred different combinations, which have been implemented. The most promising sets were taken out and tested for mutants, such as the effect of missing *Kr*, and the resulting patterns compared to experiments.

Here one faces a major problem however. If say *hb* is taken out it is not only the direct influence of missing *hb* on the rate in the Turing mechanism which appears in the computations. In the implemented gap level models *hb* has a profound effect on *Kr* as well, and it is thus the *combined* effect which emerges. This however makes the probing for reliable interactions on the zebra stripe level very much dependent on details in the implemented gap level interactions.

It is thus not possible at present to give a totally exhaustive analysis of possible interactions among the gap and zebra stripe levels. That is, the actual interactions may still hide in some untested combination. However the results won so far show that a Turing model is tenable as it provides reliable short wave length stripes, which bend near the poles and with expression in the region seen experimentally.

It should be clear that stabilization of zebra stripes is by no means trivial even with a Turing model, when the stripes as here are short wave length with respect to the characteristic length of the embryo. Thus unwanted more or less patchy patterns obtain as a rule if the stabilization of stripes from the maternal and gap level is ineffective.

Our main objective was to study zebra stripe formation and some new conclusions have emerged from extensive computer simulations. It is possible to stabilize Turing stripes with interactions from a few gap genes only, such as combinations of *hb, Kr* and *kni*. In simulations of mutants though, the removal of say *Kr* creates a haphazard Turing pattern over large parts of the embryo. This situation is not much changed if additional stabilization by *gt* is included for the short wave length Turing zebra stripe pattern. Although impressive stripes emerge in such wildtype simulations, mutants lacking one of the gap genes generate Turing patterns where the stripes break up in haphazard zones, at least in some parts of the embryo. Thus in all cases where the stabilization of zebra stripes were assumed mainly to originate from the gap level only, we did not see Turing patterns with fused stripes or missing stripes, as seen experimentally in gap mutants. Observe that such a conclusion is only possible if simulations are carried out in fully 3 spatial dimensions. In 1 dimension one always gets stripes, but in 2 dimensions patchy patterns may emerge. Plausible models based on 2 dimensional simulations however may easily create patchy patterns when simulated in fully 3 dimensions. Also we may note that some authors talk about stripes where they only obtain somewhat connected structures which are by no means equidistant stripes perpendicular to the long axis of the embryo. We call such connected (albeit none stripe) regions patchy patterns here, as they are unsuitable for generating reliably positioned zebra stripes in the embryo.

This then indicates that a possible Turing mechanism should explain experimentally recorded distorted zebra stripes in gap mutants not as altered Turing patterns, but rather with an almost intact Turing zebra stripe pattern which is then read out to the pair-rule level in a distorted manner due to the missing gap gene.

The stabilization of the Turing zebra stripes is thus mainly due to the gradients from the *bcd, nos* system. Slight addition of influence from the gap genes then has the role of enhancing the amplitude of the stripes in the central region. This model is robust towards gap gene mutants as the zebra stripe pattern is now basically undistorted.

This result may also have some relevance for segment formation in other less evolutionary advanced insects where the segments emerge sequentially. If opposing gradients exist in such embryos combined with a Turing system, they may define a narrow region at first, where the Turing system is read out with a large amplitude. Changing the position or magnitude of the posterior gradient may then gradually enhance the region in which the Turing system is read out with large amplitude. This results in a build up of peaks one after the other. The control system in *Drosophila* is thus not a totally different one, but a variant in which the gradients are sufficiently apart to allow a substantial number of stripes in the Turing system to emerge simultaneously.

What emerges is then a robust zebra stripe prepattern which is then converted to actual activated genes, quite possibly by specific combinations of gap genes acting as crude activators in their common regions. That is, the system of such cues is not the stripe *generator* but essential for activating pair-rule genes in specific regions over the Turing pattern.

Thus the model proposed on the basis of the present numerical study may be said to combine the two current alternative models, as it indicates that cues are necessary to explain the experimentally observed results on zebra stripes caused by gap mutants, but Turing stripes are necessary to provide a stable underlying stripe generator.

The present experimental state of knowledge of the proposed cue model does not specify control of more than a few of the stripes, and only in quite preliminary terms. The number of possible interactions is comparable to the above estimate (5^6) and thus impossible to exhaust in a theoretical study. However inspection of Figure 15.2 indicates some plausible global interactions which combine in a specific region to yield net activation (or repression) of response to the underlying Turing stripe generator, as proposed already by Lacalli (1990). The simulations of such simple cues have revealed however that the actual control must be more complicated. To illustrate the difficulties involved consider the following line of thought.

Figure 15.2 readily suggests that *kni* and *gt* activates *eve* read-out. A simple cue would be an additive expression ($kni + gt$), but deletion of say *kni* would *not* yield a term which reduces the *eve* rate sufficiently: the additive cue would only reduce the rate of *eve* to approximately one half of the wildtype rate, since *gt* is still present in the term. It is not feasible to reduce the influence from *gt* since this would yield a cue which is insensitive to *gt* deletions, and thus incapable of producing the reduction in *eve* rate required in *gt*- mutants. Thus to achieve the necessary deletions of *eve* expression a *combination* of *kni* and *gt* is necessary something like ($kni \times gt$) in its simplest form. This would reduce the *eve* rate appreciably in *kni*- and in *gt*- as well. The problem now arises when one continues to the influence from *Kr*. The much reduced level of *kni* in *Kr*- mutants means that in *Kr*- the level of *kni* is much lower as well. The experimental observation that *eve* expression in *Kr*- is present in the posterior part, where the proposed *kni, gt* cue is presumably now much reduced, indicates that absence of *Kr* yields overexpression of *eve* even in the absence of the *kni, gt* cue. This is difficult to envisage unless it is assumed that the still present although much

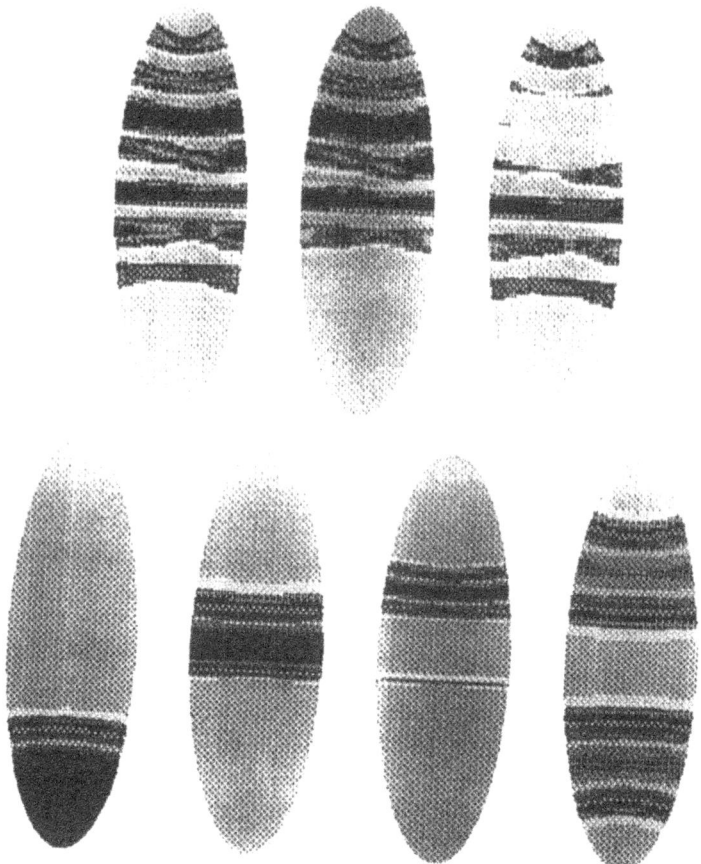

Figure 15.3. Computer simulated patterns of gap genes (left column) in the wild type (*hb, Kr, kni, gt* resp.), and of zebra stripes (right column): Gene *eve* in wild type, *hb-* and *gt-* respectively.

reduced *kni* level is sufficient to keep the *kni, gt* cue somewhat active. Similar problems arise with other suggestive proposals for simple additive cues for *hb* and *Kr*.

The study of feasible cues is thus only in a preliminary state, both experimentally and theoretically. The above example of the difficulties involved are common to both the pure cue model and the mixed Turing and cue model proposed here. What emerges so far is that the control of any pair-rule gene is highly intricate. Additionally the involved interactions in the connected control system defining the pair-rule level make interpretations of observed experimental (and even simulated) results rather complex.

Computed patterns for the most successful model so far is given in Figure 15.3. Stable zebra stripes are obtained and the result of some mutant calculations are given as well. Any model proposed to account for details in the *Drosophila* control system should be tested not only locally, in particular regions of the embryo, but should be valid globally. The present software makes such global simulations possible.

REFERENCES

Capovilla, M., Eldon, E. D., & Pirrotta, V. 1992. The *giant* gene of *Drosophila* encodes a b-ZIP DNA-binding protein that regulates the expression of other segmentation gap genes. *Development*, **114**, 99–112.

Carroll, S. B., & Vavra, S. H. 1989. The zygotic control of *Drosophila* pair-rule gene expression II. Spatial repression by gap and pair-rule gene products. *Development*, **107**, 673–683.

Goto, T., Macdonald, P., & Maniatis, T. 1989. Early and late periodic patterns of *even skipped* expression are controlled by distinct regulatory elements that respond to different spatial cues. *Cell*, **57**, 413–422.

Hafen, E., Kuroiwa, A., & Gehring, W. J. 1984. Spatial distribution of transcripts from the segmentation gene *fushi tarazu* during *Drosophila* embryonic development. *Cell*, **37**, 833–841.

Hiromi, Y., Kuroiwa, A., & Gehring, W. J. 1985. Control elements of the *Drosophila* segmentation gene *fushi tarazu*. *Cell*, **43**, 603–613.

Howard, K., Ingham, P. W., & Rushlow, C. 1988. Region-specific alleles of the *Drosophila* segmentation gene *hairy*. *Genes Dev.*, **2**, 1037–1046.

Hülskamp, M., & Tautz, D. 1991. Gap genes and gradients - the logic behind the gaps. *BioEssays*, **13**, 261–268.

Hunding, A., Kauffman, S. A., & Goodwin, B. 1990. *Drosophila* segmentation: Supercomputer simulation of prepattern hierarchy. *J. Theor. Biol.*, **145**, 369–384.

Ingham, P. W. 1988. The molecular genetics of embryonic pattern formation in Drosophila. *Nature*, **335**, 25–34.

Lacalli, T. C. 1990. Modelling the *Drosophila* pair-rule pattern by reaction-diffusion: gap input and pattern control in a 4-morphogen system. *J. Theor. Biol.*, **144**, 171–194.

Lacalli, T. C., & Harrison, L. G. 1991. From gradients to segments: models for pattern formation in early *Drosophila* embryogenesis. *Developmental Biology*, **2**, 107–117.

Murray, J. D. 1989. *Mathematical biology*. New York: Springer-Verlag.

Nicolis, G., & Prigogine, I. 1977. *Selforganization in Nonequilibrium Systems*. New York: Wiley.

Nüsslein-Volhard, C. 1991. Determination of the embryonic axes of *Drosophila*. *Development (Suppl.)*, **1**, 1–10.

Pankratz, M. J., & Jäckle, H. 1990. Making stripes in the *Drosophila* embryo. *Trends in genetics*, **6**, 287–292.

Stanjevic, D., Small, S., & Levine, M. 1991. Regulation of a segmentation stripe by overlapping activators and repressors in the *Drosophila* embryo. *Science*, **254**, 1385–1387.

Struhl, G., Johnston, P., & Lawrence, P. A. 1992. Control of *Drosophila* body pattern by the *hunchback* morphogen gradient. *Cell*, **69**, 237–249.

Turing, A. M. 1952. The chemical basis of morphogenesis. *Phil. Trans. R. Soc. Lond.*, **B237**, 37–72.

16. THE ROLE OF *HOX* GENES IN AXIS SPECIFICATION

Paul Hunt[1] and Robb Krumlauf [2]

[1]The National Institute for Medical Research
The Ridgeway
Mill Hill
London NW7 1AA, UK

[2]Development Biology Unit
Division of Cell and Molecular Biology
Institute of Child Health
30 Guildford Street
London WC1N 1EH, UK

16.1 INTRODUCTION

The discovery of vertebrate *Antennapedia* class homeobox-containing genes (the *Hox* network) has lead to a greater understanding of some of the important processes in vertebrate development. Recent analysis has revealed significant differences in the way in which the genes are expressed in different embryonic contexts, which has implications for their role in patterning each system. In this review we will summarize the evidence, both direct and indirect for the function of these genes in different regions of the developing vertebrate embryo. We will attempt to establish the common features and highlight the differences in homeobox gene function for particular systems within the vertebrate body.

16.1.1 *Antennapedia* Class Homeobox Genes in Insects and Vertebrates

Vertebrates possess four clusters of homeobox-containing genes known as *Hox* genes, related to the genes of the *Drosophila Antennapedia* and *Bithorax* complexes (Akam 1987). Furthermore the genes show general spatial restrictions in expression along the anteroposterior axis in the central nervous system and somitic mesoderm that mirror those of their *Drosophila* equivalents (Duboule & Dolle 1989; Papalopulu & Krumlauf 1989). This suggests that the gene clusters in both organisms are derived from a common ancestor. In contrast to *Drosophila*, there is extensive overlap in expression between genes of a vertebrate *Hox* cluster, so that anterior parts of the body express a few genes and posterior parts most of the genes in a given cluster. Among the genes of the four vertebrate clusters it is possible to identify families of up to four genes known as paralogous groups, referred to either by their *Drosophila* homologue where a single *Drosophila* equivalent exists or a class number (see Figure 16.1).

Recent work on the structure of the human *Hox* complexes (Simeone *et al.* 1991), has resulted in the identification of all of the genes that are likely to be present in the complexes,

Experimental and Theorietcal Advances in Biological Pattern Formation,
Edited by H.G. Othmer *et al.*, Plenum Press, New York, 1993

161

and so far no discrepancies have been found with the mouse, whose level of characterization is almost as complete. Further analysis of the expression of vertebrate genes has demonstrated a potential role in the axial patterning of a number of embryonic systems. In this review we will consider three of these systems, the branchial region of the head, the prevertebrae of the trunk and the limb bud, trying to identify the parallels and distinctions between them in their use of *Hox* genes.

16.2 THE HINDBRAIN AND BRANCHIAL ARCHES

16.2.1 The Embryological Background

The branchial region of the head has received a lot of attention in recent years, as the properties of its development suggest a particular sequence of interactions amenable to experimental manipulation. It gives rise to the hindbrain and its associated cranial ganglia, many of structures of the jaws and neck, the ears and the vessels of the outflow tract of the heart. It is also one of the parts of the vertebrate body that is dependent upon segmental structures for its development.

At an early stage, before and during the birth of the first neurons, the neural tube of the presumptive hindbrain is organized into a series of repeating bulges known as rhombomeres (Lumsden 1990). They have been shown to share developmental properties with the body segments of *Drosophila* (Fraser *et al.* 1990), where cells programmed to particular fates by gene expression are prevented from moving across segment boundaries (Ingham &

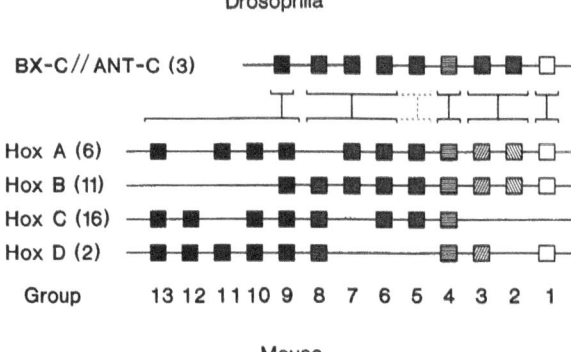

Figure 16.1. The structure of the murine *Antennapedia* class homeobox gene complexes, illustrating the relationship between members of a subfamily of vertebrate genes and their *Drosophila* equivalents. Members of a subfamily or paralogous group of genes are vertically aligned with each other, with their group number indicated at the base of the diagram. Note that some subfamilies are not represented in all *Hox* clusters. The bold brackets above the mouse genes indicate the Drosophila homeotic genes which have clear homologues in the mouse. The dashed brackets indicate that homology relationships may exist but are less distinct based on existing data. The bracketed numbers indicate the chromosomal location of their respective gene clusters.

Martinez-Arias 1992). In this way regional differences between groups of cells are maintained which could be converted into differences in final morphology (Lumsden & Keynes 1989).

The interface between the hindbrain neural plate and the surface ectoderm gives rise to neural crest cells which interact with other head tissues, contributing to a series of cranial ganglia and branchial arches (reviewed in Noden 1988). The spatial organization of crest emigration suggests that it too is organized on a rhombomeric basis (Lumsden *et al.* 1991), with the result that the crest of a ganglion and the arch it innervates originate from the same rhombomeres. The properties of the neural crest suggest that it may have a leading role in patterning structures at this axial level (Noden 1988; Richman & Tickle 1989). When premigratory neural crest from the first branchial arch is removed from an embryo and used to replace the premigratory second arch crest of a host, the grafted crest is found to migrate in a way appropriate for its new position and enter the second arch. Once there however, it makes a set of structures appropriate for its original position *i. e.* first arch jaw cartilages in the second arch. Furthermore, it is able to influence the surrounding non neural crest-derived second arch tissues, surface ectoderm and paraxial mesoderm, to form first arch structures (beaks and jaw muscles respectively). This suggests that the neural crest is imprinted with a positional value before migration that determines the form of the structures it is to make, carrying this signal with it during migration. It can also transfer this information to the tissues that surround it, influencing their development (Noden 1988). The basic relationships between origin of nerves in the hindbrain, the position of the ganglia and the specific branchial arch they innervate is conserved with a few minor variations in all vertebrates, suggesting an evolutionarily constrained system which further supports the idea of a developmental link between the structures. This and the properties described above has lead Lumsden to propose that aspects of the patterning mechanisms of hindbrain and branchial arches were linked via early regional specification of the neural plate (Lumsden 1990).

16.2.2 The Expression of *Hox* Genes During Formation of the Branchial Region

To explore the possibility that *Hox* genes are involved in early craniofacial development, the expression patterns of the genes of the four vertebrate *Hox* clusters were investigated in mouse embryos (Wilkinson *et al.* 1989; Hunt *et al.* 1991a; Hunt *et al.* 1991b). With one exception, 3' members of a *Hox* complex were found to have anterior expression restrictions in the hindbrain that corresponded to rhombomere boundaries, with successive genes in a particular cluster showing cutoffs separated by two rhombomeres. Genes of the same paralogous group shared the same rhombomeric expression cutoff. The precise rhombomere boundaries are indicated on Figure 16.2, which also emphasizes the overlapping nature of their expression. The most 3' gene in the *Hox-B* complex, *Hox-B1*, shows exceptional behavior in that its anterior expression limit is less anterior than its more 5' neighbor, *Hox-B2*, and that its expression domain in the hindbrain is confined to a single rhombomere.

Hox genes are also candidates for a role in the control of neural crest development. At 8.0 dpc (days post coitum) in mouse, before the formation of neural crest, *Hox* genes are expressed in the presumptive hindbrain with anterior limits that are thought to coincide with those they display at later states of development (Hunt *et al.* 1991a). At 8.5 dpc., *Hox* gene expression is found in the hindbrain neural plate and in two regions lateral and ventral of it, consistent with the positions of migrating neural crest cells. At this stage expression is not seen in other head mesenchyme or the surface ectoderm. By 9.5 dpc. branchial arch formation is complete, and the cranial ganglia have condensed. At this state the neural crest of the branchial arches has retained the *Hox* label of its level of origin in the hindbrain, and there is evidence for expression beginning in the surface ectoderm. The result of the retention of patterns of *Hox* expression by migrating neural crest cells and the consistent relationships between hindbrain and branchial arches documented by Lumsden *et al.* (1991) is that each branchial arch has a distinct pattern of *Hox* gene expression closely related to their level of origin in the hindbrain. The fact that the ectoderm seems to acquire a *Hox* label after it

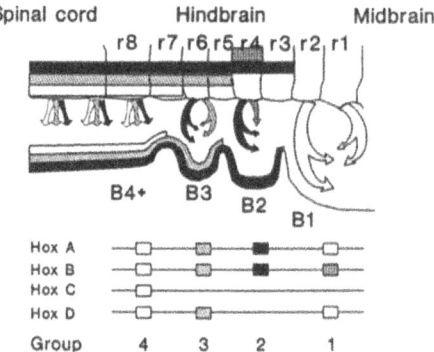

Figure 16.2. Relationship of restricted expression in the hindbrain to migrating neural crest and the branchial arches. The diagram indicates the patterns of *Hox* subfamilies expressed in the branchial region after neural crest migration when distinct rhombomeres are apparent at 9.5 dpc. The arrows indicate the migration of mesenchymal and neurogenic crest from specific rhombomeres into the branchial arches. The branchial arch ectoderm subsequently adopts an identical pattern of *Hox* subfamily expression indicated by it shading, (Hunt *et al.* 1991a; Hunt *et al.* 1991b), presumably as a result of an interaction with underlying neural crest. *Hox* subfamily expression in the hindbrain is out of phase with the branchial arches by one rhombomere as a result of the lack of contribution to branchial arch crest by r3 and r5 by analogy to the chick embryo (Lumsden *et al.* 1991). This is represented by the absence of arrows emanating from r3 and r5. The large open arrows on the right of the diagram represents first arch crest that does not have a *Hox* label. The chromosomal relationship of the relevant subfamilies is shown at the bottom of the diagram. (Based on Figure 16.4, Hunt *et al.* 1991a).

comes into contact with expressing crest (Hunt *et al.* 1991b) suggests that this information can also be transmitted between germ layers, and is consistent with evidence that the neural crest is able to influence the development of other tissues in the head. Thus the overlapping expression patterns of *Hox* genes in hindbrain and branchial arches suggest that they may be a part of the developmental system linking the two structures that has resulted in their coordinated development.

The only detectable differences between members of a paralogous group were in levels of gene expression. An example of level differences can be seen in paralogous group 3, where only two members, *Hox-A3* and *Hox-B3*, show a high level of expression in rhombomere 5. Once again the class 1 genes including *Hox-B1* were the exceptions to this rule. The fact that early stages of morphogenesis in the branchial region are characterized by very great similarities in the expression domains of paralogous genes may indicate some redundancy in function among paralogues during the formation of the branchial region.

16.2.3 Expression of *Hox* Genes at Later Stages of Head Development

At later stages of mouse development, when many of the characteristic structures of the head have formed, members of a subfamily of *Hox* genes that were expressed in identical rhombomeres and branchial arches at 9.5 dpc. now show clear differences in expression domains. While genes still show similar overall expression limits along the A-P axis, differences within the domains in relative levels, tissue or cell type specificity become superimposed upon the earlier pattern. Spatial restrictions within a tissue are shown by members of the *Hox-B* complex, which have a progressive pattern of dorso-ventral localization within the central nervous system, the time course of which reflects the final cell divisions of particular groups of neurons (Graham *et al.* 1991). In adjacent sections the *Hox-C8* gene shows a specific distribution in the dorso-ventral axis of the spinal cord, but one which does not correspond to that shown by the *Hox-B* genes.

Expression differences between paralogues have also been shown by the work of Gaunt in the mesenchymal tissues of the head (Gaunt *et al.* 1989). For example, *Hox-A4* is expressed in the thyroid gland, the mesodermal derivatives of the trachea and the thymus., while *Hox-B4* is expressed in the same tissues and in addition the surroundings of the thyroid duct. *Hox-D4* transcripts were present in the thymus and thyroid glands, but not the mesenchymal cells of the trachea. These differences suggest that *Hox* genes have distinct roles in the later events of head morphogenesis, and that this is a result of an evolution of the expression patterns of subfamily members from earlier phases of development when they are more similar.

16.2.4   Experimental Evidence for the Role of *Hox* Genes in Craniofacial Development

The assumed role for the role of *Hox* genes in craniofacial development implicit in the above discussion in based on the fact that in *Drosophila*, the anterior limit of expression of a gene correlates with the segments most strongly affected by mutations in that gene. Experimental manipulations of head morphogenesis in the mouse have been shown in produce alterations in *Hox* expression consistent with this concept.

To prove that a particular combination of *Hox* gene expression is responsible for controlling the identity of a segment, it will be necessary to directly alter the cranial *Hox* code, and show that the morphology that results is consistent with the new combinatorial code. The first experiments to attempt this have recently been performed. The technique of homologous recombination in embryonic stem cells was used to produce mice lacking functional copies of the genes *Hox A3* and *Hox-A1* (Chisaka & Capecchi 1991; Lufkin *et al.* 1991). Mice lacking these genes died shortly after birth, showing defects in the branchial region. Thus the basic assumption of many of the studies on vertebrate *Hox* genes, that their major patterning function is located in the most anterior part of their expression domain, appears to be justified in the branchial region. An implication of this is that the most 5' *Hox* gene expressed at a particular axial level is the most important in controlling the identity of that level, irrespective of the expression of other *Hox* genes. This is analogous to the "posterior prevalence" model of Lewis (1978), developed to account for the specification of abdominal identity in Drosophila. Position in a posterior prevalence model is specified by only one homeotic gene at a given axial level: the most posterior one. In such a system removal of a gene results in the conversion of units that would normally express it to the more anterior identity specified by the next most posterior gene expressed there. The expression limit of the gene normally expressed in segments posterior of those specified by the mutant gene defines the posterior extent of the domain affected by the mutation.

A significant feature of the phenotypes of both mutants was that while there were profound effects on the development of structures at a particular axial level that broadly cor-related with the anterior domain of expression of the genes, other structures derived from the same axial level and which in some cases expressed the gene in normal development were unaffected. Thus the defects in *Hox-A3* mutants were mainly in mesenchymal derivatives, while *Hox-A1* defects produced abnormal development of neural structures. Another impor-tant aspect of these two studies is that two *Hox* genes of the same cluster produced defects in different types of tissues.

16.2.5   Potential Strategies of Branchial Region Patterning Involving *Hox* Genes

In *Drosophila* a loss of function mutation in a HOM-C gene frequently results in a segment transformation, the identity of the affected segment being determined by the combination of genes that are subsequently expressed there (McGinnis & Krumlauf 1992). In the two mutants so far described other members of the paralogous groups to which the disrupted genes belong were still present in the genome. As these genes have identical expression domains at the times when specification of rhombomeres and branchial arches is thought to occur in normal animals, the situation is different from a *Drosophila* null mutant where only a single gene is potentially involved in the specification of a group of segments. This may be particularly important in the interpretation of the mechanism of action of the *Hox-A1* gene, which affects a number of structures which do not express the gene in the normal course of development. These structures are derived from axial levels that do express the gene at times before neural crest emigration has occurred, so its possible that *Hox-A1* mutations exert their effects by interfering with the establishment of the expression of other genes. The effects of a loss of function mutation in a single gene on the activities of the rest of the *Hox* network in the branchial region is not known, and until they are the potential connection between *Hox* genes and segment specification remains to be shown in vertebrates. However, we believe it is possible to infer something of the way in which *Hox* genes may be involved in regional specification in the branchial region from the present studies.

We believe that *Hox* genes are involved in two interlinked processes that occur to produce the structures of the branchial region. The first is the specification of the rhombomeres and branchial arches themselves, that is the establishment of differences between segmental units. A process results in the establishment of differential *Hox* gene expression in the neural plate, at a time when there are no other stable intrinsic differences between these populations of cells. As this is an "abstract" process there is a immediate effect in terms of production of particular structures; it is simply "putting the numbers" onto the framework of potential rhombomeres and branchial arches which are initially all equivalent. The particular combinations of *Hox* gene expression that exist in the different regions of the neural plate are then able to initiate the second process; a chain of events that gives rise to the region-specific patterns of other gene activities that will eventually involve structural proteins. By directly controlling their own target genes and by extensive interactions with other gene systems giving rise to regional identity *e. g.* the process that distinguishes odd from even rhombomeres, the unique final morphologies of the components of the brain and face are produced. Mutation of a *Hox* gene could affect both the "abstract" initial production of non equivalent embryonic structures, and the subsequent process by which these regional differences are converted into local patterns of structural gene activity, actually generating the observed final morphology.

The extensive similarities in expression domains of paralogues in the hindbrain and branchial arches argues against the use of *Hox* genes of other clusters in providing more precise spatial information than has already been described for the *Hox-B* genes (Hunt *et al.* 1991b). In addition there is no evidence that at later stages during branchial morphogenesis further spatial (as opposed to tissue-specific) differences in expression arise within a branchial arch (R. DasGupta, unpublished data); thus other developmental processes must be providing positional information within branchial arches and distinguishing odd from even rhombomeres. It is unlikely that the combined expression of all the members of a *Hox* subfamily is required for initial spatial specification of head tissues, as removing a gene from such a system would be expected to prevent the spatial specification of all *Hox*-dependent structures at an axial level. We argue that there is redundancy in the first of these processes, the "abstract" one, as other members of a subfamily *Hox* genes are able to compensate for the loss of a given gene. In mice lacking *Hox-A3* all the cranial ganglia are normal (Chisaka & Capecchi 1991), suggesting that the primary process of rhombomere and branchial arch specification upon which ganglion formation depends has occurred correctly. In a similar way, the spatial specification presumably necessary for the production of normal mesenchymal structures in the branchial arches has occurred in mice lacking *Hox-A1* (Lufkin *et al.* 1991). The presence of transcripts of other genes in the group 3 and group 1 subfamilies in the same parts of the head suggests that they may be sufficient for this first process of specification of rhombomere identity in the absence of *Hox-A3* or *Hox-A1*. To produce an effect on all the structures derived from a particular level of the neural plate, *i. e.* to alter the primary events of rhombomere specification, it may be necessary to delete all members of a subfamily of genes to prevent this compensation.

The phenotype of the mutants will result from perturbation of the first developmental process that requires either *Hox-A3* or *Hox-A1*, and which cannot be compensated for by other genes. We suggest that this process is the conversion of the "abstract" set of rhombomere and branchial arch identities into the structural differences that characterize their final morphology. In this process a specific *Hox* gene is required, and related genes can no longer compensate, while some other structures at the same level are independent of that gene and may be able to develop normally. This interpretation of regional identity to form final structures may include the emergence of particular tissue types among other processes, consistent with the observation that members of a subfamily, initially expressed at the same axial level, eventually show differences in their tissue distributions at that level.

An alternative interpretation is that while the various structures derived from the neural plate require single *Hox* genes for their normal spatial specification, the process occurs independently for each tissue, and hence some can develop in the absence of a particular *Hox* gene. As there are different numbers of genes in the various 3' subfamilies, and two genes in the same cluster affect structures which would be expected to be specified independently under this scheme, there cannot be any kind of simple relationship of the one *Hox* cluster: one set of structures type. Furthermore, the hypothesis that the primary spatial specification of groups of structures occurs independently is difficult to reconcile with the idea of a link between patterning mechanisms in the hindbrain and branchial arches that has resulted in the conserved anatomical relationships of the vertebrate branchial region (Lumsden 1990).

There is evidence to suggest that interactions between tissues of different embryonic origins are necessary for the development of much of the head (Hall 1987; Thorogood 1988). As the inability of a few cell lineages to develop normally could conceivably cause the failure of a number of structures dependent upon these interactions to develop, perturbation of a particular lineage could have a wide range of developmental effects.

Interactions between head primordia may explain another phenomenon not so far discussed, that the loss of *Hox-A3* function perturbs the development of structures which do not formally express the gene. For example, the greater wing of the hyoid bone, derived from the third arch which expresses *Hox-A3*, develops normally, while the lesser wing, derived from the second arch which never expresses *Hox-A3*, is absent. This may reflect interactions in development between third arch structures which require *Hox-A3* and second arch derivatives, or the requirement for interactions with other genes of the sort discussed earlier in the context of *Hox-A1*. If so this is further support for the idea that some of the profound developmental effects seen in *Hox* mutants may be due to secondary effects rather than alterations in primary regional specification.

16.3   THE TRUNK PARAXIAL MESODERM

The early development of the trunk proceeds by a different set of principals than does that of the head, in particular there is less evidence for primary axial specification in the neural tube derivatives of the trunk. *Hox* genes are also implicated here as part of the process of axial specification, but the way in which paralogous genes behave is different from the head, reflecting differences in the molecular processes underlying development between head and trunk.

16.3.1   The Embryological Background

The early distributions of trunk mesoderm are quite different from that of the head (Romer 1971; Noden 1988), and its developmental properties differ significantly as well. There appear to be at least two mesoderm-based patterning systems operating in the trunk, one based on the somites giving rise to the vertebral column, and another, independent system patterning the lateral plate mesoderm that gives rise to the limbs. The pattern forming ability of trunk paraxial mesoderm appears to be intrinsic, as somites will produce vertebrae and ribs appropriate to their site of origin when grafted to ectopic sites (Chevallier 1975). In normal development spinal motor nerves grow through the anterior halves of their adjacent somites. This process is also under the influence of positional values within the somites, because reversal of the A-P polarity of a group of somites causes corresponding changes in the position of nerve outgrowth, although the neural tube has remained in its original orientation (Keynes & Stern 1985). In contrast after neural tube reversal outgrowth still occurs opposite the rostral halves of somites. Replacement of head paraxial mesoderm with either segmental plate or somites results in a normal head skeleton, with normal patters of neural development (Noden 1986), consistent with the greater importance of neural plate derivatives in head development described in the previous sections.

The overtly repetitive nature of the somites raises the possibility that they, like the rhombomeres of the hindbrain, are segmental structures. However, clonal analysis of the type performed in the hindbrain does not suggest the presence of compartments in the somitic mesoderm (Stern *et al.* 1988), and observations of the behavior of somitic cells in living animals also demonstrates movement of cells across clefts between adjacent somites (D. Amanze, P. Thorogood unpublished results). The mobility of cells within the spinal cord does seem to respect boundaries, but this is thought to be due to the presence of the adjacent somites, as their removal also causes the loss of these lineage restrictions (Stern *et al.* 1991). Thus while there is evidence for positional specification of somites, it does not appear to be mediated by the same sort of lineage restriction processes that occur in the hindbrain.

16.3.2   Expression of *Hox* Genes in the Trunk

One of the earliest observations of *Hox* gene expression patterns is that they showed spatial restrictions that corresponded to somite boundaries. An extensive literature survey has been performed by Kessel, comparing the expression limits within the somites, or their prevertebral derivatives, that various *Hox* genes respect (Kessel & Gruss 1991). There has not been a direct experimental comparison of the expression limits of paralogous genes at the early stages of development, around 9 dpc. in mouse, when interpretation of axial position to establish region-specific structures would be expected to be occurring. Nevertheless it is unlikely that the somitic expression limits of the genes shown in Figure 16.3 are substantially inaccurate for this early stage of development.

In contrast to the situation in the head, there appear to be differences between paralogous genes in their expression limits in somites. These differences are small in the more anterior genes; thus *Hox-A4* shows a different prevertebral expression limit, in pv3 (prevertebra 3), from *Hox-B4* and *Hox-D4* in pv2, all of which are expressed in identical rhombomeres and branchial arches. More posterior paralogous genes show greater differences in anterior expression limits, with similar differences also being apparent in the spinal cord (Izpisua-Belmonte *et al.* 1990). This has lead Kessel to suggest that the aspects of somite identity are specified by a *Hox* code, and that in contrast to the situation in the branchial region, paralogous genes have independent roles in somite specification from the very outset.

It is clear from Figure 16.3 that not every somite has a unique *Hox* code, and that some members of paralogous groups are expressed with identical somitic cutoffs. Clearly other genetic systems must also be acting to produce distinctions between the final forms of somite derivatives, just as some alternative system in the hindbrain must be specifying the difference between odd and even rhombomeres. It is interesting to note that groups of vertebrae that are similar in their morphology, such as some of those in the cervical region, express identical *Hox* codes.

16.3.3 Experimental Evidence of the Role of *Hox* Genes in the Trunk

One group of experiments has centered on the ability of retinoic acid in vivo to produce alterations in vertebral morphology of the offspring in some cases consistent with transformations of identity (Kessel & Gruss 1991; Kessel 1992). By feeding the mother with retinoic acid from 7 to 8.5 dpc *i. e.* when gastrulation is occurring, Kessel has produced animals in which anterior vertebrae could be interpreted to have taken on an identity appropriate for more posterior parts of the body, *e. g.* cervical ribs. A number of different developmental abnormalities have been identified in mice consistent with the conversion of anterior structures to more posterior ones (Kessel & Gruss 1991), while treatment at stages later than 8.5 dpc produces different types of alteration (Kessel 1992).

It is particularly interesting that retinoic acid should produce these kinds of effects on development, as it is known to be able to modulate the expression of *Hox* genes in cultured mouse (F9) and human (NTERA-2) teratocarcinoma cells (Papalopulu *et al.* 1991; Simeone *et al.* 1991). It is therefore possible that the defects observed are a result of alterations in the domains of *Hox* gene expression, which go on to produce somites with altered morphologies. Consistent with this it has been shown that animals treated with retinoic acid at 7.4 dpc, show anterior extension in the domains of expression of particular *Hox* genes at 12.5 dpc. (Kessel & Gruss 1991). If this is occurring with other *Hox* genes, and at the stages when somite specification is occurring, then the result would be an activation of extra *Hox* genes in particular somites. A somite would thus be expressing a combination or code of *Hox* genes appropriate to a more posterior axial level, and adopt an appropriate morphology as a result.

If the structures a somite forms are controlled by the code of *Hox* genes that a somite expresses then it may be possible to change somite identity by direct alteration of the expression of *Hox* genes. This can be attempted by either inactivating a gene by homologous recombination to produce a null mutant or by expressing a gene inappropriately to produce a gain of function alteration.

The phenotypes of mice lacking a functional copy of the *Hox-C8* gene have recently been described (Mouellic *et al.* 1992). The gene was inactivated by insertion of LacZ coding sequences, thus giving a marker for activity at the disrupted locus. LacZ expression was found in the normal expression domains of the gene as determined by *in situ* hybridisation. In contrast to the effects of mutations in the branchial *Hox* genes, analysis of these mice

```
    occip. cervical   thoracic      lumbar sacral caudal
                                                              Group
    1234 1234567  1234567890123  123456 1234  1234

<-X  ....  .......  ..............  .......  ....  ....  Hox-A2    2
<-X  ....  .......  ..............  .......  ....  ....  Hox-B2

     X......  ..............  .......  ....  ....  Hox-A3    3
     X......  ..............  .......  ....  ....  Hox-B3

     X......  ..............  .......  ....  ....  Hox-D4    4
     X.....   ..............  .......  ....  ....  Hox-A4
     X.....   ..............  .......  ....  ....  Hox-B4

      X....   ..............  .......  ....  ....  Hox-A5    5
      X....   ..............  .......  ....  ....  Hox-B5
       X.    ..............  .......  ....  ....  Hox-C5

      X...   ..............  .......  ....  ....  Hox-B6    6
        X   ..............  .......  ....  ....  Hox-C6
       X............  .......  ....  ....  Hox-A6

     X ..............  .......  ....  ....  Hox-B7    7
       X+++++++++  ......  ....  ....  Hox-A7

       X+++++++++ +.....  ....  ....  Hox-C8    8
          X..  ......  ....  ....  Hox-D8
       X+++++++++++  ......  ....  ....  Hox-B9    9
          X..  ......  ....  ....  Hox-C9
         X.....  ....  ....  Hox-D9

       X ......  ....  ....  Hox-A10   10
          X...  ....  ....  Hox-D10

          X.  ....  ....  Hox-A11   11
           X  ....  ....  Hox-D11

            X ....  Hox-D12   12

            X..  Hox-D13   13
```

Figure 16.3. Summary and comparison of the anterior expression cutoffs of paralogous *Hox* genes in the prevertebrae. The location of the most anterior somite or somite derivative strongly expressing each gene is indicated by a "X". Expression extends posteriorly as indicated by the dots, but in most cases the posterior boundary is not clearly defined, or has not been investigated. In two cases, marked by the asterisks (*), but anterior and posterior boundaries of high level expression have been defined (Mouellic *et al.* 1992; Puschel *et al.* 1991) and the extent is indicated by crosses (+). The genes and their paralogous group is indicated by the right. This figure is derived from Kessel & Gruss (1991) and the references therein for patterns of expression, with modifications from our own unpublished data.

has revealed several skeletal alterations affecting different axial levels consistent with their transformation to a more anterior identity, for example ribs on the first lumbar vertebra. In each case the transformation was to the morphology of the element immediately anterior. The affected region of the body axis was more extensive than in mice lacking 3' genes, although entirely contained within the normal expression domains of the gene. The described morphological abnormalities were confined to the skeleton, although there was some evidence of abnormal motor activity in the mutants that survived to adulthood.

16.3.4   The Trunk *Hox* Code

As with the head, it is at the moment difficult to relate the effects of direct perturbation of *Hox* expression on the development of the trunk somites and their derivatives in terms of alterations in *Hox* code. It will be necessary to determine the expression patterns of the other *Hox* genes in the affected regions of the body, although the maintenance of normal *Hox-C8* expression domains in the above mutants suggests that some aspects of the *Hox* regulatory hierarchy are functioning normally.

The *Hox-C8* mutation is without profound effect on the central nervous system, and given the embryology of spinal nerve development those defects that are observed may be via perturbations of somite development. The fact that skeletal transformations can be produced by inactivation of *Hox-C8* is in striking contrast to the situation in the branchial region, and suggests that individual *Hox* genes mediate spatial specification in the trunk. This would be consistent with the observed offsets of multiple somites in the anterior expression limits of paralogous trunk *Hox* genes, as shown in Figure 16.3. It also suggests that a simple "posterior prevalence" model is not operating, as transformations of identity occur in regions which express genes 5' of *Hox-C8* (Figure 16.3). The basis of such a model is that regional specification is largely achieved by the dominant action of the most posterior homeotic gene at a given axial level; the fact that there are other genes with more anterior expression limits also present in the segment is not important in interpreting regional identity. Lack of a component in a "posterior prevalence" code would result in several elements at a specific axial level adopting one fate, rather than a greater number of elements adopting the morphology of immediately anterior ones. In mice lacking *Hox-C8*, the distribution of effects throughout the normal expression domain of the gene would be more consistent with a combinatorial code model, in which the combination of a number of genes are simultaneously necessary for spatial specification. The fact that general failure to interpret positional level does not occur suggests that final morphology is not solely due to a direct readout by each somite of its *Hox* code, as "nonsense" readouts and extensive dysmorphology would be expected over much of the area lacking *Hox-C8* if this were the case.

It is striking that direct or indirect alterations in *Hox* expression correlate with alterations of the structures formed by vertebrae, rather than simply the failure of structures to develop as is seen in the head. This suggests that the development of the skeleton from the somites may be less dependent upon external interactions to produce an interpretable morphology than the development of facial structures from a branchial arch. Thus it may be easier to determine the effects of particular combinations of *Hox* gene expression on subsequent regional identity in the trunk, and manipulation of the expression of individual genes may provide more direct evidence about mechanisms of spatial specification.

16.4   THE DEVELOPING LIMB

In the two systems using *Hox* genes so far described, their primary role is thought to be positional specification along a single dimension, the main A-P body axis. The employment of *Hox* genes in the developing limb appear to be employed in the establishment of regional identity in two dimensions, in contrast to the other two systems we have described.

16.4.1 Embryological Background

The cells of the limb bud derive from a number of sources, the most important of which appears to be the lateral plate mesoderm. This will give rise to the limb cartilages and connective tissues, and seems to be the most important tissue in defining the limb structures. Somite-derived cells also enter the limb, but the form of the muscles they contribute to appears to be controlled by the lateral-plate-derived tissues. The establishment of structures along the antero-posterior axis and the proximo-distal axis of the limb bud appears to involve two systems with different properties (reviewed in Tickle 1991).

Patterning along the A-P axis is thought to involve a positional signal which is interpreted by the cells to form structures of the appropriate type. The posterior margin of the limb bud contains an area of mesenchyme known as the polarising region. This is able to produce mirror image duplications of structures in the A-P axis when grafted to the anterior margin of another limb bud, whose tissues thus become exposed to two polarising regions. The fate of grafted cells suggests that they exert their effects by signalling to other limb cells rather than by directly contributing to the supernumerary structures. The response of cells to grafts of different amounts of polarising region tissue and grafts to different locations within the limb bud is consistent with a graded distribution of a signalling substance, the local concentration of which determines the A-P character of a particular region of limb mesenchyme. Implantation of beads soaked in retinoic acid to the anterior margin of the limb bud produces duplication effects similar to a polarising region graft, suggesting that it is able to affect the normal signalling events that occur during limb development.

A reciprocal interaction is thought to occur between limb mesenchyme and AER (Tickle 1991). Duplication of elements along the A-P axis of the limb is known to be dependent upon the continued presence of the apical ectodermal ridge, and is preceded by an extension in length of the AER dependent upon the mesenchyme (Tickle 1991). Complete removal of the AER results in cessation of limb bud outgrowth, while partial removal results in loss of pattern elements at the corresponding A-P level (Tickle 1991).

There is no evidence for compartment formation during the early events of vertebrate limb morphogenesis.

16.4.2 Expression of *Hox* Genes During Limb Morphogenesis

At an early state, before limb bud outgrowth, the lateral plate mesoderm of the presumptive limb bud expresses *Hox* genes of classes 1-9 appropriate to its level along the A-P axis. This expression decreases in intensity as development progresses, and superimposed upon this new patterns of *Hox* expression appear involving genes of classes 10-13 as the limb bud grows out. It is notable that these genes and those of class 9 all show greatest similarity to a single *Drosophila* gene, Abd-B, and that one vertebrate *Hox* cluster, *Hox-B*, does not possess any genes in classes 10-13. This suggests that either the genes of classes 10-13 have arisen by tandem duplication of an ancestral Abd-B related gene since the divergence of the lineages that lead to vertebrates and arthropods, or they are the product of an earlier duplication in an organism ancestral to both lineages and have been lost in the *Hox-B* cluster and *Drosophila* (Izpisua-Belmonte *et al.* 1991b).

The *Hox-D* genes in both chicken and mouse show progressive spatial restrictions along the A-P axis of the limb bud, consistent with their position within the *Hox-D* cluster (Dolle *et al.* 1989; Izpisua-Belmonte *et al.* 1991a; Nohno *et al.* 1991). *Hox-D9* is expressed throughout the limb bud, while *Hox-D11* is expressed in a more restricted domain which does not extend to the anterior surface of the bud. The domains of *Hox-D6* and

Hox-D13 are contained within the *Hox-D9* domain, and are progressively more posteriorly restricted. Although these spatial restrictions correspond to regions of the limb bud with consistent fates, the boundaries of gene expression do not correspond with the boundaries of particular structures in the way that expression of *Hox-B* genes corresponds to particular rhombomere boundaries. The timing of expression is such that the most anterior gene is activated first, and then successively more posterior and 5' genes are activated successively later.

Hox-A genes show spatial restrictions along the proximo-distal axis of the limb (Yok-ouchi *et al.* 1991), perpendicular to the expression domains of *Hox-D*. Successively more 5' *Hox-A* genes show progressively more distal restrictions, and there is a temporal progression in onset of expression, correlating with their spatial distribution. The expression domains are not entirely overlapping, as the *Hox-A10* gene eventually comes to be expressed in a band across the limb with both a proximal and a distal expression limit. There is no overlap in expression between this and the more distally restricted *Hox-A13* gene. The expression boundaries of *Hox-A* genes correspond to some of the points in the proximodistal axis where the number of skeletal elements changes, but this also occurs in areas with homogeneous *Hox* expression patterns.

Thus it is conceivable that the position of cells within a limb bud and hence the form of the structures they are to make is partially determined by reference to the overlapping expression domains of *Hox-A* and *Hox-D* genes, which are illustrated in Figure 16.4. The expression domain of only one 5' *Hox-C* gene, *Hox-C10*, has been characterized at the time of writing. Unlike the 5' *Hox-A* and *Hox-D* genes which show similar spatial restrictions in both fore- and hindlimbs, *Hox-C10* is not expressed in the forelimb (Peterson *et al.* 1992). The expression domain in the hindlimb shows an anterior-proximal restriction, suggesting that it could provide additional positional cues in either antero-posterior or proximo-distal axes within the limb. It is not as yet clear whether the other 5' *Hox-C* genes show different expression restrictions within the limb, and it is important to note that the more 3' *Hox-C6* gene shows a similar proximo-distal restriction within limb buds (Oliver *et al.* 1990). If all *Hox-C* genes show a similar spatial restriction within the limb bud they are unlikely to be involved in the types of regional specification within limb buds that employ *Hox-A* and *Hox-D*.

16.4.3   The Response of *Hox* Expression to Manipulation of Limb Development

The accessibility of limb buds to experimental manipulation and the extent to which the timing of processes during the development of the limb is understood have provided powerful indirect evidence for their role here.

Implantation of retinoic acid-soaked beads or grafts of polarising regions have been shown to induce ectopic domains of 5' *Hox-D* gene expression in chicken limb buds, cor-relating with the duplication of distal limb elements. Izpisual-Belmonte (1991a) were also able to investigate the timing of onset of the ectopic domains of expression. It is known that implantation of retinoic acid-soaked beads must occur for a minimum of 16 hours or greater to produce consistent mirror-image digit duplications, shorter exposure times being without effect. Limb buds exposed for up to 10 hours to retinoic acid showed no alterations in *Hox-D* gene expression immediately after bead removal; however after 16 hours an ectopic domain of *Hox-D* expression was visible at the dorsal margin of treated wings. Outgrowth of the limb bud had continued after bead implantation, and it was clear that the ectopic expression domain was separated from the implant site. At successively longer times of exposure to retinoic acid more 5' *Hox-D* genes were activated, and the fact that a 5' *Hox-D* gene was in no

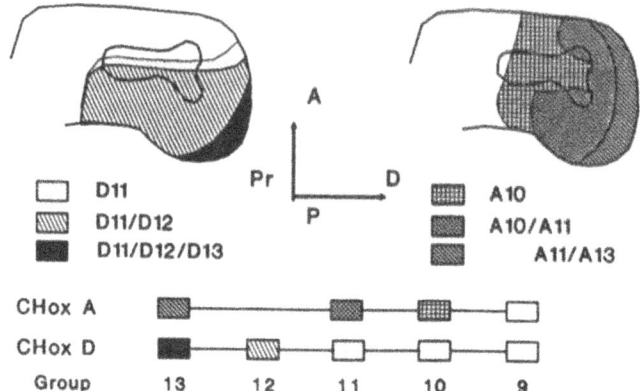

Figure 16.4. The expression domains and cluster organization of *Hox* genes expressed in the limb. A dorsal view of a stage 24 forelimb is shown, with the proximo-distal (Pr-D) and anterior-posterior (A-P) axes indicated by arrows. The structure in the centre of the limb bud represents the mesenchymal condensation that will give rise to the limb cartilages. Superimposed on this are the domains of Hox expression, with the density of shading corresponding to the combinations of gene expression indicated below. Below is shown the chromosomal organization of the genes illustrated. Adapted from alt(Yokouchi *et al.* 1991).

case activated before it more 3' neighbours suggests that the temporal and spatial sequence of gene activation are interdependent. Activation of ectopic domains of *Hox-D* expression was found to be dependent upon the presence of the AER, although a domain of expression once activated would persist if the AER was subsequently removed (Izpisua-Belmonte *et al.* 1992a). Partial AER removal resulted in the loss of gene expression domains consistent with loss of structures (Izpisua-Belmonte *et al.* 1992a).

Treatment of amniote limb buds with retinoic acid at the concentrations used in the bead assay does not produce effects on the development of the proximo-distal axis, and Yokouchi *et al.* (1991) were able to show normal proximo-distal distributions of *Hox-A* expression in limb buds which would have produced mirror-image digit duplications if allowed to develop further.

A naturally-occurring genetic perturbation of limb development, *talpid*, has been described. Proximo-distal patterning of the limb appears to occur normally, but instead of the normal pattern of digits a single, large plate of cartilage is produced. The expression of *Hox-D13* in *talpid* embryos has been found to be homogeneous across the distal parts of the limb bud, instead of the normal posterior restriction in expression (Izpisua-Belmonte *et al.* 1992b). This suggests that *talpid* results in the failure of the A-P patterning mechanism within the limb bud, resulting in the entire limb adopting a posterior identity, reflected in its pattern of Hox expression.

Thus the continuing correlation between *Hox-D* gene expression and particular morphologies in limb buds whose development has been manipulated by retinoic acid or genetically strongly suggests that the establishment of particular patterns of *Hox* gene expression is required for specific types of final morphology to be produced. The expression patterns are consistent with an involvement in the specification of regional identity by a coordinate system along the proximo-distal and anterior-posterior axes involving *Hox* genes of two clusters expressed orthogonally to each other.

16.4.4 Direct Manipulation of Limb Development with *Hox* Genes

The development of retroviral vectors able to introduce genetic material into chick embryos has enabled the direct manipulation of Hox expression in the limb bud (Morgan *et al.* 1992). A replication competent retrovirus able to express the murine *Hox-D11* gene was injected into the region of flank that will give rise to the limb bud, such that by the stage when limb mesenchyme cells are still relatively plastic in their development all cells within the bud have been infected by virus containing mouse *Hox-D11*. The expression pattern of the endogenous chicken genes, including *Hox-D11*, is unchanged, while mouse *Hox-D11* is found throughout the limb bud. This ubiquitous expression of *Hox-D11* produces two effects on the limb *Hox* code. The first is to expand the region of limb bud expressing *Hox-D11* and *Hox-D10*, which corresponds to digits II, III and IV. The second is to generate a novel *Hox* code in the anterior parts of the limb that never occurs in normal development *i. e.* the combination of *Hox-D9* and *Hox-D11* without *Hox-D10* expression.

The effects of this alteration differ between the fore- and hindlimbs. In the forelimb an extra digit is produced in the anterior portion of the wing, with a similar morphology to wing digit II. This anterior region of the wing bud normally undergoes programmed cell death. In the hindlimb bud the number of digits is unchanged, but digit I and in some cases the metatarsal from which it arises show a morphology appropriate to digit II.

16.4.5 The Limb *Hox* Code

The limb makes use of *Hox* genes in axial patterning in a way independent of the main body axis. Both fore and hind limbs use 5' genes which are also used in the patterning of very posterior parts of the body such as the genitalia (Izpisua-Belmonte *et al.* 1991b). Their onset of expression is a function of the age of the limb bud and in the case of the forelimb involves the use of "posterior" genes which are not expressed in any of the other tissues at the axial level of the forelimb. Furthermore, the limb is so far unique in using paralogous *Hox* genes from two clusters in separate axial patterning events which show distinct developmental properties from each other.

The fact that 16 hours of exposure to retinoic acid is the minimum time required to both induce digit duplications and ectopic Hox expression suggests that *Hox* genes are tightly linked to the expression of morphology of a particular axial character. The timing suggests that first of all a new set of positional values are established along the A-P axis, which now contains a duplicated set of "posterior" values and no anterior ones. Once this is established the positional values can begin to be interpreted, and an early part of this involves spatially-restricted patterns of *Hox* gene expression. The pattern of *Hox* gene expression is not itself the primary axial signal.

Retinoic acid, as has been described above, is able *in vitro* to activate *Hox* gene expression. However, the *in vitro* characteristics of 5' *Hox-D* genes are not consistent with their behavior in retinoic acid treated limb buds. *In vitro* progressively more 5' and posteriorly expressed *Hox-D* genes are increasingly repressed by retinoic acid, yet in limb buds where a retinoic acid bead is implanted anteriorly, *Hox-D* genes become activated. This implies that the signalling pathways by which *Hox-D* genes are repressed by retinoic acid *in vitro* differ from those which set up the spatial domains of expression in limb buds. Further support for this idea is the fact that the apical ectodermal ridge is required for the production of ectopic domains of *Hox-D* expression, which suggests that the signalling events that establish *Hox* expression in the limb bud are more complex than a simple direct response to retinoic acid. In addition; the timing of onset of ectopic domains of *Hox-D* expression is more rapid in response to a polarising region graft than in response to a retinoic acid bead implant, suggesting that retinoic acid activation is occurring by an indirect mechanism.

The direct manipulation of *Hox-D* expression by use of retroviral vectors has provided very clear evidence for the role of *Hox* genes in regional specification. As with the mutation of *Hox-C8* in the trunk it is clear that a simple posterior prevalence model is not operating in the limb. Ubiquitous expression of *Hox-D11* produces alterations in distal skeletal structures in regions where a *Hox-D10, Hox-D11* positive region is produced, but no apparent changes in the region where a new *Hox* code that does not normally occur in development (*Hox-D10* negative, *Hox-D11* positive) is produced. If A-P identity within the limb was the result of a simple posterior prevalence model, *Hox-D11* expression alone should be sufficient to produce posterior structures in the anterior-most parts of the limb. At the moment the precise fate of the parts of the early limb bud that express particular combinations of *Hox* genes is unclear, making it difficult to determine whether there is a direct link between particular limb structures and particular gene expression domains. The fact that there is one more digit in the basic vertebrate pentadactyl limb plan than could be specified employing *Hox-D* genes alone argues against a simple one gene: one distal element type model. In addition the development of the limb is not thought to involve the establishment of lineage-restricted compartments, further supporting the hypothesis that *Hox* genes are acting as part of an "abstract" coordinate system. It is abstract in the sense that *Hox* genes are not directly defining where particular final structures occur; the *Hox* genes are presumably part of the process that sets up the framework within which the processes which actually delineate the final morphology can occur. A prediction of this would be that animals which show extensive modifications to their limb skeleton *e. g.* the digit reductions exhibited by horses, would still delineate the same number of regions of limb bud using *Hox-A* and *Hox-D* genes. The processes of producing particular cartilages in particular places, which these genes help to define the framework for, would be the site of evolutionary changes reducing digit number.

16.5   CONCLUSIONS - *HOX* GENES IN DIFFERENT EMBRYONIC CONTEXTS

In this review we have compared the way *Hox* genes may be involved in patterning three different systems within the body. It is likely that as more detailed knowledge is obtained of *Hox* gene expression and function, other contexts will be identified which use *Hox* genes in axial patterning, and there may be yet further strategies by which Hox genes are used to achieve this. *Hox* genes are known to act in different embryonic systems in Drosophila development, and it is known that their behavior differs depending on their context.

Each context has been described as if it acted in isolation from the others, but it is more likely that the various *Hox* patterning systems overlap and interact, with positional cues from several sources being required for the appearance of regional identity. Tissues which express a *Hox* code also have to interact with those which employ different systems for regional specification. The formation and early morphogenesis of the heart tube involves very anterior mesenchyme which does not express any *Antennapedia* class *Hox* genes, yet the correct formation of the outflow tract requires a contribution from crest of the third and fourth arches that express *Hox* genes of classes3 and 4. It should be stressed that *Hox* genes may be simply one component of a spatial specification system in the regions where they are expressed; for instance differential *Hox* expression is unlikely to be involved in the formation of branchial arches and their subsequent morphogenesis involving spatial differences within them such as proximo-distal, antero-posterior or medio-lateral regional identity. Finally there are regions of the body, such as the anterior parts of the head, where *Antennapedia* class *Hox* genes are never involved in specification of regional identity.

16.5.1 Distinctions Between *Hox* Codes in Different Contexts

Although their general role has been established, the way in which Hox genes act in any context is not sufficiently understood to allow very detailed functional and regulatory comparisons to be made between systems. Certainly the systems they pattern have different embryological properties, and the differences in cellular behavior between the epithelial array of compartmentalized cells in the early neural plate and the mesenchymal cells of the limb bud would seem to imply some differences in the signalling systems that regulate the Hox genes in these two contexts. There are clear differences in the way genes in a paralogous group are thought to be employed, which may reflect differences in the downstream systems that convert codes of *Hox* expression to final morphology. Given our level of understanding, it is probably best to be cautious in extrapolating findings from one embryonic system employing *Hox* genes to another.

16.5.2 Common Themes to *Hox* Gene Function

Bearing in mind the likely differences in *Hox* function in different contexts described above, it is nevertheless possible to identify general features in common between the different embryonic systems that use *Hox* genes.

In every case described so far, *Hox* genes have been found to be involved in patterning structures along an embryonic axis. This probably reflects a general role of *Antennapedia* class genes that has resulted in their conservation in higher organisms, even if there is little other similarity between the systems they are now involved in patterning. In systems such as the branchial region there is a close relationship between *Hox* expression and repeating embryonic structures, and facial morphology may be constrained to develop via an embryonic framework of compartments specified by *Hox* genes. In others contexts such as the limb there is no simple correlation with final or embryonic morphology, and no evidence of compartments. *Hox* genes can be an abstract part of a patterning system, establishing a developmental framework which eventually causes particular structures to develop in particular places. This means that considerable flexibility is possible in developmental processes in which *Hox* genes are involved in regional specification, as evolutionary alterations in final morphology could arise from new ways of interpreting a *Hox* code, rather than requiring alterations in *Hox* gene expression.

Wherever *Hox* genes are used in spatial specification, the position of a gene within a cluster reflects its expression limits within the structure it is patterning. It is particularly interesting that this occurs in the vertebrate limb, suggesting a structure where *Hox* genes have been recruited to patterning a new structure independent of the main body axis. In the Lewis model the position of genes along the chromosome resulted in their spatially-restricted domains of action, and some genes known to encode chromatin components are known to affect the maintenance (not establishment) of HOM-C expression. This has lead to the suggestion that stable modifications in chromatin structure are the mechanism by which HOM-C expression domains are maintained, and this necessitates the observed organization of the complexes. Vertebrate homologues of the *Drosophila* genes encoding chromatin components exist, which has lead to the suggestions by Gaunt that the maintenance of vertebrate *Hox* expression occurs by a similar "open for business" model (Gaunt 1991).

A difficulty with these ideas has come from the recent demonstration that DNA sequences surrounding the murine *Hox* genes are able to direct spatially-appropriate establishment and maintenance of gene expression with the appropriate tissue specificity outside of the cluster, in an abnormal chromatin context (Puschel *et al.* 1991; Whiting *et al.* 1991). It is possible that clustered organization is necessary because the control sequences for a given gene are extensive and interspersed with the regulatory domains of other genes; while it is possible to remove a single gene from the complex and have it regulated appropriately, it may not be possible for the other genes to be expressed normally in the absence of the gene and its control sequences. This could explain the maintenance of clustered organization, but it is hard to see why this should result in strict collinearity between position of a gene in a cluster and its expression domain. A particular problem with this idea is the evolution of the limb where a group of *Hox* genes were recruited to a new role and acquired the ability to respond to a new set of positional cues.

The fact that the expression domains of *Hox* genes overlap in the three contexts described here suggests that this of general functional significance. One possibility is that the combinatorial expression of genes is required for specification of segment identity, and the distribution of defects in mice lacking *Hox-C8* would be consistent with this sort of model. In contrast, the defects in the branchial region of mice lacking *Hox-A3* or *Hox-A1* correlate with the most anterior sites of expression only, without apparent effects in the more posterior regions of expression. Thus a combinatorial code of regional specification may not be a general feature of the *Hox* genes.

One possible explanation for the general phenomenon of overlapping expression domains is suggested by the observation of the progressive activation of successively more 5' *Hox-D* genes in the limb bud (Nohno *et al.* 1991). When ectopic expression domains are induced by polarising region grafts or retinoic acid, the genes are activated in the same 3' to 5' sequence as in normal limb buds. Cross regulation between *Hox* genes is known to occur in *Drosophila*, thus it is possible that the extensive expression overlap in vertebrates could be to establish or maintain appropriate temporal and spatial expression domains rather than to provide a combinatorial code for positional specification in all cases (Izpisua-Belmonte 1992).

Retinoic acid treatment of teratocarcinoma cell lines results in modulation of expression of almost all *Hox* genes that is collinear with the position of the genes in their clusters (Simeone *et al.* 1990; Simeone *et al.* 1991; Papalopulu *et al.* 1991). The three systems we have identified are all sensitive to retinoic acid, and that treatment of embryos with retinoic acid results in changes in *Hox* expression that correlate with the induced alterations in morphology. However, it is not yet clear whether the morphological effects are directly mediated by alteration in *Hox* gene expression, or that retinoic acid alters morphogenesis by some other mechanism, and that the *Hox* genes reflect that abnormal development has been induced. The timing of *Hox-D* induction in the limb suggests the latter possibility, as does the fact that the distribution of the genes in retinoic acid treated limb buds is the precise opposite of what would be expected on the basis of their in vitro response to the compound. Studies of the mechanism of retinoic acid action may be able to identify the types of molecule that regulate *Hox* expression in the different embryonic contexts described here.

As well as operating in different spatial contexts simultaneously during development, it is likely that *Hox* genes play different roles at different times in the development in the same part of the body. Expression differences between paralogous genes have been described in the head or genes whose early expression domains are more similar, and a similar phenomenon has been found during the development of the spinal cord in the trunk. Positional specification is not necessarily their role in these later events. *Hox* genes could act as a set of transcription factors establishing the various stages in the emergence of specific cell types as well, a process which may be independent of positional cues.

REFERENCES

Akam, M. 1987. The molecular basis for metameric pattern in the *Drosophila* embryo. *Development*, **101**, 1–22.

Chevallier, A. 1975. Role du mesoderme somitique dans le developpement de la cage thoracique de l'embryon de'oiseau. I. origine du segment sternal et mecanismes de la differenciation des cotes. *JEEM*, **33**, 291–311.

Chisaka, O., & Capecchi, M. 1991. Regionally restricted developmental defects resulting from targeted disruption of the mouse homeobox gene *hox1.5*. *Nature*, **350**, 473–479.

Dolle, P., Izpisua-Belmonte, J.-C., Falkenstein, H., Renucci, A., & Duboule, D. 1989. Coordinate expression of the murine *Hox*-5 complex homeobox-containing genes during limb pattern formation. *Nature*, **342**, 767–772.

Duboule, D., & Dolle, P. 1989. The structural and functional organization of the murine *HOX* gene family resembles that of *Drosophila* homeotic genes. *EMBO J.*, **8**, 1497–1505.

Fraser, S., Keynes, R., & Lumsden, A. 1990. Segmentation in the chick embryo hindbrain is defined by cell lineage restrictions. *Nature*, **344**, 431–435.

Gaunt, S. J. 1991. Expression patterns mouse *Hox* genes: clues to an understanding of developmental and evolutionary strategies. *Bioessays*, **13**, 505–513.

Gaunt, S. J., Krumlauf, R., & Duboule, D. 1989. Mouse homeo-genes within a subfamily, *Hox*-1.4, -2.6 and -5.1, display similar anteroposterior domains of expression in the embryo, but show state- and tissue-dependent differences in their regulation. *Development*, **107**, 131–141.

Graham, A., Maden, M., & Krumlauf, R. 1991. The murine *Hox*-2 genes display dynamic dorsoventral patterns of expression during central nervous system development. *Development*, **112**, 255–264.

Hall, B. 1987. Tissue Interactions in Head Development and Evolution. In *Development and Evolutionary Aspects of the Neural Crest*, Maderson, P. F. A. (ed). pp. 215 259, New York: John Wiley.

Hunt, P., Gulisano, M., Cook, M., Sham, M., Faiella, A., Wilkinson, D., Boncinelli, E., & Krumlauf, R. 1991a. A distinct *Hox* code for the branchial region of the head. *Nature*, **353**, 861–864.

Hunt, P., Wilkinson, D., & Krumlauf, R. 1991b. Patterning the vertebrate head: murine *Hox* 2 genes mark distinct subpopulations of premigratory and migrating neural crest. *Development*, **112**, 43–51.

Ingham, P., & Martinez-Arias, A. 1992. Boundaries and fields in early embryos. *Cell*, **68**, 221–235.

Izpisua-Belmonte, J.-C. 1992. Homeobox Genes and Pattern Formation in the Vertebrate Limb. *Dev. Biol.*, **152**, 26–36.

Izpisua-Belmonte, J.-C., Dolle, P., Renucci, A., Zappavigna, V., Falkenstein, H., & Duboule, D. 1990. Primary structure and embryonic expression pattern of the mouse *Hox-4.3* homeobox gene. *Development*, **110**, 733–745.

Izpisua-Belmonte, J.-C., Tickle, C., Dolle, P., Wolpert, L., & Duboule, D. 1991a. Expression of homeobox *Hox-4* genes and the specification of position in chick wing development. *Nature*, **350**, 585–589.

Izpisua-Belmonte, J.-C., Falkenstein, H., Dolle, P., Renucci, A., & Duboule, D. 1991b. Murine genes related to the *Drosophila* abdB homeotic gene are sequentially expressed during development of the posterior part of the body. *EMBO J.*, **10**, 2279–2289.

Izpisua-Belmonte, J.-C., Brown, J. M., Duboule, D., & Tickle, C. 1992a. Expression of *Hox*-4 Genes in the chick Wing Links Pattern Formation to the Epithelial-Mesenchymal Interactions That Mediate Growth. *EMBO J.*, **11**, 1451–1457.

Izpisua-Belmonte, J.-C., Ede, D. A., Tickle, C., & Duboule, D. 1992b. The mis-expression of Posterior *Hox*-4 Genes in Talpid (ta3) Mutant Wings Correlates with the Absence of Anteroposterior Polarity. *Development*, **114**, 959–963.

Kessel, M. 1992. Respecification of vertebral identities by retinoic acid. *Development*, **115**, 487–501.

Kessel, M., & Gruss, P. 1991. Homeotic transformations of murine prevertebrae and concommitant alteration of *Hox* codes induced by retinoic acid. *Cell*, **67**, 89–104.

Keynes, R., & Stern, C. 1985. Segmentation and neural development in vertebrates. *TINS*, **8**, 220–223.

Lewis, E. 1978. A gene complex controlling segmentation in *Drosophila*. *Nature*, **276**, 565–570.

Lufkin, T., Dierich, A., Lemeur, M., Mark, M., & Chambon, P. 1991. Disruption of the *Hox*-1.6 homeobox gene results in defects in a region corresponding to its rostral domain of expression. *Cell*, **66**, 1105–1119.

Lumsden, A. 1990. The cellular basis of segmentation in the developing hindbrain. *TINS*, **13**, 329–335.

Lumsden, A., & Keynes, R. 1989. Segmental patterns of neuronal development in the chick hindbrain. *Nature*, **337**, 424–428.

Lumsden, A., Sprawson, N., & Graham, A. 1991. Segmental origin and migration of neural crest cells in the hindbrain region of the chick embryo. *Development*, **113**, 1281–1291.

McGinnis, W., & Krumlauf, R. 1992. Homeobox Genes and Axial Patterning. *Cell*, **68**, 283–302.

Morgan, B. A., Izpisua-Belmonte, J.-C., Duboule, D., & Tabin, C. 1992. Targeted expression of *Hox-4.6* in the avian limb bud causes apparent homeotic transformations. *Nature*, **358**, 236–239.

Mouellic, H. Le, Lallemand, Y., & Brulet, P. 1992. Homeosis in the mouse induced by a null mutation in the homeo-gene *Hox-3.1*. *Cell*, **69**, 251–264.

Noden, D. 1986. Patterning of avian craniofacial muscles. *Dev. Biol.*, **116**, 347–356.

Noden, D. 1988. Interactions and fates of avian craniofacial mesenchyme. *Development (Supplement; Craniofacial Development)*, **103**, 121–140.

Nohno, T., Koyama, E., Ohyama, K., Myokai, F., Kuroiwa, A., Saito, T., & Tanaguchi, S. 1991. Involvement of the chox-4 chicken homeobox genes in determination of anteroposterior axial polarity during limb development. *Cell*, **64**, 1197–1205.

Oliver, G., DeRobertis, E. M., Wolpert, L., & Tickle, C. 1990. Expression of a homeobox gene in the chick wing bud following application of retinoic acid and grafts of polarizing region tissue. *EMBO J.*, **9**, 3093–3099.

Papalopulu, A. Graham N., & Krumlauf, R. 1989. The murine and *Drosophila* homeobox clusters have common features of organization and expression. *Cell*, **57**, 367–378.

Papalopulu, N., Lovell-Badge, R., & Krumlauf, R. 1991. The expression of murine *Hox-2* genes is dependent on the differentiation pathway and displays collinear sensitivity to retinoic acid in F9 cells and *Xenopus* embryos. *N.A.R.*, **19**, 5497–5506.

Peterson, R. L., Jacobs, D. F., & Awgulewitsch, A. 1992. *Hox*-3.6 - Isolation and Characterization of a New Murine Homeobox Gene Located in the 5' Region of the *Hox*-3 Cluster. *Mech. Develop*, **37**, 151–166.

Puschel, A., Balling, R., & Gruss, P. 1991. Separate elements cause lineage restriction and specify boundaries of *Hox-1.1* expression. *Development*, **112**, 279–288.

Richman, J., & Tickle, C. 1989. Epithelia are interchangeable between facial primordia of chick embryos and morphogenesis is controlled by the mesenchyme. *Dev. Biol.*, **136**, 201–210.

Romer, A. 1971. *The vertebrate body, shorter version.* 4th edn. Philadelphia: W. B. Saunders Company.

Simeone, A., Acampora, D., Arcioni, L., Andrews, P. W., Boncinelli, E., & Mavilio, F. 1990. Sequential activation of *HOX* 2 homeobox genes by retinoic acid in human embryonal carcinoma cells. *Nature*, **346**, 736–766.

Simeone, A., Acampora, D., Nigro, V., Faiella, A., D'Esposito, M., Stornaiuolo, A., Mavilio, F., & Boncinelli, E. 1991. Differential regulation by retinoic acid of the homeobox genes of the four *HOX* loci in human embryonal carcinoma cells. *Mech. Develop*, **33**, 215–227.

Stern, C., Fraser, S., Keynes, R., & Primmett, D. 1988. A cell lineage analysis of segmentation in the chick embryo. *Development*, **104**, 231–244. (Supplement: Mechanisms of Segmentation (Eds. V. French; P. Ingham; J. Cooke and J. Smith)).

Stern, C., Jaques, K., Lim, T., Fraser, S., & Keynes, R. 1991. Segmental lineage restrictions in the chick embryo spinal cord depend on the adjacent somites. *Development*, **113**, 239–244.

Thorogood, P. 1988. The developmental specification of the vertebrate skull. *Development*, **103**, 141–153.

Tickle, C. 1991. Retinoic acid and chick limb bud development. *Dev. Suppl*, **1**, 113–121.

Whiting, J., Marshall, H., Cook, M., Krumlauf, R., Rigby, P., Stott, D., & Allemann, R. 1991. Multiple spatially-specific enhancers are required to reconstruct the patter of *Hox-2.6* gene expression. *Gene Develop.*, **5**, 2048–2059.

Wilkinson, D., Bhatt, S., Cook, M., Boncinelli, E., & Krumlauf, R. 1989. Segmental expression of *hox* 2 homeobox-containing genes in the developing mouse hindbrain. *Nature*, **341**, 405–409.

Yokouchi, Y., Sasaki, H., & Kuroiwa, A. 1991. Homeobox gene expression correlated with the bifurcation process of limb cartilage development. *Nature*, **353**, 443–445.

17. LOOKING AT EARLY DEVELOPMENT IN *XENOPUS LAEVIS* USING NMR MICRO-IMAGING

Russell E. Jacobs and Scott E. Fraser

Division of Biology
Beckman Institute (139-74)
California Institute of Technology
Pasadena, CA 91125, USA

17.1 INTRODUCTION

High-resolution magnetic resonance imaging has been applied to image individual cell lineages in developing frog embryos. By injecting a single cell with a contrast agent based on the gadolinium chelate, Gd-DTPA-dextran, its progeny could be followed uniquely over time. This technique allows the continued acquisition of three-dimensional images of the developing embryo for periods of several days, thus permitting both prospective and retrospective analyses of cell lineages and movements. As an example, we used high-resolution MRI to follow *in vivo* the relative motions and reorganizations of the different cell layers during amphibian gastrulation and neurulation.

Analysis of cell lineages and cell movements is central to understanding the processes by which an adult vertebrate, with its many types of cells, develops from a single pluripotent fertilized egg. Because the large number of indistinguishable cells in the vertebrate embryo prohibits analysis by direct observation (Sulston & Horvitz 1977; Kimble & Hirsh 1979; Hedgecock *et al.* 1983; Greenwald *et al.* 1983; Hedgecock 1985), tracing cell movements or lineages requires some means to render a cell and its progeny unique. Previous approaches have successfully employed injection of individual precursors with membrane impermeable enzymes (Weisblat *et al.* 1978; Balakier & Pedersen 1982) or fluorescent dyes (Weisblat *et al.* 1980; Gimlich & Braun 1985; Bronner & Fraser 1988; Wetts & Fraser 1988), or infection with a retroviral agent (Price *et al.* 1987; Sanes *et al.* 1986; Turner & Cepko 1987). In some cases, the labeled descendants can be imaged within the living embryo (*cf.* Kimmel & Warga 1986.) Most embryos become optically opaque at later stages of development, thus visualization of the cells is typically achieved by fixing, sectioning, and staining the specimen. The need to fix and process the tissue prohibits the direct observations of ongoing developmental events; instead, they must be inferred by comparing the results obtained from different embryos fixed at different stages.

To permit the direct observation of ongoing developmental events in living frog embryos, we have employed high-resolution magnetic resonance imaging (MRI). MRI is a non-invasive method by which three-dimensional images of optically opaque specimens may be obtained on a time scale faster than or comparable to the cell division time. Here, the descendants of individual precursors in the intact embryo are labeled by microinjection of a

Experimental and Theorietcal Advances in Biological Pattern Formation,
Edited by H.G. Othmer *et al.*, Plenum Press, New York, 1993

183

stable, nontoxic, membrane impermeable MRI lineage tracer. Because the complete time-series of high-resolution three-dimensional MR images can be analyzed forward or backward in time, one can fully reconstruct the cell divisions and cell movements responsible for any particular descendant(s). Unlike previous methods, where labeled cells are identified at the termination of the experiment, this technique allows the full kinship relationships of a clone to be determined as the clone expands. These attributes make MRI an ideal technique for examining gastrulation and neurulation (*cf.* Jacobson & Gordon 1976). In this report we show that high-resolution MRI can be used to gain new and useful insights into the spatial and temporal aspects of these phenomena.

MRI is a qualitatively different method of visualization than the light microscopy employed in previous lineage studies. In this study, the recorded signal arises from the hydrogen nuclear spin of water molecules. Spatial localization of the nuclear magnetic resonance (NMR) signal is made possible by superimposing various magnetic field gradients on the usual static magnetic field (Stehling *et al.* 1991; Morris 1986; Mansfield & Morris 1982). The use of a set of three orthogonal gradients allows the NMR signal to be parsed into a matrix of intensities, one for each volume element (voxel), yielding the 3 dimensional MR image. Contrast in the magnetic resonance image arises from voxel to voxel variations in the water concentration and local environment. Variations in the local environment (*e. g.* proximity to a paramagnetic center) and state (*e. g.* mobile versus immobile) of the water modulate the NMR relaxation times T_1 and T_2 (Moonen *et al.* 1990). We exploit the inherent T_1 contrast by employing an imaging protocol which preferentially weights the signal from those regions with spins having short T_1 relaxation times[1]. This yields an image where the intensity is a monotonic function of the relaxation rate $(1/T_1)$, all other factors being equivalent[2].

To perform cell lineage analyses, a lineage tracer with a large effect on the NMR signal is required. MRI contrast enhancement agents have such an effect, increasing the relaxation rates of nearby nuclei to provide contrast not naturally found in the specimen. For an MRI contrast agent to be a good lineage tracer it must meet three criteria: 1) induce a local signal that is characteristically different from that of the rest of the sample, 2) be physiologically inert, and 3) be membrane impermeable, thus remaining within the originally-labeled cell and its progeny. The contrast agent used in this study is a covalent conjugate of dextran with DTPA to which gadolinium has been chelated (Wang *et al.* 1990; Gibby *et al.* 1989). Although Gd is a toxic lanthanide, high affinity chelators (*e. g.* DTPA, DOTA, EDTA)[3] effectively protect living systems from its deleterious consequences (Lauffer 1987). We employ a 1-fold molar excess of the DTPA-dextran chelator to ensure that a minimal amount of free Gd ion is present. The Gd-DTPA-dextran tracer is an efficient T_1 relaxation agent (Lee 1991). Thus, our T_1 weighted imaging protocol provides MR images with enhanced intensity in those cells containing the lineage tracer. Because this tracer is a close analog of fluorescent dextran lineage tracers already in use, established techniques can be utilized for its injection into embryonic cells. Unlike the fluorescent probes which bleach and generate reactive by-products when observed,

---

[1]We employ the steady state free precession technique with a short repetition time (150ms) so that signal is observed preferentially from those spins which return to equilibrium most rapidly, *i. e.* those with short T_1's. A short echo time (4.5ms) is used to minimize T_2 effects.

[2]Inhomogeneities in water concentration, magnetic susceptibility, and diffusivity, as well as bulk flow and exchange of water between dissimilar environments can all lead to spatial variations in the MR signal. Although these issues are important in the clinical context as discussed by Mansfield & Morris (1982), the samples examined here are relatively homogeneous in all these features and the effects of the contrast agent dominate over all these possibly confounding effects.

[3]Abbreviations: DTPA, diethylenetriaminepentaacetic acid; EDTA, ethylenediaminetetraacetic acid; DOTA, 1,4,7,10-tetraazacyclododecane-N,N',N'',N'''-tetraacetic acid.

neither the MRI contrast agent nor the surrounding cytoplasm are perturbed chemically by the imaging experiment.

Magnification in the MRI experiment is achieved through the imposition of increasingly larger magnetic field gradients. To achieve the micron scale resolution necessary to follow developing cell lineages, both the static and gradient magnetic fields are several orders of magnitude larger than those employed in clinical settings (Aguayo *et al.* 1986; Johnson *et al.* 1986; Behling *et al.* 1989; Jenner *et al.* 1988). Increasing the resolution from the 1 mm^3 scale of clinical images to the $10 \mu m^3$ scale required here will decrease the voxel size by a factor of 10^6. Because the concentration of water is relatively constant, the signal strength is decreased by the same factor. NMR is an intrinsically low signal-to-noise phenomenon, thus attempts to measure signals from these small volumes are fraught with difficulties (Kuhn 1990; Callaghan 1991). Here we report images at $12 \mu m$ resolution using a 7T system with the RF coil, gradient framework, imaging protocol, and sample preparation optimized for *in vivo* micron scale imaging (Cho *et al.* 1988).

To follow the descendant of a single blastomere of a 16 cell stage embryo, the cell is injected with the Gd-DTPA-dextran tracer, and the embryo is imaged repeatedly over several days. Figure 17.1 presents a series of such images of a single embryo in which the label was introduced into a blastomere in the animal hemisphere adjacent to the prospective dorsal midline (blastomere DA in the Wetts & Fraser (1989) nomenclature) at the 16 cell stage. In the volume representations shown in Figure 17.1, labeled cells appear as the high intensity volumes. The images show the progression from blastula stages (a & b), through gastrulation (c-f), and neurulation (g-h). Progeny of the originally labeled cell (blastomere DA) appear as an intense region in the animal hemisphere. The ability of MRI to generate arbitrarily-oriented serial sections through the embryo permits more detailed observations of internal structures in living embryos. Figure 17.2 shows sections through the images shown in Figure 17.1. The true three-dimensional character of our MRI images allows both structures and labeled cells within the interior of the embryo to be clearly visualized while the animal is developing.

Gastrulation and neurulation are key steps in embryogenesis during which complex and highly-coordinated movements of a large number of cells take place. Although some species are sufficiently transparent to permit the key internal cell movements to be followed directly with light microscopy (*cf.* Gustafson & Wolpert 1967), the amphibian embryo is opaque at this stage. Thus, analyses of these processes in *Xenopus laevis* have relied upon time-lapse cinemicroscopy of surface cell movements, histological examination of fixed specimens, and a variety of explantation techniques (Wilson *et al.* 1989; Aker *et al.* 1986; Keller *et al.* 1991). These studies have demonstrated the coupling of convergence of cells towards the dorsal midline and extension of the embryonic axis (convergent extension). These insightful experiments demonstrate the importance of cell shape changes, radial intercalation, and medio-lateral intercalation in gastrulation; however, questions about the coordination and timing of these events remain unanswered. MR imaging offers the opportunity to observe directly *in vivo* the movements of both surface and deep cells while development progresses. The three dimensional nature of the data permits the display of both volume representations Figure (17.1) and single slices Figure (17.2) of an embryo from about stage 8 through stage 21. As expected from previous studies, the clone of labeled cells narrows medio-laterally (converges) and lengthens rostro-caudally (extends) during these key developmental stages.

Figure 17.1 and Figure 17.2 represent the first *in vivo* examination of the spatio-temporal relationships of internal and external morphogenic processes taking place through gastrulation and into neurulation in the frog embryo. In the late blastula stage (Figure 17.1(a) & Figure 17.2(a)) the labeled clones form a thick multilayered patch of contiguous cells. Nine hours later, shortly after the onset of gastrulation (Figure 17.1(b) & Figure 17.2(b)), the

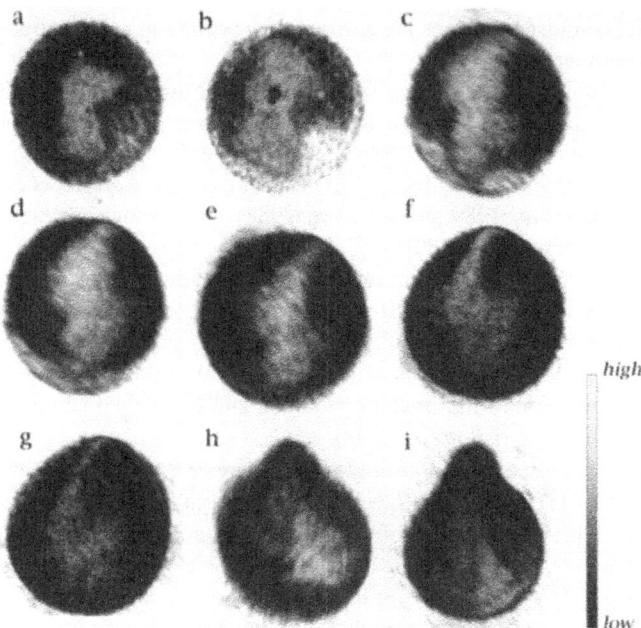

Figure 17.1. Magnetic resonance images of a single developing frog embryo taken at sixteen successive time points; **a-p** were recorded at 12, 21, 24, 27, 29, 33, 35, 45, and 47 hours after fertilization, respectively. The data sets are shown as semi-transparent volume renderings (*vide infra*) looking at the animal pole of the developing embryo. A single 16 cell stage blastomere (DA) of this embryo was injected with the contrast agent, Gd-DTPA-dextran, approximately 14 hours before image **a** was recorded. Descendants of the labeled cell appear as high intensity patch of cells, *e. g.* the contiguous bright volume in the animal hemisphere of the first row images. In **a** the embryo is in the late blastula stage. In **b** the embryo is undergoing epiboly and is experiencing gastrulation in **c-f**. It passes to an early neurula stage in **g-i**.

To introduce contrast-label, a single blastomere was pressure injected with ~20nl of a 100mg/ml solution of Gd-DTPA-dextran (17kD average mwt). Embryos were typically sealed in a 2.5mm O.D. glass NMR tube with a small amount of 10% Ringer's solution. A 1% solution of low temperature gelling agar in rearing solution is used for later stage embryos to ensure minimal movement of the animal during the imaging procedure. The MR imaging procedure had no obvious adverse effects on the development of the embryos which were routinely maintained in the spectrometer at $17°C$ for several days at a time. A three-dimensional spin echo pulse sequence with an initial $\pi/4$ radio frequency pulse, 150 ms recycle time, 4.5ms echo time, $10\mu s$ dwell time, and 256x256x(32 or 64) data array was employed. Time to record a full 3D data set was 90 minutes. Because of the short recycle time, these are T_1 weighted images; thus, significantly enhanced intensity is expected from regions in close proximity to the Gd ions as observed. The slice direction data was zero-filled to 128 slices before the three-dimensional Fourier transform. Given the strength of the magnetic field gradients employed (typically 75 G/cm for the in-plane gradients), this data collection scheme yielded an in-plane pixel resolution of $12\mu m$ and slice thickness of $36\mu m$.

For visualization purposes, we used VoxelView (Vital Images, Inc., Fairfield, Iowa) and peformed a quadratic interpolation between planes to achieve isotropic display resolution. The baseline intensity was raised to mask low intensity noise and the intensities mapped as shown in the color scale bar. In all the figures, the bottom of the color scale bar indicates low intensity and the top high intensity. To permit the labeled cells to be visualized amongst the unlabeled cells, the opacity of each voxel was adjusted so that higher voxel intensity values have exponentially higher opacity values. Thus cavities are rendered transparent, the unlabeled cells semi-transparent, and labeled cells opaque. The scale of the images is indicated by the thickness of the color scale bar = $125\mu m$

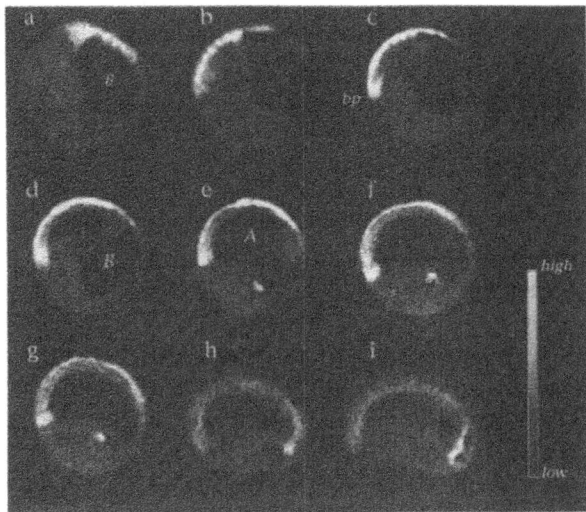

Figure 17.2. Cross-sections of the embryo shown in Figure 17.1. The single sections are taken along the dorsal midline of the embryo. Each is oriented with the animal pole to the right. The blastocoel (B), archenteron (A), and blastopore (bp) are given for orientation. As expected, this water-filled interior cavities are at essentially background intensity. These slices are derived from the same data represented in Figure 17.1.

Convergent extension of labeled neurectoderm is plainly evident along the dorsal midline in these images (see text for details). The slices in **b-d** point out the rostral extent of thickening of the labeled neurectodermal layer. The leading edge of advancing unlabeled axial mesoderm lies immediately below the transition from thick to thin labeled neurectoderm: counterclockwise from the leading edge the labeled layer is significantly thicker than clockwise. The midbrain-hindbrain junction is located just to the left (caudal) of the top of slice **e**. This is where radial intercalation driven extension begins, resulting in thinning of the labeled neurectoderm.

In **e-g** we see a separated group of labeled cells in a location consistent with presumptive heart tissue. The DA blastomere gives rise to heart tissue (Moody 1987; Dale & Slack 1987), but is thought to do so only at later stages of development. In other sections of image **e** (not shown) a trail of label can be seen connecting the labeled cells in the dorsum with the deposit of labeled cells in the blastocoel.

initially thick labeled patch has thinned and spread considerably reflecting the movements of epiboly preceding gastrulation. At this stage, the vegetal edge of the clone of labeled cells has extended around the dorsal lip highlighting the post involution side of the blastopore. In Figure 17.1(c), the labeled patch of ectodermal cells has undergone convergent extension to reach nearly the animal pole. The local increase in label intensity along the dorsal midline in the volume view of Figure 17.1(c) appears to be the product of a change in cell geometry (columnarization) in the developing neurectoderm, noticeable as the thickening of the labeled neurectoderm in Figure 17.2(c). After this geometry change, the cells continue to rearrange through intercalary motions. Interestingly, Figure 17.2(b)-(d) show a correlation between the location of the leading edge of the underlying axial mesoderm as it advances rostrally and the thickening of the neurectoderm. Radial intercalation brings about a dorso-ventral thinning of the labeled clone, which the images show is initiated at the future midbrain-hindbrain border (Figure 17.1(e) & Figure 17.2(e)) (Eagleson & Harris 1990). From this site of initiation, the thinning spreads caudally (towards the left in Figure 17.2e); four hours later, the thinning has progressed rostrally as well (Figure 17.2(f)). During this thinning of the neurectoderm, the underlying chordamesoderm undergoes medio-lateral intercalation of labeled and unlabeled cells, as evidenced by the 'salt and pepper' pattern of intensities (Keller *et al.* 1985)[4]. This progressive medio-lateral intercalation and rostro-caudal extension continues for several hours (Figure 17.2(e)-(h)). By Figure 17.2(g) the blastocoel has been completely obliterated and in Figure 17.2(h) we see the early neurula with labeled cells along the length of the neural axis. The ability of MRI to provide full *in vivo* three dimensional images augments previous *in vitro* analyses. For example, the MR images yield details ranging from the relative timing of the extension movements (ectoderm before mesoderm) to the choreography of radial convergence in the neurectoderm (initiation at the hindbrain-midbrain boundary).

The images of developing embryos shown here point out both the present limitations and future potential of magnetic resonance imaging microscopy for visualization of cell lineages in developmental biology. MRI permits structures within the living embryo, usually inaccessible to light microscopy, to be imaged clearly and non-destructively over a period of days. In light microscopy of sectioned embryos, the 3D images must be synthesized from the 2D images, requiring careful alignments of adjacent slices and corrections for distortions from fixation and sectioning. MRI generates a true 3D image with all the data points automatically in register. One may take any arbitrarily oriented 2D 'slice' through the 3D image that best allows a detailed examination of the region of interest (see Figures 17.1 & 17.2). These advantages are not without a cost; when compared to optical images of histological sections, it is clear that MRI has a smaller signal-to-noise ratio and less resolution. At later stages of development in the frog, the cells become too small to be resolved clearly in the present MR images. Of course, resolution should not be confused with the ability of an imaging technique to detect the presence of an entity smaller than the theoretical spatial resolution of the technique. In fluorescence microscopy, a small highly fluorescent feature will manifest itself as a bright pixel, even when the feature is smaller than the resolution of the light microscope. Therefore, one knows of the existence of the feature and its location to within ± 1 pixel. The same is true of MRI as it can readily detect features smaller than the MRI spatial resolution, if the feature(s) of interest is uniquely labeled.

The feasibility of MR imaging in the sub-millimeter resolution range has been demonstrated on avian (Effmann *et al.* 1988) and insect tissues (Conner *et al.* 1988). Previous studies have indicated the theoretical possibility of MR imaging in the micron resolution range (Eccles & Callaghan 1986; Ahn & Cho 1989; Maki *et al.* 1988). The images presented here demonstrate *in vivo* images of developing embryos at these resolutions, yielding information not available through optical methods. Intrinsic contrast makes internal morphology visible, while labeling specific cells with an MRI contrast agent makes the labeled cell and its descendants distinguishable from surrounding tissue. The ability to follow both the labeled and surrounding unlabeled cells permitted the examination of gastrulation and neurulation in the amphibian embryo. Not only were we able to see the relative positions of surface ectodermal cells and deep mesodermal cells over time but also the changing patterns of MRI label allowed us to infer the location and timing of specific types of cellular reorganizations (*e. g.* medio-lateral versus radial intercalation). Applications of protocols used in clinical MR imaging (*e. g.* multiple echo, diffusion, flow, chemical shift) to the realm of MR microscopy combined with labeling of specific tissues (and specific cells within tissues) at specific times during development promise new insights into the chemistry and the physiology of developing biological systems.

---

[4]This type of intercalation of deep cells has been observed by Keller *et al.* (1985) by grafting tissue with fluorescently labeled cells into unlabeled embryos.

ACKNOWLEDGMENTS We thank Z.H. Cho, C.B. Ahn, and S. Juh for their assistance and encouragement in the initial phases of this project, and J. Shih, A. Collazo and J. Ivins for helpful comments on the manuscript. Supported by NIH grant HD 25390 and a gift from the Monsanto-Searle Co.

REFERENCES

Aguayo, J. B., Blackband, S. J., Mattingly, J. Schoeniger M. A., & Hintermann, M. 1986. NMR imaging of a single cell. *Nature (Lond)*, **322**, 190–191.

Ahn, C. B., & Cho, Z. H. 1989. A Generalized Formulation of Diffusion Effects in MIcron Resolution Nuclear Magnetic Resonance Imaging. *Med. Phys.*, **16**(1), 22–28.

Aker, R. M., Phillips, C. R., & Wessel, N. K. 1986. Expression of an Epidermal Antigen Used to Study Tissue Induction in the Early *Xenopus laevis* Embryo. *Science*, **231**, 613–616.

Balakier, H., & Pedersen, R. A. 1982. Allocation of cells to inner cell mass and trophectoderm lineages in pre implantation mouse embryos. *Dev. Biol.*, **90**, 352–362.

Behling, R. W., Tubbs, H. K., Cockman, M. D., & Jelinski, L. W. 1989. Stroboscopic NMR microscopy of the carotid artery. *Nature (Lond)*, **341**, 321–323.

Bronner, F. M., & Fraser, S. E. 1988. Cell lineage analysis reveals multipotency of some avian neural crest cells. *Nature (Lond)*, **335**, 161–164.

Callaghan, P. T. 1991. *Principles of nuclear magnetic resonance microscopy*. New York: Oxford University Press.

Cho, Z. H., Ahn, C. B., Juh, S. C., Lee, H. K., Jacobs, R. E., Lee, S., Yi, J. H., & Jo, J. M. 1988. NMR microscopy with 4-μm resolution theoretical study and experimental results. *Med Phys*, **15**, 815–824.

Conner, W. E., Johnson, G. A., Cofer, G. P., & Dittrich, K. 1988. Magnetic resonance microscopy: in vivo sectioning of a developing insect. *Experientia*, **44**, 11–12.

Dale, L., & Slack, J. M. W. 1987. Fate Map for the 32-Cell Stage of *Xenopus* Embryo. *Development*, **99**, 527–551.

Eagleson, G. W., & Harris, W. A. 1990. Mapping of the presumptive brain regions in the neural plate of *Xenopus-laevis*. *J. Neurobiol.*, **21**, 427–440.

Eccles, C. D., & Callaghan, P. T. 1986. High-resolution imaging. The NMR microscope. *J. Magn. Reson*, **68**(2), 393–398.

Effmann, E. L., Johnson, G. A., Smith, B. R., Talbott, G. A., & Cofer, G. 1988. Magnetic resonance microscopy of chick embryos *in ovo*. *Teratology*, **38**, 59–65.

Gibby, W. A., Bogdan, A., & Ovitt, T. W. 1989. Cross-linked DTPA polysaccharides for magnetic resonance imaging synthesis and relaxation properties. *Invest. Radiol.*, **24**, 302–309.

Gimlich, R. L., & Braun, J. 1985. Improved fluorescent compounds for tracing cell lineage. *Dev Biol*, **109**, 509–514.

Greenwald, I. S., Sternberg, P. W., & Horvitz, H. R. 1983. The *lin*-12 locus specifies cell fates in *caenorhabditis-elegans*. *Cell*, **34**, 435–444.

Gustafson, T., & Wolpert, L. 1967. Cellular Movement and Contact in Sea Urchin Morphogenesis. *Biol. Rev.*, **42**, 442–498.

Hedgecock, E. M. 1985. Cell lineage mutants in the nematode *caenorhabditis-elegans*. *Trends Neurosci.*, **8**, 288–293.

Hedgecock, E. M., Sulston, J. E., & Thomson, J. N. 1983. Mutations affecting programmed cell deaths in the nematode *caenorhabditis-elegans*. *Science (Wash. D.C.)*, **220**, 1277–1279.

Jacobson, A. G., & Gordon, R. 1976. Changes in the shape of the developing vertebrate nervous system analyzed experimentally, mathematically, and by computer simulation. *J. Exp. Zool.*, **197**, 191–246.

Jenner, C. F., Xai, Y., Eccles, C. D., & Callaghan, P. T. 1988. Circulation of water within wheat grain revealed by nuclear magnetic resonance micro-imaging. *Nature*, **336**, 399–402.

Johnson, G. A., Thompson, M. B., Gewalt, S. L., & Hayes, C. E. 1986. Nuclear magnetic resonance imaging at microscopic resolution. *J. Magn. Reson.*, **68**, 129–37.

Keller, R., Danilchik, M., Gimlich, R., & Shih, J. 1985. Convergent extension by cell intercalation during gastrulation of *Xenopus-laevis*. In *Molecular Biology*, Edelman, G. M. (ed). New York: Alan R. Liss. New Series.

Keller, R., Shih, J., & Wilson, P. 1991. Cell Motility, Control, and Function of Convergence and Extension During Gastrulation in *Xenopus*. In Gastrulation: Movements, Patterns, and Molecules, Keller, W. Clark R., & Griffen, F. (eds). New York: Plenum Press.

Kimble, J., & Hirsh, D. 1979. The post embryonic cell lineages of the hermaphrodite and male gonads in caenorhabditis-elegans. *Dev. Biol.*, **70**, 396–417.

Kimmel, C. B., & Warga, R. M. 1986. Tissue-specific cell lineages originate in the gastrula of the zebrafish. *Science (Wash. D.C.)*, **231**, 365–368.

Kuhn, W. 1990. NMR microscopy - fundamentals, limits, and possible applications. *Angew. Chem. Int. Engl.*, **29**, 1–112.

Lauffer, R. B. 1987. Paramagnetic Metal Complexes as Water Proton Relaxation Agents for NMR Imaging: Theory and Design. *Chem. Rev.*, **87**, 901–927.

Lee, D. H. 1991. Mechanisms of contrast enhancement in magnetic resonance imaging. *Can. Assoc. Radiol. J.*, **42**, 6–12.

Maki, J. H., Johnson, G. A., Cofer, G. P., & MscFall, J. R. 1988. SNR improvements in NMR microscopy using DEFT. *J. Magn. Reson*, **80**, 482–492.

Mansfield, P., & Morris, P. G. 1982. *NMR Imaging in Biomedicine*. New York: Academic Press.

Moody, S. A. 1987. Fates of the Blastomeres of the 16-Cell Stage *Xenopus* Embryo. *Dev. Biol*, **119**, 560–578.

Moonen, C. T. W., Van, Z. P. C. M., Frank, J. A., Le, B. D., & Becker, E. D. 1990. Functional magnetic resonance imaging in medicine and physiology. *Science (Wash. D.C.)*, **250**, 53–61.

Morris, P. G. 1986. *Nuclear Magnetic Resonance Imaging in Biology & Medicine*. New York: Oxford University Press.

Price, J., Turner, D., & Cepko, C. 1987. Lineage analysis in the vertebrate nervous system by retrovirus-mediated gene transfer. *Proc. Natl. Acad. Sci. USA*, **84**, 156–160.

Sanes, J. R., Rubenstein, J. L. R., & Nicolas, J. F. 1986. Use of a recombinant retrovirus to study post-implantation cell lineage in mouse embryos. *EMBO J.*, **5**, 3133–3142.

Stehling, M. K., Turner, R., & Mansfield, P. 1991. Echo-planar imaging magnetic resonance imaging in a fraction of a second. *Science (Wash. D.C.)*, **254**, 43–50.

Sulston, J. E., & Horvitz, H. R. 1977. Post embryonic cell lineages of the nematode *caenorhabditis-elegans*. *Dev. Biol.*, **56**, 110–156.

Turner, D. L., & Cepko, C. L. 1987. A common progenitor for neurons and glia persists in rat retina late in development. *Nature (Lond)*, **328**, 131–136.

Wang, S. C., Wikstrom, M. G., White, D. L., Klaveness, J., Holtz, E., Rongved, P., Moseley, M. E., & Brasch, R. C. 1990. Evaluation of gd-DTPA-labeled dextran as an intravascular MR contrast agent: imaging characteristics in normal rat tissues. *Radiology*, **175**, 483–488.

Weisblat, D. A., Sawyer, R. T., & Stent, G. S. 1978. Cell lineage analysis by intracellular injection of a tracer enzyme. *Science (Wash. D.C.)*, **202**, 1295–1298.

Weisblat, D. A., Zackson, S. L., Blair, S. S., & Young, J. D. 1980. Cell lineage analysis by intra cellular injection of fluorescent tracers. *Science (Wash. D.C.)*, **209**, 1538–1541.

Wetts, R., & Fraser, S. E. 1988. Multipotent precursors can give rise to all major cell types of the frog retina. *Science (Wash. D.C.)*, **239**, 1142–1145.

Wetts, R., & Fraser, S. E. 1989. Slow intermixing of cells during *Xenopus* embryogenesis contributes to the consistency of the blastomere fate map. *Development (Camb)*, **105**, 9–16.

Wilson, P. A., Oster, G., & Keller, R. 1989. Cell rearrangement and segmentation in *Xenopus* direct observation of cultured explants. *Development (Camb)*, **105**, 155–166.

18. MECHANISM OF LOCAL CHAOS IN ACTIVE DISTRIBUTED MEDIA

B. S. Kerner

Research Institute
Daimler-Benz AG
Posfach 80 02 30,
7000 Stuttgart 80, Germany

18.1 INTRODUCTION

18.1.1 Active Media

The importance of the problem of pattern formation in nonequilibrium systems, especially biological pattern formation, and also the different kind of the mathematical models for their description has already been discussed in many articles, reviews and books (see references in Haken 1977; Haken 1983; Nicolis & Prigogine 1977; Murray 1977; Murray 1989; Murray & Maini 1986; Vasil'yev *et al.* 1987).

Since Turing's classic work (Turing 1952) biological pattern formation has often been described in terms of macroscopic equations of chemical kinctics (see Haken 1977; Nicolis & Prigogine 1977; Murray 1977; Murray 1989). The spontaneous formation of spatial pattern is linked with the stratification of the homogeneous state of the system at some critical value of a controlling (bifurcation) parameter $A = A_c$. The effect of stratification takes place due to the positive feedback in one of the parameters of the system — an activator. In such an *active* medium, the growth of the activator is controlled by the other parameter — an inhibitor. It was shown that biological pattern formation can in many cases be described using the models which include two equations describing the distribution of concentrations of the activator θ and the inhibitor η (Gierer & Meinhardt 1972; Meinhardt & Gierer 1974; Haken 1977; Murray 1989; Nicolis & Prigogine 1977; Vasil'yev *et al.* 1987):

$$\tau_\theta \frac{\partial \theta}{\partial t} = l^2 \triangle \theta - q(\theta, \eta, A), \qquad (18.1)$$

$$\tau_\eta \frac{\partial \eta}{\partial t} = L^2 \triangle \eta - Q(\theta, \eta, A), \qquad (18.2)$$

where A is the controlling parameter; τ_θ, τ_η and l, L are the characteristic times and variation lengths of activator θ and the inhibitor η, respectively.

Experimental and Theorietcal Advances in Biological Pattern Formation,
Edited by H.G. Othmer *et al.*, Plenum Press, New York, 1993

18.1.2 Activator-Inhibitor "Principles" and Theory of Large-amplitude Dissipative Structures

It is possible to formulate the main activator-inhibitor "principles"— some inequalities for the derivatives from the nonlinear functions q, Q (in the equations (18.1) and (18.2)) and between the values τ_θ, τ_η and l, L, — which allow to develop the *general* theory of large amplitude dissipative structures in active systems. It is convenient to write these "principles" as follows:

I "principle":

$$q'_\theta \equiv \frac{\partial q}{\partial \theta} < 0, \qquad \theta_0 < \theta_h < \theta'_0 \quad (\text{or} \quad \theta_h > \theta_0), \tag{18.3}$$

$$q'_\theta \geq 0, \qquad \theta_0 \geq \theta_h \geq \theta'_0. \tag{18.4}$$

II "principle":

$$Q'_\eta \equiv \frac{\partial Q}{\partial \eta} > 0, \tag{18.5}$$

$$q'_\theta Q'_\eta - q'_\eta Q'_\theta > 0. \tag{18.6}$$

III "principle":

$$L \gg l \tag{18.7}$$

or/and

$$\tau_\eta \gg \tau_\theta. \tag{18.8}$$

Here θ_h is the value of the activator for the homogeneous state of the system; θ_0 and θ'_0 are the values of the activator which correspond to the system of equations: $q'_\theta = 0$, $q = 0$.

The inequalities (18.3) (18.4) mean that in some range of values of the activator there is a positive feedback in the medium; due to the inequalities (18.5) and (18.6) this positive feedback is under the control of the inhibitor[1]; the inequalities (18.7), (18.8) correspond to the different space or time scale variations of the values of the activator θ and inhibitor η. These processes lead to the space or/and time competition between the variations of the activator and inhibitor and determine the main properties of the pattern in active systems.

The inequalities (18.7) or/and (18.8) in fact also imply that the critical point $A = A_c$ corresponds to a subcritical bifurcation, *i.e.* the instability of the homogeneous state of the system at $A = A_c$ results in the abrupt appearance of the *large-amplitude* dissipative structures. This situation, which is usual for many systems, will be considered in this paper.

Using the activator-inhibitor "principle", the general theory of large-amplitude dissipative structures has been developed (see references in reviews and book Kerner & Osipov 1989; 1990; 1993):

1. The possible shapes and types of the basic spatial patterns have been found.

2. The conditions of stability of static spatial patterns have been determined.

3. The processes of restructuring of the spatial patterns, their evolution and scenarios of self-organization have been investigated.

---

[1]In some active systems the inequality (18.5) (or (18.6)) can be fulfilled only in a limited range of the parameters of the medium. Pattern formation in these systems has interesting peculiarities ((Kerner & Osipov 1989),1993).

4. The classifications of the active systems and of the nonlinear phenomena in them correspondingly to the different forms of the nonlinear functions q and Q in (18.1), (18.2) and to the different quantities $\varepsilon = l/L$, $\alpha = \tau_\theta/\tau_\eta$ have been done.

From this theory it follows that the processes of self-organisation in active distributed systems are linked with the spontaneous appearance and the further evolution of the localized patterns - autosolitons.

18.1.3 Autosolitons and Local Chaos

An autosoliton is a solitary eigenstate of a dissipative medium, *i.e.* the parameters of an autosoliton (amplitude, velocity, frequency of pulsations, etc.) depend entirely on the parameters of the system. At the periphery of the autosoliton the system reaches the values of the activator and inhibitor of the homogeneous state $\theta = \theta_h$ and $\eta = \eta_h$.

An autosoliton can spontaneously form in the medium at the value of $A < A_c$. The subsequent evolution of an autosoliton can lead to the following different processes (Kerner & Osipov 1990,1993):

1. Phenomena of self-organization: The spontaneous subsequent appearance of the different stationary patterns in the form of many interacting static, pulsating or rocking autosolitons.

2. Turbulence in the whole medium: A pattern of many interacting autosolitons which appear and disappear at random in different parts of the system.

3. Local chaos: Localized nonstationary complex pattern with chaotic time/space-time behaviour.

Scenarios of self-organization and turbulence in the whole medium which are linked with the evolution of the autosolitons have been considered in the reviews and book (Kerner & Osipov 1989; 1990; 1993).

In this paper one of the possible mechanisms of the spontaneous appearance of local chaos in active one-dimensional media will be discussed. The analysis will also be restricted to the consideration of active systems for which the inequalities

$$L \gg l, \qquad \tau_\eta < \tau_\theta \tag{18.9}$$

hold. Under the conditions (18.9) the inhibition is of long range and is fast in comparison with the activator. It is well-known that in this case conditions for the appearance of uniform oscillations and travelling waves do not hold. Nevertheless there are some possibilities for the spontaneous appearance of the local chaos in such systems.

A mechanism of the appearance of local chaos in active media which will be proposed in this paper is connected with:

1. Spontaneous formation of an autosoliton.

2. "Catastrophic" processes inside the autosoliton which determine the restructuring of the autosoliton during the time of its formation.

In section 18.2 the effects of spontaneous formation of an autosoliton and effects determining the restructuring of the autosoliton will be discussed. Based on these effects, in section 18.3 the mechanism of local chaos will be considered.

18.2   EFFECTS DETERMINING APPEARANCE AND RESTRUCTURING OF AUTOSOLITON

18.2.1   Spontaneous Formation of Autosolitons

In the stable homogeneous media under consideration, *i. e.* for $A < A_c$, an autosoliton can obviously arise spontaneously due to an accidental appearance and subsequent increase of a localized, finite-amplitude fluctuation in some region of the medium. As in the well-known theory of the nucleation centre in first-order phase transitions in equilibrium systems, the probability of the appearance of the localized critical fluctuation somewhere in the active media is proportional to the size of the system. The latter follows from the formula for this probability (per unit time) P of the spontaneous formation of an autosoliton in an one-dimensional ideally homogeneous active media (Kerner & Klenov 1992):

$$P = \frac{L_x}{\triangle x_s} \frac{2\pi}{\triangle \phi_s} p_0, \qquad (18.10)$$

where L_x is the length of the medium; $\triangle x_s$ — the length of the "wandering" of the critical localized fluctuation along the system (during the some transition time) when the amplitude of the fluctuation reaches the critical value, $\triangle \phi_s$ — the corresponding value of the wandering of phase of the fluctuation; p_0 — the probability of the event that the maximum of localized fluctuation reaches the critical value.

Formula (18.10) implies that an unlimited increase of the size of the system L_x leads to the inevitable appearance of an autosoliton in a stable medium, *i. e.* at $A < A_c$, in a time on the order of the characteristic parameter change times of the active system. This result means that an autosoliton spontaneously appears somewhere in the homogeneous media before the value of the controlling parameter A reaches the critical point A_c, when the value of A is slowly increased (Figure 18.1) (Kerner & Klenov 1992).

Thus it is impossible to reach the critical point A_c of the stratification of the homogeneous state of a sufficiently extended medium: at values $A < A_c$, due to the growth of the finite amplitude localized fluctuation somewhere in the medium, an autosoliton spontaneously appears. In some cases, owing to the "catastrophic" processes inside the autosoliton (which will be discussed below), instead of the autosoliton a local chaos, *i. e.* a localized region of the time-space chaos surrounded by the homogeneous medium, spontaneously arises.

18.2.1.1   Role of Small Local Inhomogeneities

Real systems always contain small local inhomogeneities which play the role of a "seed" for the spontaneous appearance of the autosoliton (Kerner & Osipov 1989; 1990). The latter is linked with the fact that at a certain value $A = A_c^- < A_c$ in the neighbourhood of the inhomogeneity[2] a local breakdown occurs, *i. e.* a local increase (or a decrease) in the activator in an avalanche fashion (Figure 18.1(b)).

As a result of the local breakdown, different spatial patterns (autosoliton, the patterns which fill in the whole medium, local chaos) can spontaneously appear in the medium. It is necessary to emphasize that the local breakdown is a dynamic effect which is not related to the presence of fluctuations in the system. Evidently, at values $A < A_c^-$ an autosoliton can also spontaneously appear due to the growth of the finite amplitude fluctuation localized near the inhomogeneity. The probability of this process (Kerner & Klenov 1990) is exponentially

---

[2]The role of an inhomogeneity can naturally plays the boundary of the medium if the distributions of the activator and the inhibitor near the boundary are inhomogeneous.

larger than the value p_0 in the formula (18.10) for the probability of the appearance of an autosoliton in an ideally homogeneous medium. This implies that, in the systems which contain many small amplitude local inhomogeneities, the latter determines the process of the appearance of an autosoliton or a local chaos at values $A < A_c$.

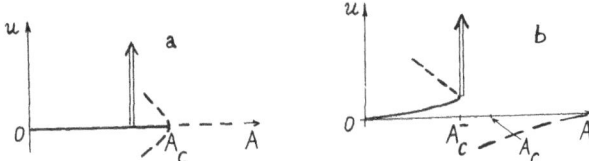

Figure 18.1. The bifurcation diagrams corresponding to a subcritical branching of the solutions: for an ideally homogeneous medium (a) and for a medium with a small amplitude local inhomogeneity (b). The arrow on Figure (a) schematically shows the increase of the amplitude of the localized fluctuation of the activator. The arrow on Figure (b) schematically shows the local breakdown near a small amplitude local inhomogeneity, *i. e.* a local increase in the activator in an avalanche fashion at $A \geq A_c^- < A_c$; u — the amplitude of the dissipative structure. The dashed lines are the parts of the curves which correspond to the unstable states.

An autosoliton in many cases can be considered as the existence of the strongly nonequilibrium local region in the (almost) homogeneous stable medium. Many catastrophic physical processes can happen inside and autosoliton. Some of them or the competition between them can lead to the spontaneous appearance in the medium of a local chaos.

18.2.2 Splitting of Autosoliton

The ffect of the splitting of the autosolitons, *i. e.* the dividing of one autosoliton into two autosolitons, is a deterministic effect which is not associated with the presence of fluctuations in the medium (for the review see Kerner & Osipov 1989; Kerner & Klenov 1990; Kerner & Osipov 1993). This effect occurs at a certain value of the bifurcation parameter $A = A_d$ and is linked with the "local breakdown" in the centre of an autosoliton. The feature of the splitting of the autosoliton depends on the form of the autosoliton. For this reason the two main forms of static autosolitons will be discussed before considering the physics of this effect.

18.2.2.1 Wide and Spike Static Autosolitons

We will consider the autosolitons of two different forms (Kerner & Osipov 1989):

1. Wide autosoliton (Figure 18.2(a)).

2. Spike autosoliton (Figure 18.2(b)).

The form of the autosolitons depends on the form of the local-coupling curve (LC curve), *i. e.* the functional dependence $\eta(\theta)$, which satisfies the equation:

$$q(\theta, \eta, A) = 0 \qquad \text{at} \quad A = \text{const.} \tag{18.11}$$

Wide autosolitons can exist in systems which have an $N-$ (Figure 18.3(a)) or $\bigvee\!\backslash-$ shaped LC curve and spike autosolitons in systems which have $\Lambda-$ (Figure 18.3(b)) or $V-$ shaped LC curve.

Such systems are correspondingly called $N-$, $\bigvee\!\backslash-$, $\Lambda-$ or $V-$ systems (for a review see Kerner & Osipov 1989).

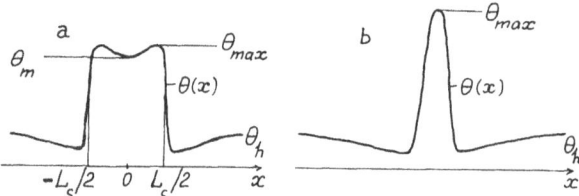

Figure 18.2. The distribution of the activator in "hot" wide (a) and spike (b) autosolitons.

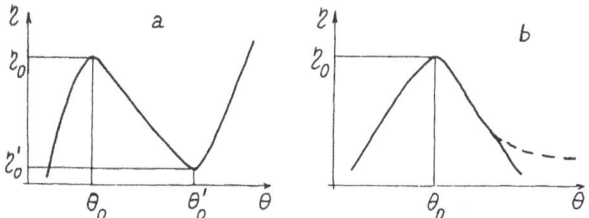

Figure 18.3. Local-coupling curves for $N-$ (a) and $\Lambda-$ systems (b).

Spike autosolitons can also exist in systems with $N-$ or $\vee\!\backslash-$ shaped LC curve for which

$$\zeta = \frac{\theta_0'}{\theta_0} \gg 1. \tag{18.12}$$

On the other hand in $\Lambda-$ or $V-$ systems with a "degenerate" LC curve (broken curve in Figure 18.3(b)) wide autosolitons with very large amplitude $\theta_{max} \gg \theta_h$ can form. The larger the value of ζ is for the systems with an $N-$ or $\vee\!\backslash-$ shaped LC curve the closer are the properties of the autosolitons realized in them to the properties of spike autosolitons (Kerner & Osipov 1989).

18.2.2.2 Physics of the Splitting of an Autosoliton

For the explanation of this effect it would be convenient to consider the real function $\theta(\eta)$ which corresponds to the distributions of the activator and the inhibitor in the autosoliton (curve d on the Figure 18.4(c)). This distribution $\theta(\eta)$ in the vicinity of the centre of the autosoliton qualitatively repeats the LC curve (curve 1 on the Figure 18.4(c)), i.e. the distribution $\theta(\eta)$ which follows from equation (18.11).

At the same time, from the shape of LC curve it follows that the corresponding function $\theta(\eta)$ is S-shaped, and at $\eta = \eta_0'$ we have

$$\frac{d\theta}{d\eta} = \infty \tag{18.13}$$

Figure (18.4(c)). When $A \to A_d$, the value of inhibitor $\eta = \eta_m$ in the centre of the autosoliton actually reaches the limiting value $\eta = \eta_0'$ for the LC curve (Figure 18.4(c)). In other words, the true function $\theta(\eta)$ in an autosoliton (curve d, Figure 18.4(c)) in fact abuts on this limiting point $\eta = \eta_0'$ for the LC curve.

Therefore, at $A = A_d$ the real function $\theta(\eta)$ which corresponds to the distribution of the activator and the inhibitor in the autosoliton (curve d, Figure 18.4(c)) in the vicinity of the autosoliton's centre also satisfies to the condition (18.13). As a result for $A > A_d$ the value of

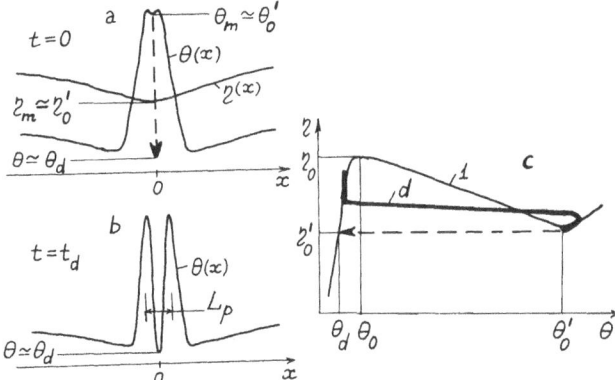

Figure 18.4. Explanation of the splitting of autosoliton: (a,b) — the distributions of activator at the initial moment time after reaching the critical point $A = A_d$ — (a) and at once after the local breakdown in the centre of the autosoliton — (b); (c) — the LC curve (curve 1) and the real $\theta(\eta)$ dependence (curve d) in the autosoliton in (a). The arrows in (a) and (c) schematically show local breakdown, *i. e.* a local decrease in the activator in the centre of the autosoliton from $\theta_m \simeq \theta_0'$ to $\theta \simeq \theta_d$ in an avalanche fashion.

activator, as shown in Figure 18.4, must decrease abruptly near the centre of the autosoliton from $\theta_m \simeq \theta_0'$ to $\theta \simeq \theta_d$. This local breakdown — that is the abrupt local change of the activator in the centre of the autosoliton — must naturally result in splitting of autosoliton during some time t_d (Figure 18.4(b)).

The condition (18.13) in the vicinity of the centre of the autosoliton can be fulfilled not only for systems with an $N-$ or $\vee\!\!\setminus-$ shaped LC curve, but also for $\wedge-$ or $V-$ systems with a "degenerate" LC curve (broken curve in Figure 18.3(b)).

Due to the local breakdown effect autosoliton splits into two interacting autosolitons, *i. e.* two striations (Figure 18.4(b)). On the other hand from the theory of autosolitons (Kerner & Osipov 1990; Kerner & Osipov 1993) one knows that two interacting autosolitons cannot be permanently situated at very small distance: owing to the effect of the pumping of activator between the autosolitons one of them disappears.

18.2.3 "Pumping"of Activator between Autosolitons

In this subsection the effect which determines the disappearance of the one of the two closely situated autosolitons will be considered. This effect is connected with the properties of the fluctuations which are dangerous, *i. e.* critical, for the autosoliton's stability.

18.2.3.1 Critical Fluctuation in Wide Autosoliton

The critical fluctuations $\delta\theta_0(x)$ (Figure 18.5(b), left) and $\delta\theta_1(x)$ (Figure 18.5(c) left) which are the most dangerous for the stability of the wide autosoliton are localized in the autosoliton's walls (Figure 18.5(a), left), *i. e.* the region of size $\sim l$ where the activator is sharply changed (see review Kerner & Osipov 1989). The increment of growth these fluctuations of the activator, *i. e.* the value $-Re\gamma$ in the expression

$$\delta\theta(x,t) = \delta\theta(x)e^{-\gamma t},$$

when changes of the inhibitor are not taken into account (*i. e.* $\delta\eta = 0$) can be estimated from the formulas:

$$Re\gamma = \lambda_0 \sim -\varepsilon\frac{L_s}{L} - e^{-\frac{L_s}{l}} \tag{18.14}$$

for the fluctuation $\delta\theta_0$ and

$$Re\gamma = \lambda_1 \sim -\varepsilon\frac{L_s}{L} \tag{18.15}$$

for the fluctuation $\delta\theta_1$; here $\varepsilon = l/L \ll 1$; L_s is the width of the autosoliton (Figure 18.5(a), left); time is measured in units of τ_θ. Formulas (18.14), (18.15) are valid for $N-$ or $\bigvee$–systems for which $\theta_0' \sim \theta_0$ (Figure 18.3(a)).

Growth of the fluctuations of the activator is damped by a corresponding change in the inhibitor value $\delta\eta$ (Figure 18.5(b),(c)). As a result the autosoliton can be stable (the value of the increment $-Re\gamma < 0$) in some range of the value of controlling parameter A (see the corresponding formulas in the review Kerner & Osipov 1989).

18.2.3.2   Critical Fluctuation in Spike Autosoliton

The critical fluctuations $\delta\theta_0(x)$ (Figure 18.5(b), right) and $\delta\theta_1(x)$ (Figure 18.5(c), right) which are the most dangerous for the stability of the spike autosoliton are localized in the spike of the autosoliton (Figure 18.5(a), right). In this case the value of λ_0 which determines the increment of the fluctuation of the activator $\delta\theta_0(x)$ at $\delta\eta = 0$ is (Kerner & Osipov 1989):

$$Re\gamma = \lambda_0 \sim -1. \tag{18.16}$$

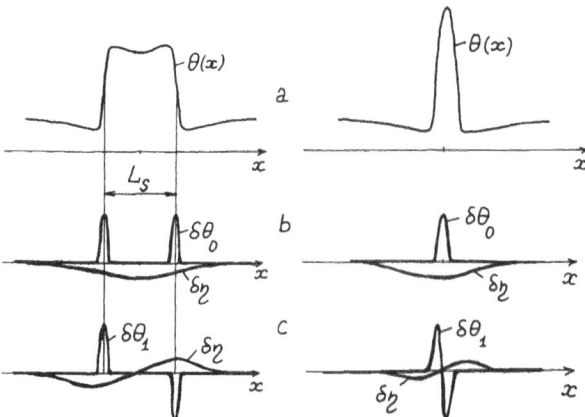

Figure 18.5. The form of the critical fluctuations of the activator $\delta\theta_0$ (b) and $\delta\theta_1$ (c) for the wide (left) and for the spike autosoliton (right) and the inhibitor perturbation $\delta\eta$ which damps the corresponding dangerous activator fluctuations.

18.2.3.3   Physics of Effect of "Pumping"of Activator between Autosolitons

Let us consider two autosolitons which are situated at some distance L_p (Figure 18.6(a)). In this case one of the dangerous fluctuations of the activator can be approximately represented as the antibinding combinations of the functions $\delta\theta_0(x)$ of each of the "isolated" autosoliton (Figure 18.5(b)):

$$\delta\theta_{1,0}(x) \simeq \delta\theta_0(x + \frac{L_p}{2}) - \delta\theta_0(x - \frac{L_p}{2}). \tag{18.17}$$

It follows from the form of this critical fluctuation $\delta\theta_{1,0}(x)$ (Figure 18.6(b)) that the growth of this fluctuation leads to an increase in the amplitude (or the width) of the one autosoliton (left on Figure 18.6(a)) and to a decrease in the amplitude (or the width) of a neighboring autosoliton (Figure 18.6(c)). In other words, it describes a pumping of activator between autosolitons. As a result of the growing of this fluctuation one of the autosolitons (the right in Figure 18.6) gradually disappears.

The growth of the fluctuation $\delta\theta_{1,0}(x)$ is damped by a sign-varying change in the inhibitor $\delta\eta$ (Figure 18.6(b)). The damping effect of this change in the inhibitor is reduced by its diffusive spreading, which is more pronounced the smaller the distance between autosolitons, L_p. So there is the critical distance between autosolitons $L_p = L_{min}$: autosolitons which are situated at the distance $L_p < L_{min}$ are unstable with respect to the growth of the fluctuation of the activator $\delta\theta_{1,0}(x)$ (Figure 18.6(b)).

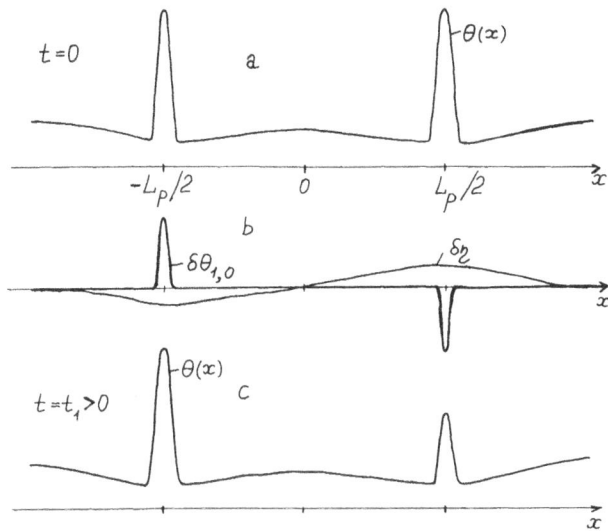

Figure 18.6. Illustrating the "pumping" of activator between two autosolitons which are situated at some distance L_p — (a): (b) — the form of the critical fluctuation of the activator $\delta\theta_{1,0}(x)$ and the corresponding disturbance of the inhibitor $\delta\eta$, damping it; (c) — some intermediate distribution of the activator realized in the process of the growth of the fluctuation $\delta\theta_{1,0}(x)$.

At very small distance between autosolitons $L_p \ll L$ (but $L_p > l$) the damping effect of the change in the inhibitor, owing to $L \ll l$, can be practically negligible small, *i. e.* $\delta\eta \simeq 0$. In this case the increment of the growth of the fluctuation $\delta\theta_{1,0}(x)$ is approximately given by formulas (18.14) and (18.16) for the wide (Figure 18.2(a)) and for the spike autosoliton (Figure 18.2(b)) correspondingly. In other words, the delay time in the disappearance of the one of the autosolitons due to the growth of the critical fluctuation $\delta\theta_{1,0}(x)$, *i. e.* due to the pumping of activator effect, for the spike autosolitons owing to $\varepsilon = l/L \ll 1$ is many times less than for the wide autosoliton.

It has been mentioned in subsection 18.2.2.1 that the autosolitons whose properties are very similar to the properties of the spike autosolitons can also exist in the active system with $N-$ or $\bigvee\!\!\backslash-$ shaped LC curve for the value $\zeta \gg 1$ (see (18.12)). The important conclusion for the active system with $N-$ or $\bigvee\!\!\backslash-$ shaped LC curve following from above is: The more the value ζ is the quicker one of the two closely situated autosolitons disappears in the system.

18.3   LOCALIZED PATTERN OF NONSTATIONARY AUTOSOLITIONS

The different processes of pattern formation which will be considered in the remainder of this paper have the following common properties:

1. All of these processes begin from the spontaneous formation of an autosoliton. The autosoliton's formation occurs when the controlling parameter A is less than the critical value $A = A_c$ for Turing instability.

2. In the process of self-formation of the autosoliton the latter splits, *i. e.* at a given value of the parameter A, a single autosoliton cannot exist in the medium.

We know from subsection 18.2.2.2 that this situation can be realized if the value of the bifurcation parameter A at which autosoliton spontaneously forms is greater than the critical value A_d at which the static autosoliton splits.

Before the consideration of the concrete mechanism of local chaos will be done (subsections 18.3.2–18.3.5), let us in subsection 18.3.1 briefly discuss the results of the experimental and numerical investigations of some consequences of the spontaneous formation of the autosoliton in the initially almost homogeneous medium.

18.3.1   Results of Experimental and Numerical Investigations

18.3.1.1   Spontaneous Appearance of Pattern of Many Stationary Autosolitons

The process of spontaneous formation of an autosoliton which has been discussed in subsection 18.2.1.1, has been investigated numerically in many concrete models of active systems. One of the possible consequences of this effect is the appearance of the periodically situated autosolitons which fill in the whole medium (Figure 18.7) (for a review see Kerner & Osipov 1989,1993). As it has been above noted, this effect can be realized if the value of A at which the formation of the autosoliton spontaneously begins is more than the critical value A_d. In this case the process of the formation of the pattern of many stationary autosolitons consists of the following stages:

1. In the process of autosoliton formation the autosoliton splits (Figure 18.7; $t = t_1$).

2. Two striations which appear owing to the splitting of the forming autosoliton move in the opposite directions and gradually transform into two almost noninteracting autosolitons (Figure 18.7; $t = t_2$).

3. Each of these autosolitons splits once more (Figure 18.7; $t = t_3$) as the autosolitons which are situated far away from one another can not exist at value $A > A_d$.

4. The periodic repetitions of the processes 2 and 3 above, finally lead to the filling of the whole medium in the interacting autosolitons (Figure 18.7; $t > t_4$)[3] which are situated on the distances $L_p < L_{max}$.

---

[3]As it has been noticed in subsection 18.2.1.1, the local breakdown is a dynamic effect which is not linked with the presence of fluctuations in the system. Owing to the sequence of the different local breakdown effects (first near the inhomogeneity (Figure 18.1(b)), then in the centre of the autosolitons (18.7; $t = t_2, t_3$)) the system in fact "chooses" one of the spatial patterns among many others which can exist in the medium at given parameters of the system. In other words, fluctuations may not play an important role in the selection of the forming spatial patterns (see review Kerner & Osipov 1990).

The critical distance L_{max} depends on the value of A and determines the critical distance between periodically-situated autosolitons: at

$$L_p > L_{max} \qquad (18.18)$$

the local breakdown effect in the centre of the autosolitons and consequently the splitting of the each of the autosolitons occur (see review Kerner & Osipov 1990).

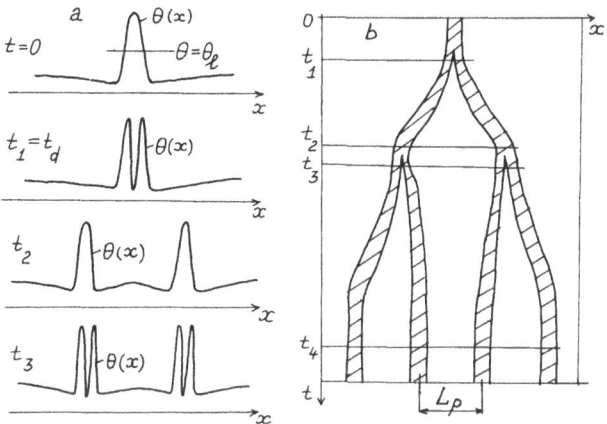

Figure 18.7. Qualitative picture of the process of the appearance of the stationary pattern of periodically situated autosolitons owing to the splitting of the autosoliton spontaneously forming in the medium: (a) — the distributions of activator in different moment of time, (b) — the same kinetics as in figure a but expressed as the location of the regions where the value of activator more than some definite value $\theta = \theta_l$ (see (a)).

Evidently, the closer the value of the controlling parameter A to the critical value A_d is the more significant the value of L_{max} is. At $A \to A_d$ we have $L_{max} \to \infty$ as at $A < A_d$ the value $L_{max} = \infty$, *i. e.* the autosolitons at $A < A_d$ can be situated any distance away from one another.

18.3.1.2 Spontaneous Appearance of Turbulence in Whole System

In the $\Lambda-$ or $V-$ systems (see subsection 18.2.2.1) the condition

$$L_{max} < L_{min} \qquad (18.19)$$

can be fulfilled. Under this condition, the processes 1, 2 and 3 of the appearance of the autosoliton and of their consecutive splitting which have been described in subsection 18.3.1.1 cannot lead to the formation of the pattern of many stationary autosolitons. Instead of this pattern, turbulence in the whole medium can spontaneously arise (for review see Kerner & Klenov 1990). The mechanism of the appearance of turbulence can be explained as follows: as a result of the processes 1–3 (subsection 18.3.1.1) the interacting autosolitons which are situated on the distances $L_p < L_{max}$ should arise. On the other side, the interacting autosolitons which are situated at the distance $L_p < L_{min}$ are unstable owing to the effect of pumping of activator between autosolitons (see subsection 18.2.3). As a result of this instability, some of the autosolitons should disappear, *i. e.* the distance between autosolitons should increase. But under the condition (18.19) the condition (18.18) is fulfilled. Therefore the local breakdown effect in the centre of the autosoliton (see subsection 18.2.2) or between them occurs. The latter processes lead to the splitting of the autosolitons and to the appearance of the new autosolitons in the medium, *i. e.* to a decrease in the distance between autosolitons. For these reasons any stationary state of many interacting autosolitons cannot exist in the medium if the condition (18.18) is fulfilled. Since the local breakdown effect and the pumping of activator effect can occur in an uncorrelated fashion in separated spatial regions, these two processes can give rise to a turbulence. The latter can look like a pattern of interacting autosolitons which appear and disappear at random at various spatial points.

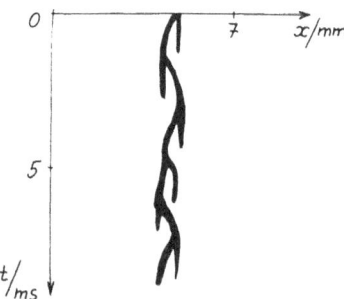

Figure 18.8. Picture of the experimentally observed local spatio-temporal oscillations during autosoliton's splitting (Willebrand *et al.* 1991)

18.3.1.3   Spontaneous Appearance of Localized Pattern of Nonstationary Autosolitons

The splitting of an autosoliton in the process of its spontaneous formation has been found experimentally in the investigations of some physical systems (Willebrand *et al.* 1991). For some parameters of the system the splitting of the autosoliton has led to the appearance of the pattern of many stationary autosolitons, as it has been described in subsection 18.3.1.1.

    For other parameters of the system, the splitting of the autosoliton has caused local spatio-temporal oscillations (Figure 18.8) (Willebrand *et al.* 1991). One can see the features of these oscillations from Figure 18.8: These oscillations are connected with the alternation of the process of the splitting of the autosoliton and of the process of the extinction of one of the two striations which appear as a result of the previous splitting of the autosoliton.

    So the splitting of the autosoliton in the process of its spontaneous formation can lead not only to the appearance of the pattern which fills in the whole medium (as it has been discussed in subsections 18.3.1.1 and 18.3.1.2) but also to spontaneous appearance of *localized* pattern of the nonstationary autosolitons. The physical mechanism which can explain this experimentally found phenomenon will be proposed below.

18.3.2   Mechanism of Repelling of Striations

    From the experimental and numerical investigations it follows that two striations which appear owing to the splitting of the forming autosoliton move in the opposite directions (Figure 18.7; $t = t_2$).

    For the explanation of this effect let us first consider the expression for the velocity v of the striation's wall (of the order of l in size) where the activator sharply changes from $\theta = \theta_1$ to $\theta = \theta_3$ (Figure 18.9). Here the quantities θ_1 and θ_3 are the values of the activator which satisfies the equation (18.11) at $\theta < \theta_0$ and $\theta > \theta_0'$ (Figure 18.3) correspondingly. The velocity of this wall one can find from the well-known formula (see, for example, Kerner &

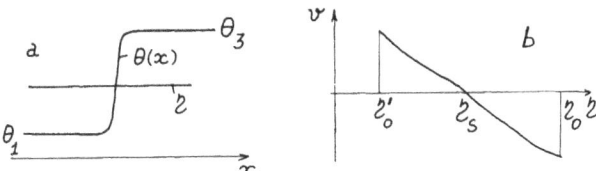

Figure 18.9. The distribution of the activator in the striation's wall (a) and the dependence of the velocity of this wall on the value of inhibitor (for $N-$ and $\Lambda-$ systems) (b).

Osipov 1989):

$$v = \tau_\theta^{-1} \left(\int_{\theta_1}^{\theta_3} q(\theta, \eta, A) d\theta \right) \left(\int_{-\infty}^{\infty} (d\theta/dx)^2 dx \right)^{-1}. \qquad (18.20)$$

The velocity of the wall (Figure 18.9) can be approximately found from formula (18.20) at $\eta = $ constant if one takes into account that the derivative $d\theta/dx$ and the value of $| q(\theta, \eta, A) |$ outside the wall and the change in the value of the inhibitor inside the wall are of the order of the value $\varepsilon = l/L \ll 1$.

The walls of the static autosoliton (Figure 18.2) have the velocity $v = 0$, and thus from (18.20) the value of the inhibitor $\eta = \eta_s$ in the walls of static autosoliton satisfies the condition

$$\int_{\theta_1}^{\theta_3} q(\theta, \eta_s, A) d\theta = 0. \qquad (18.21)$$

Owing to the local breakdown effect in the autosoliton's centre, two new walls appear (Figure 18.7; $t = t_2$). At first the value of the inhibitor in these new walls $\eta \neq \eta_s$ therefore their velocity $v \neq 0$. Correspondingly to (18.20) these two walls move in opposite directions. Besides they move from the centre of the autosoliton.

To see these conclusions easily, let us take into account that for example for $N-$ and $\Lambda-$ systems the derivative $q'_\eta < 0$ (see Kerner & Osipov 1989). From Figure 18.4(c) and the inequality $q'_\eta < 0$ follow that under the LC curve (see subsection 18.2.2.1) the function $q > 0$. On the other hand, from Figure 18.4(b) one can also see that in two new walls which appear due to the local breakdown effect in the autosoliton's centre the value of the inhibitor

$$\eta = \eta_w \simeq \eta'_0 < \eta_s,$$

and therefore the values of the activator which correspond to these new walls are almost fully situated under the LC curve where the function $q > 0$.

From formula (18.20) and inequality $q > 0$ it follows that the velocity of the right new wall where the value of the activator is the increasing function of x is more than zero and the velocity of the left new wall is less than zero. The same conclusion is valid for the process of the splitting of the autosoliton in $V-$ and $\bigvee-$ systems.

Therefore the mechanism of the repelling of the two striations which appear owing to the splitting of the forming autosoliton is linked with the considerable difference in values of the inhibitor in the new walls of these striations from the characteristic value $\eta = \eta_s$ (see (18.21)). Immediately after the splitting of the autosoliton the value of the inhibitor in the new striation's walls η_w approximately satisfies the condition:

$$\eta_w \simeq \eta'_0$$

(Figure 18.4,c), that is this value considerably differs from the characteristic value η_s:

$$| \, \eta_w - \eta_s \, | \, \eta_s^{-1} \sim 1. \tag{18.22}$$

This means that at the first moment of time the velocity of the repelling striations is

$$v \sim \frac{l}{\tau_\theta}. \tag{18.23}$$

The condition (18.23) follows from (18.20) and (18.22) if one takes into account that the width of the activator's wall (Figure 18.9) is of the order of value of l.

As further as two striations which have appeared due to the splitting of the autosoliton move away from one another the value of the inhibitor in their walls approaches the characteristic value $\eta = \eta_s$ for which the velocity of the striation's walls $v = 0$ (18.20). In other words the velocity of the repelling striations sharply decreases as the distance between them increases. In time these striations can transform into almost independent autosolitons. The latter split once more and finally the medium fills in the interacting autosolitons correspondingly to the scenario which has been considered in subsection 18.3.1.1 (Figure 18.7).

In this consideration of the process of the repelling of two striations it has not been taken into account the pumping of the activator effect (subsection 18.2.3). This effect can totally change the process of the further evolution of two striations which appear due to the splitting of the forming autosoliton.

18.3.3 Disappearance of One of the Striations

From subsection 18.2.3 it follows that if two striations are situated at the distance

$$L_p < L_{min}$$

they are unstable with respect to the growth of the fluctuation of the activator $\delta\theta_{1,0}(x)$ (Figure 18.5(b)). This pumping of the activator effect finally leads to the disappearance of one of the striations.

At first when only two striations appear due to the local breakdown effect in the autosoliton centre (Figure 18.4(b)) the distance between these striations L_p is small as the distance between their new walls is of the order of the l. At such small distances between striations the increment of the growth of the fluctuation $\delta\theta_{1,0}(x)$, i. e. the value $-Re\gamma_{1,0}$ in the expression

$$\delta\theta_{1,0}(x,t) = \delta\theta_{1,0}(x)e^{-\gamma_{1,0}t}, \tag{18.24}$$

is approximately equal to the value $-\lambda_0$ which is given by the formula (18.14) for the autosolitons in $N-$ and $\bigvee-$ systems and by formula (18.16) for the autosolitons in $\Lambda-$ or $V-$ systems.

Let us first consider the $N-$ or $\bigvee-$ systems for which applies $\theta_0' \sim \theta_0$ (Figure 18.3(a)). The value of the increment $-Re\gamma_{1,0} \simeq -\lambda_0$ (18.14) of the growth of the critical fluctuation $\delta\theta_{1,0}(x)$ in these systems to the extent of the value of $\varepsilon = l/L \ll 1$ is close to zero. For this reason the critical fluctuation $\delta\theta_{1,0}(x,t)$ cannot grow enough to make one of the striations disappear during the relatively short time when the striations are situated at a small distance from one another. Due to the repelling of these striations the distance between them increases and the fluctuation $\delta\theta_{1,0}(x)$ (18.17) can not grow any more. Therefore in these types of the active systems the process of the splitting of the autosoliton leads to the filling in the whole medium in the interacting autosolitons as it has been described in subsection 18.3.1.1 (Figure 18.7).

One can find another picture (Figure 18.10) for $\Lambda-$ or $V-$ systems, $\Lambda-$ or $V-$ systems with the "degenerated" LC curve or active systems with $N-$ or $\mathsf{\vee\!\vee}-$ shaped LC curve for which applies the ratio (18.12). Below only these types of active systems will be considered. In these systems:

Spike autosolitons of very large amplitude can exist as has been discussed in subsection 18.2.2.1. For such autosolitons the value of the increment $-Re\gamma_{1,0} \simeq -\lambda_0$ of the growth of the critical fluctuation $\delta\theta_{1,0}(x)$ corresponding to (18.16) is of the order of value τ_θ^{-1}, *i. e.* large enough (note that in (18.16) the time is measured in units of τ_θ). This increment characterizes the growth of the fluctuation $\delta\theta_{1,0}(x)$ (Figure 18.6(b)) if the distance between the striations is very small: $L_p \ll L$. On the other hand corresponding to (18.23) during the time of the order of τ_θ the distance between repelling striations increases in the small value of the order of l.

There is some critical amplitude (or critical width) of the autosoliton (striation). If the amplitude (or width) of striation becomes less than this critical value the striation self-disappears.

In other words if the amplitude (or width) of one of the striations due to the increase in the amplitude of critical fluctuation $\delta\theta_{1,0}(x)$ decreases below some critical level, this striation self-disappears independently on the distance between striations.

Therefore the decrease in the amplitude (or width) of one of the striations below some critical level owing to the pumping of the activator effect leads to the gradual disappearance of one of the striations as far as these two striation move in the opposite directions ($t = t_2$, Figure 18.10).

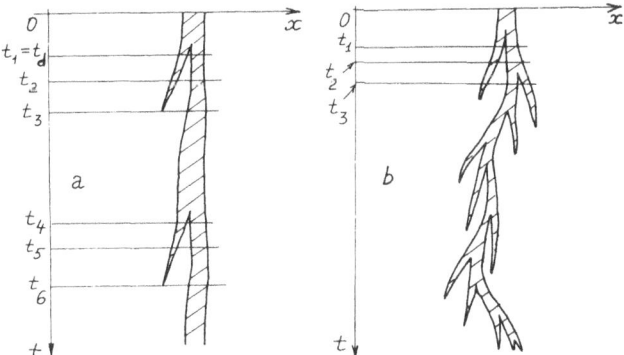

Figure 18.10. Qualitative picture of the simplest periodic localized nonstationary pattern (a) and of the local chaos (b). The kinetics of the spatial pattern expressed here as the location of the regions where the value of activator more than some definite value $\theta = \theta_l$ is (see Figure 18.7(a)).

After one of the repelling striations has disappeared ($t = t_3$, Figure 18.10(a)) the remaining striation after delayed time has to split once more ($t = t_4$, Figure 18.10(a)) as at the given value $A > A_d$ the localized striation, *i. e.* the autosoliton, can not exist in the system.

After the splitting of the latter autosoliton two new closely situated striations once more appear in the medium ($t = t_5$, Figure 18.10(a)), then they repel one another and one of the striations gradually disappears ($t = t_6$, Figure 18.10(a)). The reason for this process has been explained above. So successive processes of splitting of the autosoliton and disappearance of one of the repelling striations can periodically repeat causing the periodic space-time oscillations in the system.

18.3.4   Simplest Localized Nonstationary Patterns

The successive processes of the splitting of autosoliton and of the disappearance of one of the repelling striations can periodically repeat if the process of the splitting of the autosoliton begins in some sufficiently large time after the one of the striations has disappeared. This time ($t_3 - t_4$; Figure 18.10(a)) should be large enough to reconstruct the almost symmetrical distributions of the activator and the inhibitor in the autosoliton before the autosoliton's splitting. In this case the successive of the processes of the splitting of the autosoliton will repeat each other in the all details. As a result the processes of the appearance of two striations and the disappearance of the one of them will be periodical with a period which is determined by the time interval between successive splittings of the autosoliton.

This simplest localized pattern (Figure 18.10(a)) can spontaneously appear if the value of the controlling parameter A is more than the critical value A_d but is very close to A_d:

$$\xi = \mid A - A_d \mid A_d^{-1} \ll 1. \tag{18.25}$$

When (18.25) is satisfied the process of the autosoliton's splitting can begin after the re-constructing of the almost symmetric form of the autosoliton which is close to the form of the autosoliton at $A = A_d$. Indeed, only for this symmetric form of the distributions of the activator and the inhibitor the value of the inhibitor $\eta = \eta_m$ in the autosoliton's centre in fact reaches the limiting value $\eta = \eta_0'$ for the LC curve and the process of local breakdown occurs (for more details see subsection 18.2.2). On the other hand the effect of pumping of the activator between striations breaks the symmetrical distribution of the activator and the inhibitor in the localized pattern as the fluctuation $\delta\theta_{1,0}(x)$ whose growth causes the pumping of the activator effect is asymmetrical (Figure 18.5(b)). For this reason some time is needed ($t_3 - t_4$; Figure 18.10(a)) to reconstruct the symmetrical distribution of the activator and the inhibitor in the localized pattern.

18.3.5   Mechanism of Appearance of Localized Complex Nonstationary Patterns

If the condition (18.25) is not fulfilled, i. e. the value of the controlling parameter A at which the autosoliton spontaneously forms sufficiently exceeds the critical value $A = A_d$ (see subsection 18.2.2), the symmetric form of the autosoliton is no longer necessary for the beginning of the process of the autosoliton's splitting. In this case one can expect the localized complex nonstationary pattern or local chaos to appear spontaneously in the medium.

For the explanation of these conclusions first recall that the solution in the form of the static autosoliton exists only at $A \leq A_d$ (subsection 18.2.2). At $A = A_d$ the real function $\theta(\eta)$ which corresponds to the distribution of the activator and the inhibitor in the autosoliton (curve d, Figure 18.4(c)) in the vicinity of the autosoliton's centre satisfies to the condition (18.13). For $A > A_d$ there are no solutions corresponding to the static autosoliton, i. e. to a solitary striation in the case of the one-dimensional systems under consideration. It is evident from here that the more the value of A exceeds the critical value A_d, the more different nonstationary distributions in the form of the one striation for which the real function $\theta(\eta)$ (curve d, Figure 18.4(c)) in the vicinity of the striation's centre satisfies the condition (18.13) there are, and the process of the local breakdown begins.

Therefore if the controlling parameter A sufficiently exceeds the critical value of A_d, there are a lot of different nonstationary distributions of the activator and the inhibitor in the form of one striation which allow the beginning of the process of the splitting of the striation due to the local breakdown effect. Among these distributions there can be many asymmetrical distributions.

On the other hand, the more A exceeds the critical value A_d the less is the delay time between two subsequent processes of the autosoliton's splitting. Therefore the striation can split *before* the process of the disappearance of the other striation finishes ($t = t_3$; Figure 18.10(b)). In this case (in contrast to the case which has been described in subsection 18.3.4) the successive processes of the striation's splitting cannot repeat each other.

The latter is linked with the fact that the process of the disappearance of one of the striations is caused by the growth of the asymmetrical fluctuation $\delta\theta_{1,0}(x)$ (Figure 18.5(b)). For this reason, in the successive processes of the striation's splitting one can expect the asymmetrical distribution of the activator and the inhibitor in the localized pattern to appear directly before the splitting process begins. As a result the local breakdown effect occurs not exactly in the autosoliton's (striation's) centre. Therefore after the striation's splitting two *different* (in the width and the amplitude) striations appear. These striations repel from one another with different velocities, and what is more the direction of the moving of the striation which will disappear depends now on the process of the disappearance of the other striation which has begun after the previous splitting of the striation.

So in the complex process of the formation of the localized pattern under consideration one can distinguish three basic processes:

1. The splitting of the autosoliton into two striations.

2. The repelling of these striations.

3. The disappearance of one of the striations due to the pumping of the activator effect.

The strong competition between these three different processes can lead to the spontaneous appearance of local chaos. Indeed, if the delay time between two subsequent processes of the striation's splitting is noticeably less than the characteristic time of the disappearance of one of the striations then, as it has been shown above, the distributions of the activator and the inhibitor in the localized pattern (immediately after the autosoliton's splitting process has finished) is asymmetrical. Moreover these asymmetrical distributions in the successive splitting processes cannot repeat each other.

Two different types of localized complex nonstationary pattern can appear:

1. A periodic in time pattern whose period consists of two, three or more different aggregates of the three processes mentioned above.

2. A localized pattern with chaotic space-time behaviour, *i. e.* the local chaos.

18.3.6 Peculiarities of Kinetics of Localized Patterns

From the results of subsections 18.3.2 - 18.3.4 it follows that both the type and the behaviour of the spatial patterns which can appear owing to the competition between the three nonlinear effects described above (the splitting of the autosoliton, the repelling of the striations, the pumping of the activator between striations) depends especially on two values:

1. $|\lambda_0|$ (see subsection 18.2.2).

2. $\xi = |A - A_d|\, A_d^{-1}$.

In turn the value $|\lambda_0|$ which determines the increment of the growth of the critical fluctuation $\delta\theta_{1,0}(x)$ (Figure 18.5(b)) depends on the shape of the LC curve (more precisely, depends on the nonlinear function $q(\theta, \eta, A)$). This is connected with the fact that the gradual change in the shape of the LC curve allows $\mid \lambda_0 \mid$ to change gradually from the value of the order of $\varepsilon = l/L \ll 1$ (see(18.14)) to the value of the order of 1 (see (18.16)). This applies to a huge variety of the types of the localized nonstationary patterns which can be observed as the successive processes of the splitting and the extinction of the autosolitons.

From the results of subsections 18.3.3 and 18.3.4 it follows that the peculiarities of the kinetics of the localized patterns under consideration sharply depend also on the value ξ. Notice two effects which can be found if the value $\xi = \mid A - A_d \mid A_d^{-1}$ increases:

1. The transformation of the local chaos into the turbulence regime in the whole medium.

2. The transformation of the localized complex nonstationary pattern into the stationary pattern of the interacting autosolitons.

The first effect can be more typical for the systems which nonlinear properties are similar to the properties of $\Lambda-$ or $V-$ systems, *i. e.* the value of $\mid \lambda_0 \mid$ for the autosoliton in these systems is of the order of 1 (see (18.16)). In this case as far as the value ξ increases the initial velocity of the repelling striations which appear after the autosoliton's splitting also increases. As a result two striations can run away from one another so quickly that their parameters (the amplitude and the width) cannot reach the critical values. In other words despite the fact that the amplitude of the fluctuation $\delta\theta_{1,0}(x)$ (Figure 18.5(b)) at the first moment after the autosoliton's splitting increases, none of the striations can disappear. After scattering each of these striations splits once more. The process of the splitting goes on until the distance between striations, *i. e.* the spike interacting autosolitons, L_p reaches the critical value L_{min}. After it the pumping of the activator between the neighbouring autosolitons starts (see subsection 18.2.2). As the consequence the turbulence regime in the whole medium appears. The reason for it has been described in subsection 18.3.1.2.

The second effect can be more typical for the active systems which properties are closer to the properties of the $N-$ or $\vee\!\!\backslash-$ systems, more precisely the value of $\mid \lambda_0 \mid$ for the autosoliton in these systems is less than 1 (but large enough for the appearance of the localized patterns under consideration at $\xi \ll 1$). If the value of ξ increases two striations which appear in the process of the autosoliton's splitting can run away from one another quickly enough without the disappearance of one of them. As the consequence of the splitting of these new striations the medium fills finally in the stationary interacting autosolitons. The reason for it has been described in subsection 18.3.1.1.

18.4   CONCLUSION: SCENARIOS FOR SPONTANEOUS FORMATION AND EVOLUTION OF LOCALIZED COMPLEX NONSTATIONARY PATTERNS

The properties of the localized complex nonstationary patterns which have been examined in subsections 18.3.2 - 18.3.6 allow to formulate some scenarios for the spontaneous formation and the evolution of these patterns in the active systems under consideration:

1. As the controlling parameter A increases the process of the spontaneous formation of an autosoliton starts. After some time delay the incompletely-formed autosoliton splits into two striations. These striations repel from one another. During the process of the striation's scattering due to "the repumping of the activator" effect the amplitude (or the width) of one of the striations decreases below some critical value. After it this striation gradually disappears independently on the distance between striations. The remaining striation splits once more. As a result the successive in time of three

competing nonlinear processes (the autosoliton's splitting, the repelling of the striations and the pumping of the activator between striations) arises leading to the spontaneous appearance of the localized nonstationary patterns of one of the following types:

(a) A periodic in time localized patterns whose period consists of the one aggregate of three nonlinear processes mentioned above (Figure 18.10(a)).

(b) A periodic in time localized pattern whose period consists of several different aggregates of three nonlinear processes mentioned above.

(c) Local chaos, *i. e.* a localized pattern with chaotic space-time behaviour (Figure 18.10(b)).

2. As the controlling parameter A further increases then:

(a) The periodic in time localized pattern can spontaneously transform into local chaos.

(b) The periodic in time localized patterns or the local chaos of the active system can spontaneously transform into:

   i. The turbulence regime in the whole medium.

   ii. The stationary spatial pattern of the interacting autosolitons (Figure 18.7), depending on the nonlinearities.

(c) As the controlling parameter A decreases (but it remains more than the critical value $A = A_d$ at which static autosoliton splits) local chaos can spontaneously transform into the periodic in time localized pattern.

REFERENCES

Gierer, A., & Meinhardt, H. 1972. A theory of biological pattern formation. *Kybernetik*, **12**, 30–39.

Haken, H. 1977. *Introduction to synergetics.* New York: Springer-Verlag, Berlin.

Haken, H. 1983. *Advanced synergetics. instability hierarchies of self-organizing systems and devices.* New York: Springer-Verlag, Berlin.

Kerner, B. S., & Klenov, S. L. 1990. Spontaneous formation of autosoliton in stable nonequilibrium systems. *Sov. JETF Lett.*

Kerner, B. S., & Klenov, S. L. 1992. Spontaneous formation of localized highly nonequilibrium regions in stable media. In *Abstracts of Reports at Congress of German Physical Society.* Weinheim: Physik-Verlag GmbH.

Kerner, B. S., & Osipov, V. V. 1989. Autosolitons. *Sov. Phys. Usp.*, **32**, 101–138.

Kerner, B. S., & Osipov, V. V. 1990. Self-organization in active distributed media: Scenarios for the spontaneous formation and evolution of dissipative structures,. *Sov. Phys. Usp.*, **33**, 679–719.

Kerner, B. S., & Osipov, V. V. 1993. *Autosolitons: A new approach to the problems of self-organization and turbulence.* Dordrecht, Boston, Lancaster, Tokyo: Kluwer Academic Publishers. (In press).

Meinhardt, H., & Gierer, A. 1974. Application of a theory of biological pattern formation based on lateral inhibition. *J. Cell. Sci.*, **15**, 321–346.

Murray, J. D. 1977. *Nonlinear differential-equation models in biology.* Oxford: Clarendon Press.

Murray, J. D. 1989. *Mathematical biology.* New York: Springer-Verlag, Berlin.

Murray, J. D., & Maini, P. K. 1986. A new approach to the generation of pattern and form in embryology. *Sci. Prog. Oxf.*, **70**, 539.

Nicolis, G., & Prigogine, I. 1977. *Self-organization in Nonequilibrium Systems.* New Nork: Wiley.

Turing, A. M. 1952. The chemical basis of morphogenesis. *Phil. Trans. Roy. Soc. Lond.*, **B237**, 37–72.

Vasil'yev, V. A., Romanovskii, Yu. M., Chernavskii, D. S., & Yakhno, V.G. 1987. *Autowave processes in kinetic systems*. Dordrecht, Boston, Lancaster, Tokyo: D. Reidel.

Willebrand, H., Niedernostheide, F.-J., Ammelt, E., Dohmen, R., & Purwins, H.-G. 1991. Spatiotemporal oscillations during filament splitting in gas discharge systems. *Phys. Lett. A.*, **153**, 437–445.

19. PATTERNS FORMED THROUGH CELL-CELL INTERACTIONS: SPONTANEOUS SELECTION OF DOMINANT DIRECTIONS

Leah Edelstein-Keshet

Department of Mathematics
University of British Columbia
Vancouver, BC V6T 1Z2, Canada

19.1 INTRODUCTION

This paper presents an example of patterns formed through the direct interactions of cells. After a brief review of classical ideas from pattern formation, we introduce the idea that the selection of a dominant direction in an initially isotropic medium is analogous to a type of pattern formation, not in physical space, but rather in angle-space. The pattern forms on a unit circle, *i.e.* on a range of angles $0 < \theta < 2\pi$. It is shown that as a result of cell-cell interactions, uniform angular distributions of cells are unstable and that peaks in these distributions form spontaneously. These peaks represent dominant directions that arise in the cell population as a result of clustering and alignment of cells with one another. (See Figure 19.1). The paper will concentrate on alignment of populations of fibroblasts *in vitro*, and on analysis of typical equations that arise in modelling angular distributions. Applications of similar models to formation of preferred orientations in populations of organisms and in macromolecular networks will be discussed.

Figure 19.1. This paper describes how initially uniform angular distributions of cells (a) spontaneously become oriented as in (b).

19.2 CLASSICAL TURING PATTERN FORMATION

Turing (1952) was among the pioneers in pattern formation theories. His paper in 1952 laid the ground for theories of morphogenesis. Turing suggested that spontaneous formation of chemical "morphogen" prepatterns could stem from reaction and diffusion of substances in the cellular milieu. The key ideas in his work can be summarized as follows:

1. The mechanism is driven by local interaction and spatial diffusion.

Experimental and Theorietcal Advances in Biological Pattern Formation,
Edited by H.G. Othmer *et al.*, Plenum Press, New York, 1993

2. A minimum of two different species must be present and their rates of diffusion must differ.

3. Formation of pattern is a type of self-activation that results from instability. A slight deviation away from a uniform distribution will lead to spontaneous formation of pattern.

Linear stability theory, while limited in predictive powers, is often applied in analyzing Turing-type pattern formation. It predicts that in the patterns which tend to grow, features such as stripes or spots, and the relative sizes and spacings of the pattern elements depend on the shape of the domain, the parameters in the reaction-diffusion equation, and in particular, on the relative "domains of dominance" of the interacting species. (Domains of dominance are, roughly speaking, the ratios of chemical rate-constants to diffusion rates, and are dimensionally equivalent to area). Typically, a Turing system involves a chemical pair in which one substance is an activator and the second is an inhibitor. The domain of dominance of the inhibitor must be larger than that of the other chemical for patterns to form, *i. e.* inhibition must extend out laterally farther from a point source than activation. The term **lateral inhibition** has been used to describe this local activation and long-range inhibition (See Figure 19.5(a)).

In more recent years, numerous theories for pattern formation have been developed. We refer the reader to extensions of Turing models in Gierer & Meinhardt (1972); Meinhardt (1978); Meinhardt (1982); Bard (1981); Bard & French (1984); Murray (1981a); Murray (1981b); Murray (1982); Murray (1988); Murray (1989); Young (1984); Harrison (1987). Other mechanisms that create spatial patterns have also been introduced. (See review in Levin & Segel 1985). The mutual activation and inhibition of neurons, and their interactions through distant axons and dendrites have been modelled in several theories for pattern formation in biological neural networks. Examples date back to Ermentrout & Cowan (1979); Swindale (1980), and more recently, Ermentrout *et al.* (1986). In these models, the spatial interactions are not just local, as in the Turing diffusion-based models, but rather long-ranged. To depict such interactions, it is necessary to integrate over a whole region. The probability of interaction at a given distance is used as a weighting factor in such integrals. (These probabilities are called the kernels of the integral operators, and have a typical shape shown in Figure 19.5(a)).

The mechanochemical theories have formed a major new school of thought in which mechanical forces, such as elastic stresses are considered. See Murray & Oster (1984); Murray *et al.* (1983); Odell *et al.* (1981); Oster *et al.* (1983); Weliky & Oster (1990). However, interestingly enough, it has been shown that many seemingly unrelated mechanisms are somehow linked formally by a disguised version of lateral inhibition, or similar dispersion relations (*i. e.* relations linking the spacing between patterns to the parameters of the system). See Murray (1989) and Oster & Murray (1989) for good summaries.

As described by several papers in this volume, recent efforts at testing the pattern-formation mechanisms experimentally have provided new challenges and new ideas in some directions, and disappointment in others. Turing's idea of the morphogens that make up a reacting-diffusing system and form standing concentration patterns has remained elusive, as no single biological system has, as yet, yielded a Turing morphogen pair. However, recent advances in the chemical realization of Turing-type structures has come with the CIMA reaction, in which, by careful manipulation of substrate flow rates, temperature-sensitive patterns obeying the Turing type of instability are shown to occur. (See Lengyel & Epstein 1992; Ouyang & Swinney 1991; Winfree 1991).

In this paper a type of pattern formation is described which is not directly related to the classical idea of Turing. Nevertheless, as described below, much of the analysis and several resulting predictions share formal similarity.

19.3 PATTERNS IN ANGLE

We can think of many examples in physics where a set of particles, molecules or other bodies become polarized so that some directions of motion, of alignment, or of rotation become dominant. Magnetic spins, liquid crystals, and other types of polymers often exhibit such behaviour. In many biological situations a similar effect occurs in a population of macromolecules, of cells, or of organisms. In this paper an example of cellular dynamics that leads to polarized tissue structure is stressed, but many other examples of the same idea occur in nature: ants milling about in an early stage of swarming are able to collectively select one direction to head out on a mission of exploration. Other migratory herds, such as wildebeest, as well as flocks of birds and schools of fish are able to move in a coherent direction, even though no single bird or fish acts as leader. How a group collectively determines a preferred direction, or set of directions is the question we attempt to address here. We restrict attention to cases in which orientation is selected endogenously, not imposed artificially. The goal is to explain population behaviour based on the behaviour and interactions at the level of the individual.

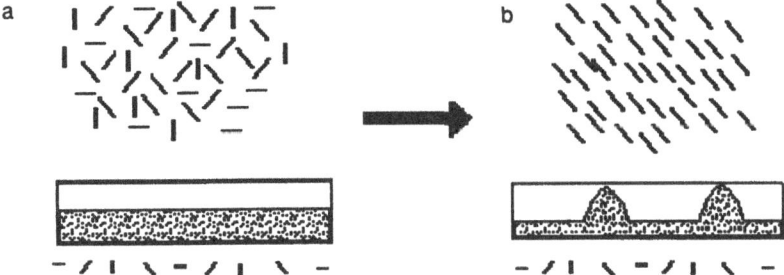

Figure 19.2. The transition from isotropic (a) to anisotropic (b) distributions can be represented graphically by density distributions on angles $0 < \theta < 360$ degrees.

The hypothesis proposed here is that a preferred direction can arise in a population of interacting units as an instability. For example, a uniform angular distribution of individuals (as shown in Figure 19.2(a)), may be subject to disruption by small random deviations. These deviations are heightened and reinforced by certain interactions between individuals that are angle-dependent. Eventually one, two, or more directions dominate. (See Figure 19.2(b)).

The above hypothesis places angle-patterns in the same framework as spatial patterns. A difference is that angles are periodic, with period 2π. (Equivalently, directions can be represented by densities on a unit circle as shown in Figure 19.3. An isotropic distribution is shown in Figure 19.3(a), and a nonisotropic one in Figure 19.3(b)). For this reason, an angular distribution is to be viewed as a function on a periodic domain: the value of the function at 0 matches exactly its value at 2π radians = 360 deg. This leads to the requirement that the number of peaks in an angle pattern in the interval $0 < \theta < 2\pi$ has to be an integer. This number is called the wavenumber of the pattern, and will be represented by the symbol k.

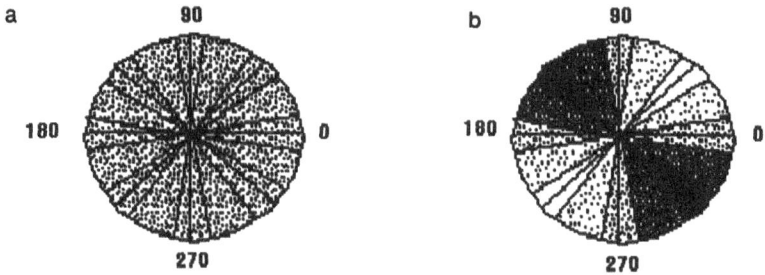

Figure 19.3. Angular density distributions can also be represented as densities on a unit circle. An isotropic distribution is shown in (a), and a nonisotropic one in (b).

19.4   CELL-CELL INTERACTIONS IN AN ARTIFICIAL 'TISSUE'

Fibroblasts grown artificially on a flat culture dish undergo a creeping movement. In the absence of collagen fibers to pull on, and despite an isotropic environment, they eventually form a set of parallel arrays, groups of cells oriented in a common direction or set of directions. Unlike the situation described by N. Hill in this volume, fluid dynamic effects play a negligible role in cell motion, which results from extrusion, adherence, and pulling by **lamelipodia** on the surface of the dish.

In a model whose detailed formulation appears in Edelstein-Keshet & Ermentrout (1990) (here abbreviated EKE-90), rules governing the behaviour of fibroblasts in culture are described. Briefly, cells undergo occasional random turning, but when bound in groups, motion is restricted, and cells can only slide past each other along their common axis. Of particular importance is the fact that when cells meet, they have a high likelihood of binding in parallel provided the angle of contact is small enough. This angle-dependent contact response of cells has been documented in the biological literature and forms a basic component of the mechanism.

19.5   A MODEL FOR ANGULAR DISTRIBUTIONS OF CELLS

The model describes changes in the angular distributions of free and bound cells, defined by:

$B(\theta, t)$ = density of bound cells oriented in direction θ at time t,

$C(\theta, t)$ = density of free cells oriented in direction θ at time t.

We also define the **contact response kernel** $K(\theta)$ as follows:

$K(\theta)$ = the probability of alignment of cells that come into contact at an angle θ with one another.

The equations governing these distributions are:

$$B_t(\theta, t) = \beta_1 CK * C + \beta_2 BK * C - \gamma B,$$

$$C_t(\theta, t) = \mu C_{\theta\theta} - \beta_1 CK * C - \beta_2 CK * B + \gamma B.$$

Here $K * C$ is a convolution integral of the form:

$$K * C(\theta) = \int_{-\pi}^{\pi} K(\theta - \theta')C(\theta', t)d\theta'.$$

The terms in these equations represent the rate at which pairs of free cells bind ($\beta_1 CK * C$), the rate at which the density of bound cells at angle θ increases by trapping free cells from other directions ($\beta_2 BK * C$), the rate at which cells are shed from clusters and become free cells (γB), the random turning of free cells ($\mu C_{\theta\theta}$), and the rate at which free cells at angle θ are trapped by cells bound at any other angle. (Observe that the terms multiplied by β_2 are not identical in the two equations.)

Analyzing the case $\beta_1 = \beta_2 = \beta$, it was shown in EKE-90 that these equations admit a uniform steady state solution in which the densities of cells at any angle is the same, namely $\bar{C} = \bar{B} = \beta M/\gamma$ where $M = \bar{C} + \bar{B}$ is the total density of free and bound cells. However this isotropic (angle independent) solution can be unstable to small angle-dependent disturbances. A small pattern whose wavenumber is k will grow provided a **dispersion relation** is satisfied. The relation is given by the inequality

$$Ak^2 < \hat{K}(k)(1 - \hat{K}(k)),$$

where

$$A = \frac{\mu}{\gamma}\left(\frac{\gamma}{\beta M}\right)^2$$

is a dimensionless combination of nonnegative parameters appearing in the equations, and $\hat{K}(k)$ is the Fourier transform of the contact response kernel. This Fourier transform appears as a consequence of assuming that the small angle-dependent perturbations are proportional to $e^{ik\theta}e^{\sigma t}$. It is worth noting that the parameters appearing in the dispersion relation represent experimentally measurable properties of the cells, as well as the total size of the population. The inequality is most likely to be satisfied if the random motility of the cells, μ, or the rate of dissociation of cells from clusters, γ are small. Increasing the affinity of cells to form pairs or clusters, β, or the total density of cells also promotes this instability to occur.

The wavenumber k for which the instability will first occur, *i. e.*the pattern spacing which will start to develop depends on details of the shape of the function of k shown on the right of the inequality. These details depend on assumptions about the contact response K. Symmetry properties of the kernel $K(\theta)$ have particularly strong effects on the shape of its Fourier transform $\hat{K}(k)$, including values of k for which it is positive. Note that only values of k that lead to $0 < \hat{K}(k) < 1$ can satisfy the dispersion relation inequality since the expression Ak^2 is always positive.

In EKE-90 it is shown that if the contact response is symmetric about zero (same turning response for turns to the right or to the left), and for angles θ and $\pi - \theta$, then wavenumbers $k = 1, 3, 5, 7 \ldots$ *i. e.* all odd integers cannot represent patterns that grow.

19.6   WHAT THE MODEL TELLS US

To interpret the predictions of the model and its analysis, we need to first take note of the role that the wavenumber, k, plays in the type of patterns. For the traditional Turing-type pattern formation in space, the wavenumber is simply the number of peaks of concentration that fit into a unit length of the domain. We usually then think of spacing between the spots or the stripes (for example in a pattern on mammalian fur described by Murray) as distances proportional to L/k where k is the wave number and L a typical domain size.

In the theory discussed in this paper, spatial patterns have been replaced by angular patterns, so k represents the number of peaks in the angular distribution that fit into the domain $0 < \theta < 2\pi$. Thus here k represents the **number of dominant directions** in the final cell distribution, not the spatial size or separation between pattern elements. For example, the mathematical analysis predicts that $k = 2$ is the first wave number to grow. This means that an angular distribution with two peaks (180 degrees apart) will form, so that cells will have one dominant axis of orientation, but will align both in parallel and anti-parallel with this axis. This type of behaviour is seen to occur in fibroblasts grown artificially in culture.

The model predicts that parallel arrays are less likely to occur if the density of cells is low, if they tend to break away from clusters easily, or if their affinity to other cells is low. Thus, one of the useful roles of a model is to give an appreciation of how phenomena depend on parameters in the problem. However, linear stability analysis has a limited power of predictability, since only the behaviour of small deviations away from a uniform steady state are governed by linear theory. Thus, the model cannot conclusively predict the final patterns except in a small range of parameter values close to the onset of pattern formation (*i. e.* close to a bifurcation, in the mathematical terminology).

The model has not been designed to describe spatial distributions, and therefore cannot tell us how the parallel arrays are distributed in space. It is not difficult to include both the spatial and angular variables in the same model, but the difficulties in analyzing the model make this of limited interest. The model also does not tell us how the average cell density translates into detailed cell configurations. This is a drawback of all continuum-based models, but is a sacrifice that permits the techniques of analysis to be applied.

The details of the cell configurations, the timescale and dynamics of pattern formation, and the nature of the fully nonlinear system are best explored in conjunction with this model by numerical techniques and computer simulations. A brief description of such techniques is given in the next section.

19.7   COMPUTER SIMULATIONS OF CELL BEHAVIOUR

The formulation of a model is based on some simplified description of the cells, with a limited number of "rules" that capture the essence of the phenomenon. These rules, or some slight modification of these rules can also be applied to a computer simulation of a cellular population. A cellular automata simulation written for this purpose by Ermentrout appears in EKE-90.

Briefly, the positions and motion of cells are represented by line segments that can move forward, reorient randomly, and turn and stick when contacting other "cells". A population of cells and their dynamic interactions can be followed on a computer monitor in "real time". The values of parameters governing cell motion or cell contact can be adjusted, and the rules can easily be changed to explore consequences. The simulations allow one to follow individual cell configurations and spatial as well as angular distributions of the cells, in a way that is not possible with the analytic model on its own. The formation of parallel arrays, with one dominant axis of orientation is well illustrated by such simulations, provided the density of the cells is high enough, and that the other parameters associated with cell binding and unbinding are appropriately chosen.

19.8   APPLICATION OF SIMILAR IDEAS TO OTHER PROBLEMS

Some of the features of this model are shared with a number of other models for collective behaviour of interacting units, specifically for cases in which an angular distribution is of interest. We mention three such cases below:

19.8.1 Branching Networks

In a paper by Edelstein-Keshet & Ermentrout (1989) the interactions of branches in a growing filamentous network was described. In surface growth of fungi, the hyphae (individual branches) often interact by forming anastomoses (connections), or by inhibiting growth in their vicinity. The spread of a network in 2D and the orientations of the branches can be analyzed by a model that shares formal similarity to the above. The orientations of neighboring branches influences the likelihood that they will grow close enough to one another to intersect and crosslink. Thus, an angle dependent contact rate, similar to the contact response kernel of the cells also appears in the model for angular distributions of branches and their directions of growth. Analysis similar to the linear stability analysis detailed above was the basis for understanding the branching model.

The model predicts that filaments would tend to form 1D strands, with parallel branches, provided branching angles were sufficiently small. Branching in a network is essentially equivalent to random reorientation in cell populations, as it tends to disperse (*i. e.* broaden) the angular distribution. This therefore agrees with a similar prediction in the cellular model, that parallel arrays occur provided that the rate of cell random reorientation is not very large.

19.8.2 Swarming in Social Insects

Social insects such as Army ants form massive migration and exploration swarms in seeking food, or in moving to a new site. The temporal dynamics of swarm formation as a colony of Doryline ants emerges from a nest was recorded by Raignier & van Boven (1955). After some random milling and loose uncoordinated motion in the neighborhood of the nest, (shown in Figure 19.4(a)) the ants form a single column of traffic, (see Figure 19.4(b)) and march out to find food, to be transported back to the nest. The selection of a single direction out of an initially confused collective motion is similar to the selection of a dominant direction in a cell population. As in the case of cells, it is believed that ants are not being directed by leaders, but rather deciding on the strategy by a group effect. One of the mechanisms involved in this coordination is trail marking and trail following.

Models for the motion of ants, their trail marking, trail following, and random exploratory motion were described in Edelstein-Keshet (1992). An angular distribution model similar to the one in this paper yields similar predictions about circumstances, rules and parameter values consistent with the ability to collectively select a dominant direction of motion.

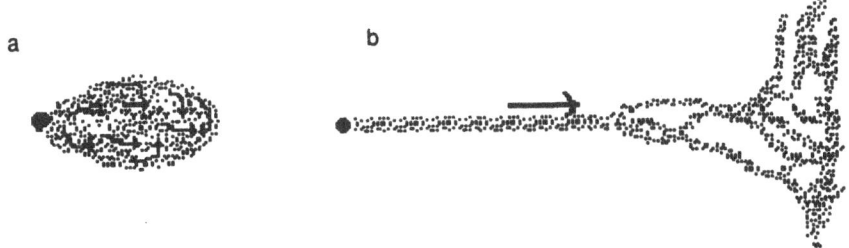

Figure 19.4. Swarming behaviour in Doryline ants emerging from their nest (black dot) illustrates direction-selection. In (a), the ants have just emerged and are milling around randomly. In (b), a prominent direction of motion has emerged.

In following pheromone trails, ants turning onto a trail do so with a probability that depends on angle of incidence with the trail. (This stems from the fact that ants detect trails by comparing concentrations of pheromone perceived by the two antennae. If the concentrations are very similar, the ant may not turn.) This angle dependent turning onto trails plays a role analogous to the contact response kernel discussed in the cell model.

19.8.3   Alignment of Actin Bundles and Networks

The shape and motility of cells is intimately linked to the structure of macromolecular cytoskeleton of the cell. Actin bundles and actin meshworks form an integral part of this assembly. Actin is a protein found ubiquitously inside cells, and it undergoes rapid polymerization from monomers into filaments. The filaments are then arranged in a 3D structure by Actin Binding Proteins (ABP). Many different types of binding proteins are known, some promoting parallel bundling of actin filaments, others promoting orthogonal, or close to orthogonal meshworks.

How the cell locally controls the type of actin structures formed is a question that can be addressed by a simple model for the angular distribution of actin filaments. In the work by Civelekoglu (1992) it has been shown that a transition between parallel and orthogonal binding can occur by locally titrating the fraction of parallel and orthogonal binding proteins. The basic features of the model are analogous to those already described in this paper. Unlike the treatment by Sherratt and Lewis in this volume, the effect of mechanical forces and stresses are not explicitly considered.

19.9   DISCUSSION

The examples cited above share the common feature that direction, rather than spatial position is considered as a primary attribute. Subsequently, the equations of the models for these phenomena describe angular, rather than spatial distributions. Peaks in the patterns represent directions along which cells or branches, or individual motion tends to align.

It is of interest to compare the ideas and results of these models with those of the Turing reaction-diffusion and other theories for pattern formation. As points of comparison, we mention the following specifics:

1. The phenomena involve at least two species or two forms of a given species (*e. g.* bound and free cells, branches and tips of branches, follower and lost ants, free and bound actin).

2. These species interact locally in some non-linear way (by binding, by crosslinking, or by recruiting one another through trails).

3. Random turning causes angular distributions to broaden, which is formally equivalent to the way that diffusion causes spatial spread.

4. Aside from the local interactions and the local diffusive spread, the angular distribution models also admit "long-ranged" interaction between angles that may be "far apart".

The long-ranged angular interaction is in distinct contrast to the Turing models (that contain only local diffusion). Some of the mechanochemical models proposed by Oster, Murray, and co-workers also contain long-ranged interactions, but these are generally expressed by including higher derivatives in the continuum equations. Othmer (1969) showed that this is formally equivalent to assuming integral kernels (as we have done), and retaining the first few moments of these operators. (See also Murray (1989) for a brief summary.)

A closer analogy exists between the long ranged angular interactions (depicted by angle dependent kernels) and the long-ranged spatial interactions between neurons (depicted by space-dependent synaptic connection kernels) in the neural network models for pattern formation. The analogy stems from the fact that neurons can have long processes that impinge on, and activate or inhibit other neurons.

We have discussed the fact that lateral inhibition is a recurring theme in models for pattern formation. In Turing models, this shows up in the nature of the chemical substances (activator and inhibitor, for example) and in the fact that one spreads laterally by diffusion at a faster rate. In the neural models, the idea of lateral inhibition shows up through a "Mexican hat" shape of interaction kernels; neurons are assumed to excite their local neighbors, but inhibit the more distant neurons. Thus, the effect is positive close to the center of an excitation but negative beyond some distance. (See papers by Ermentrout, Swindale (1980) and, for a review, Murray 1989).

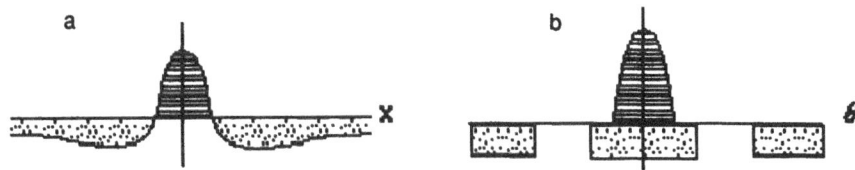

Figure 19.5. Lateral inhibition is often depicted by the "Mexican Hat response" shown in (a). Activation occurs close to some center, and inhibition extends laterally outwards. In the cell-cell interaction model, the effect is similar (b). A cluster of bound cells at some angles will contribute positively to free cells due to shedding (striped peak), and negatively due to binding (dotted function).

A similar feature exists in the contact response kernels of the models described in this paper, provided cells can bind over a reasonably wide range of angles. We reason as follows: given that a cell cluster occurs at some angle, it contributes positively to the density of free cells at very "close angles" by shedding (the free cells shed off a cluster will have an angular distribution centered about the cluster direction.) This is represented in Figure 19.5(b) by the striped positive peak. The cluster also contributes negatively to free cell density for some finite range of angles due to binding. (Represented in Figure 19.5(b) by the dotted negative rectangles. Free cells at angles close to zero and to 180 degrees can bind to the cluster.) Provided these effects are in proper balance, *i. e.* that the range for angles admitted for binding exceeds the range of angles created by shedding during a given interaction-time, an effect similar to lateral inhibition is created.

As a final comment, we mention that much of the analysis carried out on traditional models for pattern formation is also applicable to angle-dependent models, as argued above. Linear stability theory, which tests the stability to small perturbations makes a number of predictions. These predictions should be viewed with caution, since they hold close to the homogeneous steady states, and may not always describe the emergence of the patterns created by the full nonlinear interactions. A second tool, the computer simulation, takes up where analysis falls short. Because the phenomena described here involve discrete individuals interacting in (large) populations, the simulation of precise individual behaviour and group interaction via cellular automata is a particularly revealing technique. It would appear that such techniques may allow one to describe and analyze other cases in which a dominant direction or set of directions is formed through some biological interactions, whether at the level of a population of organisms, of cells, or of molecules.

ACKNOWLEDGMENTS This work has been carried out under support from the National Sciences and Engineering Research Council of Canada, grant number OGPIN 021. Initial stages of this work were funded by NSF under grant number DMS-86- 01644. The author would like to thank G.B. Ermentrout for collaboration in many aspects of this research.

REFERENCES

Bard, J. B. L. 1981. A model for generating aspects of zebra and other mammalian coat patterns. *J. Theor Biol.*, **93**, 363–385.

Bard, J. B. L., & French, V. 1984. Butterfly wing patterns: how good a determining mechanism is the simple diffusion of a single morphogen? *J. Embryol. exp. Morph.*, **84**, 255–274.

Civelekoglu, G. 1992. *Actin alignment mediated by actin binding proteins.* Ph.D. thesis, UBC, Vancouver.

Edelstein-Keshet, L. 1992. Trail following as an adaptable mechanism for population behaviour. In *3D Animal Aggregations*. (Submitted).

Edelstein-Keshet, L., & Ermentrout, G. B. 1989. Models for branching networks in two dimensions. *SIAM J. Appl. Math.*, **49**(4), 1136–1157.

Edelstein-Keshet, L., & Ermentrout, G. B. 1990. Models for contact-mediated pattern formation: cells that form parallel arrays. *J. Math. Biol.*, **29**, 33–58.

Ermentrout, G. B., & Cowan, J. D. 1979. A mathematical theory of visual hallucination patterns. *Biol. Cybernet*, **34**, 137–150.

Ermentrout, G. B., Campbell, J., & Oster, G. 1986. A model for shell patterns based on neural activity. *Veliger*, **28**, 369–388.

Gierer, A., & Meinhardt, H. 1972. A theory of biological pattern formation. *Kybernetik*, **12**, 30–39.

Harrison, L. G. 1987. What is the status of reaction-diffusion theory thirty four years after Turing? *J. Theor. Biol.*, **125**, 369–384.

Lengyel, I., & Epstein, I. R. 1992. A chemical approach to designing Turing patterns in reaction diffusion systems. *Proc. Natl. Acad. Sci.*, **89**, 3977–3979.

Levin, S. A., & Segel, L. A. 1985. Pattern generation in space and aspect. *SIAM Review*, **27**(1), 45–67.

Meinhardt, H. 1978. Models for the ontogenetic development of higher organisms. *Rev. Physiol. Biochem. Pharmacol.*, **80**, 47–104.

Meinhardt, H. 1982. *Models of biological pattern formation.* New York: Academic Press.

Murray, J. D. 1981a. On pattern formation mechanisms for lepidopteran wing patterns and mammalian coat markings. *Phil. Trans. Roy. Soc. (London)*, **B295**, 473–496.

Murray, J. D. 1981b. A pre-pattern formation mechanism for animal coat markings. *J. Theor. Biol.*, **88**, 161–199.

Murray, J. D. 1982. Parameter space for Turing instability in reaction-diffusion mechanisms: a comparison of models. *J. Theor. Biol.*, **98**, 143–163.

Murray, J. D. 1988. How the leopard gets its spots. *Sci. Amer*, **258**(3), 80–87.

Murray, J. D. 1989. *Mathematical biology.* New York: Springer Verlag.

Murray, J. D., & Oster, G. F. 1984. Cell traction models for generating pattern and form in morphogenesis. *J. Math. Biol.*, **19**, 265–279.

Murray, J. D., Oster, G. F., & Harris, A. K. 1983. A mechanical model for mesenchymal morphogenesis. *J. Math. Biol.*, **17**, 125–129.

Odell, G. M., Oster, G., Alberch, P., & Burnside, B. 1981. The mechanical basis of morphogenesis, I. epithelial folding and invagination. *Dev. Biol.*, **85**, 446–462.

Oster, G. F., & Murray, J. D. 1989. Pattern formation models and developmental constraints. *J. exp. Zool.*, **251**, 186–202.

Oster, G. F., Murray, J. D., & Harris, A. K. 1983. Mechanical aspects of mesenchymal morphogenesis. *J. Embryol. exp. Morph.*, **78**, 83–125.

Othmer, H. G. 1969. *Interactions of reaction and diffusion in open systems.* Ph.D. thesis, Chemical Engineering Dept., Univ. of Minnesota.

Ouyang, Q., & Swinney, H. L. 1991. Transition from a uniform state to hexagonal and striped Turing patterns. *Nature*, **352**, 610–612.

Raignier, A., & van Boven, J. 1955. étude taxonomique, biologique, et biometrique des *Dorylus* du sous-genre *Anomma* (hymenoptera: Formicidae) *Annals du Musee Royal du Congo Belge* n.s. 4. *sciences zoologiques*, **2**, 1–359.

Swindale, N. V. 1980. A model for the formation of ocular dominance stripes. *Proc. R. Soc. Lond.*, **B208**, 243–264.

Turing, A. M. 1952. The chemical basis of morphogenesis. *Phil. Trans. Roy. Soc. Lond.*, **B237**, 37–72.

Weliky, M., & Oster, G. 1990. The mechanical basis of cell rearrangement I epithelial morphogenesis during fundulus epiboly. *Development*, **109**, 373–386.

Winfree, A. T. 1991. Crystals from dreams. *Nature*, **352**, 568–569.

Young, D. A. 1984. A local activator-inhibitor model of vertebrate skin patterns. *Math. Biosci.*, **72**, 51–58.

20. CONTROLLING REACTION-DIFFUSION PATTERN WITH GRADIENTS: LESSONS FROM *DROSOPHILA*, AND TRAJECTORIES THROUGH PARAMETER SPACE

Thurston C. Lacalli

Biology Department
University of Saskatchewan
Saskatoon, Sask. S7N-0W0, Canada

20.1 INTRODUCTION

Since it was first reported by Hafen *et al.* (1984), the beautifully simple periodic pattern of pair-rule stripes in *Drosophila* embryos has intrigued theoretical and experimental biologists alike. A primary concern among theoreticians has been the question of whether pair-rule stripes are, to any extent, Turing structures, sustained by far-from-equilibrium kinetic processes. Current opinion among experimentalists tends to the view that they are not. It is instead proposed that the precise boundaries of all seven pair-rule stripes are each independently specified by the underlying expression patterns of gap genes in a concentration-dependent fashion, in the same manner that the gap genes are themselves activated by the broader maternal gradients. This means the iterative operation of a process of gradient reading. The problems with this view have already been discussed at this meeting at some length by Axel Hunding. His message is that it is difficult if not impossible to test either view critically on the basis of currently available data. Models that depend solely on iterative gradient reading require, in their extreme form, many untested suppositions. A Turing mechanism avoids some of the apparent problems, but there is as yet little direct evidence to support the idea that a Turing mechanism is operating, or if it is, what the key components, *i. e.* the morphogens, might be.

Regardless of how this issue is eventually resolved, simply asking the question has led theoreticians to fruitful new areas of inquiry, specifically concerning how stripes might be stabilized, and the role graded signals might have as inputs to a Turing-type reaction system (*e. g.* see Lyons & Harrison 1993). I want to deal here with two of the more interesting issues that arise. First is the question of precision in hierarchical control systems. Suppose a graded signal at one level of a hierarchy is to be used to provide cues for the next. One then must consider whether the errors, or noise, inherent in the signal are propagated along with the signal itself, and if not, what is the mechanism by which they are removed. This is important in relation to *Drosophila* because the process of gene activation, no matter how precisely controlled, is still a molecular event subject to the random fluctuations and uncertainties of real molecular dynamics. Precise concentration-dependent response by the transcriptional machinery to a graded signal requires some means of time or spatial averaging to counter

Experimental and Theorietcal Advances in Biological Pattern Formation,
Edited by H.G. Othmer *et al.,* Plenum Press, New York, 1993

223

these effects. How this might be achieved in a developing system like *Drosophila* is far from clear.

A second issue concerns the question of how robust particular classes of patterning mechanisms are against parameter change, since gradients are essentially a way of arranging for key developmental parameters to vary over distance. To explore this question requires an understanding of parameter spaces. Here, briefly, I summarize some features of the particular form of parameter space that allows for most ready comparison between models. On this basis, a distinction is proposed between two broad classes of models, depending on whether they are "tight" or "loose" in stoichiometric terms. By the former, I mean models with a strict chemical stoichiometry in which the morphogens are reactants or reaction intermediates. These are widely used as models for oscillatory phenomena. The latter refers to models with a looser functional relationship between components in kinetic terms, with the substances involved being catalysts or modulators of their own and each others' synthesis. Such models are especially relevant to complex synthetic control processes of the type found in living systems, of which transcriptional regulation is a prime example. As a class, they appear to be especially well controlled and thus suitable for robust use by living systems for making patterns in embryos.

20.2 SIGNAL-TO-NOISE AND THE PROBLEM OF ERROR PROPAGATION

This issue arises naturally in the context of reaction-diffusion models because they enable patterning to be controlled in an especially precise way. While they are only one of many ways of actually generating pattern, they characteristically produce beautifully regular ones, a consequence of the intrinsic pattern wavelength imposed by the reaction-diffusion mechanism. So even if a preexisting pattern is used to provide positional cues, a significant proportion of the positional variation or error in these cues is removed when the Turing pattern is imposed over them. In this respect, reaction-diffusion systems act as signal amplifiers, extracting a particular signal from a noisy input. The boundaries of pair-rule stripes, once they have formed, show irregularities in width on the order of 1 - 2 cell diameters. This suggests the mechanism of positional specification has an inherent uncertainty on the order of 1%. If iterative gradient reading is the means by which each stripe boundary is specified, we need to demonstrate that it can achieve this degree of precision.

To read a gradient precisely, the cell's molecular machinery faces a number of practical problems. They can be summarized briefly. For a more extensive treatment see Lacalli & Harrison (1991). First is the question of how effectively a gene promoter can operate as a monitor of concentration. This is basically a sampling problem, similar in nature to the problem of how a chemotactic cell might monitor concentration in its surroundings. The latter is examined from a theoretical standpoint by Berg & Purcell (1977). Precision seems to be achieved mainly by increasing the number of surface receptors on each cell. By analogy, this may explain why the promoters of important control genes have multiple binding sites for each of a number of regulatory factors. But the effectiveness with which a given promoter will be able to act as a concentration-dependent transcriptional switch also depends crucially on kinetic factors like binding constants, turnover times and rates of transcript initiation, and it is by no means clear how appropriate it is to view gene activation in a simple deterministic way (Ko 1991; Ko 1992). There is, in addition, the question of how reliably the gradient itself can be reproduced from embryo to embryo. The maximal concentration achieved by the gradient at its source depends on rates of synthesis, which may include both transcription and RNA translation. Its slope will depend on diffusion, but also on rates of hydrolytic degradation, which in turn depend on the steady state concentrations of other synthesized proteins. Various concentrations and rates of reaction presumably have to be constrained within certain limits to maintain the resulting gradient to a given degree of precision.

In short, while the structural complexity of the cell's synthetic machinery can often be effectively portrayed in simple, cartoon-like diagrams, these mask a host of kinetic and statistical complexities. These need to be understood quantitatively before such diagrams can be accepted as accurate representations of reality.

It is probably correct to assume that however *Drosophila* controls pattern, it has pushed the control system to its limits. Flies are specialized, compared with other insects, for rapid and simultaneous specification of embryonic structures that primitive insects have the leisure to generate sequentially. *Drosophila*'s compressed developmental schedule may test the limits of the ability of the transcriptional system to respond to graded positional cues. Whether this means we are effectively at the limit of precision for iterative gradient reading or whether a Turing mechanism has been interpolated as a corrective device, the errors inherent in the process deserve attention in themselves, as clues to the underlying mechanism.

20.3 TRAJECTORIES IN PARAMETER SPACE

The second issue arises when we examine how reaction-diffusion models behave as one or more of the input parameters is varied over time or space. A spatial gradient in a reaction precursor, or in an activator of one of the reaction steps, would be examples. The latter imposes, in effect, a gradient in reaction rate. If a pattern of repeated structures is to be produced, we can ask whether these will be regularly positioned along the gradient, or whether they are less well behaved in some respect, varying in, say, size, spacing, or time of appearance. Likewise with time: as the patterning mechanism is progressively turned on by increasing one or more parameters past some critical value, does pattern scale change with time? One could have, for example, patterns that start large and get small, or *vice versa*. These are all features of the patterning process that could potentially be observed in real organisms, so discovering how particular classes of models behave is of some practical significance.

Dealing with the above requires an understanding of parameter spaces. Murray (1982) discusses how these can be devised for particular models, but there is also a canonical form of the parameter space that allows diverse models to be compared directly. Each model then gives a characteristic set of trajectories. The analysis requires that the basic rate equations first be converted to linear form, *i. e.* Turing's equations:

$$\partial X/\partial t = k_1 X + k_2 Y + \mathcal{D}_X \partial^2 X/\partial s^2;$$
$$\partial Y/\partial t = k_3 X + k_4 Y + \mathcal{D}_Y \partial^2/\partial s^2.$$

Here the four k's are combinations of the rate constants that appear in the original rate equations, and the D's are diffusivities. To construct the parameter space we look at three expressions, n, k_1' and k_4', defined as follows:

$$n = \mathcal{D}_Y/\mathcal{D}_X; \qquad k_1' = k_1/k_{cr}; \qquad k_4' = k_4/k_{cr}; \qquad k_{cr} = (-k_2 k_3)^{\frac{1}{2}}.$$

Six conditions define the linear behavior of Turing's equations:

$$k_1' + k_4' = 0; \tag{20.1}$$

$$k_1' k_4' = -1; \tag{20.2}$$

$$k_1' - k_4' = 2; \tag{20.3}$$

$$nk_1' - k_4' = 2n^{\frac{1}{2}}; \tag{20.4}$$

$$k_1' - k_4' = 4n^{\frac{1}{2}}/(n+1); p \tag{20.5}$$

$$k_1' - k_4' = (n+1)/n^{\frac{1}{2}}. \tag{20.6}$$

The key conditions are shown graphically in the figures. Further details of the analysis are discussed by Lacalli & Harrison (1979). To understand Figure 20.1, note that equations (20.1)-(20.3) do not depend on n, the ratio of diffusivities. If we consider only the reaction terms in Turing's equations, and ignore diffusion, conditions (20.1)-(20.3) are equivalent to setting to zero the trace, determinant and discriminant, respectively, of the Jacobian obtained. These three conditions (solid lines, Figure 20.1) divide the fourth quadrant ($k_1' > 0$, $k_4' < 0$) into various domains. There are three main ones: above the hyperbola defined by (20.2) the reaction system will be stable to the left of (20.1), but oscillates to the right of (20.1). Line (20.1) is the Hopf bifurcation. Below the hyperbola the system is spontaneously unstable, which means chain reactions, explosions or similar will occur.

Including diffusion adds conditions (20.4)-(20.6), the key one being (20.4), a line of slope n (dashed line, Figure 20.1). This opens a roughly wedge-shaped domain to the immediate left of (20.1) within which a diffusion-driven Turing instability can occur, producing stable pattern. For large n, this domain is relatively large. For n approaching 1, line (20.4) inclines to the right until it is coincident with (20.3), and the Turing domain disappears. For the mathematically inclined, Figure 20.2 shows the dispersion relations for the four main subdomains in Figure 20.1.

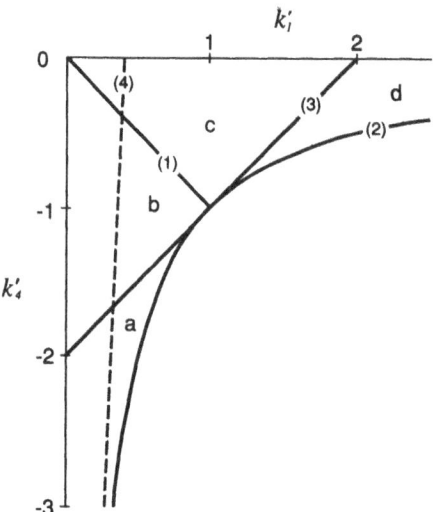

Figure 20.1. Parameter space for 2-component Turing models, drawn for diffusivity ratio (n) = 20. Linear stability conditions are shown as heavy lines; numbers correspond to equations in the text, letters a - d to the dispersion relations in Figure 20.2. The roughly wedge-shaped domain bounded by lines (20.1), (20.2) and (20.4) defines the domain within which stable pattern is formed. To the right of (20.1), the Hopf bifurcation, oscillations or related time-dependent phenomena occur.

There is a characteristic expression for the product of k_1' and k_4' for each model, typically of the following form:

$$k_1' k_4' = -q \left(1 - f(a, b \ldots)\right), \tag{20.7}$$

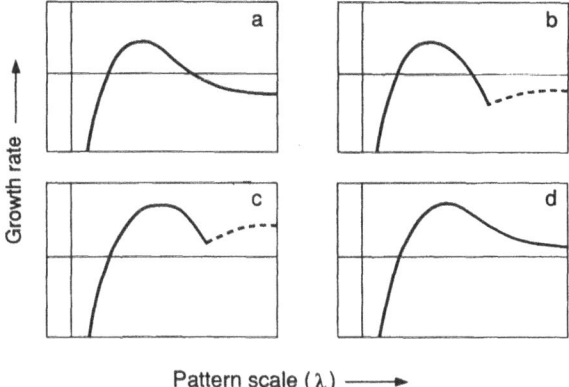

Growth rate

Pattern scale (λ) ———▶

Figure 20.2. Dispersion relation for Turing's equations for various regions of parameter space as indicated by corresponding letters in Figure 20.1. Plots are of spatial scale (wavelength) of the growing pattern versus the exponential growth rate for that pattern. The dashed parts of the curves show the real part of the eigenvalue solutions where these are complex.

where q is a fraction less than or equal to 1, and f is a sum of terms involving various combinations of rate constants, a, b, etc. Thus, the parameter trajectories move along hyperbolic trajectories lying between the origin and the limiting hyperbola defined by q. For q close to 1, even moderate values of n give a fair-sized Turing domain. For q much smaller than about $\frac{1}{2}$, which occurs in some models, the Turing domain will be quite limited for any but very large n.

In very general terms, two basically different types of trajectories are encountered. First the simple case: trajectories cross the domain on a roughly hyperbolic track with the k's changing monotonically with change in the parameter being varied, from left to right as A in Figure 20.3, or the reverse, from right to left. In other cases, however, the trajectories reverse direction (as B in Figure 20.3), leaving the Turing domain soon after entering it. They often do not cross the Hopf bifurcation at all. The first type of behavior is characteristic of a number of simple chemical reaction systems designed to model oscillating reactions. Most, but not all of the Schnakenberg and Brusselator schemes commonly used by theoreticians behave this way. For a fairly wide choice of parameters these models oscillate, which is of course desirable if one is modelling oscillations. To achieve this, however, the models must usually traverse some distance across parameter space, taking on a range of k_1' and k_4' values. Both of these parameters are important determinants of λ, the pattern wavelength, and other linear properties that govern how the model will behave (see Lacalli & Harrison (1987) for further details). A model that crosses parameter space readily in this way thus has considerable potential to be badly behaved in various ways, including exhibiting a continuously varying pattern size.

The connection between this type of behavior and strict stoichiometry, in very simple terms, can be seen as follows: in simple models, the X's and Y's are typically reactants or reaction intermediates. Changes in X and Y concentration are strongly linked because specified amounts of Y are needed to form X, or are destroyed when X is formed. In consequence, terms of identical form appear in the kinetic equations for rates of change of both X and Y. These simplify the mathematical analysis by cancelling at various points, which leads to very simple expressions for k_1' and k_4', and simple trajectories in parameter space.

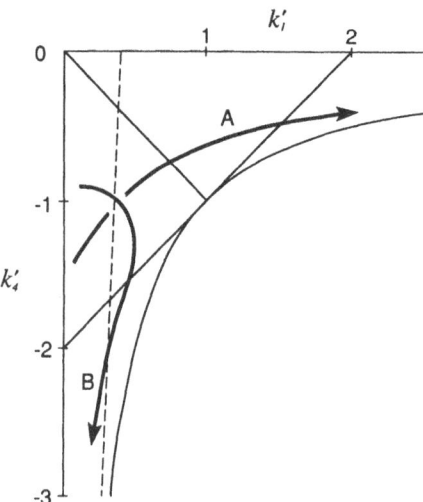

Figure 20.3. Parameter space as in Figure 20.1 showing two typical types of parameter trajectories, A and B. See text for discussion.

Model reaction systems can be altered in various ways that lead to trajectories like B in Figure 20.3, and the type of alterations required are easily justified in biological terms. Consider a complex biological synthetic process involving DNA transcription. Transcription rates are under the control of a multitude of protein factors which are themselves produced via DNA transcription. They can act to enhance or inhibit their own and each others' rates of synthesis, but they are acting as catalysts or modulators of the underlying transcriptional process, not as reactants. In kinetic terms, in a model with X and Y as synthetic regulators, the best way to express the effect of X on Y formation, and Y on X, will be to use various kinds of generalized functions. These will typically involve a high degree of cooperativity, since transcriptional regulators can act as both homotypic and heterotypic dimers that either activate or inhibit (Ptashne 1988; Lamb & McKnight 1991). The resulting kinetic terms for X and Y formation are unlikely to be identical, and the mathematical analysis will be more complicated. This leads to complicated expressions for the Turing constants and more complex trajectories in parameter space. The synthetic steps also must saturate, since there are clearly upper limits on the overall rate at which transcriptional processes can occur. As well, it is useful to include zero-order input and first-order decay terms for all components, to represent the basic rates of unactivated synthesis. By way of example, a model satisfying some of these criteria is given below: X synthesis is activated by homodimers of X with the help of Y, but saturates, while both X and Y are removed as inactive XY heterodimers.

$$\partial X/\partial t = bY + \frac{cX^2Y}{(1+KX^2)} - gXY - dX + \mathcal{D}_X\partial^2 X/\partial s^2;$$

$$\partial Y/\partial t = a - gXY + \mathcal{D}_Y\partial^2 Y/\partial s^2.$$

Meinhardt (1982) has long advocated the inclusion of features like those just described in biological models. He emphasizes that modellers should be thinking in terms of components that "help" each other, even if the details are poorly understood, and processes that saturate. So long as the kinetic order of the reaction is high enough, such models can enter the Turing domain. Typically they then leave it as saturation effects come into play, along trajectories like B in Figure 20.3, which is what happens in the heterodimer model just described. Models of this type make it relatively easy to stay within the Turing domain so long as the system is kept near saturation, which may be the way such systems operate in reality. More importantly, such models are quite restricted in terms of the values of k'_1 and k'_4 they can achieve. They are consequently well behaved in terms of sustaining, for example, constant pattern size, or their tendency to avoid oscillatory regimes. This is a kind of robustness achieved by "loosening up" the mechanism, freeing it from the constraint of strict stoichiometry.

20.4 BIOLOGICAL RELEVANCE

The previous section distinguishes between models that are basically designed to oscillate, and a second class of models, based on a much oversimplified conception of how transcriptional regulation might work, that do not. The latter show, in principle, more promise of being well behaved in various ways, including the ability to maintain constant spacing in the face of parameter change. How relevant is this type of distinction in biological terms? It is clear that some developing systems use control mechanisms that can oscillate. We have seen a beautiful example in this symposium in the session on the slime mold *Dictyostelium*. This appears to be a case in which an organism has co-opted an oscillatory mechanism for use in patterning, controlling it in part by imposing a gradient (P. Schaap & M. Wang, this volume). It is then quite reasonable to think in terms of mechanisms that can produce both oscillations and stable spatial pattern depending on parameters or other imposed constraints. Not all organisms will have evolved their pattern control mechanisms by this same route, however. Mobilizing the autocatalytic capacities of transcriptional control machinery and allowing for diffusion or diffusion-like spread of some signal from cell to cell would produce non-oscillatory control systems of the type discussed above. *Drosophila* provides an example of a well-controlled pattern generator that gives constant spacing without evidence, so far, for any underlying oscillatory behavior. The *Drosophila* blastoderm embryo is not, however, a typical multicellular system. It is instead syncytial when the pattern first forms. Most embryonic patterns are formed by groups of individual cells in which communication over distance may involve cel-to-cell signalling mechanisms rather than simple diffusion. If the machinery of intracellular signal transduction is involved, oscillatory behavior might be important, because we know these systems can generate oscillations and propagated waves. But communication may also occur by diffusion in the extracellular space, via gap junctions, or by other diffusion-like processes. We need to know more about how communication over distance actually occurs in developing systems. Dr. Müller emphasised this earlier in this meeting in his discussion of how models for hydra morphogenesis might be improved. If it can be established that communication over distance is by means of diffusion-like processes, the question of whether the reaction systems operating in each cell are oscillatory or non-oscillatory becomes important as a criterion for assessing the merits of competing models.

A key feature of Turing's model is the differential mobility required of the two components; one must be effectively less mobile than the other, by roughly an order of magnitude at least if the model is to be very robust. In the chemical systems now being studied that yield Turing structures, *e. g.* the CIMA reaction (Ouyang & Swinney 1991), this is achieved by having the reaction occur in a gel to which one of the components binds. Binding reduces the effective mobility of the one substance relative to the other. In *Drosophila*, there are gene products that could potentially be morphogens that bind either to cytoskeletal structures or chromatin in nuclei. This could provide the basis for the required differential mobility. In multicellular systems, the same could be achieved by virtue of differential permeability through gap junctions, or by having reduced exchange with the extracellular space, which would effectively "bind" components within individual cells. In brief, there is no shortage of means by which the mobility difference required by the theory could be achieved.

Jim Murray has commented, during this meeting, that one ideally wants to design models that are structured, from the start, to embody the kinds of behaviors and robustness seen in the real biological system. Here I have used parameter analysis to show how the choice of reaction kinetics, whether oscillatory or non-oscillatory, may be important in governing the overall behavior of any Turing mechanism in which they participate. This is only a small step towards understanding how better, more realistic models might be devised. It is as significant step, however, given how difficult it is to make real progress with the theoretical analysis when we have as yet so little precise information on the substances and processes that are actually involved.

ACKNOWLEDGMENTS Supported by NSERC Canada. I thank L. G. Harrison, M. J. Lyons, and Axel Hunding for extensive discussion.

REFERENCES

Berg, H. C., & Purcell, E. M. 1977. Physics of chemoreception. *Biophys. J.*, **20**, 193–219.

Hafen, E., Kuriowa, A., & Gehring, W. J. 1984. Spatial distribution of transcripts from the segmentation gene *fushi tarazu* during *Drosophila* embryonic development. *Cell*, **37**, 833–841.

Ko, M. S. H. 1991. A stochastic model for gene induction. *J. Theor. Biol.*, **153**, 181–194.

Ko, M. S. H. 1992. Induction mechanism of a single gene molecule: stochastic or deterministic? *Bioessays*, **14**, 341–346.

Lacalli, T. C., & Harrison, L. G. 1979. Turing's conditions and the analysis of morphogenetic models. *J. Theor. Biol.*, **76**, 419–436.

Lacalli, T. C., & Harrison, L. G. 1987. Turing's model and branching tip growth: relation of time and spatial scales in morphogenesis, with application to *Micrasterias*. *Can. J. Bot.*, **65**, 1308–1319.

Lacalli, T. C., & Harrison, L. G. 1991. From gradient to segments: models for pattern formation in early *Drosophila* embryogenesis. *Seminars in Devel. Biol.*, **2**, 107–117.

Lamb, P., & McKnight, S. L. 1991. Diversity and specificity in transcriptional regulation: the benefits of heterotypic dimerization. *Trends Biochem. Sci.*, **16**, 417–422.

Lyons, M. J., & Harrison, L. G. 1993. Stripe selection: an intrinsic property of some pattern-forming models with nonlinear dynamics. *Developmental Dynamics*. (In press).

Meinhardt, H. 1982. *Models of biological pattern formation*. New York: Academic Press.

Murray, J. D. 1982. Parameter space for Turing instability in reaction-diffusion mechanisms: a comparison of models. *J. Theor. Biol.*, **98**, 143–163.

Ouyang, Q., & Swinney, H. L. 1991. Transition from a uniform state to hexagonal and striped Turing patterns. *Nature*, **352**, 610–612.

Ptashne, M. 1988. How embryonic transcriptional activators work. *Nature*, **335**, 683–689.

21. STEADY-STATE PATTERNS IN A REACTION DIFFUSION SYSTEM WITH MIXED BOUNDARY CONDITIONS

Philip K. Maini[1], Robert Dillon[2], and Hans G. Othmer[2]

[1]Centre for Mathematical Biology
Mathematical Institute
24–29 St. Giles'
Oxford OX1 3LB, UK

[2]Department of Mathematics
University of Utah
Salt Lake City, UT 84112, USA

A number of models for pattern formation and regulation are based on the hypothesis that a diffusible morphogen supplies positional information that can be interpreted by cells. Such models fall into two main classes: those in which pattern arises from distributed sources and/or sinks of the morphogens, and those which can spontaneously produce pattern via the interaction of reaction and transport. In source–sink models, specialized cells maintain the concentration of the morphogen at fixed levels, and given a suitable distribution of sources and sinks, a tissue can be proportioned into any number of cell types with a threshold interpretation mechanism. However, the spatial pattern established is strongly dependent on the distances between the sources and sinks, and additional hypotheses must be invoked to ensure that the pattern is invariant under changes in the scale of the system. This is most easily seen in a one-dimensional system with a source at one end and a sink at the other. If the ends are held at c_0 and c_1 respectively, then the morphogen distribution is given by $c(x) = (c_1 - c_0)(x/L) + c_0$, and so the flux through the system must vary as $1/L$. Thus the homeostatic mechanism that maintains the boundary concentrations at fixed levels must be able to vary the production or consumption of morphogen over a wide range.

Turing models (Turing 1952) are an example of systems in which the pattern can arise spontaneously. These involve two or more morphogens that react together and diffuse throughout the system. In Turing's analysis no cells are distinguished *a priori*; all cells can produce or degrade the morphogens. Moreover, Turing only considered periodic systems or closed surfaces, in which case no boundary conditions are needed. More generally, we call any system of reaction-diffusion equations for which the boundary conditions are of the same type for all species, and such that the elliptic system which governs the steady state admits a constant solution, a Turing system. For an appropriate choice of parameters, it is well known that a spatially–homogeneous stationary state of a Turing system can become unstable with respect to small non–uniform disturbances. Such instabilities, which Turing called symmetry-breaking because the homogeneous locally-isotropic stationary state becomes unstable and therefore physically inaccessible, can lead to either a spatially non–uniform stationary state or to more complicated dynamical behavior. Such transitions from uniform stationary states to spatially– and/or temporally-ordered states might in turn lead, via an unspecified 'interpretation' mechanism, to spatially–ordered differentiation. For mathematical simplicity most

Experimental and Theorietcal Advances in Biological Pattern Formation,
Edited by H.G. Othmer *et al.,* Plenum Press, New York, 1993

analyses of Turing models deal with instabilities of uniform stationary states, since numerical analysis is generally required for more general reference states. However, Turing himself recognized the biological unreality of this in stating that 'most of an organism, most of the time is developing from one pattern to another, rather than from homogeneity into a pattern'.

Reaction diffusion systems have been proposed to account for spatial pattern formation in several other biological systems and in chemical systems, but in many of these cases experimental evidence is lacking. Recently, however, Turing-type structures have been found in the chlorite-iodide-malonic acid reaction (Castets *et al.* 1990; Ouyang & Swinney 1991). Aside from the difficulty of identifying morphogens and the reactions in which they participate in a biological context, there are several general properties of Turing systems that limit their applicability.

- The spatial patterns in a Turing system typically arise via an instability, and thus the parameters must be tightly controlled to obtain the onset of the instability at the desired point in parameter space. In particular, for a given kinetic mechanism, the diffusion coefficients must have the proper relative magnitudes.

- Because the instabilities result from the interaction of reaction and diffusion, the patterns that arise are sensitive to the overall scale of the system. As a result, it is difficult to obtain the degree of scale-invariance that is observed in various biological systems. However, modifications of Turing's model can circumvent this difficulty (Othmer & Pate 1980).

- Frequently there are multiple stable solutions that coexist in a Turing system, which raises the problem of pattern selection. Generally tight control of the initial conditions is needed to select the desired pattern.

We have analyzed the spatial pattern formation properties of a two-component reaction-diffusion system in one-dimension, in which the two species are subject to different boundary conditions (Dillon *et al.* 1993). For example, one species may be subject to Neumann conditions, whereas the other species may satisfy Dirichlet conditions. We refer to these as non-scalar boundary conditions. We have concentrated on a simplified version of a model for glycolysis, which is obtained from a biochemical model in the limiting case in which the enzymes are far from saturation (Ashkenazi & Othmer 1978). The governing equations are

$$u_t = \nu u_{\zeta\zeta} + \beta - \kappa u - uv^2, \; v_t = \nu \delta v_{\zeta\zeta} + \kappa u + uv^2 - v, \tag{21.1}$$

where $u(\zeta, t)$ and $v(\zeta, t)$ are nondimensionalised chemical concentrations at position ζ and time t; $\nu = D_1/\omega L^2$, $\delta = D_2/D_1$, where D_1 and D_2 are the diffusion coefficients of u and v respectively, ω^{-1} is a typical reaction time scale and L a measure of the domain length; $\zeta \in [0, 1]$, and β and κ are parameters that we set to 1.0 and 0.001, respectively.

The time evolution system is analysed by a combination of linear analysis, which is non-trivial for the case of non-scalar boundary conditions, bifurcation analysis and numerical integration. The steady state system is analysed using the numerical package AUTO (Doedel 1981). In particular, we consider the properties of solutions as the length scale L is varied. We find that for non-scalar boundary conditions, qualitatively new phenomena arise. For example, stable, non-uniform solutions exist for small values of L. It is well known that for Turing systems all solutions converge to a spatially uniform solution for sufficiently small L (Othmer 1977).

Furthermore, patterns are less sensitive to both the length parameter and the initial conditions. In particular, for certain combinations of boundary conditions we find smooth

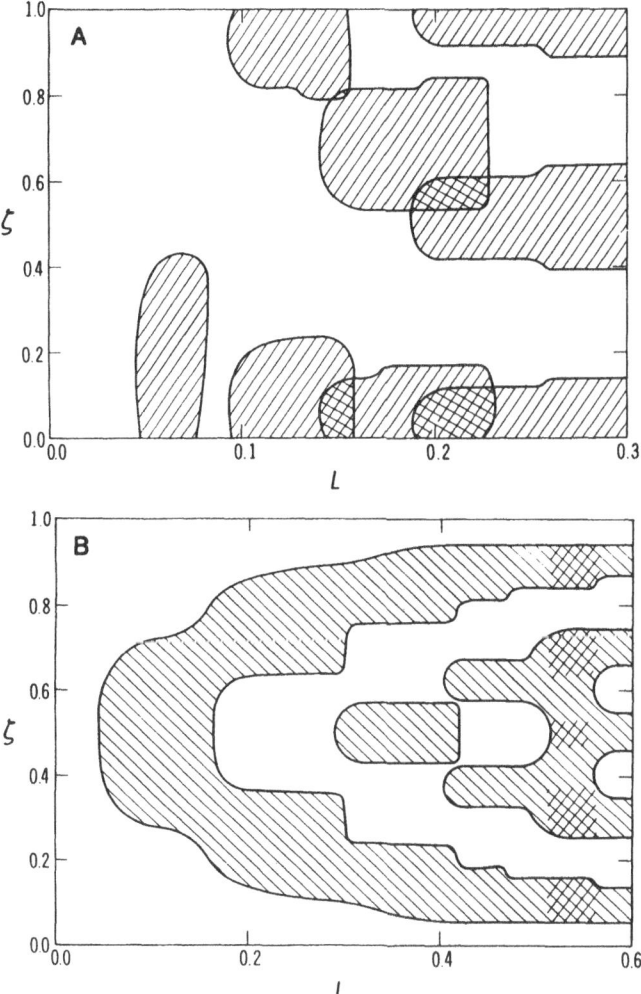

Figure 21.1. Comparison of steady states of the modified glycolysis model for (a) scalar boundary conditions and (b) homogeneous Neumann conditions on u and zero Dirichlet conditions on v (*cf.* Dillon *et al.* 1993).

transitions between different spatial patterns, and these transitions do not involve bifurcations. For example, we find a transition from 1 to 2 to 3 stable pattern elements in a one-parameter family parameterized by the length. This type of transition is similar to that observed in skeletal development in the tetrapod limb. Moreover, these solutions are apparently the only stable solutions. By contrast, for Turing systems a tortuous path in parameter space would be required, because different stable patterns may coexist under certain conditions (see Figure 21.1). In this analysis we have used the domain length L as the bifurcation parameter. However, as this parameter occurs in the model equations as the non-dimensional grouping that involves diffusion coefficients and reaction rate, the sequence of transitions illustrated in Figure 21.1(b) could be generated by changes in the diffusion coefficients. For instance, it is known that the gap junction permeability of cells can be modulated by other species (Mehta *et al.* 1989; Brümmer *et al.* 1991), and such changes would be reflected in the diffusion coefficients in the continuum model used here (Othmer 1983). This is incorporated in the model described by Dillon and Othmer elsewhere in this volume.

Note that our model solutions capture neither the anterior-posterior spatial asymmetries observed in the skeletal elements of the limb nor their temporal sequence of development along this axis. Recently, Benson *et al.* (1992) have shown that a spatially varying diffusion coefficient can produce such spatial asymmetry. The temporal sequence of pattern formation may be due to cells responding to the spatial pattern in a time-specific fashion.

Imposing non-scalar boundary conditions also results in pattern formation occurring over a larger ratio of diffusion coefficients, thereby enlarging the parameter domain over which certain patterns exist and hence lowering pattern sensitivity to small changes in the environment.

Clearly, therefore, boundary conditions have a marked affect on the patterns exhibited by reaction-diffusion models in one-dimension. We would expect this effect to be even more pronounced in two- and three- dimensions, where one has an even wider choice of different types of boundary conditions.

ACKNOWLEDGMENTS This work (RD and HGO) was supported in part by NIH Grant No. GM29123.

REFERENCES

Ashkenazi, M., & Othmer, H. G. 1978. Spatial patterns in coupled biochemical oscillators. *J. Math. Biol.*, **5**, 305–350.

Benson, D. L., Sherratt, J. A., & Maini, P. K. 1992. Diffusion driven instability in an inhomogeneous domain. *Bull. Math. Biol.*, **55**, 365–384.

Brümmer, F., Zempel, G., Buhle, P., Stein, J.-C., & Hulser, D. F. 1991. Retinoic acid modulates gap junction permeability: a comparative study of dye spreading and ionic coupling in cultured cells. *Exp. Cell Res.*, **196**, 158–163.

Castets, V., Dulos, E., & Kepper, P. De. 1990. Experimental evidence of a sustained standing Turing-type nonequilibrium chemical pattern. *Phys. Rev. Letts.*, **64**(24), 2953–2956.

Dillon, R., Maini, P. K., & Othmer, H. G. 1993. Pattern formation in generalized Turing systems. *J. Math. Biol.* (To appear).

Doedel, E. J. 1981. AUTO: A program for the automatic bifurcation and analysis of autonomous systems. In *Proc. 10th Manitoba Conf. Num. Anal. and Comp.*, pp. 265–284.

Mehta, P., Bertram, J. S., & Loewenstein, W. R. 1989. The actions of retinoids on cellular growth correlate with their actions on gap junctional communication. *J. Cell Biol.*, **108**, 1053–1065.

Othmer, H. G. 1977. Current problems in pattern formation. In *Lectures on Mathematics in the Life Sciences 9*, pp. 57–85. AMS.

Othmer, H. G. 1983. A continuum model for coupled cells. *J. Math. Biol.*, **17**, 351–369.

Othmer, H. G., & Pate, E. F. 1980. Scale invariance in reaction-diffusion models of spatial pattern formation. *Proc. Natn. Acad. Sciences*, **77**, 4180–4184.

Ouyang, Q., & Swinney, H. L. 1991. Transition from a uniform state to hexagonal and striped Turing patterns. *Nature*, **352**, 610–612.

Turing, A. M. 1952. The chemical basis of morphogenesis. *Phil. Trans. R. Soc. Lond.*, **B237**, 37–72.

22. PATTERN CONTROL IN *HYDRA*: BASIC EXPERIMENTS AND CONCEPTS

Werner A. Müller

Zoological Institute
University, INF 230
D 6900 Heidelberg, Germany

When Alan Turing (1952) coined the term *'morphogen'* and proposed his famous *'reaction-diffusion systems'* of chemical pattern generation, he pointed to *Hydra* as an example, showing that the tentacle pattern in this animal might be created by prepatterns of such morphogens. Since those days *Hydra* has been one of the most frequently addressed organisms in theoretical papers on biological pattern formation. What is peculiar to *Hydra*?

First of all, even non-zoologists may know that the fresh-water polyp *Hydra* is endowed with an amazing capacity to regenerate any lost body part, and with reference to our zoological textbooks, can quickly learn that this polyp is among the simplest multi-cellular organisms in the animal kingdom.

On first inspection we see a tube-like body about 5 mm in length, with a whorl of tentacles surrounding a mouth at the upper end and a disk-shaped organ for adhesion at the lower end. To facilitate communication with non-zoologists, we subdivide the longitudinal pattern into 'head', 'gastric' region, budding zone (where new animals are generated by a process of natural cloning), stalk and 'foot' (Figure 22.1).

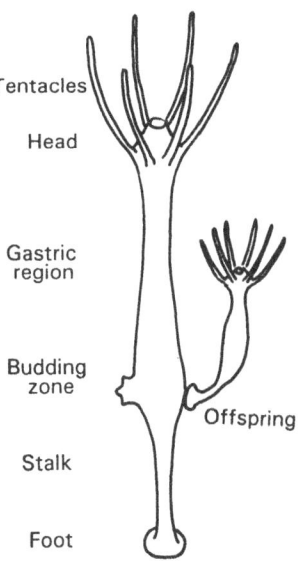

Figure 22.1. *Hydra*

Experimental and Theorietcal Advances in Biological Pattern Formation,
Edited by H.G. Othmer *et al.,* Plenum Press, New York, 1993

However, in spite of this simple overall architecture the cellular composition reveals internal richness: the polyp is composed of about a dozen cell types, including sensory cells and nerve cells embedded in two concentric epithelial layers. In terms of the history of life, *Hydra* and its relatives are perhaps the oldest organisms equipped with those typically animal cells.

Yet, all these attributes most likely would not prompt us to develop theories of biological pattern formation if the hydra did not display some peculiar features.

22.1   *HYDRA* IS IMMORTAL; ITS CELLS CONTINUOUSLY NEED AND USE POSITIONAL INFORMATION

Hydra is immortal not because it consists of immortal cells but because cells of advanced age and used-up cells are unceasingly replaced by newly-generated cells. In this process of perpetual renewal all cells are exchanged, even the nerve cells (for review see, for example, Gierer 1977).

Nerve cells, for instance, are generated by multipotent 'interstitial stem cells' which also give rise to other cell types, including the various types of stinging cells. These stem cells are found in the middle of the body. Their derivatives, *e. g.* the nerve cell precursors, first have to migrate to their final position before they can take over the function of their dying ancestors.

Likewise, the epithelial cells, which constitute the body wall, are born in the middle and are subsequently shifted to the terminal poles where they become committed to their final function and eventually die. During their displacement appropriate functional roles are reassigned to the cells according to the region they cross.

The overall body pattern is in a state of equilibrium, just like a population in which the aging members are displaced into the outskirts whilst the rising generation occupies the city.

However, in *Hydra* it is not merely the age of the cells but their location which determines their social function and place of residence. Young cells adopt the function of adults if they are shifted to the periphery. This can be shown by very simple cutting experiments: after a transverse cut cells of the gastric region find themselves at the upper end of the lower fragment, their former neighbours at the lower end of the upper fragment. Within 2-3 days they form a head or a foot, respectively. Moreover, overlapping cut levels show that the same cells can form either gastric region, or head, or foot, according to their position along the body axis (Figure 22.2). The cells change their function in response to positional cues. Positional information is available all the time.

22.2   PATTERN FORMATION IN *HYDRA* IS BASED ON SELF-ORGANIZATION OF INTERACTING CELLS

Hydra can be 'dissociated', *i. e.* separated into a suspension of free floating, single cells. The cells sink to the bottom, reestablish adhesive contact and form aggregates. Within two days the aggregates develop into multiheaded monsters. Over the course of about 7-21 additional days the monsters give rise to one or more normal animals depending on the initial size of the aggregate. Axes of asymmetry are established with no need for maternal positional cues like those laid down in the egg of animals such as *Drosophila*.

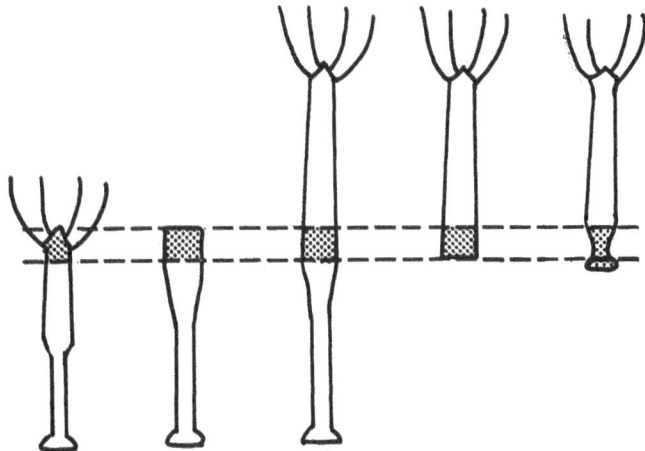

Figure 22.2. Cutting experiment showing the availability of positional cues: The same cells form a foot, a gastric segment or a head depending on their position along the body column.

The reorganization of intact animals starts from a random distribution of the various cell types. Reorganization of the longitudinal body pattern is not the result of a spatial rearrangement, but of concerted changes in the functional state of the cells. The cells do not sort with respect to their original position along the body axis (Sato *et al.* 1992; Technau & Holstein 1992). However, classical experiments (Gierer *et al.* 1972) indicate that new centers of head formation preferentially arise from patches of cells derived from high levels in the body (*i. e.* near the head). When by experimental manipulation the cells are not randomly mixed but grouped according to their origin (Figure 22.3), a bias in the behaviour of the cells becomes evident. A group of cells derived from the upper gastric region dominates over a group of cells derived from the lower gastric region in organizing a new head successfully. The higher the axial origin is, the higher is the inherent bias to form head structures.

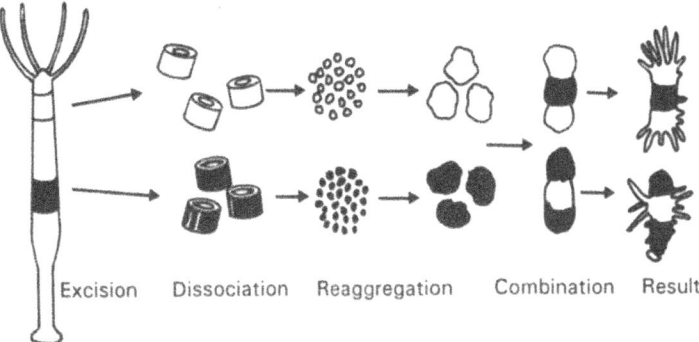

Excision   Dissociation   Reaggregation   Combination   Result

Figure 22.3. Reaggregation experiment. Aggregates derived from the upper gastric region (white) dominate over aggregates derived from a lower position (black) in organizing head structures (tentacles).

The experiment just described also showed that in regeneration the **polarity is tightly linked to a gradient and not determined by vectorial forces**. From such studies the following general concept of pattern reconstitution arose:

1. The cells are endowed with some kind of *positional memory* which survives in dissociated cells for a significant period of time. This memory is associated with a *gradient
   in the potential for head formation* in the intact animal. This potential has been termed
   'positional value' (Wolpert 1971) or **'head activation potential'** (Sugiyama 1982;
   MacWilliams 1983a).

2. Non-uniformities in this potential are amplified by processes involving non-linear *positive feedback loops* (Gierer 1977).

3. *Lateral inhibition* restricts the range of the amplification and suppresses competing
   centers in the vicinity (Gierer 1977).

No attention has been paid to foot formation in explaining pattern reconstitution in
aggregates. Additional properties of the pattern forming system can be deduced from studies
of pattern control.

22.3 PATTERN REGULATION: *HYDRA* CONTROLS THE INTEGRITY OF ITS BODY AND AVOIDS SUPERNUMERARY HEADS AND FEET

Hydra replaces a missing head and a missing foot. But not only terminal body parts
are replaced. When the gastric region is cut out and head and foot are directly joined together,
a new gastric segment is intercalated (for a review on pattern control without intercalation see
Bode & Bode 1984; with intercalation see Müller 1982).

In excised midgastric segments, the original polarity is strictly maintained. The
gradient in the 'head activation potential' ensures head formation at the end which was closest
to the original head. (By experimental interference, however, *Hydra* can be stimulated to
produce heads or feet at opposing ends and elsewhere, as will be shown below. But we will
also see that even multiheaded hydras are able to correct their body structure.)

The replacement of missing structures is preceded by position-dependent changes in
positional value.

22.3.1 Positional value: a graded property enabling the cells to collectively make structures in conformance with position

To avoid misunderstandings it is emphasized from the outset that *positional values*
do not, or not only, refer to the finished longitudinal body pattern but indicate potentials or
capacities to *develop* structures in conformance with position. When a head is removed, the
remaining body still has a latent capacity for head formation, the highest capacity being found
at its upper end. To a first approximation, decreasing positional values mean a decreasing
capacity to form head structures. However, the term 'positional value' (Wolpert 1971) is
more comprehensive than the recently preferred term 'head activation' (MacWilliams 1983a),
as tissue possessing a high potential for head formation always has a low potential for foot
formation. Both these potencies are invariantly and inversely related to each other and change
simultaneously during regeneration.

Moreover, a medium positional value such as exists in the middle of the body column,
not only indicates a moderate potential for head and foot formation, but enables the cells
actually to form the gastric region. During regeneration of a head from the gastric region,
positional value has to increase to its maximum *before* a head can be formed. Likewise,
foot formation presupposes a *previous* decrease in positional value down to a minimum. The

increase or decrease takes time, from minutes to many hours, depending on the level of the cut. Thus, a first, though indirect, measure of the positional value is the *speed of regeneration*. At high cut levels head regeneration is fast, foot regeneration is slow. At low cut levels head regeneration is slow, foot regeneration is fast.

The classical, more direct method of determining positional values is the *transplantation experiment* (*e. g.* MacWilliams 1983b; 1983a; Müller 1990). Pieces of tissue are grafted from one animal into another (Figure 22.4), and one of three outcomes is observed.

(1) If the former and the new place of residence are at corresponding positions along the body column, the piece is integrated into the host and nothing else happens. (2) If the positional origin of the graft is higher than its new place, it forms a head. Moreover, a graft too small to form a complete head itself induces the surrounding host tissue to participate in head formation. (3) A piece taken from a lower body level forms a foot.

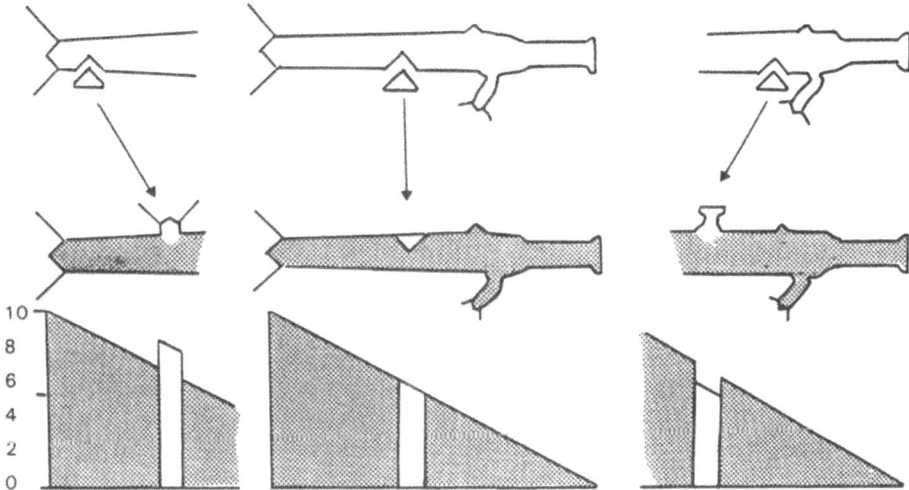

Figure 22.4. Determination of relative positional values by transplantation. Disparities between the positional value of the transplant and the surrounding host tissue result in head formation or foot formation, respectively. Quantitative measurements are based on repetitions; the percentage of head or foot formation is scored.

To summarize the outcome of many series of experiments, we may assign an index of positional value ranging from 10 to 0, to the various levels of the body column, starting with 10 for the head. The following results are of particular significance.

● Whenever the index of the graft is higher than the index of its surroundings the index in the graft increases to 10 and a head is formed. Whenever the index of the graft is lower it decreases to 0 and a foot is formed.

● Probability and speed of head and foot formation by a graft are higher, the larger the difference is between the positional values of the graft and the adjacent host tissue (Figure 22.6).

● After removal of the upper part of the body, the positional value at the upper end of the remaining stump increases rapidly. It reaches the final value of 10 in a few hours, while tentacles do not appear for an additional 28-30 h (Figure 22.5). The same rates of change are found at the lower end of a fragment where the positional value rapidly decreases.

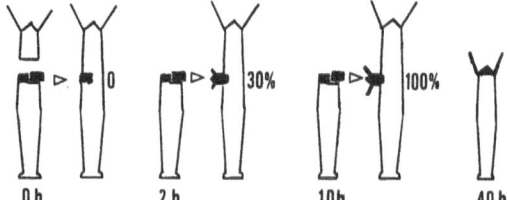

Figure 22.5. Increase in positional value (head-inducing potency) at the upper end of a decapitated body column

All these observations may be summarized as follows:

1. *An increase or decrease in positional value is initiated*

    (a) *when body regions with higher or lower values are removed,* as it is done in the traditional regeneration experiment, or

    (b) *when cells with disparate positional values are placed next to each other.*

    The direction of the change is determined by the sign of the difference; the probability and speed of the change are determined by the height of the difference.

2. *The increase and decrease imply mechanisms of self-enhancement.*

22.3.2   Lateral inhibition: how to establish a hierarchy of dominance

Self-enhancement (positive feedback, autocatalysis) must be limited. In intracellular biochemical pathways three general mechanisms of restriction are known: limited supply of precursors, negative feedback by metabolites or end products, and saturation.

But now we have to consider the question on how cells impose constraints of their neighbors and how a hierarchy of dominance is set up along the body column. A term often used to address this problem is *'lateral inhibition'*, (a term borrowed from neurobiologists). The imagination of almost all scientists occupied with biological pattern formation quickly and inevitably focused on that mechanism which is believed to be the simplest: the head is thought to release an inhibitory substance that diffuses widely and prevents competitive head formation elsewhere. How good is the evidence for this assumption?

Grafts have been taken from a high, subtentacular position in donors and inserted into lower levels of hosts at three positions, one-quarter, one-half, and three-quarters of the way from the head of the host (Figure 22.6). The grafts form and induce head structures more frequently, the farther the site of insertion is from the host head (MacWilliams 1983b). The conventional interpretation of the data is that the existing head exerts an inhibitory influence the strength of which declines with distance (Sugiyama 1982; MacWilliams 1983b; Takano & Sugiyama 1983).

However, what actually is scored is not inhibition as such but lack of a change. The probability of head formation by the graft primarily is a function of the disparity in the positional values between the graft and its new surroundings. Slight disparities may result in integration without change rather than head formation; large differences frequently result in head formation.

There is, however, indirect evidence pointing to active inhibition. If the head of the host is removed, grafts located along the body column form heads with higher probability than in intact hosts (Figure 22.6). The yield in head formation is highest when the graft is

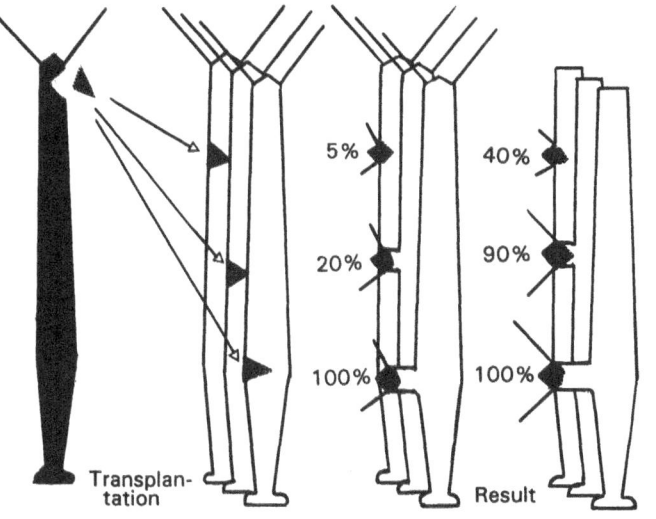

Figure 22.6. Experiment demonstrating (a) the effect of disparities in positional value on the probability of head formation and (b) the inhibiting influence of the extant host head

inserted 6 h after decapitation. The apparent inhibition decays with a half-life of 2-3 h and does not regain its original level until the host has regenerated its own head. By careful and comprehensive transplantation studies it has even been possible to determine an apparent diffusion constant of $2 \times 10^{-6} cm^2$/sec (MacWilliams 1983b) (*cf.* Figure 22.13).

Conversely, when a donor head is grafted laterally onto the body column, a removed host head is not, or only rarely, regenerated depending on the distance between the grafted head and the upper end of the decapitated host. Likewise, when a young bud with its own head is arising, the regeneration of the parental head is inhibited.

The set of data summarized in Figure 22.6 can be interpreted in several ways. The conventional interpretation, however, does not discriminate between the lack of stimulation of head formation by a lateral graft, whose own positional value does not exceed, or only slightly exceeds, that of the surrounding host tissue, and the additional long-range suppressing effect of the head of the host. Both effects are assumed to be due to a diffusible inhibitory molecule which spreads from cell to cell through intercellular channels known as gap junctions (Fraser *et al.* 1987 *cf.* Figure 22.13).

Attempts to isolate the putative inhibitory morphogen were promising (Berking 1979), but the candidate molecule has not yet been identified. If inhibitors exist, how many do we expect? One for the entire body column? One (or more) for head formation, one (or more) for foot formation? Why not additional inhibitors for the gastric region, or for any other structures? These are open questions.

22.3.3 Long-range interactions: the head assists the foot formation

Long-range interactions do exist, whatever the physical basis of these interactions may be. Removal of the head slows down foot formation. Moreover, as *Hydra* allows surgical manipulations of all kinds, instead of removing a head we may graft additional heads taken from beheaded donors. In such multiheaded animals, not only does the apparent inhibition increase, but the polyp responds to the graft by forming additional feet. If the additional heads are grafted near the upper end, the additional feet appear near the lower end. If the additional heads are periodically placed along the body column, additional feet also appear periodically at intervals half way between the heads (Figure 22.7; Müller 1989, 1990). Hence *suppression of competitive head formation is correlated with promotion of foot formation.* In

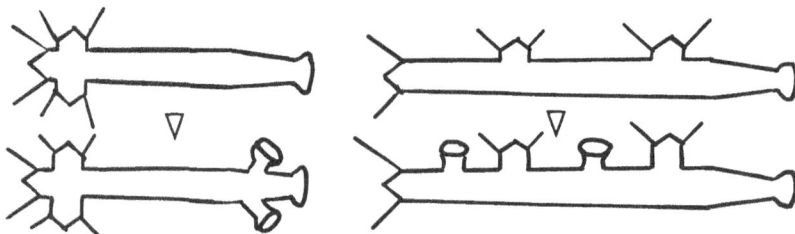

Figure 22.7. Additional heads evoke additional feet.

the reciprocal experiment, however, additional feet do not evoke additional heads. The long-range assistance is unidirectional, not mutual. Moreover, the head *supports* foot formation, but its presence is not an essential prerequisite, for under certain experimental conditions feet, and only feet, are formed between and at both ends of excised segments, even in the absence of any head (Hassel *et al.* 1993).

A relationship between head and foot formation has also been shown in a curious but ingenious experiment: tandem grafts consisting of many body rings all excised from donor animals at the same axial level developed into monsters with multiple heads and feet. Tubes made of rings taken from high levels formed many heads but also many feet, and heads were formed before feet (Ando *et al.* 1989).

In the long term, the systems of pattern control will correct even those monsters, in part by reducing supernumerary structures, in part by segregation of head-bearing branches. The final separation into several individuals is accomplished by the fission of common feet (*cf.* Figure 22.11).

22.4 HOW TO MAKE POLYPS WITH MULTIPLE HEADS AND FEET WITHOUT SURGERY

It is an old proposal that a signal molecule, termed 'polarizing factor', 'head in-ducer' (Lesh & Burnett 1966) or 'head activator' (Schaller 1973), may exist. The idea is plausible, but efforts to identify such a molecule have not yet succeeded, in spite of promising initial indications. However, absence of proof is no argument. The isolation and identification of biological signal molecules is not a very easy task. One has to find traces of a substance among many thousands of various other molecules, most of which are present in much larger quantities, and one has to develop an appropriate bioassay.

In animal cells many external signals are received by specific receptors on the cell surface and processed by means of signal transducing systems. A transducing system is a molecular device to translate an external message into various internal messages. These are generated and channeled through biochemical pathways into the various compartments of the cell to evoke coordinated responses. In the present context the *PI-PKC system* (phosphatidylinositol-protein kinase C system) is of particular interest (Figure 22.8).

Starting from the receptor on the cell surface, subroutes run along the inner side of the cell membrane, leave the membrane and lead to various compartments within the

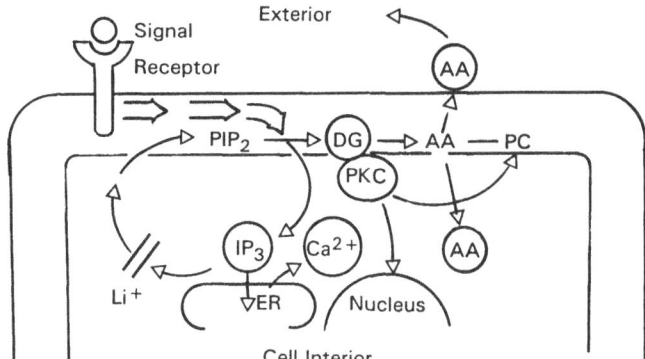

Figure 22.8. The PI signal-transducing system. The encircled members of biochemical pathways are of particular interest: diacylglycerol (DG), protein kinase C (PKC), inositol triphosphate (IP_3), arachidonic acid (AA). Binding of an extracellular signal to the receptor leads to activation of the PKC; externally applied DG and AA directly stimulate PKC; the receptor is bypassed. Li^+ blocks the PI system by preventing the formation of the DG precursor PIP_2.

cell including the nucleus; other subroutes, termed arachidonic acid cascades, lead to the release of signal molecules which are able to cross the membrane and to convey messages to neighbouring cells (Figure 22.8).

What can we do if the putative external signal is not available? Under favourable circumstances an unknown signal and its corresponding receptor can be bypassed by directly stimulating subordinate key enzymes which open decisive subroutes of intracellular pathways or by importing secondary signal molecules. Furthermore, one can try to block the whole system or distinct subroutes. We have done this and the results were dramatic.

(1) Prolonged exposure of the animals to lithium ions causes a drop of positional value and eventually the appearance of additional, ectopic feet along the lower body column (ectopic = not in the normal position). Lithium is known to block several enzymes which keep the PI-cycle going, and to cause accumulation of certain metabolites (*e. g.* inositol phosphates) within the cell. Pilot measurements have verified lithium-induced shifts in the concentration of such metabolites, correlated with ectopic foot formation (Hassel *et al.* 1993).

(2) On the other hand, stimulation of the key enzyme PKC by diacylglycerol induces an increase in positional value and eventually the appearance of ectopic heads (Müller 1989; 1990).

(3) New evidence points to a supporting role of the arachidonic acid cascade, a subordinate signal relaying system, in mediating ectopic head formation. While diacylglycerol as well as arachidonic acid evoke only a slow increase in positional value, coadministration of both inducers strongly speeds up the increase (Müller *et al.* 1993). Our present working hypothesis is that PKC-channeled intracellular pathways and arachidonic acid-mediated transcellular pathways bring about an increase in positional value cooperatively.

However, at present it is not possible to propose a coherent hypothesis of the underlying cellular and biochemical events and, therefore, it would not be appropriate to discuss details. Instead, I will show what actually happens when diacylglycerol and/or arachidonic acid are used to increase the positional value.

To induce a significant increase in positional value, the animals have to be treated repeatedly, once a day. After 2-4 pulse treatments the outward appearance of the animals is still unchanged, but the behaviour in regeneration is altered. Excised segments progressively

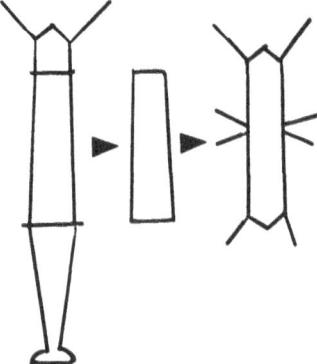

Figure 22.9. Change in regeneration upon repetitive treatment of the intact animals with diacylglyc- erol. Excised gastric segments fail to regenerate a foot and form a head instead at their lower end. Additional tentacles may appear in the middle.

fail to regenerate feet, and form heads even at their lower end and eventually over the entire segment (Figure 22.9). Tentacles appear first and are later supplemented by mouth cones.

With increasing frequency of treatment, intact animals also begin to form ectopic tentacles, the first whorl sprouting about two-thirds of the body length away from the existing head. Later more and more clusters of tentacles are intercalated, more or less regularly spaced (Figure 22.10). The extending body column produces head structures as long as the treatment is continued.

The temporal emergence of ectopic structures belonging to higher positional levels reflects the normal spatial sequence of these structures along the body column (Figure 22.10). Head formation is heralded by the appearance of nerve cells of a particular type which normally occupy the upper body, forming a net of graded density. Then tentacles begin to sprout and finally cones bearing a mouth are inserted between the clusters of tentacles. Eventually what looks like a known multiheaded mutant develops. Yet these animals are not mutants, but are genetically intact and still able to correct irregularities and to convert into normal polyps. The time course of the correction again reflects the change in positional value. To show this, we start with a biheaded, mirror-symmetrical form (Figure 22.11) which arose from a gastric region excised from a diacylglycerol-treated animal. Within weeks, the missing lower body parts are intercalated in a sequence reflecting the normal spatial order. The visible appearance of place-dependent structures is a function of the rise or fall of positional value.

22.5   CONCEPTS AND MODELS

22.5.1   What has been proposed?

Basic classical hypothesis: *The intactness of the body is recognized by the congruency of actual gradient systems with nominal systems.*

Actual gradients ought to have a defined maximum, a defined minimum and a monotonous slope in between. Steps have to be smoothed out. Only the two most stud- ied models are referred to.

(1) Positional information (Wolpert)

At the head end a morphogen S is emitted (and perhaps an S′ at the foot end). The morphogen spreads by diffusion and is broken down in the body. The resulting slope

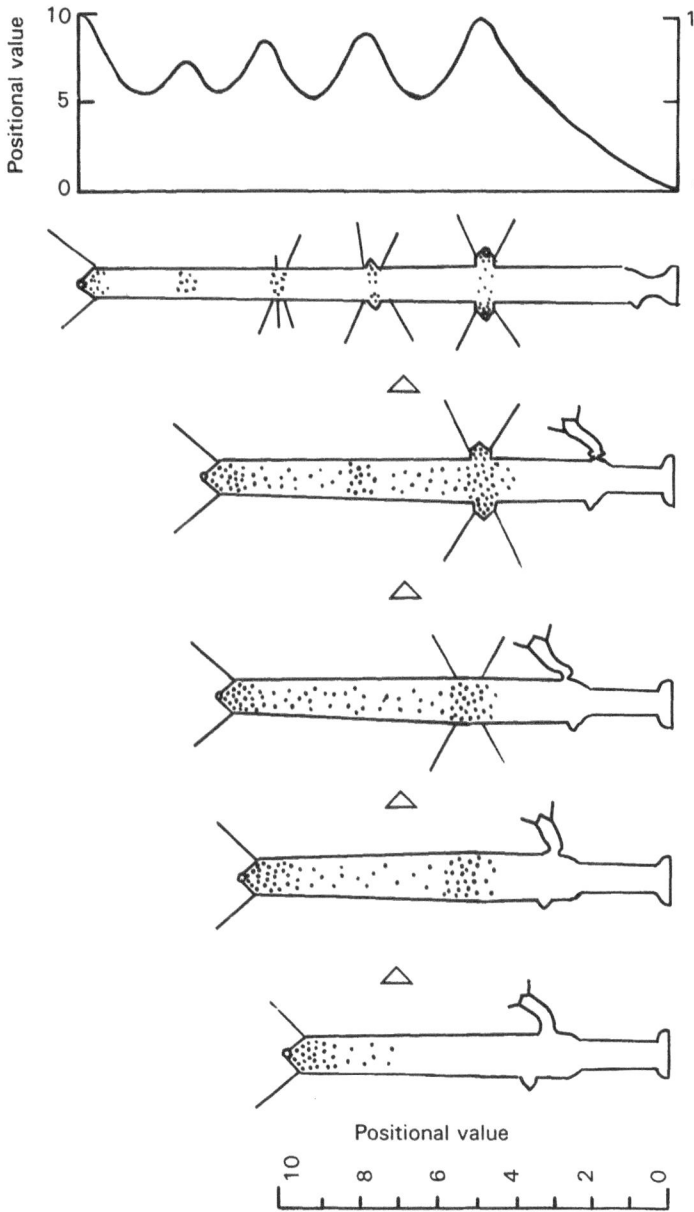

Figure 22.10. Periodic emergence of supernumerary head structures upon repetitive treatment of intact animals with diacylglycerol. Dots represent a particular type of nerve cell (RF-amide positive). Note: The temporal sequence with which additional head-specific structures arise reflects the normal spatial order. The positional value appears to increase in the elongating gastric region in a wavelike manner.

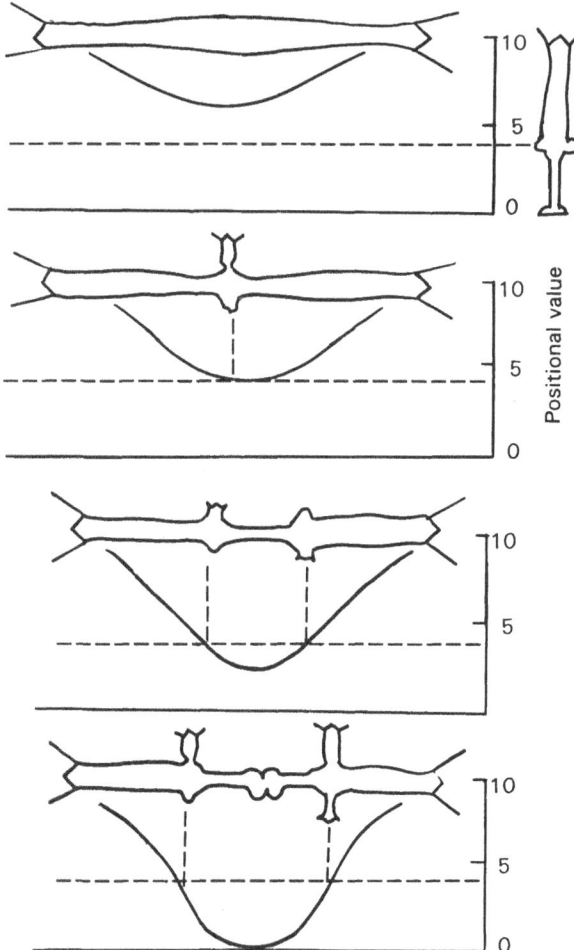

Figure 22.11. Normalization of the body pattern after the daily treatment with diacylglycerol has finished. In the middle of the mirror-symmetrical biheaded segment positional value decreases. This is reflected by the successive appearance of structures belonging to the lower body column. Note: Positional value starts to decrease at the site where the distance to the two heads is largest. If only one head were present, the decrease would take place at the end opposing the head.

of S represents the nominal gradient. It provides lateral inhibition as well as 'positional information' and is used to adjust a second, more stable gradient, the 'positional value P' (Figure 22.12, Wolpert 1971; Wolpert *et al.* 1971). No suggestions are made as to how *de novo* patterns are created and why positional values increase or decrease when the sources of S (or S') are removed.

(2) Turing type models (*e. g.* **Gierer & Meinhardt**)

Gradients of morphogens are formed, even *de novo*, by reaction-diffusion mechanisms. A minimum gradient system comprises a short-range *activator* and a long-range *inhibitor* (Gierer & Meinhardt 1972). The positional value is interpreted in this model as a gradient in the density of morphogen sources (Figure 22.12). The density gradient represents the positional memory and ensures that the morphogen concentration peaks at one end and hence the head always appears at the same end (Meinhardt & Gierer 1974; Meinhardt 1982).

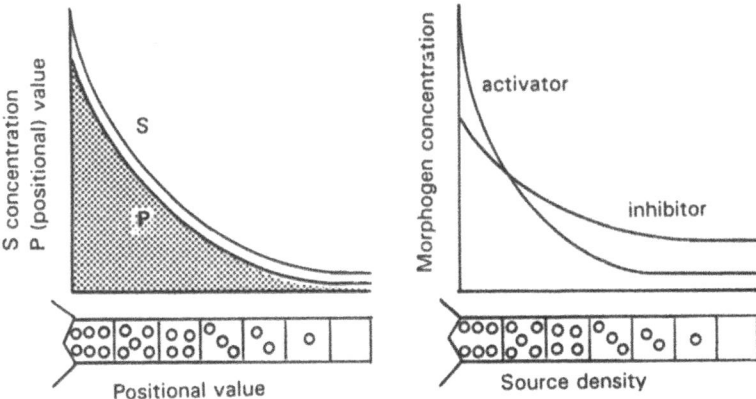

Figure 22.12. Basic models of pattern control in the hydra. Left: Wolperts concept of positional information and positional value. Right: The Gierer-Meinhardt model as an example of a Turing-type model.

Refined models incorporate features for proportion regulation (Meinhardt 1982; MacWilliams 1982).

It is beyond the scope of the present review to discuss the accuracy of the various models. Only two points, which may be worth reconsidering, are mentioned. (1) Both models operate with long-range morphogens. The proposed morphogens are still purely hypothetical and long-range gradients are not easily established and maintained in *Hydra*. (2) No published model covers all the features of the real pattern-forming system outlined here. For instance, no model takes into account that the long-range inhibition of competitive head formation and the long-range assistance of the head in foot formation are correlated, and probably causally related to each other.

22.5.2   Some proposals and constraints for refined or new models

Models based solely on reaction-diffusion systems are probably not very realistic.

In an attempt to classify known mechanisms of intercellular communication operating in animal development, three categories may be distinguished with respect to the range of signal transfer (Figure 22.13).

(1) *Very short-range interactions based on the presentation of stationary signal molecules* on the cell surface, or based on signal molecules fixed on solid extracellular structures. Adjoining cells may pick up the information via cell surface receptor systems. The involvement of this type of information transfer in *Hydra* regeneration has been suggested quite recently (Sarras *et al.* 1991).

(2) *Short-range interactions based on molecules which spread by diffusion.* Pattern control in *Hydra* very likely also implies signal transfer by diffusion. Two types of channels can be envisaged:

(a) Small, hydrophilic messengers such as calcium ions and inositol phosphates may spread from cell to cell via a type of transcellular channel known as 'gap junctions'.

(b) Lipid mediators such as arachidonic acid and its metabolites might move within, along and between cell membranes.

interstitial space
spread of large signal
molecules by convection   Gap junction, allowing diffusion
of small signals (e.g. Ca²⁺, IP₃)

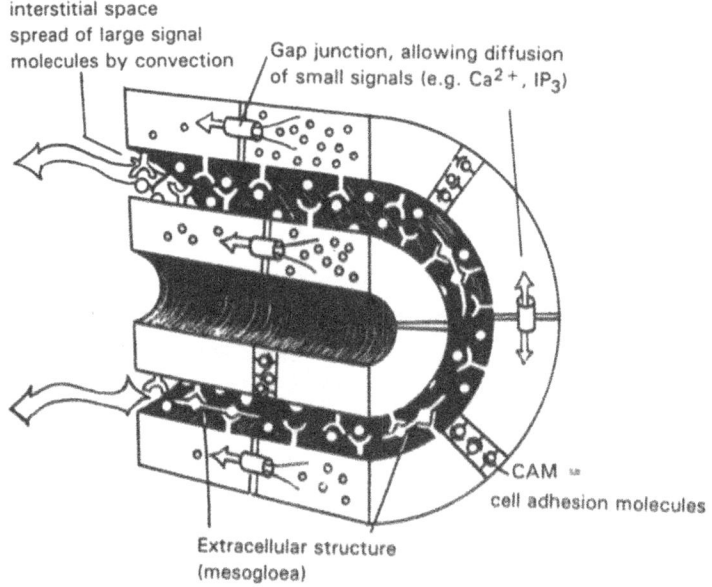

CAM
cell adhesion molecules

Extracellular structure
(mesogloea)

Figure 22.13. Suggested models and routes of signal transfer in *Hydra*.

Diffusion may mediate interactions in fields comprising some hundreds or some thousands of cells. *Hydra* is composed of about 100,000 cells and the distance between head and foot can surpass 1 cm.

(3) *Long-range interactions through convection.* Proteins and other large molecules which may act as growth factors and/or hormones cannot pass through gap junctions but most probably are released into the network of interconnected 'interstitial spaces' which run through the epithelial layers.

The fluid in the interstitial spaces is not motionless. The hydra often bends, contracts and expands. Dyes injected into this fluid are rapidly dispersed by the mixing effect of convection and movement (Fraser *et al.* 1987). I propose that the long-range assistance of the head in foot formation and also the long-range suppression of the competitive head formation is mediated by molecules which spread by convection rather than diffusion.

Conceptually it may be useful to distinguish between local position-dependent developmental capacities and long-range inhibition in terms of active suppression. When cells in the middle of the body form gastric tissue and not head or foot structures this is not necessarily due to some long-range inhibition exerted by the distant head and foot, but is the expression of the local positional value. Likewise, whether a transplant forms a head or a foot, is primarily a function of the height and sign of the disparity between the positional values of the transplant and the adjacent host tissue, and is not necessarily the result of too weak a long-range inhibition by the head or foot, respectively, of the host. The apposition of disparate tissues might generate a short-range stimulus for a change in positional value in the one or the other direction. Two observations, however, appear to require the introduction of particular equations for long-range interactions: the *increase in the probability of head formation by a lateral transplant upon removal of the host head* (Figure 22.6) *and the assistance of the head in foot formation.* Both these effects are *long-range* and closely correlated, being stronger the more additional heads are grafted onto the body column. Removal of the original head, on the other hand, not only increases the probability of competitive head formation, but also slows down foot regeneration. I propose that both effects derive from the same molecular mechanism.

This proposal leaves open two alternative possibilities:

(a) The head may emit a substance which suppresses head formation but promotes foot formation, or

(b) the head may remove from the interstitial fluid some substance, the uptake of which is needed to gain, and to maintain, a level of high positional value.

Since the substance is present in only limited quantity, its concentration is kept at a low level by extant heads. Convection leads to a uniform decline of the concentration along the body column. But in the head region the cells may have more receptors to pick up the molecules than cells in distant regions (Figure 22.14) and, therefore, are at an advantage over distant cells having fewer receptors. At the lower end of the body the cells possess the smallest number of receptors and, therefore, run short of the essential molecules first.

What is positional value? Nowadays many developmental biologists believe that positional values are established by the activity of some selector genes coding for proteins which control the expression of whole sets of other, subordinate genes. Since selector genes often contain a homeobox, they are often collectively referred to as *homeotic genes.* Such genes are also present in the hydra and are preferentially expressed in head regeneration (Schummer *et al.* 1992).

By definition, true homeotic genes are not used to produce a particular type of cell, such as a nerve cell, but to provide all cells of a given body region with position-dependent characteristics. Different positional values along the body axis might be established by a varying expression pattern of such homeotic genes - more and more genes being activated with increase in distance from a reference boundary. In addition to, or as a consequence of the expression of homeotic genes, the cells expose a *molecular code on their surface* which serves as a positional cue and which can be read by neighbouring cells by means of their receptors.

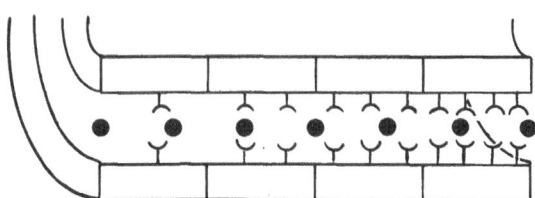

Figure 22.14. 'Positional value' might include the density of receptors used to pick up a growth-factor like molecule which is needed to gain and to maintain positional value including receptor density itself.

In any event, pattern formation in *Hydra* is a process of self-organization based on *cell-cell interactions.* For coordinated responses of many cells, two routes of signal transfer are required: (a) in cells emitting signals, or presenting signals on their surface to adjacent neighbours, we need a route from the genome in the center of the cell to the periphery of the cell, and (b) in cells receiving signals we need a route from the periphery into the interior. In a developmental field perhaps not all cells have to emit signals, but all cells have to receive signals and to respond to them.

In *Hydra*, signal molecules of pivotal significance appear to be picked up by cell surface receptors linked to the PI-PKC transducing system. Therefore, it will be of significance to determine whether or not cells are equipped with such receptors and how many receptors they possess. The molecular complex which constitutes the *positional value might include place-dependent capacities to produce not only signals but also place-dependent variations in the quality and quantity of receptors to receive signals* (Figure 22.14). Why should we not try to incorporate feedback loops in our models in order to change and control the density of receptors on the cell surface? However, the availability of receptors can only be one of several parameters which are of significance in pattern formation.

In conclusion, almost all of the detailed information needed by designers of models is still lacking. Our knowledge is scant, our ignorance huge. On the other hand, since the constraints imposed by our knowledge are so modest, designers of models have almost every freedom to employ their creativity.

REFERENCES

Ando, H., Sawada, Y., Shimizu, H., & Sugiyama, T. 1989. Pattern formation in *Hydra* tissue without developmental gradients. *Dev. Biol.*, **133**, 405–414.

Berking, S. 1979. Analysis of head and foot formation in *Hydra* by means of an endogenous inhibitor. *W. Roux Arch. Dev. Biol.*, **186**, 189–210.

Bode, M. P., & Bode, H. R. 1984. Patterning in *Hydra*. In *Pattern formation*, Malacinski, G. M. (ed). New York: Macmillan Publishing Co.

Fraser, S. E., Green, C. R., Bode, H. R., & Gilula, N. B. 1987. Selective disruption of gap junctional communication interferes with patterning process in *Hydra*. *Science*, **273**, 35–49.

Gierer, A. 1977. Biological features and physical concepts of pattern formation by *Hydra*. In *Current Topics in Dev. Biol.*, Moscona, A. Monroy A. A. (ed). pp. 17–58, New York: Academic Press.

Gierer, A., & Meinhardt, H. 1972. A theory of biological pattern formation. *Kybernetik*, **12**, 30–39.

Gierer, A., Berking, S., Bode, H., Flick, C. N. David K., Hansmann, G., Schaller, H., & Trenkner, E. 1972. Regeneration of *Hydra* from reaggregated cells. *Nature, New Biol.*, **239**, 98–101.

Hassel, M., Albert, K., & Hofheinz, S. 1993. Pattern formation in *Hydra vulgaris* is controlled by lithium-sensitive processes. *Dev. Biol.*, **155**. (In press).

Lesh, G. E., & Burnett, A. L. 1966. An analysis of the chemical control of polarized form in *Hydra*. *J. Exp. Zool.*, **163**, 55–78.

MacWilliams, H. 1982. Numerical simulations of *Hydra* head regeneration using a proportion-regulating version of the Gierer-Meinhardt model. *J. Theor. Biol.*, **99**, 681–703.

MacWilliams, H. 1983a. *Hydra* transplantation phenomena and the mechanism of *Hydra* head regeneration. II. Properties of the head activation. *Dev. Biol.*, **96**, 239–257.

MacWilliams, H. 1983b. *Hydra* transplantation phenomena and the mechanism of *Hydra* head regeneration. I. Properties of the head inhibition. *Dev. Biol.*, **96**, 217–238.

Meinhardt, H. 1982. *Models of Biological Pattern Formation*. New York: Academic Press, London.

Meinhardt, H., & Gierer, A. 1974. Application of a theory of biological pattern formation based on lateral inhibition. *J. Cell Sci.*, **15**, 321–346.

Müller, W. A. 1982. Intercalation and pattern regulation in hydroids. *Differentiation*, **22**, 141–150.

Müller, W. A. 1989. Diacylglycerol-induced multihead formation in *Hydra*. *Development*, **105**, 309–316.

Müller, W. A. 1990. Ectopic head and foot formation in *Hydra:* diacylglycerol-induced increase in positional value and assistance of the head in foot formation. *Differentiation*, **42**, 131–143.

Müller, W. A., Leitz, T., Stephan, M., & Lehmann, W. D. 1993. Arachidonic acid and the control of body pattern in *Hdra*. *Roux's Arch. Dev. Biol.*, **202**. (In press).

Sarras, M. P., Meador, D., & Zhang, X. 1991. Extracellular matrix (mesoglea) of *Hydra vulgaris*. II, Influence of collagen and proteoglycan components on head regeneration. *Dev. Biol.*, **148**, 495–500.

Sato, M., Tashiro, H., Oikawa, A., & Sawada, Y. 1992. Patterning in *Hydra* cell aggregates without sorting of cells from different axial origins. *Dev. Biol.*, **151**, 111–116.

Schaller, H. C. 1973. Isolation and characterisation of a low-molecular weight substance activating head and bud formation in *Hydra. J. Embryol. Exp. Morph*, **29**, 27–38.

Schummer, M., Scheurlen, I., Schaller, C., & Galliot, G. 1992. HOM/HOX homeobox genes are present in *Hydra (Chlorohydra viridissima)* and are differentially expressed during regeneration. *The EMBO J.*, **11**, 1815–1823.

Sugiyama, T. 1982. Roles of head-activation and head-inhibition potentials in pattern formation of *Hydra:* Analysis of a multi-headed mutant strain. *Amer. Zool.*, **22**, 27–34.

Takano, J., & Sugiyama, S. 1983. Genetic analysis of developmental mechanisms in *Hydra.* VIII. Head-activation and head-inhibition potentials of a slow-budding strain (L4). *J. Embryol. Exp. Zool.*, **78**, 141–168.

Technau, U., & Holstein, T. W. 1992. Cell sorting during the regeneration of *Hydra* from reaggregated cells. *Dev. Biol.*, **151**, 117–127.

Turing, A. M. 1952. The chemical basis of morphogenesis. *Phil. Trans. R. Soc. Lond.*, **B237**, 37–72.

Wolpert, L. 1971. Positional information and pattern formation. *Curr. Top. Dev. Biol.*, **6**, 183–224.

Wolpert, L., Hicklin, J., & Hornbruch, A. 1971. Positional information and pattern regulation in *Hydra. Symp. Soc. Exp. Biol.*, **25**, 391–415.

23. PREDICTED AND OBSERVED SPATIAL PREPATTERNS IN (HAIR) WOOL FOLLICLE BULBS

B. N. Nagorcka,[1] D. L. Adelson,[1] J. R. Mooney,[2] and B. A. Kelley[1]

[1]CSIRO, Division of Animal Production
P.O. Box 239
Blacktown, NSW 2148, Australia

[2]CSIRO, Division of Mathematics and Statistics
P.O. Box 1965
Canberra, ACT 2601, Australia

23.1 INTRODUCTION

It is now well established both theoretically (see the review by Murray 1989) and experimentally (Ouyang & Swinney 1991) that reaction-diffusion (RD) systems, of the type originally proposed by Turing (1952), are able to spontaneously produce stable spatial wave-like patterns in an otherwise completely homogeneous medium. It has been argued that these wavelike spatial patterns have considerable potential to function as spatial prepatterns controlling the differentiation of groups of cells during biological development (Nagorcka 1989). The particular interest of this paper is the proposal that a 2-component RD system present in the epidermis and epithelium, the two chemical components being denoted here as U and V, is responsible for the regulation of many aspects of wool (hair) follicle initiation and development as well as the formation of the fibre in the mature wool (hair) follicle bulb (Nagorcka & Mooney 1989; Nagorcka 1989). Our research is currently aimed at experimentally testing this proposal. To date these tests have included:

(i) experiments designed to observe prepatterns predicted to arise but not yet observed. For example, particular spatial patterns of mitotic activity have been predicted to arise in early stage follicles during follicle development (Mooney & Nagorcka 1985), and in follicle bulbs during fibre formation (discussed in this paper), but these patterns have not previously been observed or reported.

(ii) experiments to alter prepatterns by changing the diffusion rates and hence the characteristic wavelength of the RD system.

(iii) experiments to alter the prepattern by mechanical manipulation of tissue size and shape.

This paper concentrates on testing the RD theory as it applies to fibre formation. The aim of the paper is to present a preliminary report of the results of an experiment, belonging to category (i) above, designed to observe prepatterns in the follicle bulb as a function of size and shape of the bulb, and to relate these observations to the structure of the fibre produced

Experimental and Theoretcal Advances in Biological Pattern Formation,
Edited by H.G. Othmer *et al.,* Plenum Press, New York, 1993

in the bulb. The paper also aims to compare these observations and relationships with the predictions of the RD theory of fibre formation. The relevant features of the RD theory of fibre formation are briefly outlined in the next section. The calculated prepatterns produced by an RD system in realistically shaped follicle bulbs are presented in the third section. The calculations presented here remove the approximations used earlier by Nagorcka & Mooney (1982) regarding the shape of the follicle bulb. A 2-dimensional approximation previously used to obtain numerical estimates of the prepatterns is also removed. The numerical solutions obtained here contain some differences from the earlier results but largely confirm the sequence of prepatterns expected in small, medium and large follicle bulbs (as defined by Nagorcka & Mooney 1989) on the basis of a linear analysis of the RD system equations. Some of the prepatterns predicted to arise in this range of bulb sizes are shown in detail in the third section. In the fourth section the prepatterns predicted to arise in the bulb are compared with observations of spatial patterns of mitotic activity and cellular differentiation in follicle bulbs of various sizes. A comparison is also made with the spatial patterns of cell type observed in the fibres produced by those follicle bulbs.

23.2 THE RD THEORY OF FIBRE FORMATION

During fibre growth there is a high level of mitotic activity in the follicle bulb. Epithelial cellsthe follicle bulb migrate out of the bulb moving along paths similar to those shown in Figure 23.1(a). The epithelial cells are envisaged to arise from a population of "stem" cells attached to the basal lamina. A proportion of the progeny of the stem cells becomes committed to follow a path of differentiation leading ultimately to cell death. Following commitment, cells are still able to undergo a limited number of divisions. The committed epithelial cells choose which particular differentiation path they will follow while they are still in the follicle bulb. They differentiate as one of a range of different cell types, some of which are indicated in Figure 23.1(a): a full list of all cell types arising from the follicle bulb is given in table 1 of Nagorcka & Mooney (1982).

The RD theory of fibre formation proposes that some time after commitment, say T_4 days, epithelial cells are assumed to have differentiated to a state where they are impervious to U and V, and therefore such cells impose an upper boundary to the diffusion of U and V in the follicle bulb (Figure 23.1(a)). Prior to this time the cells make a number of molecular decisions which determine their cell type. It is proposed that epithelial cells aged T_1 days (Figure 23.1(a)) (*i. e.* T_1 days after commitment occurs) decide whether they will differentiate as inner root sheath (IRS) cells or as fibre cells. Subsequently the presumptive fibre cells differentiate as orthocortical or paracortical cells during the period T_2 to T_3 days after commitment, where $T_1 \approx T_2 < T_3 < T_4$. (Other fibre cell types are also produced but they will not be discussed here.) Paracortical cells differ from orthocortical cells in that they produce relatively large amounts of proteins with a high content of sulphur-amino acids (Bradbury 1973). The RD theory of fibre formation proposes that these two stages of cellular differentiation (*i. e.* the decisions between IRS and fibre cells, and between paracortical and orthocortical cells) are regulated by the spatial distribution of one chemical component of the RD system, namely U, along with the spatial distribution of a factor of dermal origin, denoted as Z. Z is produced in the dermal papilla and diffuses radially outwards. The two RD system components, U and V, on the other hand, are produced within the follicle bulb (and outer root sheath (ORS)) where they diffuse and react with each other.

The particular form chosen here for the two-component RD system is the same as that used to account for both follicle initiation and dermatoglyphics by Nagorcka & Mooney

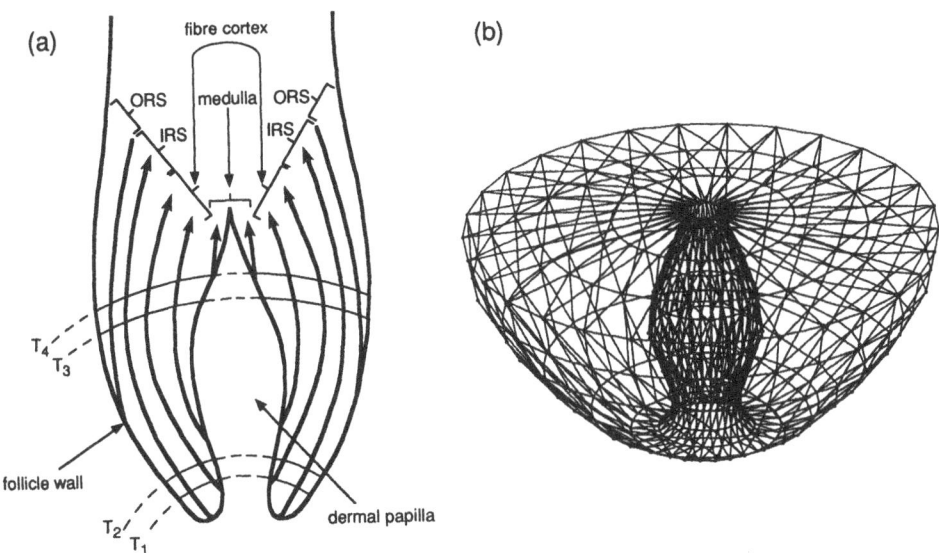

Figure 23.1. (a) A longitudinal section through a wool (hair) follicle bulb which is radially symmetric about the vertical axis of symmetry. Epithelial cells divide rapidly in the base of the bulb and adjacent to the dermal papilla. They migrate out of the bulb along paths indicated by the solid arrows. Epithelial cells aged T_1 days reach a level in the bulb indicated by the T_1 curve. It is assumed that epithelial cells aged T_1 days make a decision whether to differentiate as fibre cells, or as inner root sheath (IRS) cells, or as outer root sheath (ORS) cells. Subsequently fibre cells differentiate further as paracortical, orthocortical or medulla cells (see text). (b) The set of triangular elements displayed represents the surface of the epithelial tissue of the follicle bulb (*i. e.* the follicle wall plus the membrane separating the epithelial tissue from the centrally located dermal papilla) with an upper limit defined by the T_4 boundary. This representation of the surface is required to solve numerically the diffusion equations (23.1) (see text) describing the diffusion of Z away from the dermal papilla.

(1992). In that case the RD system components, U and V, were considered to be transportable across the basement membrane separating the epidermis (epithelium) and dermis. Therefore U and V diffuse and react with each other in both the dermis and epidermis. However, the diffusion rates of the RD system in the dermal tissue are set at values such that the RD system in the dermis does not obey the Turing conditions (Nagorcka & Mooney 1992; Murray 1989). It follows that wavelike spatial patterns in U and V will not arise spontaneously in the dermis; instead, the uniform steady state (defined in the next section) will remain stable. On the other hand, the parameters defining the diffusion and reaction rates of U and V in the epithelium are set at values which lie within the Turing space (Nagorcka & Mooney 1992). Hence wavelike spatial patterns in the distribution of U and V are expected to arise spontaneously in the epithelium of the follicle bulb. Therefore the same approximation used by Nagorcka & Mooney (1992) is also employed here; *i. e.* the terms defining the transport of U and V across the basal lamina surrounding the epithelium of the follicle bulb are incorporated into the non-linear reaction terms of the RD system equations, and the RD system equations are solved only in the epithelium of the follicle bulb. This approximation simplifies the numerical problem. The solution of the RD system equations is discussed in the third section.

The RD theory of fibre formation (Nagorcka & Mooney 1982; Nagorcka 1984) states that inhomogeneous distributions of the RD system component, U, along with the dermal factor Z, regulates cellular differentiation according to the following rules:

Stage 1 Cells aged T_1 days which experience $[Z][U] > \phi_1$ where ϕ_1 is a constant threshold value, differentiate as fibre cells (as do cells aged T_1 days still in contact with the membrane separating bulb and dermal papilla), otherwise cells aged T_1 days differentiate as IRS cells. In addition to cellular differentiation it is also proposed that the quantity $[Z][U]$ also regulates the rate of cell division of all committed cells in the follicle bulb.

Stage 2 Fibre cells and IRS cells in contact with each other differentiate as fibre cuticle and IRS cuticle cells, respectively. (This stage will not be considered here.)

Stage 3 Fibre cells (other than fibre cuticle cells) aged between T_2 and T_3 days differentiate as paracortical cells if $[Z][U]^{-1} < \phi_2$, where ϕ_2 is a constant. Alternatively, if $[Z][U]^{-1} > \phi_2$ then the expression of genes coding for proteins containing relatively high amounts of sulphur amino acids is suppressed and the fibre cells differentiate as orthocortical cells.

The differentiation of fibre (cortical) cells was originally assumed, for the sake of simplicity, to be dependent on $[Z][U]$, *i. e.* the same functional dependence on $[Z]$ and $[U]$ as at stage 1 (Nagorcka & Mooney 1982). However, since it is known that paracortical cells arise from the constricted side of asymmetrical follicle bulbs, as for example in the case of zig-zag fibres in mice, it can be shown that differentiation at stage 3 must be dependent on a functional form of $[Z]$ and $[U]$ which is different to that employed at stage 1: the function $[Z][U]^{-1}$ was therefore used at stage 3 (Nagorcka 1984).

In fact, if Z and U act independently in suppressing the expression of genes coding for the high sulphur proteins, such as those associated with paracortical cells, then a more appropriate functional form would involve the addition of independent functions of $[Z]$ and $[U]^{-1}$ (instead of the product $[Z][U]^{-1}$). Although the independent functions would in general be nonlinear, we consider, as a first approximation, the consequences of a linear dependence on $[Z]$ and $[U]^{-1}$ of the form $a[Z] + b[U]^{-1}$, where a and b are constants.

The consequences of the differentiation rules defined above are that any spatial inhomogeneities (patterns) in $[U]$ and $[Z]$ should be reflected in

(a) the spatial patterns of mitotic activity in the follicle bulb,

(b) the fibre cross-sectional shape, and

(c) the spatial pattern of ortho- and paracortical (O/P) cells as observed in fibre cross-sections.

The spatial prepatterns of U and Z expected in the follicle bulb are discussed in the next section. The predicted patterns of mitotic activity and O/P cells in the fibre, as well as the cross-sectional shape of the fibre, are calculated and presented in section 23.4.

23.3   PREPATTERNS PREDICTED IN THE FOLLICLE BULB

23.3.1   The Spatial Distribution of Z

Z is produced in the dermal papilla of the follicle bulb (stippled region in Figure 23.2(a)). Z diffuses through dermal tissue at a rate D_d and through epithelial tissue at $D_d/10$. The steady state concentration of Z at a point $\mathbf{r}$ in these tissues must satisfy the following equations:

$$\nabla \cdot (D_Z(\mathbf{r}, t) \nabla Z(\mathbf{r}, t)) = \rho(\mathbf{r}, t) \tag{23.1}$$

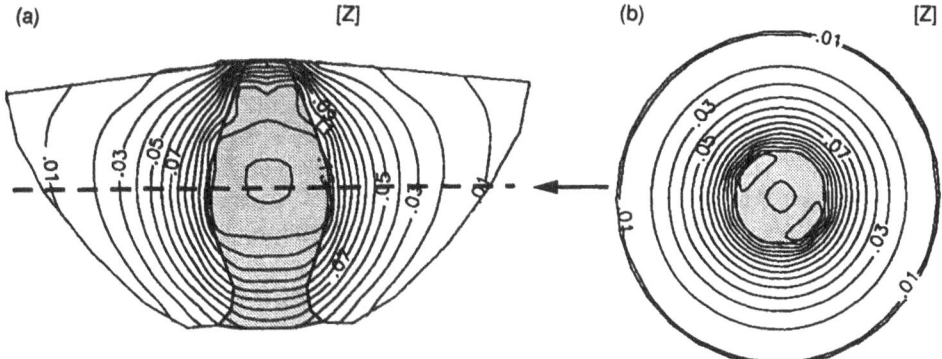

Figure 23.2. The dermal factor Z is produced in the dermal papilla of the follicle bulb (stippled region) and diffuses away from its source. The diffusion equation (23.1) is solved numerically to obtain the steady state distribution of Z. The distribution of Z is shown here as a contour plot in (a) a longitudinal section through the follicle bulb (Figure 23.1(b)), and in (b) a cross-section at half height through the follicle bulb. (Concentration of Z is in arbitrary units and the contour interval is 0.01).

$$\text{where} \quad D_Z(\mathbf{r}, t) = \begin{cases} D_d & \text{in dermal tissue} \\ D_d/10 & \text{in epithelial tissue} \end{cases}$$

$$\text{and} \quad \rho(\mathbf{r}, t) = \begin{cases} \rho_0 & \text{in dermal papilla} \\ 0 & \text{elsewhere} \end{cases}$$

The diffusion equation (23.1) may be solved numerically using a boundary integral method (Brebbia & Dominguez 1989). The boundary is taken here to be a 3-dimensional surface representing the membrane surrounding the epithelial tissue of the follicle bulb, with an upper limit defined by the T_4 boundary in Figure 23.1(a). In Figure 23.1(b) the boundary is shown constructed of triangular elements; it encloses a domain denoted here as G. The 3-dimensional domain G represents the epithelial tissue of the follicle bulb which encloses the dermal papilla.

Solving equation (23.1) for ρ_0 equal to unity throughout the dermal papilla gives the stable steady state spatial distribution of Z as it diffuses away from the dermal papilla. The distribution (or pattern) of Z obtained in a symmetrical follicle bulb is shown in Figure 23.2. It is displayed as a contour plot in a longitudinal section through the follicle bulb, and in a cross-section at approximately half-height through the bulb. The pattern of Z is clearly radially symmetric about the dermal papilla and its concentration declines rapidly with distance away from the papilla.

23.3.2 The Spatial Distribution of U and V

It has been proposed that the RD system components diffuse and react with each other while being confined to the follicle bulb (Nagorcka & Mooney 1982), in which case the RD system equations are given by the mass balance equations for U and V at a point $\mathbf{r}, t$ in G, namely,

$$\frac{\partial U(\mathbf{r}, t)}{\partial t} = D_U \nabla^2 U(\mathbf{r}, t) + F_U(U(\mathbf{r}, t), V(\mathbf{r}, t))$$

$$\frac{\partial V(\mathbf{r}, t)}{\partial t} = D_V \nabla^2 V(\mathbf{r}, t) + F_V(U(\mathbf{r}, t), V(\mathbf{r}, t))$$

$$(23.2)$$

subject to zero flux boundary conditions on the surface, ∂G, of the domain in Figure 23.1(b). F_U and F_V are, in general, nonlinear functions of $U(\mathbf{r}, t)$ and $V(\mathbf{r}, t)$ defining the net rates of production of U and V, respectively. It is now well known that the coupled nonlinear equations (23.2) spontaneously give rise to spatial wavelike patterns in the distribution of U and V, provided the diffusion rates D_U, D_V and the first order differentials of F_U and F_V with respect to U and V obey a set of conditions called the Turing conditions (Murray 1989).

More recently it has been suggested that if U and/or V are transportable across the basement membrane, separating epithelium and dermis, then the form of the nonlinearities in (23.2) would produce spatial patterns ranging from stripes to spots in response to changes in a single parameter (Nagorcka & Mooney 1992). It is this approach which we adopt here.

Nagorcka and Mooney assumed that the ratio D_U/D_V in the dermis is such that the parameter values of the RD system (23.2) lie outside the Turing space. It follows that the uniform steady state solution of equations (23.2) will remain stable in the dermis. By contrast the ratio D_U/D_V in the epithelium is larger than that in the dermis, so that the parameter values now lie within the Turing space (see Figure 3 – Nagorcka & Mooney 1992). Wavelike spatial patterns in the distribution of U and V are, therefore, expected to arise spontaneously in the epithelium (*i. e.* within domain G). Given these conditions, and since the precise form of F_U and F_V remains unknown, the same approximation as that used earlier by Nagorcka & Mooney (1992) is also employed here; namely, to explicitly incorporate the transport of U and V into the net production functions F_U and F_V, and to apply zero flux boundary conditions. This approximation is not expected to alter either the basic type of spatial pattern produced, or the sequence of patterns produced with increasing size of the follicle bulb (discussed below). The approximation also allows a direct comparison to be made with earlier predictions (Nagorcka & Mooney 1982).

F_U and F_V are defined as follows,

$$F_U(U, V) = AU - BV + P_U - q_U(U, V)[\tau_{U1}(U - U_0) + \tau_{U3}(U - U_0)^3 + \ldots] \quad (23.3)$$

$$F_V(U, V) = CU - DV + P_V - q_V(U, V)[\tau_{V1}(V - V_0) + \tau_{V3}(V - V_0)^3 + \ldots] \quad (23.4)$$

where the reaction rates A, B, C and D are constants and P_U and P_V are constant source terms. τ_{ij}, where $i = U, V$ and $j = 1, 3$, are constant coefficients defining the transport (loss/gain) terms dependent on the gradients $(U - U_0)$ and $(V - V_0)$. $U = U_0$, $V = V_0$ is, in fact, a uniform steady state solution of equations (23.2), so that $F_U(U_0, V_0) = F_V(U_0, V_0) = 0$. q_U and q_V are monotonic functions of U and V representing the effects of blocking or saturation on the transport of U and V. Values used for these parameters are given in the caption of Figure 23.3. They are the same values as those used earlier to produce spotted prepatterns to account for follicle initiation (Nagorcka & Mooney 1992). In fact, it can be shown that the wavelength required in the follicle bulb (*i. e.* the distance between adjacent maxima — which is a characteristic of the RD system) is the same as that required to predict prepatterns for the early stages of follicle initiation (Nagorcka & Mooney 1992; Nagorcka & Mooney 1985).

The RD system equations (23.2) are solved numerically in 3-dimensions using a finite element method (Bathe 1978). For this purpose the domain G (Figure 23.1(b)) is reconstructed using 20-node rectangular elements (not shown). In all solutions the spatial pattern of the distribution of V was found to be similar to that of U, hence only spatial patterns in U are presented below. As in earlier work (Nagorcka & Mooney 1989), three ranges of follicle bulb size are considered, namely,

I	small		$d_b < 90\mu$m		$d_f < 29\mu$m
II	medium	90μm $<$	$d_b < 100\mu$m	29μm $<$	$d_f < 33\mu$m
III	large	100μm $<$	$d_b < 118\mu$m	33μm $<$	$d_f < 40\mu$m

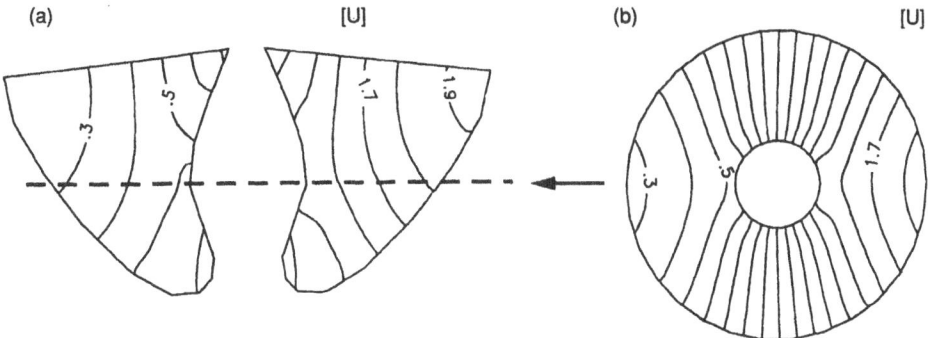

Figure 23.3. The stable steady state distributions of U and V are obtained by solving the RD system (23.2) (see text). The pattern in the distribution of V is similar to that of U, and therefore only the distribution of U is shown here. A small follicle bulb (defined in text) was used to obtain the solution for U displayed here as a contour plot in (a) a longitudinal section through the follicle bulb (Figure 23.1(b)) and in (b) a cross-section taken at half height through the bulb. (Concentration of U is units chosen such that $U_0 = 1$, contour interval is 0.1). Parameter values defining the RD system are $D_U = 16$, $D_V = 1$, $A = 3.6$, $B = 4.6$, $C = 8.6$, $D = 9.6$, $P_U = 1.0$, $P_V = 1.0$, $\tau_{U1} = 1.0$, $\tau_{U3} = 0.5$, $q_U(U, V) = (.75/(.15UV^2 + 1)) + .25$, $q_V(U, V) = 0$.

where d_b is the diameter of the follicle bulb and d_f is the diameter of the fibre produced by the bulb (Nagorcka & Mooney 1989). The shape of the bulb remains unchanged.

The pattern of U obtained in small bulbs is shown in both a longitudinal section and a cross-section in Figures 23.3(a) and (b), respectively. The wavelike variation in $[U]$ around the dermal papilla is referred to as a "bilateral" pattern (Figure 23.3(b)). The amplitude of the bilateral pattern increases with distance from the base of the follicle bulb Figure 23.3(a) (also displayed in 23.4Ia). A much lower amplitude bilateral pattern, referred to as a "small gradient" in $[U]$, is predicted to arise in medium sized follicle bulbs (Figure 23.4IIa). The pattern of U produced in large bulbs, shown in Figure 23.4IIIa, is referred to as a "bilobed" pattern in $[U]$. As in the case of the bilateral pattern, the amplitude of the bilobed pattern also increases with distance from the base of the follicle bulb (not shown).

It follows from the RD theory of fibre formation (outlined in the second section) that the spatial pattern of the quantity $[Z][U]$ is a major factor determining the fibre diameter and the fibre cross-sectional shape. The spatial pattern in mitotic activity is also expected to be determined by the pattern in $[Z][U]$. The patterns of $[Z][U]$ calculated in small, medium and large follicle bulbs, are shown in Figures 23.4Ib, IIb and IIIb, respectively. The three patterns are referred to here as "asymmetric" (Figure 23.4Ib), "near-symmetric" (Figure 23.4IIb) and "elliptical" (Figure 23.4IIIb). The RD theory also states that the spatial distribution of ortho-/paracortical (O/P) cells in the fibre is determined by a threshold level in the spatial pattern of $a[Z] + b[U]^{-1}$. These patterns have also been calculated in small, medium and large bulbs and are shown in Figure 23.4Id, IId and IIId.

It is emphasized that the calculated patterns displayed in Figure 23.4 are derived from 3-dimensional numerical solutions of equation (23.1) and equations (23.2). The 2-dimensional approximations of earlier calculations have, therefore, been removed. The major difference between the patterns in Figure 23.4 and the earlier calculations (Figure 6 of Nagorcka & Mooney 1989) is in medium sized follicle bulbs where a small amplitude bilateral pattern in $[U]$ (and $[V]$) is expected in place of the uniform steady state values of $[U]$ and $[V]$. As a consequence the fibres produced in medium sized follicle bulbs are expected to consist predominantly of orthocortical cells with a small group of paracortical cells (possibly) being located on one side (edge) of the fibre cross-section (Figure 23.4IId). Previously such fibres were expected to contain a core of orthocortical cells.

23.4   SPATIAL PATTERNS OBSERVED IN THE FOLLICLE

Romney Marsh sheep were maintained indoors in metabolism cages receiving a high level (1200g/day) of a high protein diet (consisting of equal proportions of ground and pelleted lucerne, and wheaten hay) for 7.5 weeks. The fleeces produced contain fibres having a wide range of fibre diameters throughout the fleece. Since fibre diameter is known to be proportional to follicle bulb size (Henderson 1965), all regions of the skin were expected to contain follicles having a wide range of follicle bulb sizes (diameters). At the end of the feeding period, when the wool growth rate had reached an equilibrium, skin samples 1cm in diameter were taken from the midside along with fibre samples from the fleece.

23.4.1   The Spatial Pattern of Mitotic Activity in the Follicle Bulb

Two hours prior to sampling the skin, the animals were injected with 5–Bromo–2'– Deoxyuridine (BrdU). The skin samples were fixed, sectioned and subsequently stained with an antigen to BrdU, and an antigen, known as Hit-96 (Hewish & French 1986), specific to orthocortical cells. The experimental technique is described by Adelson *et al.* (1990; 1992). The anti-BrdU staining clearly shows the spatial pattern of mitotic activity in the follicle bulb. The most frequent patterns observed were asymmetrical (Figure 23.4Ic) and elliptical (Figure 23.4IIIc) corresponding closely to those patterns in $[Z][U]$ (Figure 23.4Ib and 23.4IIIb) predicted on the basis of the bilateral and bilobed patterns in $[U]$ (Figure 23.4Ia and 23.4IIIa). Less frequently observed was a slightly asymmetrical (or near symmetrical) pattern (Figure 23.4IIc) which is consistent with the pattern of $[Z][U]$ in Figure 23.4IIb expected on the basis of the small gradient in $[U]$ (Figure 23.4IIa).

23.4.2   The Spatial Patterns of Ortho-/Paracortical (O/P) Cells in the Fibre

As summarized in Figure 23.4, the RD theory of fibre formation predicts that follicle bulbs in which the mitotic activity is asymmetrical should produce fibres containing a bilateral pattern of O/P cells in the fibre cross-section (Figure 23.4Id). Similarly, elliptical patterns of mitotic activity (Figure 23.4IIIc) must, according to the theory, produce fibres containing a bilobed pattern of O/P cells in the fibre cross-section (Figure 23.4IIId), while the near symmetrical patterns of mitotic activity (Figure 23.4IIc) are associated with fibres containing a small gradient of orthocortical cells in the fibre (Figure 23.4IId).

To test these predictions the skin samples were serially sectioned into $8\mu m$ thick sections and stained using both the BrdU antibody and Hit-96. The protein, specific to orthocortical cells, which binds Hit-96, is strongly expressed in the keratogenous zone (Hewish and French, 1986). About fifteen follicles were then traced all the way from the base of the bulb to the keratogenous zone, a distance of about $500\mu m$. In this way it is possible to associate the O/P cell pattern in the fibre cross-section with the spatial pattern of mitotic activity which gives rise to the fibre in the follicle bulb. Approximately equal numbers of the three different patterns in the follicle bulb were chosen. In all cases it was observed that asymmetrical, near-symmetrical and elliptical patterns of mitotic activity were associated with fibres containing bilateral, a small (orthocortical) gradient and bilobed patterns, respectively, of O/P cells in the cross-section. Typical results are shown in Figure 23.4 Ic and Ie, IIc and

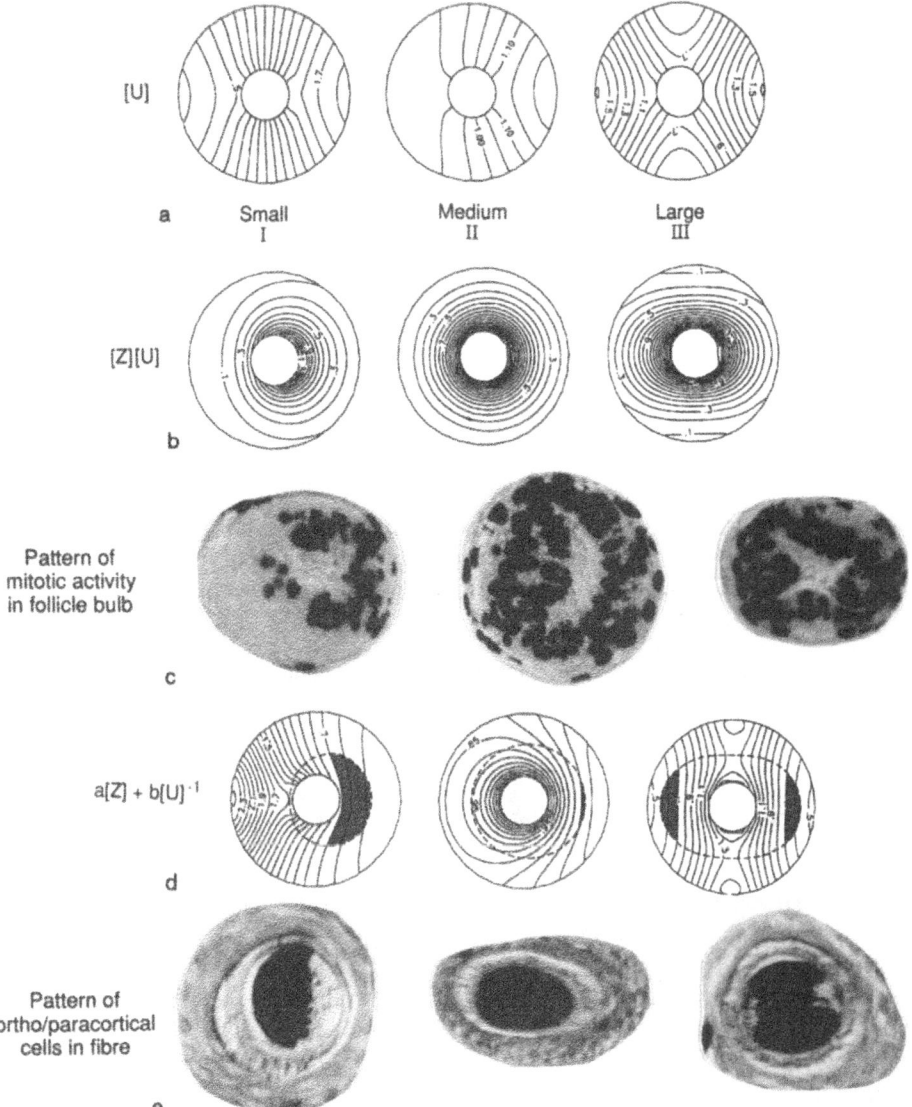

Figure 23.4. The figure is a concise summary of some of the main predictions of the RD theory of fibre formation and related observations in small (I), medium (II) and large (III) follicle bulbs (defined in text). Predicted spatial patterns are shown for (a) the distribution of U, and for (b) the function $[Z][U]$, in cross-sections at half height through different sized bulbs. According to the RD theory of fibre formation the quantity $[Z][U]$ regulates mitotic activity and hence the spatial pattern in $[Z][U]$ should reflect the spatial pattern of mitotic activity in the bulb. Patterns of mitotic activity in follicle bulbs observed in Romney Marsh sheep are shown in (c). In addition, epithelial cells which experience $[Z][U] > 0.4$ are considered to differentiate as fibre cells. The threshold $[Z][U] = 0.4$ is also plotted as a dashed curve in (d). The quantity $a[Z] + b[U]^{-1}$ is also plotted in follicle bulb cross-sections (using the relative weightings $a = .3$ and $b = .7$) in (d). A threshold value of 0.7 in this quantity is chosen to determine which fibre cells will differentiate as orthocortical cells (light stippled region) and paracortical cells (dark stippled region). The predicted ortho-/paracortical cell distributions in (d) may be compared (e) with the O/P cell distributions observed in fibres sectioned higher in the same follicles from which the bulb sections shown in (c) were taken. The mitotic distributions in (c) produce the fibres and O/P cell distributions shown in (e). The dark region in (e) indicates the location of the orthocortical cells in the fibre. The paracortical cells of the fibre are unstained. The fibre is the central region surrounded by layers of IRS cells, which in turn are surrounded by layers of ORS cells.

IIe, IIIc and IIIe. The results were found to be consistent with the predictions, which are also shown in Figure 23.4.

23.4.3   Fibre Cross-sectional Shape and the O/P Cell Pattern

The calculated patterns of $[Z][U]$, shown in Figure 23.4(d) for a particular threshold, also give us a prediction of the fibre cross-sectional shape and its relationship to the O/P cell pattern in the fibre. However, it is important to note that these predictions are for the symmetrical follicle bulb shape shown in Figure 23.1(b); in general, other shapes including asymmetrical shapes should also be considered. Nevertheless, the predictions for the follicle bulb shape in Figure 23.1(b) are that:

(1) a bilateral pattern in $[U]$ (Figure 23.4Ia) is expected to produce a slightly elliptical fibre cross-section, being flatter on the orthocortical side of the fibre (Figure 23.4Id), so that the major axis is approximately aligned with the internal O/P boundary of the bilateral O/P pattern,

(2) a gradient in $[U]$ (Figure 23.4IIa) is expected to produce a round fibre composed almost entirely of orthocortex, but may also contain a small amount of paracortex localized on one side (edge) of the fibre (Figure 23.4IId) (referred to earlier as a small orthocortical gradient),

(3) a bilobed pattern in $[U]$ (Figure 23.4IIIa) is expected to produce a fibre with an elliptical shaped cross-section containing small regions of paracortex localized at the extremes of the major axis (Figure 23.4IIId),

(4) the frequency of observation of the three different O/P cell patterns in the fibre is expected to be correlated with fibre diameter.

Image analysis software is currently being tested and developed to analyze fibre cross-sections stained for O/P cells. It is already known that, despite considerable variation in the O/P cell pattern and in the fibre cross-sectional shape, prediction (23.1) above is correct. On average, bilateral O/P cell patterns do have their internal boundary aligned with the major axis of an elliptical shaped fibre (Ahmad & Lang 1957). The remaining predictions have not been tested and the available published information is not sufficient to provide an adequate test. The observations of Ahmad & Lang (1957) lend some support to the predictions (23.3) and (23.4), although, the predominant bilobed O/P cell pattern which they draw (*i. e.* the type F pattern Figure 23.1 Ahmad & Lang 1957) is not consistent with that predicted here in relation to fibre shape (Figure 23.4IIId). Our observations of O/P cell patterns in the keratogenous zone of the follicle, by contrast, appear to be in substantial agreement with prediction (23.3) in the significant proportion of fibres whose cross-section is elliptical. Further testing of these predictions using image analysis techniques is in progress.

23.5   SUMMARY AND DISCUSSION

Three-dimensional solutions have now been obtained for equation (23.1), providing a steady state distribution of Z as it diffuses away from the dermal papilla of a (hair) wool follicle bulb. Three-dimensional solutions have also been obtained for the RD system (23.2) predicting that one of several different spatial patterns in the distribution of U (and V) will arise in the epithelial tissue of the follicle bulb depending on the size of the bulb. The patterns expected in the distribution of U in small, medium and large sized follicle bulbs are referred to as bilateral, small gradient and bilobed, respectively.

On the basis of these calculated distributions in Z and U, the RD theory of fibre formation further predicts that asymmetric, near-symmetric and elliptical spatial patterns of mitotic activity should be observed in follicle bulbs of different sizes. We have carried out detailed measurements of the spatial patterns of mitotic activity in order to test for the existence of the range of patterns predicted on the basis of an RD system present in the follicle bulb. Observations of mitotic patterns in the wool follicle bulbs of Romney Marsh sheep, a breed containing a wide range of follicle bulb sizes throughout the skin, were carried out and the results have confirmed that the predicted patterns do indeed occur in wool follicle bulbs. It is significant that neither this classification of patterns, nor the presence of elliptical patterns of mitotic activity has been suggested or reported previously.

The RD theory also predicts that the asymmetric, near-symmetric and elliptical patterns of mitotic activity must be associated with fibres containing bilateral, small gradient and bilobed patterns, respectively, of O/P cells in the cross-sections of the fibres produced in the follicle bulbs. The results of serially sectioned follicles reported here have confirmed this association of mitotic pattern with fibre cortical cell pattern. Once again, such an association has not previously been reported.

Initial observations reported here along with some existing reports (Ahmad & Lang 1957) also provide some support for the predicted orientation of O/P cell patterns in fibres relative to the cross-sectional shape of the fibres. Further analysis of the measurements is required to provide an adequate test of this prediction.

Another important prediction is that the frequency of occurrence of these patterns, both in the bulb and in the fibres, must be correlated with the size of the follicle bulb and with the diameter of the fibre. Analysis of the fibre and skin samples is continuing in order to test the predicted size dependence. Confirmation, along with confirmation of the other predictions noted above, would provide strong support for the suggestion that a prepattern mechanism, having the spatial pattern forming properties of an RD system, is present in the follicle bulb where it regulates the formation of the fibre.

ACKNOWLEDGMENTS This work was carried out with financial support from the Australian Wool Research and Development Corporation. We also thank Mr. D. Tunks and Ms. S. Munro for assistance with animal maintenance.

REFERENCES

Adelson, D. L., Nagorcka, B. N., Mooney, J. R., & Hewish, D. R. 1990. Evidence for a reaction-diffusion system in the control of cell proliferation and differentiation in the wool follicle. *Proc. Aust. Soc. Anim. Prod.*, **18**, 132–135.

Adelson, D. L., Kelley, B. A., & Nagorcka, B. N. 1992. Increase in dermal papilla cells by proliferation during development of the primary wool follicle. *Aust. J. Agric. Res.*, **43**, 843–856.

Ahmad, N., & Lang, W. R. 1957. Ortho-para cortical differentiation in "anomalous" merino wool. *Aust. J. Biol. Sci.*, **10**, 118–124.

Bathe, K. J. 1978. *ADINAT— A finite element program for automatic dynamic incremental nonlinear analysis of temperatures*. Tech. rept. *Report No. 82448-5* . M.I.T., Acoustics and Vibration Laboratory, Mechanical Engineering Dept.

Bradbury, J. H. 1973. The structure and chemistry of keratin fibres. *Adv. Prot. Chem.*, **27**, 111–211.

Brebbia, C. A., & Dominguez, J. 1989. Boundary elements an introductory course. *Computational Mechanics Publications, Southampton*.

Henderson, A. E. 1965. Relationship and wool follicle and wool fibre dimensions. In *Biology of the Skin and Hair Growth*, Lyne, A. G., & Short, B. F. (eds). pp. 447–460, Sydney: Angus and Robertson.

Hewish, D. R., & French, P. W. 1986. Monoclonal antibodies to a subfraction of merino wool high-tyrosine proteins. *Aust. J. Biol. Sci.*, **39**, 431–451.

Mooney, J. R., & Nagorcka, B. N. 1985. Spatial patterns produced by a reaction-diffusion system in the development of primary hair follicles. *J. of Theor. Biol.*, **115**, 299–317.

Murray, J. D. 1989. *Mathematical Biology*. New York: Springer-Verlag.

Nagorcka, B. N. 1984. Evidence for a reaction-diffusion system as a mechanism controlling mammalian hair growth. *Biosystems*, **16**, 323–332.

Nagorcka, B. N. 1989. Wavelike isomorphic prepatterns in development. *J. Theor. Biol.*, **137**, 127–162.

Nagorcka, B. N., & Mooney, J. R. 1982. The role of a reaction-diffusion system in the formation of hair fibres. *J. Theor. Biol.*, **98**, 575–607.

Nagorcka, B. N., & Mooney, J. R. 1985. The role of a reaction-diffusion system in the initiation of primary hair follicles. *J. Theor. Biol.*, **114**, 243–272.

Nagorcka, B. N., & Mooney, J. R. 1989. The reaction-diffusion system as a spatial organizer during the initiation and development of hair follicles and the formation of the fibre. In *The Biology of Wool and Hair*, G.E. Rogers, P.J. Reis, K. A. Ward, & Marshall, R. C. (eds). pp. 365–379, London; New York: Chapman and Hall.

Nagorcka, B. N., & Mooney, J. R. 1992. From spots to stripes. *IMA J. Maths. Appl. Med. Biol.* (In press).

Ouyang, Q., & Swinney, H. L. 1991. Transition from a uniform state to hexagonal and striped Turing patterns. *Nature*, **352**, 610–612.

Turing, A. M. 1952. The chemical basis of morphogenesis. *Phil. Trans. R. Soc. Lond.*, **B237**, 37–72.

24. THE SPEED OF A BROWNIAN RATCHET

George F. Oster[1] and Charles S. Peskin[2]

[1]Departments of Molecular and Cellular Biology, and Entomology
University of California
Berkeley, CA 9472, USA

[2]Courant Institute of Mathematical Sciences
251 Mercer Street
New York, NY 10012, USA

24.1 INTRODUCTION

Some cellular motile processes do not appear to involve molecular motors. These include many types of cellular extensions, including filopodia, lamellipodia, and acrosomal extension. These processes transduce chemical bond energy into directed motion by a mechanism quite different from motor molecules. They do not operate in a direct mechanochemical cycle and need not depend directly upon nucleotide hydrolysis. In this paper we describe several such processes and present simple formula for the velocity and force they generate. We shall call these machines "Brownian Ratchets" (BR) to distinguish them from the usual class of protein motors-although motors may be Brownian Ratchets as well (Cordova *et al.* 1991; Meister *et al.* 1989; Mitsui & Ohshima 1988; Vale & Oosawa 1990).

Consider a particle diffusing in one dimension with diffusion coefficient D. The approximate time, τ, it takes a particle to diffuse from the origin, $x = 0$, to the point $x = \delta$ is: $\tau = \delta^2/2D$. Now, suppose that a domain extending from $x = 0$ to $x = L$ is subdivided into $N = L\delta$ subintervals, and that each boundary, $x = n\delta, n = 1, 2, \ldots, N$ is a "ratchet", that is, it is absorbing from the left, but reflecting from the right. The mechanism of the ratchet depends on the application; for example, the particle may be prevented from reversing its motion by a polymerizing fiber to its left. Then the time to diffuse a distance $L = N\delta$ is simply $NT\delta : T = NT_\delta = N\delta^2/2D = L\delta/2D$. The average velocity of the particle is $< v >\equiv L/T$, and so the average speed of a particle that is "ratcheted" at intervals δ is

$$\langle v \rangle = \frac{2D}{\delta}.$$

This is the speed of a perfect BR. Note that as the ratchet interval, δ, decreases, the ratchet velocity increases. This is because the frequency of smaller Brownian steps grows more rapidly than the step size shrinks (when δ is of the order of a mean free path, then this formula obviously breaks down).

Several ingredients must be added to this simple expression to make it useful in real situations. First, the ratchet may not be perfect: a fraction of the sites crossing a ratchet boundary may be able to cross back. Second, in order to perform work, the ratchet must work

Experimental and Theorietcal Advances in Biological Pattern Formation,
Edited by H.G. Othmer *et al.*, Plenum Press, New York, 1993

against a force resisting the motion. Below we will give explicit expressions for the velocity of a BR; they will have the form $< v >= 2D/\delta F(w, \alpha_i), 0 < F < 1$, where $w = f\delta/k_BT$ is the dimensionless work the ratchet performs against the load force, f, and α_i are rate constants characterizing the chemical reaction that effects the rectification.

24.2 FILOPODIAL PROTRUSION

Janmey was able to load actin monomers into liposomes and trigger their polymerization (Janmey *et al.* 1992). He observed that the polymerizing fibers extruded long spikes resembling filopodia from the otherwise spherical liposomes. A similar phenomenon was described by Miyamoto & Hotani (1988) using tubulin. This appears to demonstrate that polymerization can exert an axial force capable of overcoming the bending energy of a lipid bilayer. Using a bilayer bending modulus of $B = 2 \times 10^{-12}$ dyne-cm (Bo & Waugh 1989; Duwe *et al.* 1990), the energy required to elongate a lipid cylinder of radius 50 nm from zero length to $5\mu m$ long is $\sim 10^4 \ k_BT$. Since we are dealing with thermal motions, we will henceforth we will express all energetic quantities in terms of $k_BT \approx 4.1 \times 10^{-14}$ dyne-cm, where k_B is Boltzmann's constant and T the absolute temperature. The free energy change accompanying actin polymerization is $\triangle G \approx -14 \ k_BT$/monomer (Gordon *et al.* 1976). At a concentration of 20 μM, actin polymerizes *in vitro* at about 0.6 μm/second ≈ 280 monomers/second (Pollard 1986), yielding 3800 k_BT/second. So polymerization can provide sufficient free energy to drive membrane deformation (Hill & Kirschner 1982; 1983); the BR model provides an explanation for how this free energy is transduced into an axial force.

Consider the ratchet shown in Figure 24.1. An actin rod polymerizes against a barrier, B (*e. g.* a membrane) whose mobility we characterize by its diffusion coefficient, D. We model a polymerizing actin filament as a linear array of monomers; here, the ratchet mechanism is the intercalation of monomers between the barrier and the polymer tip. Denote the gap width between the tip of the rod and the barrier by ξ, and the size of a monomer by δ. When a sufficiently large fluctuation occurs the gap opens wide enough to allow a monomer to polymerize onto the end of the rod. The polymerization rate is given by $R = k_{on}(\delta) \ M - \beta$, where M is the local monomer concentration and the polymerization rate constant; $k_{on}(\delta)M$, reflects the *conditional* probability of adding a monomer when the gap width is ξ. We set $k_{on}(\xi) \ M = \alpha$ when $\xi \geq \delta$, and $k_{on}(\xi) \ M = 0$ when $\xi < \delta$. If no barrier were present, actin could polymerize at a maximum velocity of $\delta R \approx 0.75\mu$m/second at 25 μM concentration of actin monomers (Pollard 1986). Cellular filopodia protrude at velocities about 0.16 μm/second (Argiro *et al.* 1985), well below the maximum polymerization velocity. We can derive a formula for the polymerization BR by writing a diffusion equation for the gap width, ξ, (Peskin *et al.* 1992)

$$\frac{\partial c}{\partial t} = D\frac{\partial^2 c}{\partial \xi^2} + \left(\frac{fD}{k_bT}\right)\frac{\partial c}{\partial \xi} + \alpha\left[c(\xi + \delta, t) - H(\xi - \delta)c(\xi, t)\right]$$
$$+\beta\left[H(\xi - \delta)c(\xi - \delta, t) - c(\xi, t)\right] \qquad (24.1)$$

where D is the diffusion coefficient of the load, $-f$ is the load force (*i. e.* to the left, opposing the motion), $H(\xi - \delta)$ is the Heaviside step function $H(\xi - \delta) = 0$ for $\xi < \delta$, and $H(\xi - \delta) = 1$ for $x > \delta$). The boundary conditions are that $\xi = 0$ is reflecting and $c(\xi, t)$ is continuous at $\xi = \delta$. The steady state solution to equation (24.1) gives the force-velocity relation if we define the ratchet velocity by $v(w) = \delta\left(\alpha\int_\delta^\infty c(x)dx - \beta\int_0^\infty c(x)dx\right)$ (*i. e.* we weight the polymerization velocity by the probability of a δ-sized gap). When depolymerization can be neglected, *i. e.* $\beta << \alpha$ – which is the case for actin polymerization - we obtain:

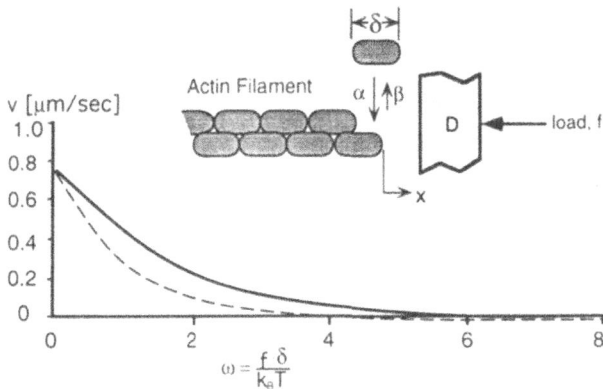

Figure 24.1. The speed of the polymerization ratchet, driven by a single actin filament, as a function of dimensionless load force. $v[\mu m/\sec]$, $w = f\,\delta/k_BT$. The solid line is based on equation (24.2), the formula for the ratchet speed when depolymerization is negligible ($\beta \to 0$). The dashed line is based on equation (24.3), valid when polymerization is much slower than diffusion, $\alpha\delta^2/D << 1$. The rate constants were taken from Pollard (1986) for actin polymerization: $\alpha = k_{on}\,M = 11.3[1/\sec \text{ molar}]\,10[\mu\text{molar}]$, $\beta = 1.6[1/\sec]$, $\delta = \text{(monomer size)}/2 \approx 2.7 \times 10^{-7}$ cm, since actin is a double helix, We used a load diffusivity of $D = 1 \times 10^{-9}$ cm/sec, corresponding to a disk of diameter ≈ 1 μm. Filopodial velocities are below 0.16 μm/sec (Argiro *et al.* 1985), which is about 20% of the maximum polymerization velocity, $\delta\,(\alpha - \beta) = \delta\,(k_{on}\,M - k_{off}) \approx 0.76\mu$m/sec (Cooper 1991; Pollard 1986). From equation (24.4), the stall force for a single actin fiber is $f_o \approx 7.8$ pN. A filopod composed of 20 filaments presumably could exert 20 times this force.

$$v = \frac{2D}{\delta}\left[\frac{(\mu - w)\frac{w^2}{2}}{w^2 + (e^\infty - w - 1)\mu}\right] \qquad (24.2)$$

where w is the dimensionless work done against the load: $w = f\,\delta/k_BT$, and $\mu(w, \delta, \alpha, D) > w$ is given by solving a transcendental equation, $\alpha\delta^2\,(1 - e^{-\mu})\,/\mu D + \mu - w = 0$. Figure 24.1 shows a plot of $v(w)$. If the polymerization and depolymerization velocities are much slower than the ideal ratchet velocity, *i. e.* $\alpha\,\delta$, $\beta\,\delta << 2D/\delta$, then the ratchet equation can be solved explicitly for $\beta \neq 0$. The result is a startlingly simple formula:

$$v \approx \delta[\alpha e^{-\infty} - \beta] \qquad (24.3)$$

That is, the polymerization rate, $\alpha = k_{on}\,M$, is weighted by the probability of the load allowing a monomer-sized gap, δ. Note that in this limit the ratchet velocity does not depend on the diffusion coefficient of the load. The force required to stall the ratchet is found by setting $v = 0$ in equation (24.3), which yields the familiar thermodynamic relationship $\beta/\alpha = \exp(-f\,\delta/k_BT)$, or

$$f_0 = \frac{k_BT}{\delta}\ln\left(\frac{\alpha}{\beta}\right) \qquad (24.4)$$

This formula for the stall force is exact; it remains valid for all parameter values, even those that violate the assumptions used in deriving equation (24.3).

    Two observations support the BR model for filopodial growth. Firstly the velocity of extension is almost constant with length (Argiro *et al.* 1985), unlike the acrosomal extension of *Thyone* sperm, in which length grows as the square root of time (Inoué & Tilney 1982; Oster & Perelson 1988; Oster *et al.* 1982; Perelson & Coutsias 1986; Tilney & Inoué 1982,1985). The BR mechanism produces a constant velocity providing the polymerization affinity is constant.

Eventually, the filopod may grow long enough so that the diffusion of actin monomers to the tip is limiting, in which case the velocity will decrease. Secondly, experiments by Bray *et al.* (1991) demonstrated that filopodial extension velocities actually increased somewhat with external osmolarity. This is consistent with the BR mechanism, since pulling water out of the cell will concentrate the actin monomers, thus increasing for a time the affinity, and hence the ratchet velocity. This contrasts with acrosomal protrusion of *Thyone* wherein increasing external osmolarity decreases protrusion velocities (Inoué & Tilney 1982; Tilney & Inoué 1985; Tilney & Inoué 1982). Membrane tensions fall in the range 0.035–0.039 dyne/cm, which amount to a load force of about 25 pN. A filopod of 20 filaments could produce a thrust 20 times as strong as a single filament, or about 200 pN. However, once a filopod grows long enough so that diffusion limits the concentration of actin monomers at the tip, the protrusion velocity will fall to zero quite quickly.

The BR formula omits an important feature: proteins are flexible, elastic structures, whose internal fluctuations significantly affect their motions. In the ratchet formula (24.2) the rod is assumed to be stiff and the gap width depends solely on the diffusion of the barrier. However, since the actin monomers are themselves flexible, Brownian motion will induce thermal "breathing" modes which will contribute to the gap width. There is no simple way to include this in the derivation; however, we can use numerical simulations to investigate elastic effects in particular situations. We have performed a molecular dynamics simulation of this situation using the parameters for actin; the details of this computation will be published elsewhere. We find that for rod lengths of more than 50 monomers the elastic fluctuations of the rod can compress the rod enough to permit polymerization even if the barrier is too large to diffuse appreciably. In this situation the elastic compression energy generated by thermal motions is the proximal origin of the force.

24.3   THE PROPULSION OF *LISTERIA*

The bacterium *Listeria monocytogenes* moves through the cytoplasm with velocities between 0.05–1.5 μm/second. As is moves, it trails a long tail of polymerized actin (Dabiri *et al.* 1990; Tilney & Portnoy 1989) consisting of many short fibers cross linked into a meshwork; the fibers are oriented predominantly with the barbed end in the direction of motion. Using fluorescent photoactivation, Theriot *et al.* (1992) were able to visualize the tail as the bacterium moved (Sanger *et al.* 1992; Theriot *et al.* 1992). They found that the tail remained stationary, and that actin inserted into the tail meshwork adjacent to the bacterial body. Taken together, these observations suggest that actin polymerization may drive bacterial movement.

We propose that *Listeria* is driven by a BR mechanism: the polymerizing tail rectifies the random thermal motions of the bacterium, preventing it from diffusing backwards, but permitting forward diffusion. In this view the tail does not actually push the bacterium; propulsion is simply Brownian diffusion, rendered unidirectional by the polymerization of the actin tail. We assume that (i) the *Listeria* diffuses as a Stokes particle of size $\sim 1\,\mu$m (Berg 1983), (ii) the ratchet distance is half the diameter of an actin monomer, $\delta \approx 2.7$ nm, and (iii) the polymerization rate constants are the same as we used in the filopod calculation (Pollard 1986,1990). The effective concentration of actin monomers near the bacterium *in vitro* is unknown, but is likely to be much higher than at the tip of a filopod. Using a local concentration of 25 μM (Cooper 1991), the stall force for a single actin fiber is $f_o \approx 7$ pN, about five times the force generated by a myosin crossbridge. Since the tail consists of many fibers, whose orientations are not collinear, we cannot directly compute the thrust of the tail without knowledge of the fiber number and orientation distributions. All we can say is that, since *Listeria* moves through the cytoplasm at a top speed of about 1.5 μm/second, the computed load-velocity curve shows that one fiber would be sufficient to drive a 1 μm bacterium at 1.5 μm/second against a load of 1 pN. The elastic resistance of the cell's dense actin gel is the major impediment to the bacterium's motion, and it may be reasonable to ascribe the load force to this elastic resistance.

According to the BR mechanism the speed of the BR depends on the polymerization rate of actin - although it is not *driven* directly by the polymerization. The faster the bacterium can recruit actin from the cytoplasmic pool - perhaps via an enzyme that liberates acting from profilin - the faster the bacterium moves and the longer the tail grows. Indeed, it can easily be shown that the observed linear relationship between velocity and tail length requires only that the polymerization rate of actin be constant, regardless of the mechanism of force generation. Using a laser trap it should be possible to measure the stall force as a function of monomer concentration, which equation (24.4) predicts should vary as $f_o \sim \ln(M)$. *In vivo* values of diffusion coefficients and monomer concentrations may be quite different from those *in vitro*; and so our computed load-velocity curve is probably not too accurate. In order to characterize the *Listeria* BR motor, it is necessary to design experiments to measure accurately the diffusion coefficient of a "dead" bacterium along with the *in situ* polymerization rates and the fiber orientations.

A possible analog of the Listeria system was recently reported by Forscher *et al.* (1992): latex beads dropped onto the surface of certain cells commenced to move in the plane of the membrane at speeds of about 0.16 μm/second. Closer inspection revealed a tail of polymerized actin streaming behind the moving bead. This resembles the tail of *Listeria*, and it is tempting to assert that this too is a manifestation of the Brownian Ratchet mechanism.

24.4 PROTEIN TRANSLOCATION

We have previously proposed that post-translational translocation of a protein across a membrane may be driven by a BR (Simon *et al.* 1992). We addressed the process that begins after the proximal tip of the protein is threaded through the translocation pore (Simon & Blobel 1991). Brownian motion causes the protein to fluctuate back and forth through the pore, but with no net displacement in either direction (analogous to a reptating polymer (de Gennes 1983)). If a chemical modification of the protein occurs on the distal side of the membrane which inhibits the chain from reptating back through the pore, the chain will be ratcheted. The model assumes that the protein is maintained in an unfolded conformation so that it is free to fluctuate back and forth through the translocation pore. There are several known chemical asymmetries that could bias the Brownian walk of a chain (Cheng *et al.* 1989; Kagan *et al.* 1981; Ostermann *et al.* 1989; Simon *et al.* 1992). As a polypeptide emerges from the translocation apparatus the chain is subjected to glycosylation, formation of disulfide bonds, cleavage of the signal sequence (which affects folding of the chain, and binding of chaperonins). Any, or all, of these can induce the asymmetry in the system required for the BR. This multiplicity of ratchet mechanisms may explain why different laboratories have attributed the translocation motor to different constituents of the translocation machinery, and why almost any protein can be translocated if given the proper signal sequence.

This ratchet is somewhat different from the polymerization BR considered above since there are many ratcheting sites rather than one. We can derive a force-velocity curve for the translocation ratchet in the case where the ratchet mechanism is the binding of chaperonins on the luminal side of the translocation pore, as shown in Figure 24.2 (Peskin *et al.* 1992;

Simon *et al.* 1992). Since the motion of each segment is equivalent we consider an ensemble of points diffusing on a circle of circumference equal to the length of a ratchet segment of the polymer, δ. As before, each point is subject to a force $-f$ which imparts a drift velocity $-f D/k_B T$. Points are in rapid equilibrium between two states: $S_O \Leftrightarrow S_1$, with rate constants k_+ and k_-. Points in state S_O pass freely through the origin in both directions, but points in state S_1 are ratcheted: they cannot cross back across the origin. Let p be the probability of finding a point in state $S_1 : p = k_+/(k_+ + k_-)$. Then we can write the net flux of points as $\phi(x, t) = -\frac{Df}{kT}c - D\frac{\partial c}{\partial x}$, where $c(x, t)$ is the density of points at position x and time t. $\phi(x)$ satisfies the steady state conservation equation $\frac{\partial \phi}{\partial x} = 0$, with boundary conditions $\phi(0) = \phi(\delta)$, and $c(\delta) = (1 - p) c(0)$. We solve for $c(x)$ and define the average velocity as $\langle v \rangle = \phi/(N/\delta)$, where $N = \int_0^\infty C(x, t)dx$ is the total number of points in the ensemble. The result is (Peskin *et al.* 1992; Simon *et al.* 1992)

$$\langle v \rangle = \frac{2D}{\delta} \left[\frac{\frac{1}{2}w^2}{\left(\frac{(e^w - 1)}{1 - K(e^w - 1)} \right) - w} \right] \tag{24.5}$$

where w is defined as before, and the parameter $K = (1 - p)/p = k_- k_+$ is the dissociation constant of the trans chaperonins. The maximum (no load) velocity is $< v >_{max} = (2D)/\delta \; (1/(1 + 2K))$, and the stall load is $f_0 = (k_B T/\delta) \, ln \, (1 + 1/k)$. Note that when $K = 1$, translocation still proceeds at a finite rate, whereas the polymerization ratchet stalls even in the no-load condition when $\alpha = \beta$. A typical force-velocity curve computed from equation (24.5) is plotted in Figure 24.2. Equation (24.5) has two important limitations. Firstly, it assumes that the rates k_+ and k_- are very fast, and secondly that the ratchet is inelastic.

Using equation (24.5) we can put some quantitative bounds on the translocation time of a protein (the calculation is simplified since there is no obvious load force, f, resisting translocation). For example, the slowest time corresponds to the situation where one end is just threaded through the TP and translocation is completed when the other end passes through the TP. Taking $\delta \approx 100$ nm as the length of an unfolded protein, and $D \approx 10^{-8} cm^2$ second as the longitudinal diffusion coefficient, the translocation time is ≈ 5 m/second; but if the chain is ratcheted every 5nm, the transit time is 0.25 m/second - faster by a factor of 20. This estimate of τ is probably too short, since the 1-dimensional formula (24.5) cannot take into account the effects of chain coiling; for this a full 3-dimensional calculation must be carried out. On the other hand, equation (24.5) neglects the effect of chain elasticity, which significantly adds to the translocation velocity. However, a full stochastic simulation of the polymer dynamics shows the quadratic dependence of translocation time on chain length implied by equation (24.5) (Simon *et al.* 1992). Thus, both our numerical and analytical calculations demonstrate that the BR mechanism is more than sufficient to account for the observed rates of translocation.

There are several other phenomena that are possibly driven by rectified diffusion. For example, the polymerization of sickle hemoglobin into rods that deform the erythrocyte membrane appear similar to filopod protrusion, (Liu *et al.* 1991) and probably derive their thrust from the same mechanism. Finally, *in vitro* model systems show that depolymerizing microtubules can drive kinetochore movements towards the minus end at velocities of $\approx 0.5 \; \mu$ m/second, and exert forces on the order of $\approx 10^{-5}$ dyne (Koshland *et al.* 1988). Koshland, (1988) describe a qualitative model for how depolymerization could drive kinetochore movement, and Hill & Kirschner (1982) have shown that such movements are thermodynamically feasible. The BR model fills in the mechanical mechanism, and equation (24.5) may apply to this phenomenon as well.

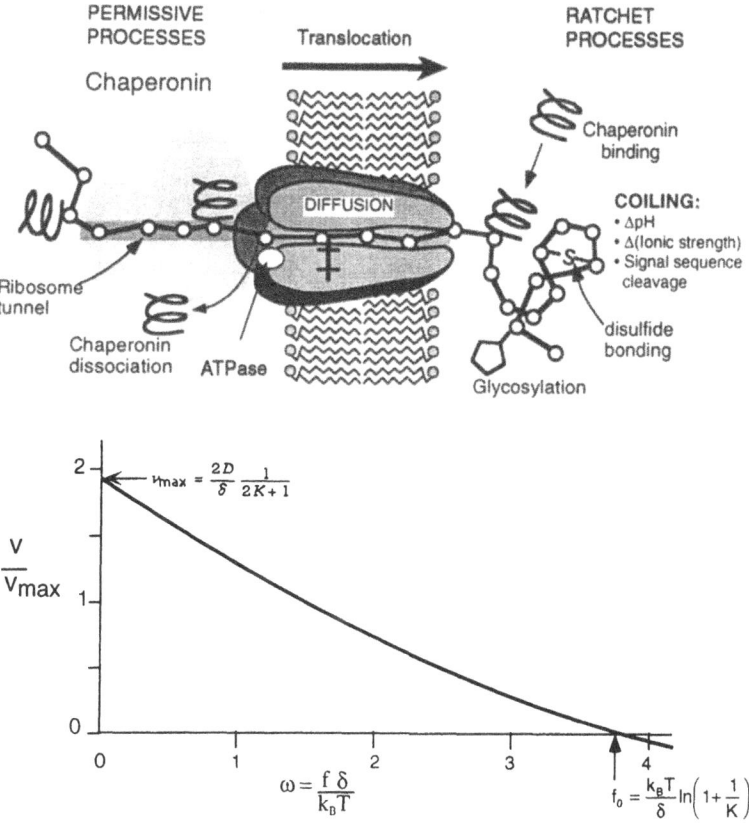

Figure 24.2. (a) The BR driven velocity of *Listeria* as a function of actin monomer concentration, M, is of Michaelis-Menten form. (b) The stall force, f_o, also saturates with increasing M. Parameters: $\delta = 2.5$ nm. $D = 2 \times 10^{-9}$ cm^2/second, K = 0.11. At physiological concentrations, and with no loading resistance, the ratchet would effectively operating at its maximum velocity, with a stall force of about 12 pN. These curves may underestimate the power of the BR since *Listeria*'s tail is composed of many short, crosslinked fibers whose orientation is distributed statistically about the long axis of the bacterium.

24.5 DISCUSSION

The notion that biased Brownian motion drives certain biological motions is not new. Huxley implied as much in his 1957 model for myosin (Huxley 1957), and later authors have proposed similar models for other molecular motors (Cordova *et al.* 1991; Leibler & Huse 1991; Meister *et al.* 1989; Mitsui & Ohshima 1988; Vale & Oosawa 1990). The model we present here is unlike these in two important respects. Firstly, we are modeling mechanisms that do not operate in the same thermodynamic cycle as do molecular motors. Rather they are "one-shot" engines; for example, after protrusion of a filopod the polymers must be disassembled and the process started anew. Secondly, we do not treat the motion as a biased random walk, as in Feynman's "thermal ratchet" machine (Feynman *et al.* 1963). Biased random walk models assume asymmetric jump probabilities in either direction at each step; in the limit of small step sizes this produces a continuous drift velocity proportional to the difference in jump probabilities (Zauderer 1989). By contrast, we assume that the jump probabilities are symmetric, and so diffusion is unbiased. Only when diffusion crosses a ratchet threshold does the motion become ratcheted. Ultimately, the free energy driving the

BR mechanism derives from the free energy of chemical reactions: actin polymerization in the case of *Listeria* and filopodial motion, and by a variety of processes in protein translocation, including binding of chaperonins, post-translational coiling, glycosylation, etc. However, the transduction is indirect: the proximal force for movement arises from random thermal fluctuations, while the chemical potential release accompanying reactions serves to rectify the thermal motions of the load. For example, the binding free energy of a monomer to the end of an actin filament ($\triangle G_b \approx -13 k_B T$/monomer) must be tight enough to prevent the load from back diffusion. If $\triangle G_b$ were $\sim k_B T$, the residence time of the monomer would be short and the site would likely be empty when the load experiences a reverse fluctuation - or, if the site is occupied, the force of its collision with the load would dislodge the monomer. Hence the concentration of monomers and the binding free energy of polymerization supply the free energy to implement the ratchet. Thus the process does not violate the 2nd Law; rather it borrows from the available thermal energy to drive the ratchet.

ACKNOWLEDGMENTS This work was supported by the National Science Foundation, the MacArthur Foundation, and The Neurosciences Institute. The authors would like to acknowledge J. Theriot and T. Mitcheson for sharing unpublished data with us, and P. Janmey, D. Lerner, and S. Simon for their valuable comments.

REFERENCES

Argiro, V., Bunge, M., & Johnson, M. 1985. A quantitative study of growth cone filopodial extension. *J. Neurosci. Res.*, **13**, 149–162.

Berg, H. 1983. *Random walks in biology*. Princeton, NJ: Princeton University Press.

Bo, L., & Waugh, R. E. 1989. Determination of bilayer membrane bending stiffness by tether formation from giant, thin-walled vesicles. *Biophys. J.*, **55**, 509–517.

Bray, D., Money, N., Harold, F., & Bamburg, J. 1991. Responses of growth cones to changes in osmolality of the surrounding medium. *J. Cell Sci.*, **98**, 507–515.

Cheng, M. Y, Hartl, F. U., Martin, J., Pollock, R. A., Kalousek, F., Neupert, W., Hallberg, E. M., Hallberg, R. L., & Horwich, A. L. 1989. Mitochondrial heat-shock protein hsp60 is essential for assembly of proteins imported into yeast mitochondria. *Nature*, **337**, 620–625.

Cooper, J. A. 1991. The role of actin polymerization in cell motility. *Ann. Rev. Physiol.*, **53**, 585–605.

Cordova, N., Ermentrout, B., & Oster, G. 1991. The mechanics of motor molecules I. The thermal ratchet model. *Proc. Natl. Acad. Sci. (USA)*, **89**, 339–343.

Dabiri, G. A., Sanger, J. M., Portnoy, D. A., & Southwick, F. S. 1990. *Listeria monocytogenes* moves rapidly through the host-cell cytoplasm by inducing directional actin assembly. *Proc. Natl. Acad. Sci. (USA).*, **87**, 6068–6072.

de Gennes, P. 1983. Reptation d'une chaine heterogene. *J. Physique Lett.*, **44**, L225–L227.

Duwe, H., Kaes, J., & Sackmann, E. 1990. Bending elastic moduli of lipid bilayers - modulation by solutes. *J Phys. France.*, **51**, 945–962.

Feynman, R., Leighton, R., & Sands, M. 1963. *The Feynman Lectures on Physics*. Addison-Wesley.

Forscher, P., Lin, C. H., & Thompson, C. 1992. Inductopodia: A novel form of stimulus-evoked growth cone motility involving site directed actin filament assembly. *Nature*, **357**, 515–518.

Gordon, D., Yang, Y.-Z., & Korn, E. 1976. Polymerization of acanthamoeba actin. *J. Biol. Chem.*, **251**, 7474–79.

Hill, T., & Kirschner, M. 1982. Bioenergetics and kinetics of microtubule and actin filament assembly and disassembly. *Intl. Rev. Cytol.*, **78**, 1–125.

Hill, T., & Kirschner, M. 1983. Regulation of microtubule and actin filament assembly-disassembly by associated small and large molecules. *Intl. Rev. Cytol.*, **84**, 185–234.

Huxley, A. F. 1957. Muscle structure and theories of contraction. *Prog. Biophys. biophys. Chem.*, **7**, 255–318.

Inoué, S., & Tilney, L. 1982. The acrosomal reaction of *Thyone* sperm I. changes in the sperm head visualized by high resolution video microscopy. *J. Cell Biol.*, **93**, 812–820.

Janmey, P., Cunningham, C., Oster, G., & Stossel, T. 1992. Cytoskeletal networks and osmotic pressure in relation to cell structure and motility. In *Swelling Mechanics: From Clays to Living Cells and Tissues*, Karalis, T. (ed). Heidelberg: Springer-Verlag.

Kagan, B., Finkelstein, A., & Colombini, M. 1981. Diphtheria toxin fragment forms large pores in phospholipid bilayer membranes. *Proc. Natl. Acad. Sci.*, **78**, 4950–4954.

Koshland, D. E., Mitchison, T. J., & Kirschner, M. W. 1988. Polewards chromosome movement driven by microtubule depolymerization *in vitro*. *Nature*, **331**, 499–504.

Leibler, S., & Huse, D. 1991. A physical model for motor proteins. *C. R. Acad. Sci. Paris*, **313**, 27–35.

Liu, S.-C., Derick, L., & Palek, S. Zhai J. 1991. Uncoupling of the spectrin-based skeleton from the lipid bilayer in sickled red cells. *Science*, **252**, 574–575.

Meister, M., Caplan, S. R., & Berg, H. C. 1989. Dynamics of a tightly coupled mechanism for flagellar rotation: bacterial motility, chemiosmotic coupling, protonmotive force. *Biophys. J.*, **55**, 905–914.

Mitsui, T., & Ohshima, H. 1988. A self-induced translation model of myosin head motion in contracting muscle. I. force-velocity relation and energy liberation. *J. Muscle Res. Cell Motil.*, **9**, 248–260.

Miyamoto, H., & Hotani, H. 1988. Polymerization of microtubules within liposomes produces morphological change of their shapes. In *Taniguchi International Symposium on Dynamics of Microtubules*, Hotani, H. (ed). pp. 220–242, Taniguchi, Japan: The Taniguchi Foundation. 14.

Oster, G., & Perelson, A. 1988. The physics of cell motility. *J. Cell Sci.*, **8**, 35–54. Suppl.: Cell Behavior: Shape, Adhesion and Motility. C.M.J. Heaysman F. Watt.

Oster, G., Perelson, A., & Tilney, L. 1982. A mechanical model for acrosomal extension in thyone. *J. Math. Biol.*, **15**, 259–65.

Ostermann, J., Horwich, A. L., Neupert, W., & Hartl, F. U. 1989. Protein folding in mitochondria requires complex formation with hsp60 and ATP hydrolysis. *Nature*, **341**, 125–130.

Perelson, A. S., & Coutsias, E. A. 1986. A moving boundary model of acrosomal elongation. *J. Math. Biol.*, **23**, 361–79.

Peskin, C., Odell, G., & Oster, G. 1992. Cellular motions and thermal fluctuations: The Brownian ratchet. *Biophys. J.* (Submitted).

Pollard, T. 1986. Rate constants for the reactions of ATP- and ADP-actin with the ends of actin filaments. *J. Cell Biol.*, **103**, 2747–2754.

Pollard, T. 1990. Actin. curr. opin. *Cell Biol.*, **2**, 33–40.

Sanger, J. M., Sanger, J. W., & Southwick, F. S. 1992. Host cell actin assembly is necessary and likely to provide the propulsive force for intracellular movement of *Listeria monocytogenes*. *Infection & Immunity*, **60**, 3609–3619.

Simon, S., & Blobel, B. 1991. A protein-conducting channel in the endoplasmic reticulum. *Cell*, **65**, 371–380.

Simon, S., Peskin, C., & Oster, G. 1992. What drives the translocation of proteins? *Proc. Natl. Acad. Sci. USA.*, **89**, 3770–3774.

Theriot, J. A., Mitchison, T. J., Tilney, L. G., & Portnoy, D. A. 1992. The rate of actin-based motility of intracellular *Listeria monocytogenes* equals the rate of actin polymerization. *Nature*, **357**, 257–60.

Tilney, L., & Inoué, S. 1982. The acrosomal reaction of thyone sperm. II. the kinetics and possible mechanism of acrosomal process elongation. *J. Cell Biol.*, **93**, 820–827.

Tilney, L., & Inoué, S. 1985. Acrosomal reaction of the thyone sperm. III. the relationship between actin assembly and water influx during the extension of the acrosomal process. *J. Cell Biol.*, **100**, 1273–83.

Tilney, L. G., & Portnoy, D. A. 1989. Actin filaments and the growth, movement, and spread of the intracellular bacterial parasite, *Listeria monocytogenes*. *J. Cell Biol.*, pp. 1597–1608.

Vale, R. D., & Oosawa, F. 1990. Protein motors and maxwell's demons: does mechanochemical transduction involve a thermal ratchet? *Adv. Biophys.*, **26**, 97–134.

Zauderer, E. 1989. *Partial differential equations of applied mathematics*. 2 edn. New York: John Wiley & Sons.

25. OSCILLATIONS AND WAVES IN A MODEL OF InsP$_3$-CONTROLLED CALCIUM DYNAMICS

Hans G. Othmer and Yuanhua Tang

Department of Mathematics
University of Utah
Salt Lake City, UT 84112

25.1 INTRODUCTION

Various types of cells use changes in intracellular calcium triggered by hormones, neurotransmitters or growth factors as a second messenger to trigger a variety of intracellular responses, including contraction, secretion, growth and differentiation. One of the earliest experiments that revealed the complexity of the calcium response is that due to Woods *et al.* (1986), who studied the response of hepatocytes to vasopressin, a hormone that mobilizes intracellular calcium. They found that over the range of 200 nM to 1 μM in the hormone concentration, stimuli evoke repetitive spikes in the intracellular calcium concentration, rather than simply elevating the level of calcium. Moreover, they found that as the hormone concentration was raised, the frequency of spiking increased, but the amplitude remained nearly constant. Thus the continuously-graded (analog) extracellular hormone signal was converted into a frequency-encoded digital signal (the number of calcium spikes). Similar dynamic behavior has been found in a large number of cell types since then, and has lead to the suggestion that calcium spiking and frequency encoding must have a physiological role. Since high calcium levels are cytotoxic, it is necessary to maintain a low average concentration of calcium, but since calcium frequently serves primarily as a trigger for other processes, a transient elevation above a low mean concentration suffices for this purpose. Moreover, using a large transient increase as the trigger permits the use of a sharper threshold for response, and hence better noise discrimination. Other possible advantages are suggested by Meyer & Stryer (1991), but to date there is little hard evidence that the spiking plays an essential physiological role.

Of potentially greater significance is the fact that the response in many cells is also spatially inhomogeneous. In a variety of cell types, waves of calcium release propagate across the cell in response to hormonal or other stimuli. For instance, in *Xenopus laevis* oocytes, penetration of a sperm into the egg triggers a localized increase in cytosolic calcium that propagates away from the point of entry at approximately 10 μ/sec, inducing cortical contraction, meiosis, and structural rearrangement (Gilkey *et al.* 1978; Nuccitelli 1991). In addition, there is evidence that at least some of the hormone-sensitive calcium stores in *Xenopus* oocytes are localized at the animal pole (Lupu-Meiri *et al.* 1988; Berridge 1990), and thus in either fertilization or hormonal stimulation, propagation of the wave is essential for inducing the entire cell to respond to a localized stimulus. In hepatocytes the oscillations

Experimental and Theorietcal Advances in Biological Pattern Formation,
Edited by H.G. Othmer *et al.,* Plenum Press, New York, 1993

appear to originate from a single locus and propagate across the cell. Moreover, the initiation site seems to be relatively constant in a given cell, even for different agonists. Thus when hepatocytes are treated with phenylephrine, followed by washout and restimulation with vasopressin, Ca^{+2} waves originate from the same site. The speed of the waves in hepatocytes is typically 20-25 μ/sec. These cells are polarized to a degree and a variety of receptors are known to be concentrated at the sinusoidal membrane, which may account for the origin of the wave. Finally, spatial variations in calcium are important at the supracellular level as well. For instance, calcium waves may be used to synchronize large cell assemblies such as ciliated epithelial cells (Meyer 1991), and diffusion of inositol 1,4,5–trisphosphate ($InsP_3$) through gap junctions appears to be necessary for the propagation of these waves (Boitano *et al.* 1992).

In cardiac and smooth muscle cells, calcium is sequestered primarily in the sarcoplasmic reticulum, whereas in cells that are not electrically excitable it is stored in the endoplasmic reticulum. Calcium in the sarcoplasmic reticulum is bound to a high-affinity protein called sequestrin, which is thought to reside in organelles called calciomes. Wong *et al.* (1992) have developed a model for calcium-induced calcium release (CICR) from the SR. In non-excitable cells at least part of the sequestered calcium can usually be released by binding of $InsP_3$ to a receptor that controls the permeability of a calcium channel in the ER membrane. This channel is a tetramer which can open to four distinct conductance levels that are multiples of a unit step of 20 pS (Watras *et al.* 1991). $InsP_3$ binding is obligatory for opening of the channel (Harootunian *et al.* 1991), as is calcium (Watras *et al.* 1991). Low calcium levels promote opening of the channels, while high levels inhibit opening (Finch *et al.* 1991; Bezprozvanny *et al.* 1991).

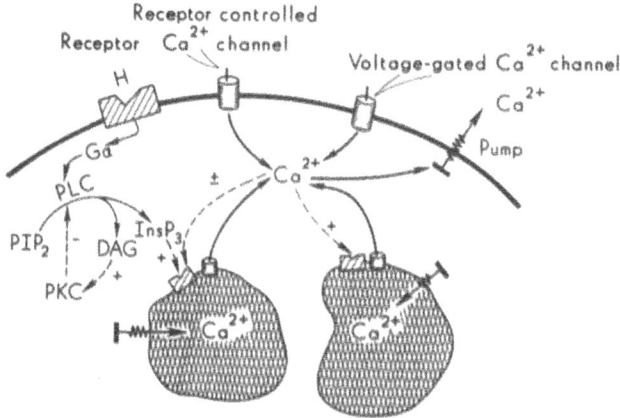

Figure 25.1. A schematic of the transduction system for hormonal stimuli, the calcium transport mechanisms, and the $InsP_3$-sensitive and $InsP_3$-insensitive stores. Key: G_α: The G protein that activates phospholipase C (PLC); PIP_2: phosphatidylinositol 4,5-bisphosphate; $InsP_3$: inositol 1,4,5-trisphosphate; DAG: diacylglycerol; PKC: protein kinase C. Solid lines indicate a material flow, and dashed lines indicate a control influence.

A typical sequence for transduction of an agonist signal into a variation in intracellular calcium is as follows (*cf.* Figure 25.1). The agonist binds to a plasma membrane receptor and the complex catalyzes the GDP/GTP exchange in a G protein. The G_α component in turn activates phospholipase C, the enzyme that catalyzes the hydrolysis of phosphatidylinositol 4,5-bisphosphate to $InsP_3$ and diacylglycerol. $InsP_3$ then binds to an $InsP_3$ receptor that is part of the calcium channel complex on the ER membrane, and if calcium is also bound at the calcium-activating site the channel opens. Frequently, sequestered calcium in cells that are not electrically-excitable can also be released, either from the $InsP_3$-sensitive store or from a distinct store, through another type of calcium channel that is $InsP_3$-insensitive. For instance, Thomas *et al.* (1991) cite evidence that in hepatocytes only 30-50% of the non-mitochondrial calcium is released from the $InsP_3$-sensitive pool, even when the calcium pump is inhibited; the remainder is released by calcium ionophores. In Figure 25.1 we indicate that these stores are distinct, but that is not known in general.

Under prolonged hormonal stimulation, or when $InsP_3$ is injected directly into the cell, calcium spiking occurs only in a range of hormonal or $InsP_3$ concentrations (DeLisle *et al.* 1990). Rooney *et al.* (1989) found that there was a correlation between the latent period and the subsequent oscillation period over a range of agonist doses and oscillation frequencies. In hepatocytes the shape of the oscillations depends on the agonist. The rate of rise of the calcium transient is more-or-less independent of the agonist, but the declining phase is highly dependent on the agonist. For example, phenylephrine-induced transients decay much more rapidly than do the transients induced by vasopressin. Thomas *et al.* (1991) cite evidence that protein kinase C may be involved in the re-uptake.

Extracellular calcium is needed in hepatocytes to sustain the calcium oscillations: in its absence the frequency of oscillation is reduced and the oscillations eventually run down. However in larger cells such as oocytes, the major features of the observed dynamics are not dependent on the presence of extracellular calcium or calcium-induced calcium release from intracellular stores. Consequently, in this paper we focus on a model in which there is no calcium transport between the interior and the exterior of the cell, and no calcium-induced calcium release; the oscillations center on the dynamics of the $InsP_3$ receptor. Our objective is to develop a model that is complex enough to reproduce the major aspects of the experimental observations, yet which is simple enough to be analyzed qualitatively in order to understand how the component processes affect the overall dynamics. The major assumptions of the model are that calcium binds to the activating site on the channel only after $InsP_3$ has bound to the receptor, and that the binding of calcium to the inhibitory site occurs only after calcium is bound to the activating site. This assumption of sequential binding leads to a model with only a few states that lends itself to qualitative analysis of the effect of various parameters. As we shall see later, the oscillations arise from the biphasic response of the $InsP_3$-sensitive calcium channel to calcium.

An outline of the remainder of the paper is as follows. In the following section we briefly review some of the previous models for calcium dynamics. In Section 25.3 we introduce a simple two-state receptor model and show analytically that under very general conditions this scheme cannot produce oscillations. We then introduce a four-state model, and in Section 25.4 we show that this model correctly reproduces the experimentally-observed bell-shaped response of the activated channel to calcium. We also show how excitability arises in this scheme, we demonstrate that it produces oscillations over a range of $InsP_3$ levels, and we show that these oscillations show frequency coding as a function of the $InsP_3$ level. Finally, in Section 25.5 we illustrate some of the wave phenomena that are possible in this model.

25.2   AN OVERVIEW OF MODELS FOR CALCIUM DYNAMICS

Calcium oscillations will in general involve many intracellular components, but in order to understand which components are essential and which are not, it is helpful to identify major subsystems and determine how they are involved in the overall process. One can identify three major subsystems in Figure 25.1, (i) the transduction subsystem whereby hormonal signals are transduced via G-proteins into an InsP$_3$ signal, (ii) a system comprising the InsP$_3$-sensitive calcium store, and (iii), a system comprising an InsP$_3$-insensitive calcium store, which may include voltage- and receptor-gated membrane channels. Of course not all cells will necessarily have all of these components.

Given this subdivision, one can identify a number of distinct ways in which calcium oscillations can be generated. Firstly, the transduction system may generate an oscillatory InsP$_3$ signal in response to a constant hormonal stimulus. This could arise from a feedback loop in which DAG activates protein kinase C, which in turn phosphorylates the G-protein that couples the receptor to phospholipase C, thereby leading to a reduction in the production of InsP$_3$ and DAG, which would reduce the inhibition of transduction, and so on. We call this a type I oscillator, and the model proposed by Cobbold *et al.* (1991) is of this type. This type produces an oscillatory input into the InsP$_3$-sensitive calcium subsystem, and in such systems the InsP$_3$ oscillations drive the calcium oscillations.

In a second class of models the oscillations are generated within the InsP$_3$-sensitive calcium store subsystem. As we remarked earlier, it is known that in a number of systems the InsP$_3$-sensitive calcium channel is not activated unless calcium occupies an activating site on the receptor. Furthermore, the channel conductivity is reduced at high calcium concentrations, probably due to the presence of a calcium-binding inhibitory site on the receptor. The model proposed by De Young & Keizer (1992); Keizer & De Young (1992) and the model we propose later incorporate these observations. In both these models InsP$_3$ is treated as a parameter; oscillations in the cytosol calcium arise without any temporal variation in the level of InsP$_3$. As we show later, such a system is excitable in the usual sense that there is a threshold level of the InsP$_3$ stimulus below which there is no significant response, but above which there is a large response. As frequently happens in such systems, tuning the system parameters slightly produces oscillatory behavior. In our model InsP$_3$ is both the stimulus that produces an excitable response, and the parameter which produces oscillatory behavior in a suitable range. This is in accord with the experimental observations, in that injection of InsP$_3$ can produce either single calcium transients or calcium oscillations, depending on the stimulus level (Dupont *et al.* 1991).

In a third class of models the oscillations are confined within the calcium subsystem. Since these do not involve transduction of hormonal stimuli, except as they involve modulation of calcium channels in the plasma membrane, they will not be treated here.

In addition to the three pure types described above, there are a variety of mixed types, in which there are feedforward or feedback interactions between two or more of the subsystems described above. For example, the model of Meyer & Stryer (1988) involves cooperative release of calcium from the InsP$_3$-sensitive store, coupled with Ca^{+2} activation of PLC. Thus the transduction subsystem and the InsP$_3$-sensitive store are mutually coupled, and oscillations arise from the feedback interaction between these two subsystems. A model due to Berridge and co-workers (Goldbeter *et al.* 1990), which is conceptually similar to a model proposed by Kuba & Takeshita. (1981), involves both an InsP$_3$-sensitive calcium store and an InsP$_3$-insensitive store. In this model InsP$_3$ is also a parameter, not a dynamic variable. A more complicated model, in which the transduction subsystem is linked to calcium dynamics has been proposed by Cuthbertson & Chay (1991).

25.3 THE MATHEMATICAL MODEL

25.3.1 Two states for the IP3 receptor do not suffice

Our interest is in systems in which a steady calcium input into the cytoplasm, either from the extracellular space or an internal pool, is not essential for generating the oscillations. In the absence of such inputs the total amount of intracellular calcium is conserved if there is no transport out of the cell, and it follows that a two-variable model in which the only dynamic variables are the concentrations of calcium in the cytosol and in the active storage compartment cannot give rise to oscillations, whether or not InsP$_3$ plays a role in the release of calcium. This applies, in particular, to the model of Goldbeter *et al.* (1990), and to that of Somogyi & Stucki (1991). It is conceivable however, that a model in which the InsP$_3$ receptor has two distinct states may give rise to oscillation. However we show in this section that oscillations cannot arise in response to hormonal stimulation in such a model under very mild restrictions on the functional form of the receptor response to InsP$_3$ and calcium. This leads us to a more complex four state model that we describe later in this section and analyze in the following section.

Suppose that the receptor can exist in one of two states, and that transitions between these states occur according to the first-order scheme

$$R \underset{k_{-1}}{\overset{k_1}{\rightleftharpoons}} R^*. \tag{25.1}$$

To incorporate the InsP$_3$ binding to the bare receptor R, as well as the activating and inhibitory effects of calcium, we suppose that

$$k_1 = k_{10} f_0(I, C), \tag{25.2}$$

where I and C are the concentrations of InsP$_3$ and cytosol calcium, respectively. The function f_0 should be monotone increasing in I, monotone increasing in C at low C, and monotone decreasing in C at high C. In addition to the calcium flux through the InsP$_3$-sensitive channel, we suppose that there is a basal calcium leakage between the store and the cytosol, and a pump from the cytosol into the store (*cf.* Figure 25.1). Finally, we suppose that the conductivity of the InsP$_3$-sensitive channel is an increasing function of the fraction of receptors in the state R^*. Let z denote the fraction of receptors in the state R^*; then the governing equations for C and z can be written

$$\frac{dC}{dt} = (1 + v_r)(\gamma_0 + \gamma_1 f(z))(C_0 - C) - \bar{g}(C)$$

$$\frac{dz}{dt} = k_{10} f_0(I, C)(1 - z) - k_{-1} z. \tag{25.3}$$

Here γ_0 is the basal permeability of the ER (which includes the channel conductance in the absence of InsP$_3$), and γ_1 is the density of InsP$_3$-sensitive channels per unit volume of ER. The function $f(z)$ represents the dependence of the channel conductivity on the fraction of channels in the conducting state RIC^+, and we assume that $f' \geq 0$. C_0 is the volume-average intracellular calcium concentration, which is defined by the relation

$$C_0 = \frac{C + v_r C_s}{1 + v_r},$$

where v_r is the ratio of the ER volume to the cytosol volume and C_s is the calcium concentration in the store. The function $\bar{g}$ represents the active pump, and thus $\bar{g}(C) \geq 0$ and $\bar{g}'(C) \geq 0$. It is necessary that $C \geq 0$, and that $z \in [0, 1]$, and it is easy to see that the set $\Omega \equiv \{(C, z) \mid C \geq 0$ and $z \in [0, 1]\}$ is invariant under the flow of (25.3).

The stable steady state of the system at low values of InsP$_3$ should correspond to low values of cytosol calcium, and we wish to determine whether this can be destabilized and lead to oscillations at a critical value of InsP$_3$. Thus consider the Jacobian of (25.3) at a steady state (C^*, z^*), which is the matrix

$$K = \begin{bmatrix} -(1 + v_r)(\gamma_0 + \gamma_1 f(z^*)) - \bar{g}'(C^*) & (1 + v_r)\gamma_1 f'(z^*)(C_0 - C^*) \\ k_{10}\dfrac{\partial f_0}{\partial C}(I, C^*)(1 - z^*) & -k_{10}f_0(I, C^*) - k_{-1} \end{bmatrix}.$$

In order to destabilize the system to growing oscillations (*i. e.* in order to produce a Hopf bifurcation) as I is increased, the sum of the diagonal terms must vanish, whereas the off-diagonal terms must have opposite signs. However, neither of these conditions can be met under the stipulated conditions on the functions f_0, f and g. Thus the steady state cannot be destabilized by increasing the hormone concentration, and hence the InsP$_3$ concentration, and we conclude that under the forgoing conditions, a single-pool, InsP$_3$-gated model with two receptor states cannot undergo a Hopf bifurcation. In fact one can prove that the system can never oscillate, but this will not be done here. Thus we must incorporate more receptor states, and we next introduce a scheme with four states that can produce oscillations. One can show that a three-state version of this model can also oscillate, but the dynamic behavior possible is limited.

25.3.2   A four-state single pool model

We suppose that the receptor has four states, and that the transitions between them occur according to the following scheme.

$$I + R \underset{k_{-1}}{\overset{k_1}{\rightleftharpoons}} RI$$

$$RI + C \underset{k_{-2}}{\overset{k_2}{\rightleftharpoons}} RIC^+ \qquad (25.4)$$

$$RIC^+ + C \underset{k_{-3}}{\overset{k_3}{\rightleftharpoons}} RIC^+C^-$$

As before, R denotes the bare receptor, I denotes InsP$_3$ and C denotes the cytosol calcium concentration. Further, RI denotes the receptor-InsP$_3$ complex, and RIC^+ (respectively, RIC^+C^-) denotes RI with calcium bound at the activating site (respectively, the activating and inhibitory sites). In addition, there is calcium diffusion between the calcium store and the cytoplasm, and a calcium pump between the cytosol and the calcium store. At present I is treated as a parameter.

Let $x_i, i = 2, \ldots 5$, denote the fractions in states R, RI, RIC^+ and RIC^+C^-, respectively. Then the governing equations are

$$\frac{dC}{dt} = (1 + v_r)(\gamma_0 + \gamma_1 f(x_4))(C_0 - C) - \bar{g}(C)$$

$$\frac{dx_2}{dt} = -k_1 I x_2 + k_{-1} x_3$$

$$\frac{dx_3}{dt} = -(k_{-1} + \bar{k}_2 C)x_3 + k_1 I x_2 + k_{-2} x_4 \tag{25.5}$$

$$\frac{dx_4}{dt} = -(k_{-2} + \bar{k}_3 C)x_4 + \bar{k}_2 C x_3 + k_{-3} x_5$$

$$\frac{dx_5}{dt} = \bar{k}_3 C x_4 - k_{-3} x_5. \tag{25.6}$$

We suppose that $f(x) = x$, *i. e.* that the channel conductivity is a linear function of the fraction of channels open. The calcium pump between the cytoplasm and the ER is known to be a tetramer with four calcium binding sites, and therefore we assume that

$$\bar{g}(C) = \frac{\bar{p}_1 C^4}{C^4 + \bar{p}_2^4}.$$

We define $x_1 = C/C_0$, and after eliminating x_5 by use of the conservation condition $\sum_{k=2}^{5} x_k = 1$, we can write the governing equations in the form

$$\frac{dx_1}{dt} = \lambda(\gamma_0 + \gamma_1 x_4)(1 - x_1) - \frac{p_1 x_1^4}{p_2^4 + x_1^4}$$

$$\frac{dx_2}{dt} = -k_1 I x_2 + k_{-1} x_3$$

$$\frac{dx_3}{dt} = -(k_{-1} + k_2 x_1)x_3 + k_1 I x_2 + k_{-2} x_4 \tag{25.7}$$

$$\frac{dx_4}{dt} = k_{-3} - k_{-3} x_2 + (k_2 x_1 - k_{-3})x_3 - (k_{-2} + k_3 x_1 + k_{-3})x_4,$$

where $\lambda \equiv 1 + v_r$, $k_2 = \bar{k}_2 C_0$, $k_3 = \bar{k}_3 C_0$, $p_1 = \bar{p}_1/C_0$, and $p_2 = \bar{p}_2/C_0$. In this scaling all x_i range between 0 and 1, and it is easy to see that the set $\{x_i \mid 0 \le x_i \le 1\}$ is invariant under the flow of (25.7). As we show in Section 25.4, this scheme is sufficiently robust to reproduce many of the observations.

25.3.3 Parameter values

In this section we give the values of parameters in the four-state model that are used in later simulations and give the rationale for their choice.

The ratio of the ER volume to the cytoplasmic volume v_r is taken to be 0.185, after Alberts *et al.* (1989). The leakage rate between the calcium store and cytoplasm is known to be small. We use the value $\gamma_0 = 0.1 s^{-1}$, but the value of this parameter does not significantly influence the dynamics as long as it is small.

The average calcium concentration is taken as $C_0 = 1.56 \mu M$. This leads to a maximal calcium concentration in the ER (with zero calcium concentration in the cytoplasm) of $10 \mu M$, which is in the physiological range (De Young & Keizer 1992).

Joseph *et al.* (1989) report that the dissociation constant for InsP$_3$ binding to microsomal ER fractions in the absence of calcium is approximately $0.15 \mu M$. We choose $k_1 = 12.0 (\mu M \cdot s)^{-1}$ and $k_{-1} = 8.0 s^{-1}$, which gives $K_1 = 0.15 \mu M$.

There are no experimental data on the kinetic constants of calcium binding to the channels. We choose the set of values of k_2, k_{-2}, k_3, and k_{-3} given in Table 25.1 so as to make activation a much faster process compared to the binding of calcium to the inhibitory site. This set of data also provides an adequate fit to the bell-shaped channel opening curve in response to calcium changes and the saturation curve for the channel opening in response to InsP$_3$ increases.

Table 25.1. Parameter values for the four-state model

Parameter	Value	Parameter	Value
v_r	0.185	k_1	$12.0 \, (\mu M \cdot s)^{-1}$
γ_0	$0.1 s^{-1}$	k_2	$15.0 \, (\mu M \cdot s)^{-1}$
γ_1	$20.5 s^{-1}$	k_3	$1.8 \, (\mu M \cdot s)^{-1}$
$\bar{p}_1$	$8.5 \, (\mu M \cdot s)^{-1}$	k_{-1}	$8.0 s^{-1}$
$\bar{p}_2$	$0.5 \mu M$	k_{-2}	$1.65 s^{-1}$
C_0	$1.56 \mu M$	k_{-3}	$0.21 s^{-1}$

γ_1 is a parameter that incorporates both the effect of calcium channel density and the channel conductance. Although the channel conductance is known for cerebellum cells (Watras *et al.* 1991), the channel concentration is not reported there. However, the value of γ_1 we use is consistent with that conductance and the channel densities reported for other cells. The calcium pumping parameters $\bar{p}_1$ and $\bar{p}_2$ are also unknown. We choose them, as well as the remaining kinetic constants, so as to qualitatively reproduce the following experimental results.

- In the absence of InsP$_3$, the equilibrium Ca^{2+} concentration in the cytoplasm should be of the order of $50 nM$ (*cf.* Figure 25.4).

- The K_d for InsP$_3$ binding to the channel increases to about $0.5 \mu M$ in the presence of $1 \mu M$ calcium (Joseph *et al.* 1989).

- The Ca^{2+} oscillations should occur in a reasonable range of InsP$_3$ concentration and the system should be excitable for InsP$_3$ below this range.

- When the InsP$_3$ concentration is in the range that produces oscillations, the system should exhibit frequency coding.

The parameter values are listed in Table 25.1. The corresponding parameters needed for the nondimensional equations can easily be calculated from these values.

25.4   ANALYSIS OF THE LOCAL DYNAMICS IN THE FOUR-STATE MODEL

25.4.1   Steady-state analysis

From the steady-state version of (25.7) we find that

$$
\begin{aligned}
x_3 &= (\frac{k_1 I}{k_{-1}}) x_2 & &\equiv K_1^{-1} x_2 \\
x_4 &= (\frac{k_2 x_1}{k_{-2}}) x_3 & &\equiv K_2^{-1} x_1 x_3 \\
x_2 &= \frac{K_1 K_2}{K_2(K_1 + 1) + x_1 + K_3 x_1^2}
\end{aligned}
$$

where $K_3 \equiv k_3/k_{-3}$. Therefore the fractions in the states RI and RIC^+ are given by

$$
x_3 = \frac{K_2}{K_2(K_1 + 1) + x_1 + K_3 x_1^2} \tag{25.8}
$$

$$x_4 = \frac{x_1}{K_2(K_1 + 1) + x_1 + K_3 x_1^2} \equiv X_4(x_1), \qquad (25.9)$$

The InsP$_3$ dependence of these fractions enters through K_1, and it follows that $x_i \to 0$, $i = 2, 3, 4$, as InsP$_3 \to 0$, as they should. Furthermore, $x_4 \to 0$ as $x_1 \to 0$.

Since the fractions in the various receptor states at steady state can be expressed in terms of the dimensionless calcium concentration, these fractions can be computed knowing the solution of the scalar equation

$$\lambda(\gamma_0 + \gamma_1 x_4)(1 - x_1) = \frac{p_1 x_1^4}{p_2^4 + x_1^4}, \qquad (25.10)$$

where x_4 is given by (25.9). Before analyzing this equation we consider the case in which calcium is clamped, as in the experiments by Bezprozvanny *et al.* (1991).

It follows from (25.9) that the fraction of channels open at a fixed calcium concentration increases with the calcium concentration at low concentrations and decreases at high concentrations. One finds that the maximum open fraction occurs at the dimensionless calcium concentration

$$x_1^* = \sqrt{\frac{K_2(K_1 + 1)}{K_3}},$$

and that the fraction open at this value is

$$x_4^* = \frac{1}{1 + \sqrt{K_2 K_3(K_1 + 1)}}.$$

Since $K_1 = k_{-1}/k_1 I$, the maximum increases as $\sqrt{I}$ for small I. The width of the graph of $X_4(x_1)$ at half the maximal open fraction is determined by the roots of the equation

$$\frac{1}{2}x_4^* = X_4(x_1),$$

from which one finds that the half-width is

$$\Delta x_1 \equiv x_1^+ - x_1^- = \frac{\sqrt{4 - 2x_4^* + (1 - 4K_3 K_2(K_1 + 1))(x_4^*)^2}}{K_3 x_4^*}$$

$$= \frac{\sqrt{3\left(1 + 2\sqrt{K_2 K_3(K_1 + 1)}\right)}}{K_3}.$$

Thus the half-width is very weakly dependent on the InsP$_3$ concentration. The graph of $X_4(x_1)$ is shown in Figure 25.2(a), and the experimentally-derived curve is shown in Figure 25.2(b). Certainly the model parameters could be tuned to reproduce the experimental results as well as desired, but because there is no complete set of data available for the system from which Figure 25.2 is obtained, we have not done this.

Now consider the steady state Equation (25.10), which reads

$$\lambda\left(\gamma_0 + \frac{\gamma_1 x_1}{K_3 x_1^2 + x_1 + K_2(K_1 + 1)}\right)(1 - x_1) = \frac{p_1 x_1^4}{p_2^4 + x_1^4}. \qquad (25.11)$$

The left-hand side of (25.11) has the value $\lambda \gamma_0$ at $x_1 = 0$, it vanishes at $x_1 = 1$, and has a unique minimum in $(0,1)$. On the other hand, the right-hand side vanishes at $x_1 = 0$, is monotone increasing in x_1, and attains its half-maximal value at $x_1 = K_4$. Depending on the

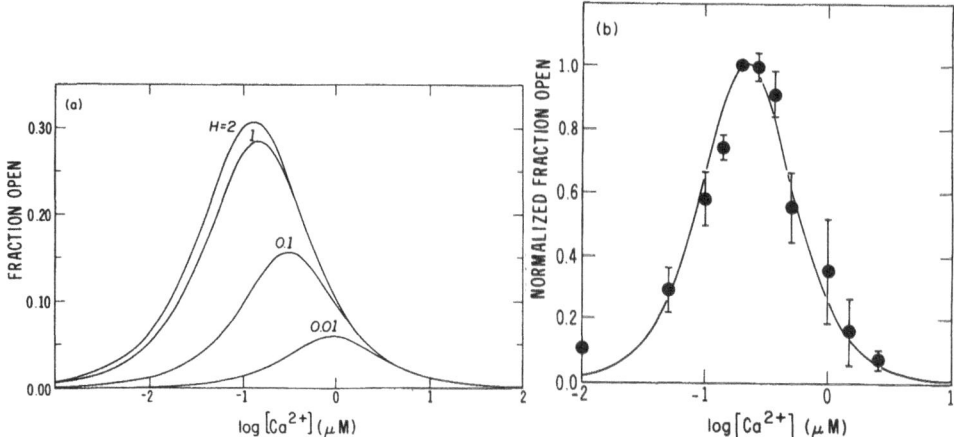

Figure 25.2. (a) The theoretically-predicted graph of the fraction of activated channels as a function of cytosol calcium. (b) The experimentally-measured curve for InsP$_3$ $= 2.0\mu M$ (from Bezprozvanny *et al.* 1991.)

choice of parameters, there may be one or three steady states at a fixed InsP$_3$ concentration. However, if K_4 is chosen sufficiently large there is a unique positive steady state for all values of I. It is easy to show that the steady-state calcium concentration is a monotone increasing function of I. It then follows from (25.9) that the fraction of channels open at the steady state is monotone increasing in I provided that the steady state level of calcium is such that $x_1 < x_1^*$ for all $I \in (0, \infty)$. A plot of the fraction open computed for the standard parameters is shown in Figure 25.3(a), and the experimentally-observed curve for cerebellar cells is shown in Figure 25.3(b). One can formally compute $\partial x_1 / \partial I$ from (25.9) and (25.10) to determine the asymptotic dependence of the fraction open on InsP$_3$, but the results are too complex to be useful in the range of interest. However, one finds numerically that the theoretical curve given in Figure 25.3(a) can be fit moderately well using a Hill function in InsP$_3$ with an exponent of 1.75. Experimental values of the Hill coefficient for InsP$_3$-induced calcium release range from 1.0-1.3 (Volpe *et al.* 1990) to greater than 3.7 (Delisle 1991).

25.4.2   Excitability and Oscillations

As we observed earlier, the experimentally-observed dynamics of the InsP$_3$-triggered calcium release have the hallmarks of a classical excitable system as the phrase is used, for example, in the context of nerve conduction equations (Alexander *et al.* 1990; Othmer 1991). To understand the origin of excitability in this system it is helpful to represent the kinetic mechanism schematically as follows.

$$R \underset{8}{\overset{12}{\rightleftharpoons}} RI \underset{1.65}{\overset{15}{\rightleftharpoons}} RIC^+ \underset{0.21}{\overset{1.8}{\rightleftharpoons}} RIC^+C^- \tag{25.12}$$

The first forward reaction involves binding of InsP$_3$, while the last two forward reactions involve binding of calcium. The numerical values of the rate constants are indicated, and from these one sees that binding of calcium to the activating site is an order of magnitude faster than binding of calcium to the inhibitory site at all concentrations. Similarly, InsP$_3$ binding is fast as long as InsP$_3$ is $\geq \mathcal{O}(0.1)$.

There are two aspects of excitability that are of interest here: (i) the response to pulses of InsP$_3$ and (ii), the response to a pulse of calcium in a system with fixed InsP$_3$. We

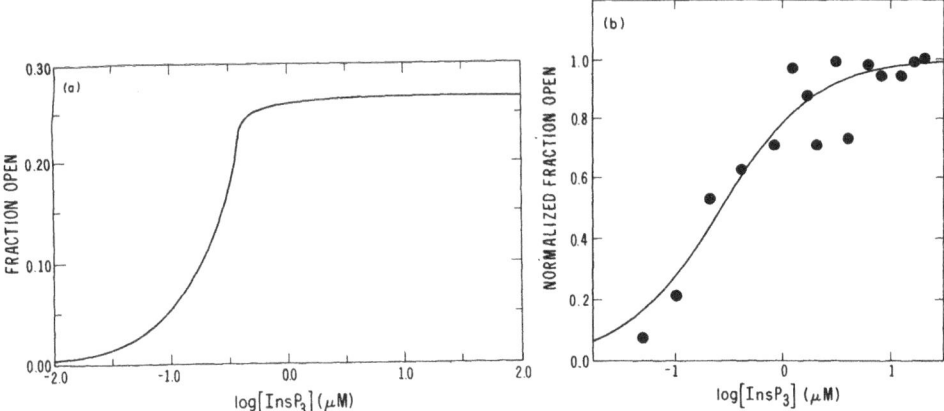

Figure 25.3. (a) The theoretically-predicted fraction of channels open as a function of the InsP$_3$ concentration. In this figure the calcium is not clamped, but rather, is obtained from the solution of the full set of steady state equations. (b) The experimentally-measured curve for cerebellar cells. 100% corresponds to 14% open. (From Watras *et al.* 1991)

consider the former case first. In the absence of InsP$_3$, $x_2 = 1$, $x_i = 0, i = 3, 4, 5$, and the calcium concentration is low. If a step change in InsP$_3$ is made, InsP$_3$ binds rapidly with R, and calcium rapidly binds with the RI complex at the activating site. Since RIC^+ opens the calcium channels, the cytosol calcium increases, which in turn produces more activated receptor. Thus the initial rapid response only involves $x_1, \cdots x_4$, and one of the receptor fractions can be eliminated by the conservation condition. More precisely, if one rescales the full system (25.5) and makes use of the difference in kinetic coefficients in (25.12) to eliminate x_5, and if one uses the conservation condition to eliminate x_2, then one arrives at the system

$$\frac{dx_1}{dt} = \lambda(\gamma_0 + \gamma_1 x_4)(1 - x_1) - \frac{p_1 x_1^4}{p_2^4 + x_1^4} \tag{25.13}$$

$$\frac{dx_3}{dt} = -(k_{-1} + k_2 x_1 + k_1 I)x_3 + (k_{-2} + k_1 I)x_4 + k_1 I$$

$$\frac{dx_4}{dt} = k_2 x_1 x_3 - (k_{-2} + k_3 x_1)x_4.$$

On the surface $\dot{x}_3 = 0$ one has

$$x_3 = \frac{k_1 I + (k_{-2} - k_1 I)x_4}{k_{-1} + k_2 x_1 + k_1 I}. \tag{25.14}$$

Therefore, on the intersection of $\dot{x}_4 = 0$ with this surface one has that

$$x_4 = \frac{k_1 k_2 x_1 I}{k_{-1} k_{-2} + k_3 x_1(k_2 + k_{-1}) + k_1 I(k_{-2} + k_3 + k_2 x_1)} \tag{25.15}$$

Similarly, on $\dot{x}_1 = 0$ one has

$$x_4 = \left(\frac{g(x_1)}{\lambda(1 - x_1)} - \gamma_0\right) / \gamma_1. \tag{25.16}$$

In Figure 25.4(a) we show the intersections of the null surfaces $\dot{x}_1 = 0$ and $\dot{x}_4 = 0$ with the null surface $\dot{x}_3 = 0$, projected into the $x_1 - x_4$ plane. One sees from (25.15) that on

$\dot{x}_4 = 0$, x_4 is an increasing function of x_1, while on $\dot{x}_1 = 0$, x_4 is independent of x_1. It is easy to see that depending on the value of I, there may be one or three steady states in the fast dynamics. For instance, when $I = 0.36$ there are three, and the significance of this value will become clear later. In Figure 25.4(a) we also show the x_1 and x_4 components of the full system (25.5), starting at the steady state for $I = 0$, and applying a square-wave stimulus of amplitude $I = 0.36$ and duration either 3 or 4 seconds. One sees there that a 3-second stimulus is subthreshold, while a 4-second stimulus is superthreshold. It is clear in that figure that the unstable manifold of the intermediate steady state of the fast system, which is a saddle point, serves as a threshold, and stimuli which carry the state above this manifold produce a significant amplification of cytosol calcium, while stimuli that leave the state below this manifold can be termed subthreshold. The time course of the calcium component of the response for the foregoing stimuli is shown in Figure 25.4(b), and these illustrate that this system is excitable in the sense used in Othmer (1991).

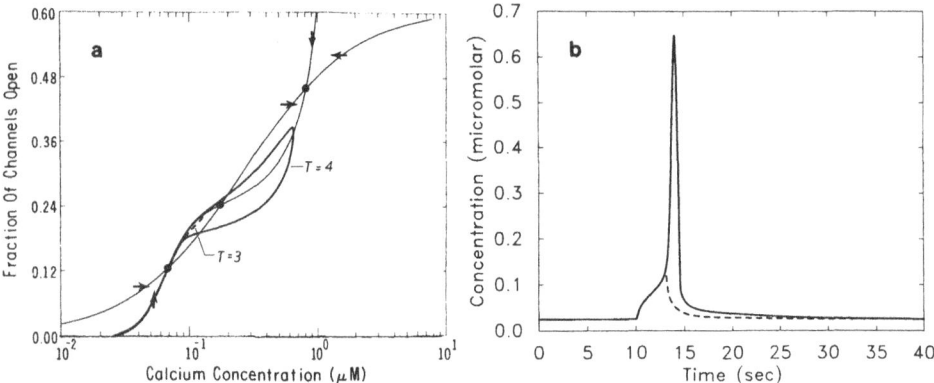

Figure 25.4. (a) The intersection of the null surfaces $\dot{x}_1 = 0$ and $\dot{x}_4 = 0$ with the null surface $\dot{x}_3 = 0$, projected into the $x_1 - x_4$ plane. The dashed curve represents the response of a system at steady state with $I = 0$ to a square-wave impulse in InsP$_3$ of amplitude $I = 0.36$ and of 3 seconds duration, and the heavy solid curve is the response to a 4 second stimulus of the same duration. (b) The time course of the calcium component for these stimuli.

The response for an InsP$_3$ stimulus of fixed duration and varying amplitude is shown in Figure 25.5. This figure also shows that the latency in the response is inversely-related to the amplitude of the stimulus, as is observed experimentally. It is clear that one could compute the threshold stimulation for a variety of amplitude-duration pairs and thereby construct an amplitude-duration curve for this excitable system, but this is not done here. A similar analysis can be done to understand excitability with respect to calcium perturbations at fixed InsP$_3$. In this case we freeze x_5 at its steady-state value for the given value of InsP$_3$ and analyze the fast dynamics as before. In Figure 25.6(a) we show the time course of calcium and in Figure 25.6(b) the time course of the fraction of open channels for two initial perturbations in calcium. It is clear from these figures that a sufficiently large pulse triggers an excitable response, and in view of this, one expects that these dynamics can generate propagating calcium waves when InsP$_3$ is spatially-uniform and a spatially-localized perturbation of calcium is introduced. It should be noted in Figure 25.6(a) that the initial pulse triggers a secondary response, which is what is observed experimentally when REF2 cells that are treated with vasopressin are exposed to a calcium pulse (Harootunian *et al.* 1991). In the experimental context the secondary response is broad (*cf.* Figure 2(b) in Harootunian *et al.* (1991)), but this is the result of cell-to-cell differences in the population.

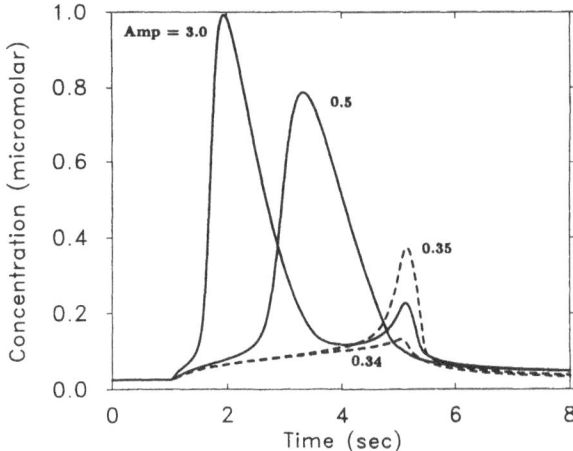

Figure 25.5. The response of a system at steady state to a square-wave stimulus in InsP$_3$ of 4 seconds duration and of the amplitude indicated on the figure. Note that a sufficiently large stimulus generates a secondary response that coincides time-wise with the response generated to a stimulus of smaller amplitude.

As I is increased, one finds that oscillations set in at $I \sim 0.366$ and persist until $I \sim 2.57$. We show the amplitude of the calcium component of the periodic solution in this range in Figure 25.7. It should be noted that the maximum amplitude of the oscillations decreases somewhat over the oscillatory range of InsP$_3$, but the period changes by an order of magnitude. Thus the system exhibits the experimentally-observed frequency coding. In Figure 25.8 we show the calcium concentration and the conducting channels at $I = 0.4$, which is in the oscillatory regime. Note that there are two phases in the early part of each cycle, a phase in which calcium rises slowly and the fraction of channels open begins to rise, followed by a phase in which the channels open very rapidly and calcium rises rapidly.

In the preceding figures the concentration of InsP$_3$ has been held fixed. However, under normal circumstances InsP$_3$ is degraded, and thus the foregoing results only strictly apply when a non-hydrolyzable analog of InsP$_3$ is used. However, it is clear that if InsP$_3$ is degraded, and if the initial concentration is larger than the upper limit of the oscillatory range of InsP$_3$, then as InsP$_3$ decreases the system sweeps through the entire range of oscillatory dynamics. The details of the dynamics will of course depend on the rate at which InsP$_3$ is degraded. By controlling the InsP$_3$ level experimentally one can control the passage through various dynamical regimes. This is illustrated in Figure 25.9(a), where we show the calcium concentration as a function of time when the time course of InsP$_3$ is as given in the figure caption. The experimentally-observed behavior for endothelial cells under the same stimulus protocol (albeit on a different time scale) is shown in Figure 25.8(b), which is reproduced from R. Jacob, J. E. Merritt & Rink (1988).

25.5 WAVES

One of the most striking aspects of calcium dynamics in oocytes is the wide variety of wave patterns that are observed (Lechleiter *et al.* 1991; Lechleiter & Clapham 1992). A model based on calcium-induced calcium release that can qualitatively reproduce these observations is proposed in Girard *et al.* (1992). This model uses the model of Goldbeter *et al.* (1990)

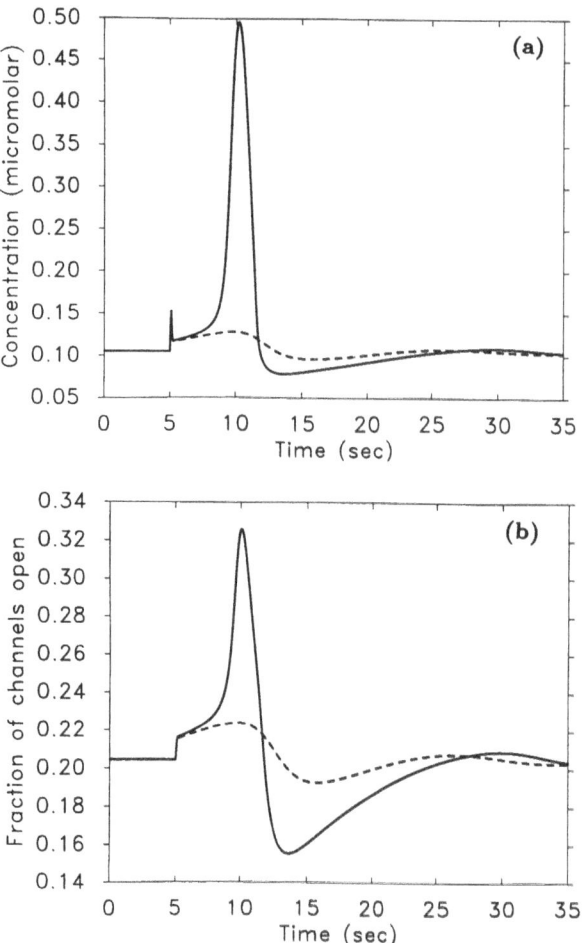

Figure 25.6. The time course of calcium (a) and the fraction of open channels (b). The system is in a steady state for fixed $I = 0.36$ when $T \in (0, 5)$, and a square wave of calcium of amplitude 1.2 (solid lines) or amplitude 1.15 (dashed lines) is imposed for 0.1 second at $T = 5$.

for the local dynamics. However, as we noted earlier, calcium transport across the plasma membrane plays an essential role in this model, yet the major features of the calcium dynamics in oocytes do not depend on this, at least on a short time scale. Furthermore, it is known that there are no ryanodine receptors, and therefore there is no InsP₃-insensitive calcium pool, in *Xenopus* oocytes (D. Clapham, personal communication). In hamster eggs the InsP₃-sensitive pool plays an essential role in the calcium waves, for when the eggs are treated with antibodies to the InsP₃ receptor, the waves are blocked (Miyazaki *et al.* 1992).

In this section we briefly describe some of the wave patterns that are predicted by the InsP₃-controlled model developed here. We include diffusion of both calcium and InsP₃, and therefore the governing equations in a spatially distributed system are as follows.

$$\frac{\partial x_1}{\partial t} = D_{CA}\Delta x_1 + \lambda(\gamma_0 + \gamma_1 x_4)(1 - x_1) - \frac{p_1 x_1^4}{p_2^4 + x_1^4}$$

$$\frac{\partial y}{\partial t} = D_I \Delta y + H - y$$

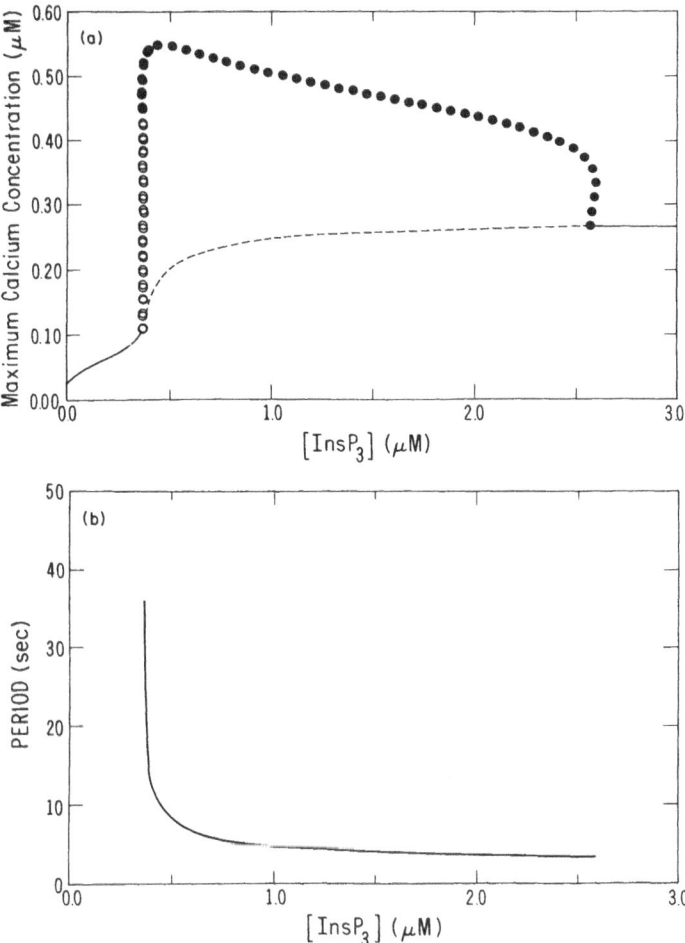

Figure 25.7. (a) The amplitude at steady state (solid and dashed lines) and the maximum amplitude of a periodic solution (open and filled circles) of calcium as a function of the InsP$_3$ concentration. Solid lines and circles indicate stable solutions; dashed lines and open circles indicate unstable solutions. (b) The period in seconds of the periodic solutions in (a). These results were obtained using the software package AUTO (Doedel 1986).

$$\frac{dx_2}{dt} = -k_1 I x_2 + k_{-1} x_3 \tag{25.17}$$

$$\frac{dx_3}{dt} = -(k_{-1} + k_2 x_1)x_3 + k_1 I x_2 + k_{-2} x_4$$

$$\frac{dx_4}{dt} = k_{-3} - k_{-3} x_2 + (k_2 x_1 - k_{-3})x_3 - (k_{-2} + k_3 x_1 + k_{-3})x_4.$$

Here y is the dimensionless concentration of InsP$_3$, Δ denotes the Laplace operator, D_{CA} and D_I are the diffusion coefficients of calcium and InsP$_3$, respectively, and H is the dimensionless hormone concentration. Note that we have augmented the previous equations (25.7) to incorporate source terms of InsP$_3$ and first-order decay. In a spatially-uniform system InsP$_3$ relaxes exponentially to the value H. We can thereby mimic experiments in which there

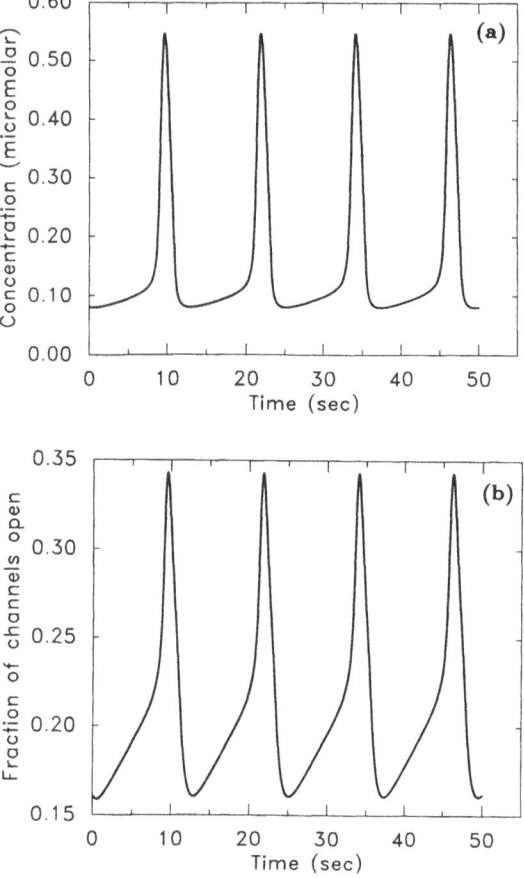

Figure 25.8. The time course of the calcium concentration (a) and the fraction of conducting channels (b) for the periodic solution at $I = 0.4$.

are spatially-distributed sources of InsP$_3$ by specifying the spatial variation of H. We shall describe the solutions of these equations for one- and two-dimensional domains, and for a variety of source distributions. In all cases we impose homogeneous Neumann (*i. e.* zero-flux) boundary conditions on the boundary of the domain. The numerical procedure used is based on a fully-implicit time-stepping algorithm and finite difference approximations to the spatial derivatives. Unless otherwise stated, we use the value $5. \times 10^{-4}\ mm^2/sec$ for both D_{CA} and D_I, and the spatial extent of all systems is $1\ mm$.

We first solve these equations in a one-dimensional domain. In Figure 25.10 we show the contour lines for waves that originate at a pacemaker located at the center of the domain (by a pacemaker region we mean that if InsP$_3$ in the equations for a spatially-uniform system is fixed at the specified H, then the dynamics in that system would be oscillatory). In Figure 25.10(a) the period of the pacemaker is approximately 12.2 seconds, and 1:1 locking with a period of 14.4 seconds results. The speed of the resulting waves is approximately 16.5 μ/sec, which is in the experimentally-observed range for various systems. However, much more complex patterns result when the pacemaker region has a shorter period, particularly when several pacemakers interact. In Figure 25.10(b) we set

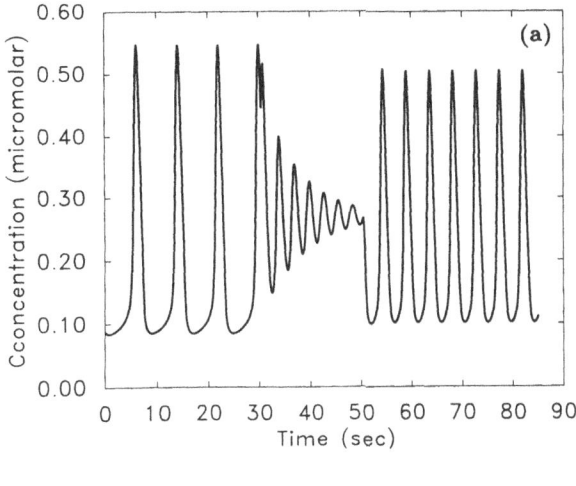

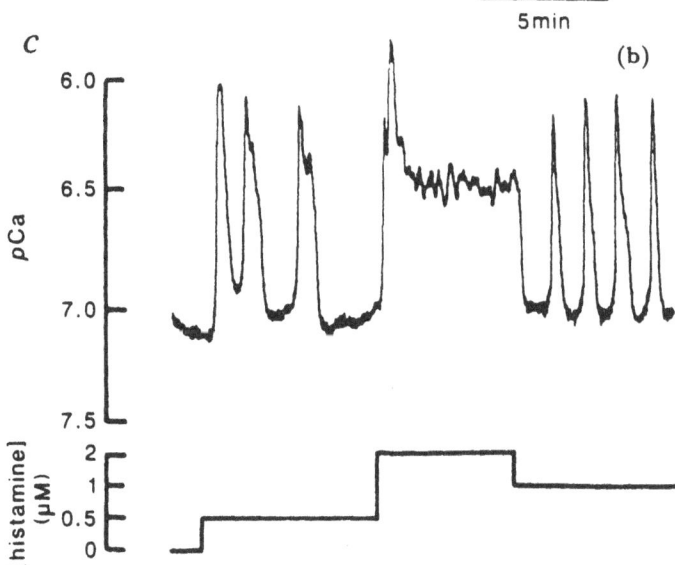

Figure 25.9. (a) The theoretically-predicted calcium concentration, and (b) the experimentally-observed results for endothelial cells. In (a) the InsP$_3$ concentration is held at 0.5 for $T \in (3, 30)$, then is increased to InsP$_3$ = 5. for 20 seconds, and held at InsP$_3$ = 1 thereafter.

$H = 0.5$ near $x = 0.25$, which gives rise to a period of approximately 8 seconds, and $H = 1.0$ near $x = 0.5$, which produces a period of approximately 4.6 seconds. Note that every other wave that emanates from $x = 0.25$ is blocked, whereas only 2 of every 5 waves that emanate from $x = 0.5$ propagate. What is noteworthy here is that the faster pacemaker does not entrain the slower one; they coexist stably for as long as we have continued the computations.

A similar phenomenon exists in higher dimensions. In Figure 25.11 we show two snapshots of the contour pattern generated by three interacting pacemakers. In this figure $H = 0.4$ within a disk of radius 0.0707 centered at $(0.5, 0.75)$, $H = 1.0$ within a disk of

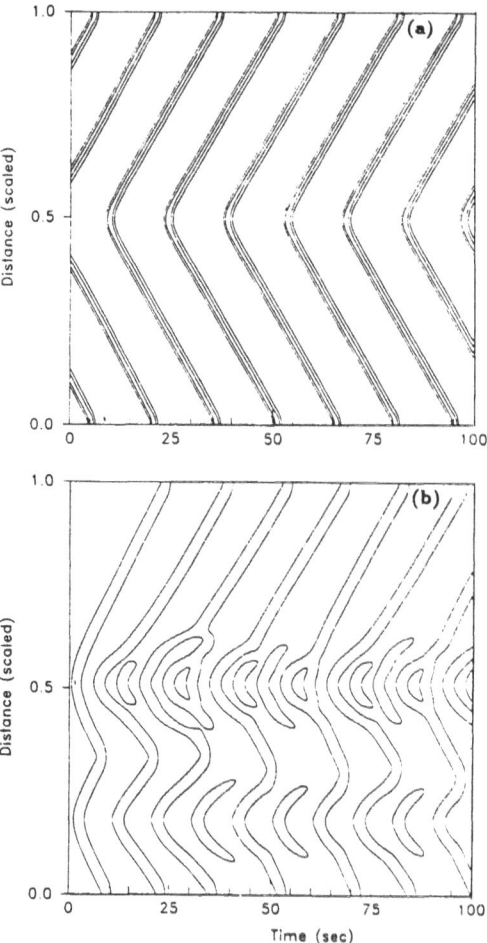

Figure 25.10. (a) The contour lines for waves generated by a pacemaker at the center of the domain ($H = 0.4$ for $x \in (.475, .525)$). (b) The wave patterns when two pacemakers interact. In both (a) and (b) $H = 0.36$ outside the pacemaker regions.

radius 0.0707 centered at $(0.25, 0.25)$, and $H = 1.5$ within a disk of radius 0.0707 centered at $(0.5, 0.25)$ (the last of these produces a period of approximately 3.95 seconds). Again the conclusion is similar to that in one space dimension; pacemakers of widely-disparate periods can coexist stably, apparently indefinitely. In the two-dimensional system we have not established that the overall pattern is in fact periodic.

Other types of waves are also possible. In Figure 25.12 we show a spiral wave that exists when $H = 0.36$ throughout the medium. By examining a sequence of time snapshots of the wave, one finds that the period is approximately 14.9 seconds and the wavelength is 260 μ, which yields a speed of 17.45 μ/sec. Just as multiple pacemakers can coexist stably, so too can a spiral and a pacemaker. In Figure 25.13 we show two time snapshots that illustrate this coexistence. In this figure $H = 0.5$ in a disk of radius 0.0707 centered at $(0.5, 0.5)$, and $H = 0.35$ elsewhere. The coexistence of these waves is more sensitive to parameter values than the multiple pacemakers. If one slightly alters the combination of H values either the pacemaker or the spiral will dominate the asymptotic behavior, depending on the alteration.

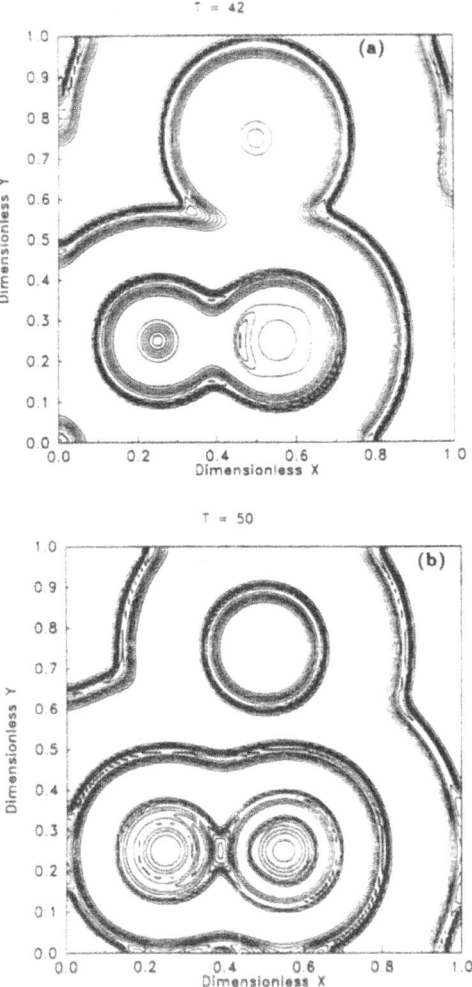

Figure 25.11. The contour pattern of waves produced by three pacemaker regions. $H = 0.36$ outside the pacemaker regions.

However, such solutions are more than a passing curiosity, for they are seen experimentally in oocytes.

25.6 DISCUSSION

As we noted in Section 25.2, there are a number of distinct classes of models for calcium dynamics. Models based on calcium-induced calcium release developed by Berridge and co-workers can reproduce many aspects of the local dynamics, and it has recently been shown that they generate traveling waves (Girard *et al.* 1992; Dupont *et al.* 1991). Similarly, models that involve calcium-activated PLC can also reproduce the observed behavior in some systems. Although it is known that some systems have several types of calcium stores (Brundage *et al.* 1991; Bazotte *et al.* 1991), it is also known that the InsP$_3$-sensitive

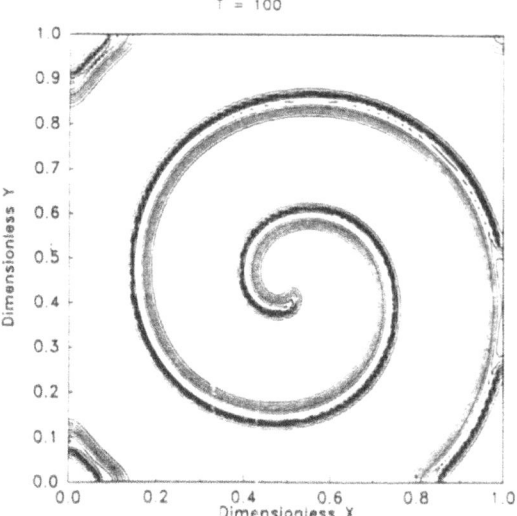

Figure 25.12. A spiral wave propagating through a medium in which $H = 0.36$ everywhere.

store plays an essential role in some systems, and that some systems do not have a calcium-sensitive store (Wakui *et al.* 1989). Within the context of models for a a single InsP$_3$-sensitive pool, one can consider schemes in which InsP$_3$ oscillates and those in which it does not, but oscillation of InsP$_3$ is not essential, as shown by experiments using the non-metabolizable InsP$_3$ analogue InsP$_3$S$_3$ (Wakui *et al.* 1989). This leads to a model based on one calcium pool with the InsP$_3$ signal as a parameter which controls the various types of response.

The bell-shaped channel opening in response to the calcium levels, which emerges from the four-state model developed here, has been observed in cerebellum cells, using reconstituted membrane containing the channels (Bezprozvanny *et al.* 1991). Moreover, activation of channels at low calcium concentrations and inhibition at high concentrations are seen in a variety of cell types (Zhao *et al.* 1990; Wakui & Petersen 1990; Parker & Ivorra 1990; Meyer & Stryer n.d.). As we have seen, single-pool models based on an InsP$_3$-controlled channel must incorporate more than two states for the InsP$_3$ receptor in order to generate oscillations. This is true even when the activating and inhibitory effects of calcium on the channel dynamics are incorporated. The reason is that the bell-shaped dependence of openings on calcium reflects different time scales, that of the fast activation and that of the slow inhibition, that must be incorporated in the dynamics in order to generate oscillations. The two-state model is too severe a reduction to reflect these scales.

A multi-state model for the InsP$_3$ receptor has been developed by DeYoung and Keizer (1992; 1992). Our model differs from theirs in the following respects: (i) The 4-state model can be studied analytically to understand the origin of the excitable and oscillatory behavior, whereas this is not possible in the 8-state model they propose. (ii) In the DeYoung-Keizer model a channel is assumed to be open only if three subunits in the channel are in the state RIC^+, whereas we assume that a channel can conduct with only one subunit in this state. This implies that the channel in our model can open to 4 different levels, while in theirs it can only open to one level. A major advantage of our four-state model is that we can obtain the bell-shaped distribution directly, and thus show that it arises naturally when there are sequential activation and inhibition calcium binding steps. Moreover, parametric analysis is relatively easy, and the effects of changing on– and off–rates are easy to predict.

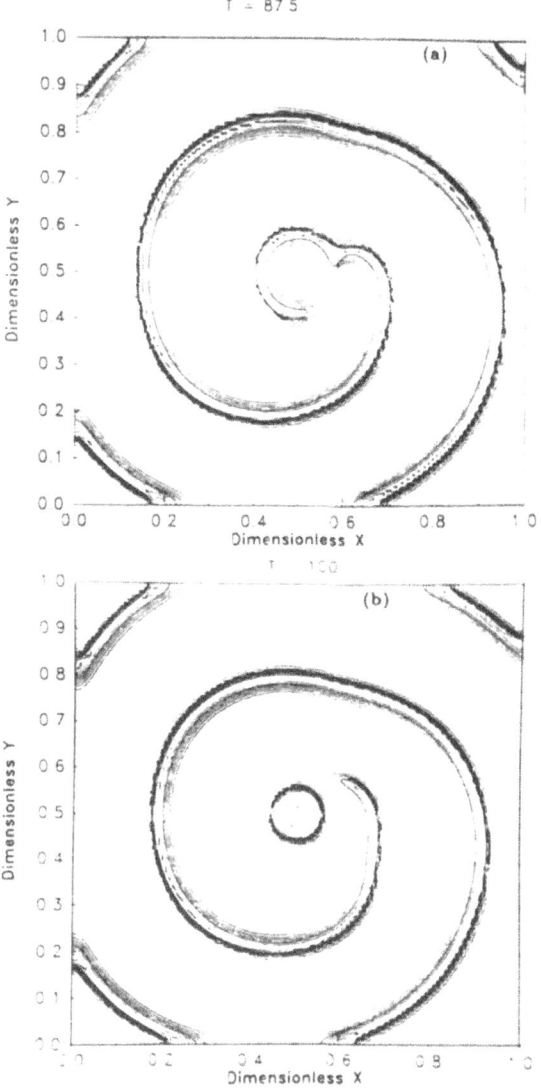

Figure 25.13. The stable coexistence of a spiral and a pacemaker.

The local dynamics of our model show four different types of response to different levels of InsP$_3$ concentrations, namely; a subthreshold response, an excitable response, an oscillatory response with frequency coding, and the over-saturated response. Our model can generate either an oscillatory or an overdamped approach to a high calcium level, depending on the parameters. The second type of response is observed in experiments when InsP$_3$ is very high. The parameter values we used in this paper do not show this type of response, but it is possible with another set of parameters. The different types of responses cover essentially all the local aspects of calcium dynamics observed to date. The distributed versions of our model show traveling waves in one spatial dimension, and target patterns and spiral waves in two spatial dimensions. The new types of wave interaction patterns predicted from numerical calculations are confirmed by experimental results.

Further numerical simulations are needed on the fertilization waves in the *Xenopus* oocyte, the sea urchin egg (Swann & Whitaker 1986), and endothelial cell populations (Boitano *et al.* 1992), where the stimuli are localized. By incorporating positive feedback of Ca^{2+} on InsP$_3$ generation in our model, we have shown that spatially-localized hormone stimuli can generate constant amplitude propagating waves. These results will be reported elsewhere.

Exchange of calcium between the cytoplasm and the extracellular medium is not included in our model, but it can easily be incorporated. If it were, the model would predict that the oscillations die out when calcium is removed from the extracellular medium. Similarly, we could include a calcium-sensitive store, and other types of interactions between the subsystems identified in Section 25.2.

ACKNOWLEDGMENTS This research was supported in part under NIH grant #GM29123.

REFERENCES

Alberts, B., Bray, D., Lewis, J., Raff, M., Roberts, K., & Watson, J. D. 1989. *Molecular biology of the cell.* second edn. New York: Garland Publishing, Inc.

Alexander, J. C., Doedel, E. J., & Othmer, H. G. 1990. On the resonance structure in a forced excitable system. *SIAM J. Appl. Math.,* **50**(5), 1373–1418.

Bazotte, R. B., Pereira, B., Higham, V. Shoshan-Barmatz S., & Kraus-Friedmann, N. 1991. Effects of ryanodine on calcium sequestration in the rat liver. *Biochemical Pharmocology,* **42**(9), 1799–1803.

Berridge, M. J. 1990. Inositol 1,4,5-trisphosphate-induced calcium mobilization is localized in *Xenopus* oocytes. *Proc. Roy. Soc. B,* **238**, 235–343.

Bezprozvanny, I., Watras, J., & Ehrlich, B. E. 1991. Bell-shaped calcium-response curves of Ins(1,4,5)P$_3$- and calcium-gated channels from endoplasmic recticulum of cerebellum. *Nature,* **351**, 751–754.

Boitano, Scott, Dirksen, Ellen R., & Sanderson, Michael J. 1992. Intercellular propagation of calcium waves mediated by inositol trisphosphate. *Science,* **258**(Oct.), 292–295.

Brundage, Rodney A., Fogarty, Kevin E., Tuft, Richard A., & Fay, Fredric S. 1991. Calcium gradients underlying polarization and chemotaxis of eosinophils. *Science,* **254**(Nov.), 703–706.

Cobbold, P. H., Sanchez-Bueno, A., & Dixon, C. J. 1991. The hepatocyte calcium oscillator. *Cell Calcium,* **12**, 87–95.

Cuthbertson, K. S. R., & Chay, T. R. 1991. Modelling receptor-controlled intracellular calcium oscillations. *Cell Calcium,* **12**, 97–109.

De Young, G., & Keizer, J. 1992. A single-pool inositol 1,4,5-trisphophate-receptor-based model for agonist-stimulated oscillations in Ca^{+2} concentration. *Proc. Nat. Acad. Sci.,* **89**, 9895–9899.

Delisle, S. 1991. The four dimensions of calcium signalling in *Xenopus* oocytes. *Cell Calcium,* **12**, 217–227.

DeLisle, S., Krause, K. H., Denning, G., Potter, B. V. L., & Welsh, M. J. 1990. Effect of inositol trisphosphate and calcium on oscillating elevations of intracellular calcium in *Xenopus* oocytes. *J. Biol. Chem.,* **265**, 11726–11730.

Doedel, E. 1986. *AUTO: Software for continuation and bifurcation problems in ordinary differential equations.* Tech. rept. California Institute of Technology.

Dupont, G., Berridge, M. J., & Goldbeter, A. 1991. Signal-induced Ca^{2+} oscillations: Properties of a model based on Ca^{2+}-induced Ca^{2+} release. *Cell Calcium,* **12**, 73–85.

Finch, E. A., Turner, T. J., & Goldin, S. M. 1991. Calcium as a coagonist of inositol 1,4,5-triphosphate-induced calcium release. *Science,* **252**, 443–446.

Gilkey, J. C., Jaffe, L. F., Ridgeway, E. B., & Reynolds, G. T. 1978. A free calcium wave traverses the activating egg of the medaka, *Oryzias latipes. J. Cell Biol.,* **76**, 448–466.

Girard, S., Luckhoff, A., Lechleiter, J., Sneyd, J., & Clapham, D. 1992. Two-dimensional model of calcium waves reproduces the patterns observed *Xenopus* oocytes. *Biophys. J.,* **61**, 509–517.

Goldbeter, A., Dupont, G., & Berridge, M. J. 1990. Minimal model for signal-induced Ca^{2+} oscillations and for their frequency encoding through protein phosphorylation. *Proc. Natl. Acad. Sci. USA*, **87**, 1461–1465.

Harootunian, A. T., Kao, J. P. Y., Paranjape, S., & Tsien, R. Y. 1991. Generation of calcium oscillations in fibroblasts by positive feedback between calcium and IP_3. *Science*, **251**, 75–78.

Joseph, S. K., Rice, H. L., & Williamson, J. R. 1989. The effect of external calcium and pH on inositol trisphosphate-mediated calcium release from cerebellum microsomal fraction. *Biochem. J.*, **258**, 261–265.

Keizer, J., & De Young, G. W. 1992. Two roles for Ca^{2+} in agonist stimulated Ca^{2+} oscillations. *Biophys. J.*, pp. 649–660.

Kuba, K., & Takeshita., S. 1981. Simulation of intracellular Ca^{2+} oscillations in a sympathetic neurone. *J. of Theor. Biol.*, **93**, 1009–1031.

Lechleiter, J., & Clapham, D. 1992. Molecular mechanisms of intracellular calcium excitability in *X. laevis* oocytes. *Cell*, **69**, 283–294.

Lechleiter, J., Girard, S., Peralta, E., & Clapham, D. 1991. Spiral calcium wave propagation and annihilation in *Xenopus laevis* oocytes. *Science*, **252**, 123–126.

Lupu-Meiri, M., Shapira, H., & Oron, Y. 1988. Hemispheric asymmetry of rapid chloride responses to inositol trisphosphate and calcium in *Xenopus* oocytes. *FEBS Letts.*, **240**, 83–87.

Meyer, T., & Stryer, L. Transient calcium release induced by successive increments of inositol 1,4,5-trisphosphate. *Proc. Nat. Acad. Sci. USA*.

Meyer, T., & Stryer, L. 1988. Molecular model for receptor-stimulated calcium spiking. *Proc. Natl. Acad. Sci. USA*, **85**, 5051–5055.

Meyer, Tobias. 1991. Cell signaling by second messenger waves. *Cell*, **64**, 675–678.

Meyer, Tobias, & Stryer, Lubert. 1991. Calcium spiking. *Ann. Rev. Biophys. Biophys. Chem.*, **20**, 153–174.

Miyazaki, S., Yuzaki, M., Nakada, K., Shirakawa, H., Nakanishi, S., Nakade, S., & hiba, K. Mikos. 1992. Block of Ca^{2+} wave and Ca^{2+} oscillation by antibody to the inositol 1,4,5-trisphosphate receptor in fertilized hamster eggs. *Science*, **257**, 251–255.

Nuccitelli, R. 1991. How do sperm activate eggs? *Curr. Top. Dev. Biol.*, **25**, 1–16.

Othmer, H. G. 1991. The dynamics of forced excitable systems. In *Nonlinear Wave Processes in Excitable Media*, Holden, A. V., Markus, M., & Othmer, Hans G. (eds), pp. 213–232. NATO, Plenum Press.

Parker, I., & Ivorra, I. 1990. Inhibition by Ca^{2+} of inositol trisphosphate-mediated Ca^{2+} liberation: A possible mechanism for oscillatory release of Ca^{2+}. *Proc. Natl. Acad. Sci USA*, **87**, 260–264.

R. Jacob, J. E. Merritt, T. J. Hallam, & Rink, T. J. 1988. Repetitive spikes in cytoplasmic calcium evoked by histamine in human endothelial cells. *Nature*, **335**(Sept.), 40–45.

Rooney, Thomas A., Sass, Ellen J., & Thomas, Andrew P. 1989. Characterization of cytosolic calcium oscillations induced by phenylephrine and vasopressin in single fura-2-loaded hepatocytes. *J. Biol. Chem.*, **264**(29), 17131–17141.

Somogyi, R., & Stucki, J. W. 1991. Hormone-induced calcium oscillations in liver cells can be explained by a simple one pool model. *J. Biol. Chem.*, **266**(17), 11068–11077.

Swann, Karl, & Whitaker, Michael. 1986. The part played by inositol trisphosphate and calcium in the propagation of the fertilization wave in sea urchin eggs. *J. Cell Biol.*, **103**(6 Pt. 1), 2333–2342.

Thomas, A. P., Renard, D. C., & Rooney, T. A. 1991. Spatial and temporal organization of calcium signalling in hepatocytes. *Cell Calcium*, **12**, 111–126.

Volpe, P., Alderson-Lang, B. H., & Nickols, G. A. 1990. Regulation of inositol 1,4,5-triphosphate-induced Ca^{2+} release. I. effect of Mg^{2+}. *Am. J. Physiol.*, **258**, C1077–C1085.

Wakui, M., & Petersen, O. H. 1990. Cytoplasmic Ca^{2+} oscillations evoked by acetylcholine or intracellular infusion of inositol trisphosphate or Ca^{2+} can be inhibited by internal Ca^{2+}. *FEBS Lett.*, **263**, 206–208.

Wakui, M., Potter, B. V. L., & Petersen, O. H. 1989. Pulsatile intercellular calcium release does not depend on fluctuations in inositol trisphosphate concentrations. *Nature*, **339**, 317–320.

Watras, J., Bezprozvanny, I, & Ehrlich, B. E. 1991. Inositol 1,4,5-trisphoshate-gated channels in cerebellum: Presence of multiple conductance states. *J. Neuroscience*, **11**(10), 3239–3245.

Wong, Alan Y. K., Fabiato, Alexandre, & Bassingthwaigthe, J. B. 1992. Model of calcium-induced calcium release mechanism in cardiac cells. *Bulletin of Mathematical Biology*, **54**(1), 95–116.

Woods, N. M., Cuthberson, K. S. R., & Cobbold, P. H. 1986. Repetitive transient rises in cytoplasmic free calcium in hormone-stimulated hepatocytes. *Nature*, **319**, 600–602.

Zhao, Hong, Loessberg, Peggy A., Sachs, George, & Muallem, Shmuel. 1990. Regulation of intracellular Ca^{2+} oscillation in AR42J cells. *J. Biol. Chem.*, **265**(34), 20856–20862.

26. FOUR SIGNALS TO SHAPE A SLIME MOLD

Pauline Schaap[1] and Mei Wang[2]

[1]Department of Biology, University of Leiden
Kaiserstraat 63, 2311 GP Leiden, The Netherlands

26.1 INTRODUCTION

The *Dictyostelium* life cycle is a fascinating example of the generation of highly ordered patterns of cell movement and tissue specification by a population of initially identical cells. Notably its spiral waves of chemotactic movement and the regulative aspects of its differentiation patterns have attracted considerable attention. Oscillatory cell movement was first visualized by Arndt (1937) in time lapse films, who compared this phenomenon to gusts of wind passing over a cornfield, transforming into rhythmic wave patterns emanating from a few dominant centers. The regulative capacity of *Dictyostelium* slugs was demonstrated by a series of elegant experiments by Raper (1940), showing that dissected isolates from every part of a migrating slug would regulate to form a new slug and fruiting body. Bonner (1957) demonstrated that the proportion of stalk and spore cells in the fruiting body is practically independent of the total size of the organism over three orders of magnitude.

Dictyostelium development consists of two major processes: (i) coordinated movement of individual cells, which guides both the formation of multicellular aggregates and the characteristic shape changes leading to slug and fruiting body formation, and (ii) establishment of a system of intercellular communication, which provides spatio-temporal clues for regulation of stage- and cell-type specific gene expression and subsequent formation of tissue patterns. Morphogenetic movement and pattern formation are interdependent processes and often share the same signals.

A general outline of *Dictyostelium* development is presented in Figure 26.1. Under appropriate conditions, amoebae emerge from spores and start feeding on soil bacteria to which they are chemotactically attracted. The amoebae grow and multiply until the bacterial food source is depleted and the aggregation phase of development starts. In primitive species such as *D. minutum* and *D. lacteum*, some amoebae start to secrete the attractant of their food source, folate and pterins, and cells then aggregate by simply moving towards the highest attractant concentration. Other species as *D. discoideum* use a more sophisticated aggregation mechanism. After a few hours of starvation a few cells in the population spontaneously secrete pulses of attractant, in many species this attractant is cAMP. Surrounding cells respond with a short period of chemotactic movement and by secreting cAMP themselves. By means of this relay response, the original oscillation propagates through the entire population, and induces

---

*Present address: Center for Phytotechnology, University of Leiden, Wassenaarseweg 64, 2332 XZ Leiden, The Netherlands

Experimental and Theorietcal Advances in Biological Pattern Formation,
Edited by H.G. Othmer *et al.*, Plenum Press, New York, 1993

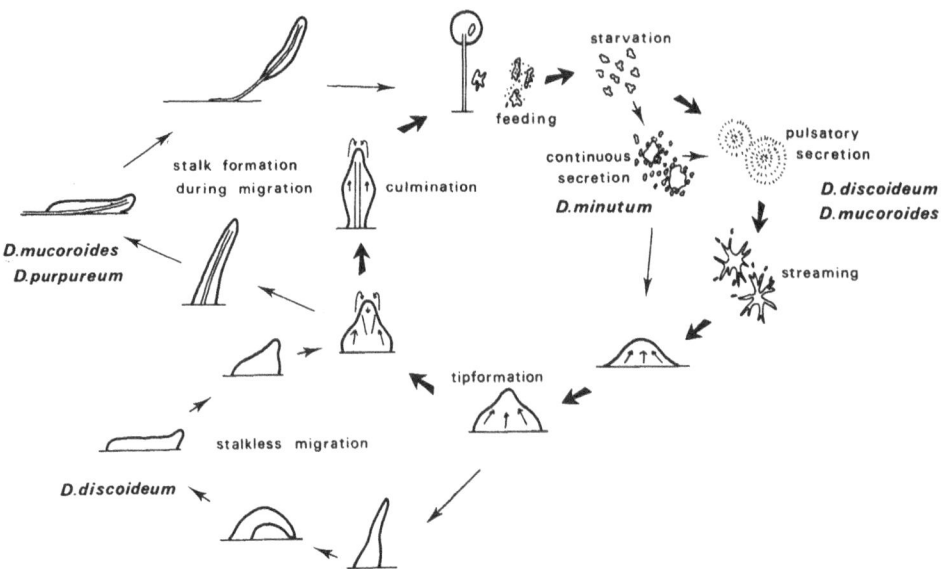

Figure 26.1. Generalized *Dictyostelium* life cycle.

chemotactic movement in a large field of cells. This mode of attractant secretion typically results in streams of amoebae moving towards the oscillating center, because amoebae are here attracted to a local source, rather than to the center itself.

After aggregation, the cell mass performs a series of shape changes, which ultimately results in the formation of one or more fruiting bodies. This morphogenetic movement is guided by a small group of cells called tip cells. The number and time of appearance of initial and accessory tips determines the species-specific maximal size of the fruiting body as well as its specific pattern of side branches. The tip is first evident as a small elevation on top of the aggregate and initiates formation of a slug-shaped structure, which in some species migrates freely over the substratum. About concomitant with tip formation, the antero-posterior pattern of prestalk and prespore cells is being established. After a period of migration, the slug erects itself to start the process of culmination. Cells at the tip move into a central prefabricated cellulose tube, and meanwhile differentiate into highly vacuolated stalk cells. This involution continues until the stalk has reached a certain length and has carried the bulk of the cell mass aloft; the remaining cells then turn into spores,

26.2   SPATIO-TEMPORAL PATTERNS OF GENE EXPRESSION

Developmental gene expression has been examined in great detail in the species *D. discoideum*. Starvation initiates the rapid transient expression of a class of "early" genes (Figure 26.2), followed after a few hours of starvation by expression of aggregative genes coding for cAMP receptors, adhesive contact sites A and a G-protein α-subunit (Gα2). Expression of these genes is strongly enhanced by oscillatory cAMP stimulation (Darmon *et al.* 1975; Gerisch *et al.* 1975; Noegel *et al.* 1986; Kumagai *et al.* 1989). During aggregation, two subpopulations of prestalk genes are expressed. Expression of one subpopulation is induced by the Differentiation Inducing Factor, DIF (see below), while the expression of the second subpopulation is induced by micromolar concentrations of cAMP. Cells expressing these genes initially appear at random, but accumulate later at the prestalk region of the slug (Jermyn *et al.* 1987; Esch & Firtel 1991; Traynor *et al.* 1992). Shortly after aggregates have formed, prespore specific genes are transcribed (Barklis & Lodish 1983; Mehdy *et al.* 1983; Morrissey *et al.* 1984). Expression of these genes can also be induced by micromolar cAMP concentrations (Kay 1982; Mehdy & Firtel 1985; Schaap & Van Driel 1985). Cells containing prespore transcripts first appear at the basal/central region of the late aggregate (Krefft *et al.* 1984; Haberstroh & Firtel 1990). During slug formation, the prespore phenotype becomes very distinct in the posterior 60-80%, and can be readily identified by the presence of prespore vesicles, which contain highly antigenic spore coat proteins (Takeuchi 1963; Devine *et al.* 1983).

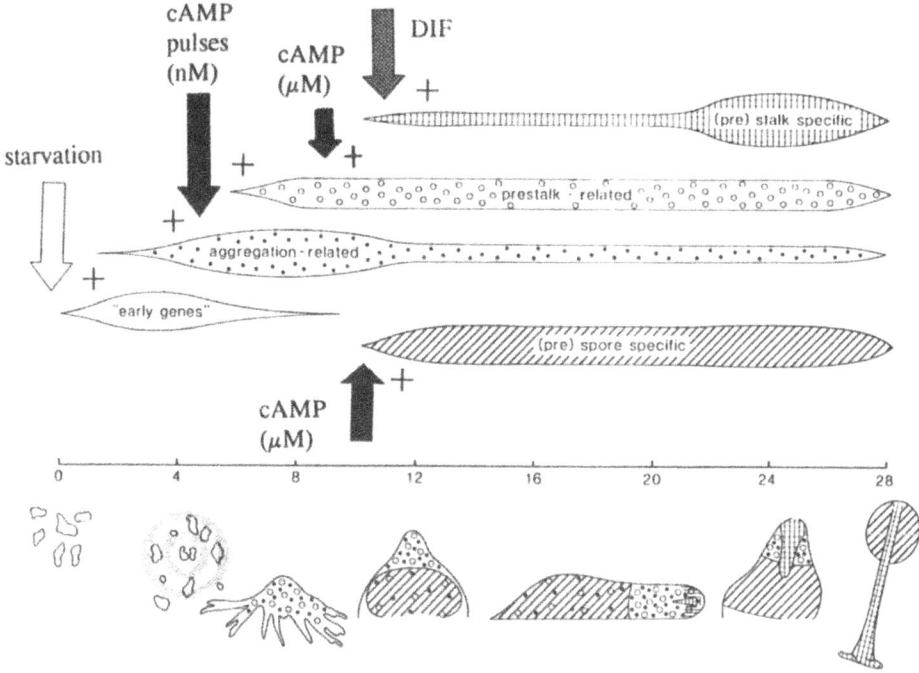

Figure 26.2. Gene regulation and pattern formation during development.

Cells at the anterior 20-40% of the slug do not show major changes in phenotype and retain some of the functions and gene products of aggregative cells. This region is however rather heterogeneous with respect to expression patterns of individual genes. The DIF-induced prestalk gene *ecm*B is expressed at the central core of the tip region (Jermyn *et al.* 1989; Williams *et al.* 1989), the DIF-induced *ecm*A gene as well as the cAMP-induced *ras* gene are highly expressed at the most anterior 10% of the slug and only weakly at the more posteriorly located prestalk cells (Esch & Firtel 1991; Jermyn *et al.* 1989). A third class of cells is located at the posterior region, intermixed with prespore cells; these so-called anterior-like cells express both the DIF- and cAMP-induced prestalk genes, but no prespore genes (Sternfeld & David 1981; Sternfeld & David 1982; Devine & Loomis 1985; Esch & Firtel 1991; Jermyn & Williams 1991).

Anterior prestalk cells pass through major changes in gene transcription and phenotype at the early culmination stage. About 10-20 stalk-specific proteins are being synthesized (Kopachik *et al.* 1985; Morrissey *et al.* 1984) and starting from the tip, the anterior cells vacuolate and synthesize a cellulose cell wall. Spore maturation is the final phase of development and is accompanied by minor changes in gene transcription (Morrissey *et al.* 1984; Richardson *et al.* 1991).

26.3   MORPHOGENETIC SIGNALING DURING DEVELOPMENT

Four intercellular signaling molecules have been identified in *Dictyostelium*. Their regulation and function is summarized in Table 26.3.   The major signalling molecule is cAMP.   cAMP oscillations mediate cell aggregation in *D. discoideum, D. mucuroides* and related species and several lines of evidence indicate that postaggregative cell movement in *D. discoideum* and highly divergent species as *D. minutum* and *Polysphondylium* is uniformly controlled by oscillatory cAMP signaling (Durston & Vork 1979; Schaap *et al.* 1984; Schaap & Wang 1984; Fukushima & Maeda 1991). The tip cells essentially take over the function of the aggregation center and act as autonomous oscillators.   Recent advanced imaging techniques showed that prespore cells show coherent periodic movement along the longitudinal axis of the slug, while anterior prestalk cells rotate in spirals around the tip.   This suggests that cAMP oscillations are shaped as three dimensional scroll waves in the anterior region, which transform to planar waves in the posterior region (Siegert & Weijer 1992).

In addition to its function as a chemoattractant, cAMP induces expression of several classes of developmentally-regulated genes.   Pharmacological studies show that gene regulation is mediated by surface cAMP receptors and is therefore controlled by extracellular cAMP (Schaap & Van Driel 1985; Oyama & Blumberg 1986).   cAMP is degraded extracellularly to adenosine by two surface-associated enzymes, cAMP phosphodiesterase and 5'nucleotidase.   Adenosine also functions as a signal molecule; it reduces the density of autonomously oscillating centres in preaggregative fields and aggregates (Newell & Ross 1982).   Adenosine furthermore inhibits the cAMP-induced expression of prespore genes, while promoting the cAMP-induced expression of a subclass of prestalk genes (Weijer & Durston 1985; Schaap & Wang 1986; Spek *et al.* 1988).

Two entirely distinct *Dictyostelium* signal molecules are DIF, a chlorinated alkylphenone (Morris *et al.* 1987) and its antagonist ammonia (Gross *et al.* 1983). DIF can induce stalk cell differentiation in cell monolayers *in vitro* and induces expression of the prestalk genes *ecm*A and *ecm*B as well as synthesis of stalk specific proteins (Kay & Jermyn 1983; Williams *et al.* 1987; Kopachik *et al.* 1985).   DIF inhibits the cAMP-induced expression of prespore genes (Wang *et al.* 1986; Early & Williams 1988).   DIF levels increase during aggregation and reach their maximum in the tipped aggregate stage.   DIF levels are higher in the prespore than in the prestalk region (Brookman *et al.* 1982; Brookman *et al.* 1987). This is a possible consequence of the fact that the DIF degrading enzyme, DIF-1 dechlorinase, is almost exclusively active at the anterior prestalk region (Insall *et al.* 1992). Ammonia is produced abundantly by *Dictyostelium* cells as the end product of protein degradation (Schindler & Sussman 1977). It inhibits DIF-induced gene expression and cell differentiation. Ammonia may also regulate slug and fruiting body orientation (Bonner & Dodd 1962; Bonner *et al.* 1986; Feit & Sollitto 1987). A recent study indicates that ammonia may inhibit anterior-like cells from sorting to the prestalk region during normal development (Feit *et al.* 1990).

To conclude, considerable knowledge has been gained of spatio-temporal expression of cell-type specific genes and putative signal molecules regulating expression of these genes. A major challenge lies in combining these data in a model that can account for both the qualitative and regulative aspects of *Dictyostelium* pattern formation and regeneration.

Table 26.1. Morphogenetic Signals in *Dictyostelium*

FUNCTION	REGULATION
cAMP -Chemoattractant -Nanomolar pulses enhance aggregative gene expression -Micromolar concentrations induce expression of prespore genes and a subclass of prestalk genes	Autonomous oscillatory secretion by aggregation centers and tips Propagation by relay mechanism.
adenosine -Inhibits cAMP-induced prespore gene expression -Promotes cAMP-induced prestalk gene expression -Inhibits initiation of oscillatory signaling	cAMP degradation by cAMP-PDE and 5'nucleotidase
DIF[2] -Induces expression of a subclass of prestalk genes and expression of stalk genes -Inhibits cAMP-induced expression of prespore genes	Synthesis induced by cAMP Preferentially present in prespore region.
ammonia -Inhibits DIF-induced expression of prestalk and stalk genes -Inhibits sorting of anterior-like cells -Orients multicellular structures	End product of protein degradation

[2] 1-[3,5-dichloro-2,6-dihydroxy-4-methoxy-phenyl]hexane-1-one

26.4   MECHANISMS CONTROLLING PATTERN FORMATION IN *DICTYOSTELIUM*

Considerable controversy exists in the *Dictyostelium* literature concerning the elementary nature of the mechanisms controlling pattern formation. Extreme sorting models attribute all aspects of *Dictyostelium* pattern formation to random homeostatic regulation of the correct proportion of prestalk and prespore cells, followed by sorting to the correct positions. Alternatively, several aspects of *Dictyostelium* pattern formation indicate that cells respond to positional cues, such as graded morphogen distributions.

The following evidence emphasizes the importance of sorting in development. (i) Before aggregation, cells are to some extent predetermined for a specific fate. This predetermination depends on the phase of the cell cycle a cell has reached at the onset of starvation. Cells starved in early cell cycle phase form slugs with a relatively large percentage of prestalk cells and sort to the prestalk region when mixed with nonsynchronized cells (Weijer *et al.* 1984; McDonald & Durston 1984). (ii) At the posterior prespore region about 5-10% of the cells exhibit prestalk characteristics. When slugs are bisected in prestalk and prespore regions, these anterior-like cells (ALC) sort chemotactically to form a new prestalk region. The posterior part then regulates homeostatically to the original proportion of prespore cells and ALCs (Sternfeld & David 1981). In intact slugs, sorting of ALCs is inhibited by a diffusible molecule secreted by the anterior region, which is possibly ammonia (Sternfeld & David 1982; Feit *et al.* 1990). During culmination this inhibitory effect is relaxed and ALCs move to both the anterior and basal region to participate in the formation of the stalk and in a lower cup supporting the spore mass (Sternfeld & David 1982; Jermyn & Williams 1991).

Other studies have contradicted the importance of sorting in pattern regulation. Pogge-Von Strandmann & Kay (1990) showed that in prespore isolates, the new prestalk zone is recruited from the most anteriorly located prespore cells and not from anterior-like cells. They argue that sorting may only occur when the original polarity of the cell mass is disrupted. In prestalk isolates, prespore cell differentiation initiates at the most posterior location and then moves forward (Gregg & Karp 1978). Also in favor of positional cues are observations that expression of prespore genes is first evident in the basal part of late aggregates and is not observed in the anterior tip region (Krefft *et al.* 1984; Haberstroh & Firtel 1990). Prespore differentiation appears to be actively inhibited in the anterior prestalk region, since prespore cells rapidly dedifferentiate when they cross the posterior-anterior boundary (Gregg & Davis 1982; Schaap 1986). The differentiation of stalk cells is typically and exclusively initiated at the extreme apex of culminating structures and apparently depends on some signal specific for that location.

In the following paragraphs we develop a hypothesis which incorporates both sorting and positional mechanisms in *Dictyostelium* pattern formation.

26.5   AN ADENOSINE GRADIENT MAY CONTROL PATTERN FORMATION IN SLUGS

A very simple and straightforward mechanism for pattern regulation in slugs was suggested by the antagonistic effects of cAMP and adenosine on prespore gene expression (Weijer & Durston 1985; Schaap & Wang 1986), combined with observations that cAMP degrading enzymes are preferentially active in prestalk cells (Armant & Rutherford 1979; Tsang & Bradbury 1981; Schaap & Spek 1984). Figure 26.3A,B show that cAMP-phosphodiesterase (cAMP-PDE) and 5'nucleotidase are at least 10 times more active in prestalk cells than in prespore cells. As discussed above, cAMP oscillations are generated in slugs by the tip region

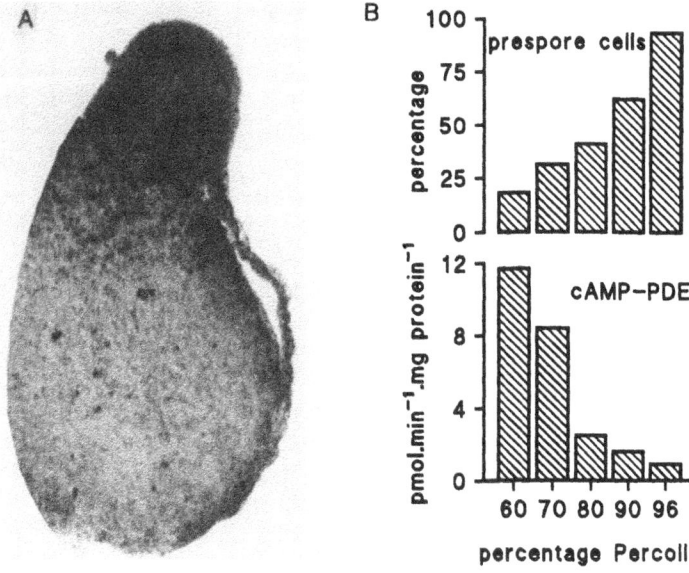

Figure 26.3. Distribution of cAMP degrading enzymes in *Dictyostelium* slugs. A: Slug section stained histochemically for 5'nucleotidase activity. The enzyme activity (dark precipitate) is almost exclusively localized at the anterior prestalk region (M. Wang, unpublished results). B: Prestalk and prespore cells from dissociated slugs were separated using discontinuous Percoll density gradients. The percentage of cells showing positive reaction to spore specific antiserum and cell surface cAMP-PDE activity was measured in each fraction. cAMP-PDE activity is highest in fractions that contain almost no prespore cells and is therefore specifically associated with prestalk cells (data retrieved from Schaap & Spek 1984).

and propagated along the length of the slug by means of the cAMP relay mechanism (Kesbeke *et al.* 1986; Siegert & Weijer 1992).

Our hypothesis simply states that at the anterior, where cAMP-PDE and 5'nucleotidase activity is high, cAMP will be effectively degraded to adenosine. At the posterior, cAMP degradation is much lower and little adenosine is formed. In this region prespore differentiation is induced and maintained, while at the anterior, prespore differentiation is inhibited. The following evidence supports this model: (i) cAMP depletion from slugs induces dedifferentiation of prespore cells (Figure 26.4A,B), while adenosine depletion causes prespore cells to differentiate in the anterior prestalk region (Figure 26.4C). (ii) In slugs of mutant HL101, which is defective in 5'nucleotidase activity (MacLeod & Loomis 1979), prespore cells differentiate up to the utmost tip region (Figure 26.4D).

We recently determined whether intercellular cAMP and adenosine levels in anterior and posterior regions of the slug are in quantitative agreement with concentration requirements of either compound for, respectively, prespore induction or inhibition. The average cAMP level at the anterior is 5.8 ± 1.7 μM and at the posterior 4.7 ± 2.2 μM. The average adenosine level at the anterior is 980 ± 210 μM and at the posterior 230 ± 63 μM (unpublished results). Dose-response relationships for effects of cAMP and adenosine on prespore gene expression show that 2 μM cAMP is required for half-maximal induction of prespore gene expression, while 200 μM adenosine is required for half-maximal inhibition of this process (Figure 26.5). When comparing these data with measured cAMP and adenosine levels, it appears that throughout the whole slug sufficient cAMP is present to induce prespore gene

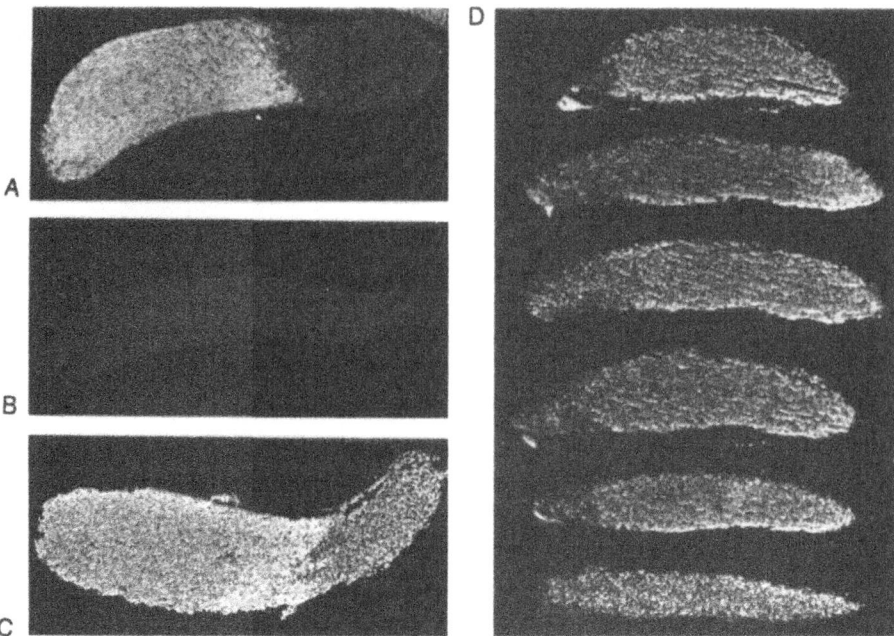

Figure 26.4. Effects of cAMP and adenosine depletion on pattern formation in slugs. Intact slugs were alternatively submerged in phosphate buffer (A) or in buffer containing 0.25 U/ml beef heart cAMP- phosphodiesterase, to inactivate endogenous cAMP (B), or containing 400 U/ml adenosine-deaminase to inactivate endogenous adenosine (C). After 4 hour of incubation, slugs were sectioned and stained with prespore specific antiserum. (Schaap & Wang 1986; Wang *et al.* 1988b). (D). Complete serial sections of slugs of the 5'nucleotidase defective mutant HL101, stained with prespore specific antiserum. Degradation of 1 mM 5'AMP is 10-fold reduced in mutant compared to wild-type slugs (M. Wang, unpublished data).

expression. At the anterior, but not at the posterior region, adenosine levels are sufficiently high to inhibit prespore gene expression.

These data strongly suggest that the graded distribution of adenosine in slugs may give rise to pattern, but do not make clear why cAMP-degrading enzymes are actually more active at the anterior region.

26.6   ESTABLISHMENT OF INITIAL POLARITY BY CHEMOTACTIC SORTING

The graded distribution of adenosine in slugs depends on initial heterogeneity in the distribution of cAMP-degrading enzymes along the longitudinal axis of the slug. A number of observations suggested that initial polarity may be established by sorting of cells differing in cell cycle phase. (i) Cells entering starvation while in early cell cycle phase (E-cells) sort preferentially to the prestalk region, while cell starved in late cell cycle phase (L-cells) will accumulate in the prespore region (Weijer *et al.* 1984; McDonald & Durston 1984). (ii) E-cells exhibit a much stronger potency to initiate autonomously oscillating centers than L-cells (McDonald 1986). (iii) E-cells compared to L-cells exhibit chemotactic cAMP receptors, cAMP-PDE and 5'nucleotidase earlier and to higher levels than L-cells (Figure 26.6). Combining these data, it appears that E-cells not only have the highest potency

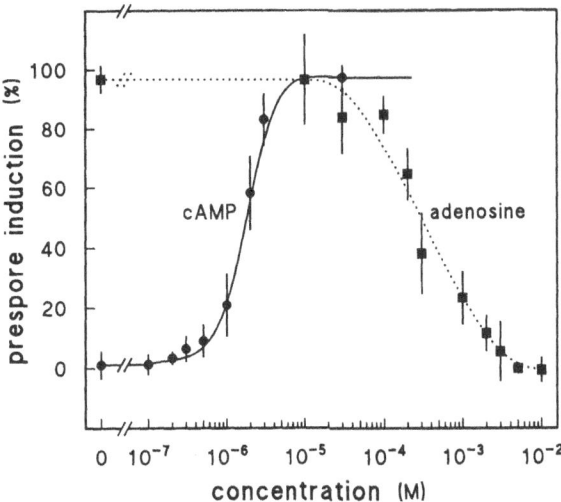

Figure 26.5. cAMP and adenosine concentration requirements. Differentiation competent *D. discoideum* transformants containing a gene fusion of the D19 prespore promoter and the *E. coli lacZ* gene (Dingermann *et al.* 1989) were incubated at 10^5 cells/ml for 6 hours with the indicated cAMP concentrations added at 60 minute intervals (○———○), or with 3 μM cAMP/60 min and the indicated adenosine concentrations added at t=0 h (■ ... ■). ß-galactosidase activity in freeze-thawed lysates was measured and expressed as percentage of maximal induction.

to initiate autonomous oscillators, but because of their higher levels of chemotactic cAMP receptors are also most responsive to the chemotactic signal. It is easy to comprehend that these cells will preferentially accumulate at the aggregation center and later at the apical region, thus explaining the cell cycle related sorting behaviour. At the same time sorting of E-cells with relatively high cAMP-PDE and 5'nucleotidase activity to the anterior region of the slug, will generate the initial conditions to form the adenosine gradient and establish the antero-posterior pattern (Figure 26.7). The observation that E-cells form slugs with a relatively large proportion of prestalk cells is a logical consequence of the fact that E-cell slugs contain a higher level of cAMP degrading activity than unsynchronized populations.

26.7   REGULATION OF SIZE AND SHAPE BY ADENOSINE GRADIENTS

The size of *D. discoideum* slugs is determined by the number of cells entering the aggregate, but is limited to about 100,000 cells. Aggregates containing more cells will break up to form several slugs. Formation of multiple tips occurs as a rule in *Dictyostelium* species with small fruiting body size. Secondary tip formation results in bifurcation of the cell mass during slug migration and in formation of side branches during the culmination process; each side branch consisting of a separate stalk and spore head. The most striking example is the species *Polysphondylium*, which forms highly regular whorls of side-branches from globular cell masses left behind on the stalk by the main slug.

Transplantation experiments show that anterior tissue displays the largest potency to form secondary tips when transplanted on host slugs (MacWilliams 1982). Additionally, an antero-posterior gradient of resistance to formation of secondary tips appears to be operative (Durston 1976; MacWilliams 1982). This tip inhibition gradient is imposed by a diffusible molecule secreted by the tip region (Durston 1976; Kopachik 1982).

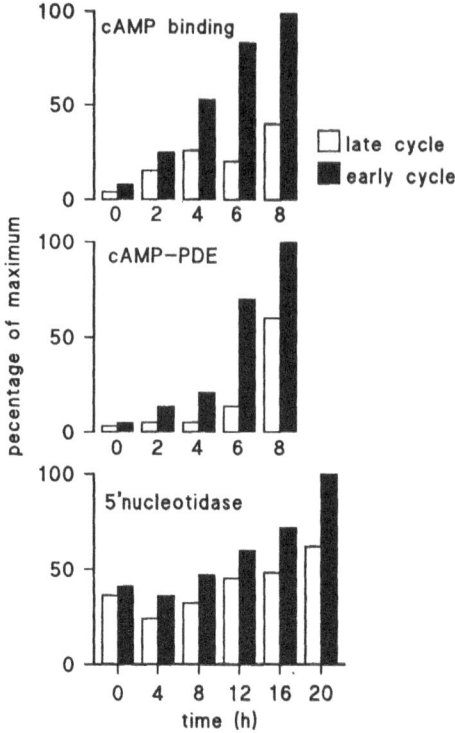

Figure 26.6. cAMP receptors and cAMP degrading enzymes during development of early and late cell cycle cells. Cells were synchronized in either late G2 or in S and early G2 phase of the cell cycle (The *Dictyostelium* cell cycle shows no G1 phase) and incubated on non-nutrient agar. At the indicated time periods the levels of chemotactic cAMP receptors, cAMP-phosphodiesterase and 5'nucleotidase were assayed (data retrieved from Wang *et al.* 1988a).

Considering the evidence that tips control morphogenetic movement by functioning as cAMP oscillators, the tip activation gradient most likely represents the intrinsic capacity of cells to initiate autonomous oscillations. A gradient of tip activation along the slug corresponds perfectly with the sorting gradient of cells differing in cell cycle phase, since the early cell cycle cells, which sort to the anterior display the highest potency to initiate autonomous oscillators (McDonald 1986).

Adenosine is a good candidate for the diffusible tip inhibitor. It reduces the density of signaling centers in aggregative fields by inhibiting cAMP signaling (Newell 1982; Newell & Ross 1982; Theibert & Devreotes 1984), and is preferentially produced in the prestalk region by degradation of cAMP. When adenosine levels in aggregates are artificially reduced multiple tips are formed while, on the other hand, an increase of adenosine levels results in larger slugs (Schaap & Wang 1986). The volume of slugs of the 5'nucleotidase defective mutant HL101 (which cannot produce sufficient adenosine) is 15 fold lower than the volume of wild type slugs (M. Wang, unpublished data). Interplay between cAMP signaling and adenosine production could therefore not only regulate formation of differentiation patterns, but also determine the size of the organism and the species-specific variations in fruiting body shape.

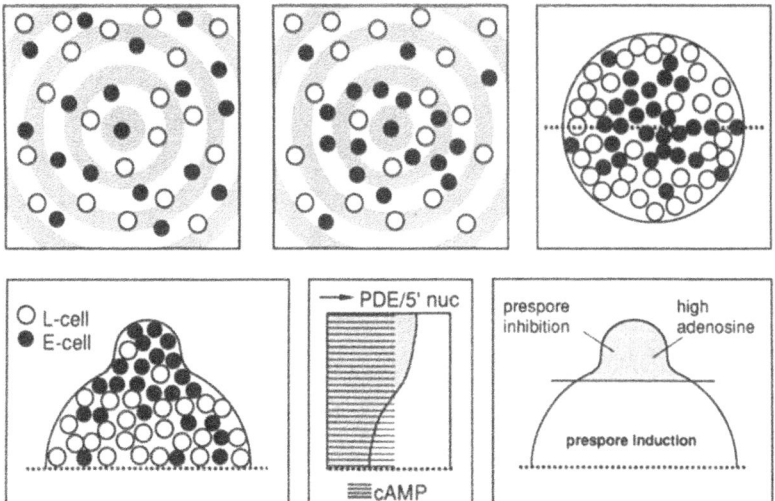

Figure 26.7. Establishment of an adenosine gradient by sorting. Cells starved while in early cell cycle phase (E-cells) initiate oscillating centers and preferentially accumulate at aggregation centers and tips, due to high chemotactic responsiveness compared to cells starved in late cell cycle phase (L-cells). Because E-cells also exhibit higher levels of cAMP hydrolysing enzymes, this results in the establishment of an antero-posterior gradient of cAMP degrading activity.

26.8 INDUCTION OF STALK CELL DIFFERENTIATION BY AMMONIA DEPLETION

We described a hypothesis for formation of the antero-posterior pattern in terms of prespore induction by cAMP and prespore inhibition by adenosine. The differentiation of stalk cells does not enter this hypothesis, because neither cAMP nor adenosine induces this process. Expression of stalk specific genes is effectively induced by DIF (Williams *et al.* 1987). However, the spatio-temporal regulation of DIF production does not suggest an obvious mechanism for formation of the stalk. DIF synthesis starts during aggregation and is maximal at the tipped aggregate stage. Thereafter, a moderate decline in DIF levels is evident, which is most pronounced in the anterior prestalk region, which contains 2-3 fold lower DIF levels than the posterior prespore region (Brookman *et al.* 1982; 1987). DIF can induce differentiation of mature stalk cells in low density cell monolayers (Kay & Jermyn 1983), but not in intact slugs (Wang & Schaap 1989), although DIF-treated slugs show a considerably reduced proportion of prespore cells (Table 26.2). This suggests that in intact slugs an inhibitor is active, which prevents stalk cell differentiation.

A good candidate for this inhibitor is ammonia, which is abundantly produced as the end product of protein degradation and inhibits DIF-induced stalk differentiation *in vitro* (Gross *et al.* 1983; Wang *et al.* 1990). When intact slugs are incubated with DIF and an enzyme mixture which uses ammonia as substrate, more than 20% of cells in the slug differentiate to mature vacuolated stalk cells, surrounded by the characteristic cellulose matrix. cAMP was also reported to inhibit stalk cell differentiation *in vitro* (Berks & Kay 1988). However, cAMP depletion in combination with DIF does not induce stalk cell differentiation, indicating that cAMP does not control stalk formation *in vivo*.

A possible mechanism for stalk induction is presented in Figure 26.8. During migration, slugs move around in a thin water layer above the substratum. Slugs and the surrounding

Table 26.2. Effect of DIF, ammonia depletion and cAMP depletion on prespore and stalk cell differentiation in slugs

Addition	Stalk cells (%)	Prespore cells (%)
None	0.2	61.8
DIF (15,000 U/ml)	2.8	36.9
ammonia depletion	3.2	60.2
DIF + ammonia depletion	21.9	36.7
cAMP-PDE (0.25 U/ml)	0.5	14.6
DIF + cAMP-PDE (0.25 U/ml)	2.0	10.1

slugs were incubated in buffer containing the indicated amounts of DIF and beef-heart cAMP phosphodiesterase or containing 0.03 U/ml L-glutamate dehydrogenase, 0.1 mM NADH and 70 mM α-ketoglutarate (ammonia depletion mix) or combinations of the three variables. After 10 h, slugs were dissociated into single cells and stained with prespore antiserum and Calcofluor, a dye that stains the cellulose wall of stalk cells. The proportion of prespore and stalk cells to total cells was determined. (Data retrieved from Wang & Schaap 1989).

water layer contain ammonia levels sufficiently high to inhibit DIF induced stalk cell differentiation. When slugs start to culminate and the tip moves upwards, it can effectively loose ammonia by evaporation. The relatively large surface to volume ratio of the tip compared to the rest of the cell mass aids to specifically reduce ammonia levels in this part of the structure. At the tip, inhibition of DIF-induced stalk gene expression is lost and stalk formation initiates.

26.9   HOMEOSTATIC REGULATION OF CELL TYPE PROPORTIONS

Another aspect of *Dictyostelium* pattern formation which cannot be readily explained by interactions between cAMP and adenosine is the regulation of a fixed proportion of anterior-like cells in the prespore region. Regulation of correct proportions of prespore and prestalk/anterior-like cells also occurs under submerged conditions where pattern formation does not take place (Oyama *et al.* 1983; Weijer & Durston 1985; Inouye 1989). The observations that (i) DIF inhibits prespore differentiation (Wang *et al.* 1986; Early & Williams 1988), (ii) DIF synthesis is stimulated by cAMP (Brookman *et al.* 1982) (iii) DIF is preferentially present in the prespore region (Brookman *et al.* 1987), suggest that regulation of the correct proportion of prespore and anterior-like cells could result from feed back interactions of cAMP and DIF on cell type conversion as indicated in Figure 26.9A. cAMP generated by oscillatory signaling induces anterior-like cells to differentiate into prespore cells. These cells in turn produce DIF, which induces the opposite conversion. The DIF antagonist ammonia may also contribute to proportion regulation. Prestalk and anterior-like cells may be expected to produce larger amounts of ammonia because lysosomal activity is relatively

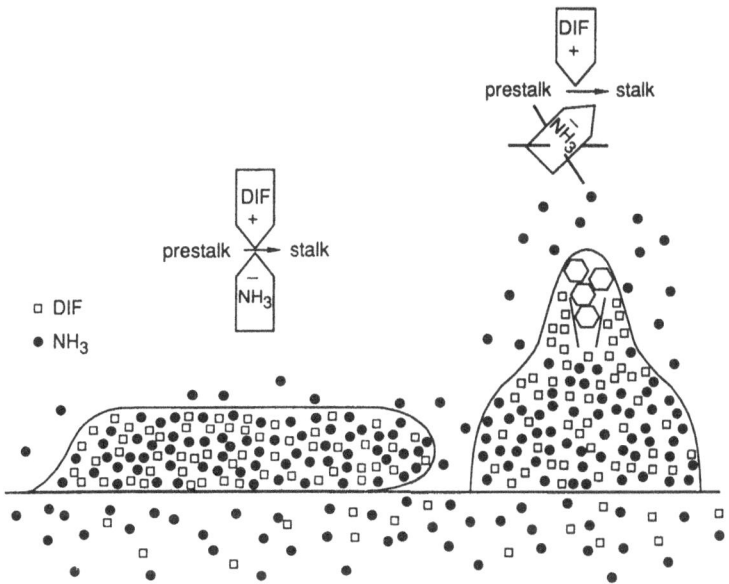

Figure 26.8. Model for induction of stalk cell differentiation by ammonia depletion.

high in these cells (Oohata & Takeuchi 1977; Fong & Rutherford 1978; Yamamoto *et al.* 1981). Ammonia itself does not induce prespore gene expression (Van Lookeren Campagne *et al.* 1989) but may influence cell type proportions by inhibiting the prespore to anterior-like conversion (Figure 26.9B). The data presented in Table 26.2 show that ammonia depletion does not affect the proportion of prespore cells in slugs, however, recent studies suggest that ammonia may control the conversion of anterior-like cells into prestalk cells. During slug migration, anterior-like cells express low levels of the DIF-induced *ecm*A and *ecm*B genes.

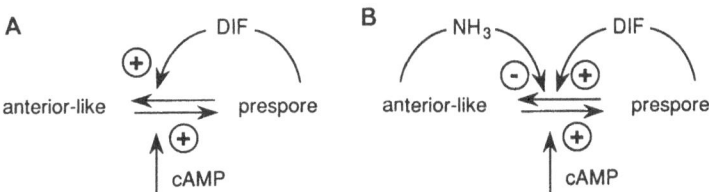

Figure 26.9. Possible feedback interactions controlling cell-type proportions

When the cell mass elongates upwards during culmination, levels of expression of both genes in anterior-like cells increase strongly. Additionally, anterior-like cells now start to sort to the prestalk region of the organism (Jermyn & Williams 1991). Combining the evidence of Feit *et al.* (1990), that ammonia inhibits sorting of anterior-like cells, and the antagonistic effect of ammonia on DIF-induced prestalk gene expression with the assumption that ammonia evaporation occurs when culminating structures rise from the substratum, it can be envisaged that ammonia depletion results in the three-fold response of (i) stalk cell differentiation at the apex, where the most extreme drop in ammonia levels can be expected, (ii) enhanced expression of prestalk genes and (iii) sorting of anterior-like cells. To validate this hypothesis, it should be demonstrated that the supposed decline in ammonia concentration during culmination indeed occurs.

26.10 CONCLUSIONS

Experimental data suggest a sequence of regulatory mechanisms, which yield a qualitative explanation for the generation of pattern in *Dictyostelium*. In short, sorting of cells with cell cycle correlated variations in cAMP detection and degradation generates an antero-posterior concentration gradient of the prespore antagonist adenosine. cAMP produced by oscillatory signaling induces prespore differentiation; at the anterior prestalk region, prespore induction is inhibited by high adenosine levels. cAMP also induces synthesis of the stalk-inducing factor DIF, but during slug migration stalk formation is inhibited by high ambient ammonia levels. Local ammonia depletion at the tip of culminating structures initiates stalk cell differentiation. Homeostatic regulation of prespore and anterior-like cells in slug posteriors can be understood in terms of two feedback interactions, in which DIF produced by prespore cells promotes the conversion of prespore cells into anterior-like cells, while ammonia produced by anterior-like cells inhibits this conversion.

Several quantitative aspects of pattern formation and regulation are not yet understood. (i) The boundary between prestalk and prespore cells in slugs is very distinct. This requires either a virtually stepwise gradient of adenosine over this boundary or extreme sensitivity of cells to respond to small concentration differences. The former would not be predicted from a gradient in cAMP degrading activity established by sorting of cells with stochastically determined variations in cell cycle phase, the latter is not evident from the dose-dependency of prespore inhibition by adenosine. An element of autocatalytic feedback on adenosine production could generate a more pronounced discontinuity in adenosine concentrations. Such a mechanism is plausible since, *e. g.* expression of the cAMP-PDE gene is stimulated by cAMP pulses, but repressed by high cAMP concentrations (Peters *et al.* 1991). (ii) In our model all adenosine is supposed to be formed through cAMP degradation. However, measured adenosine levels are 200 fold higher than cAMP levels. This could only occur if both synthesis and degradation rates of cAMP far exceed the degradation rate of adenosine. Further analysis of cAMP and adenosine metabolism in slugs is required.

The present hypothesis is essentially a hybrid between a sorting and a positional information type model. One might ask whether formation of a simple bipolar pattern requires such complex regulation. Could pattern be generated and regulated by either cAMP/adenosine gradients or homeostatic proportion regulation and sorting? The regulative capacity and scale-invariance of pattern formation are intrinsic properties of the latter mechanism. However, schemes for homeostatic regulation, as presented in Figure 26.9, indicate that in a pure population of anterior cells, the anterior to prespore conversion is strongly stimulated. Since prespore cells were observed to dedifferentiate in the anterior region, an inhibitory rather than a stimulatory signal appears to be operative, which may represent adenosine in the alternative scheme. On the other hand, since generation of the adenosine gradient requires sorting, a model not involving any sorting can presently not be formulated. We therefore conclude that a hybrid mechanism incorporating four morphogens yields the best explanation for the known aspects of pattern formation and regulation in *Dictyostelium* at present.

REFERENCES

Armant, D. R., & Rutherford, C. L. 1979. 5'-AMP nucleotidase is localized in the area of the cell-cell contact of prespore and prestalk regions during culmination of *Dictyostelium discoideum*. *Mech. Ageing Dev.*, **10**, 199–217.

Arndt, A. 1937. Rhizopodienstudien. III Untersuchungen über *Dictyostelium mucoroides* Brefeld. *Wilhelm Roux Arch. Entwicklungsmech. Org.*, **136**, 681–747.

Barklis, E., & Lodish, H. F. 1983. Regulation of *Dictyostelium discoideum* mRNAs specific for prespore or prestalk cells. *Cell*, **32**, 1139–1148.

Berks, M., & Kay, R. R. 1988. Cyclic AMP is an inhibitor of stalk cell differentiation in *Dictyostelium discoideum*. *Dev. Biol.*, **126**, 108–114.

Bonner, J. T. 1957. A theory of the control of differentiation in the cellular slime molds. *Quart. Rev. Biol.*, **32**, 232–246.

Bonner, J. T., & Dodd, M. R. 1962. Evidence for gas-induced orientation in the cellular slime molds. *Dev. Biol.*, **5**, 344–361.

Bonner, J. T., Suthers, H. B., & Odell, G. M. 1986. Ammonia orients cell masses and speeds up aggregating cells of slime moulds. *Nature*, **323**, 630–632.

Brookman, J. J., Town, C. D., Jermyn, K. A., & Kay, R. R. 1982. Developmental regulation of stalk cell differentiation-inducing factor in *Dictyostelium discoideum*. *Dev. Biol.*, **91**, 191–196.

Brookman, J. J., Jermyn, K. A., & Kay, R. R. 1987. Nature and distribution of the morphogen DIF in the *Dictyostelium* slug. *Development*, **100**, 119–124.

Darmon, M., Brachet, P., & da Silva, L. H. Pereira. 1975. Chemotactic signals induce cell differentiation in *Dictyostelium discoideum*. *Proc. Nat. Acad. Sci. USA*, **72**, 3163–3166.

Devine, K. M., & Loomis, W. F. 1985. Molecular characterization of anterior-like cells in *Dictyostelium discoideum*. *Dev. Biol.*, **107**, 364–372.

Devine, K. M., Bergmann, J. E., & Loomis., W. F. 1983. Spore coat proteins of *Dictyostelium discoideum* are packaged in prespore vesicles. *Dev. Biol.*, **99**, 437–446.

Dingermann, T., Reindl, N., Werner, H., Hildebrandt, M., Nellen, W., Harwood, A., Williams, J., & Nerke, K. 1989. Optimation and *in situ* detection of *Escherichia coli* ß-galactosidase gene expression in *Dictyostelium discoideum*. *Gene*, **85**, 353–362.

Durston, A. J. 1976. Tip formation is regulated by an inhibitory gradient in the *Dictyostelium discoideum* slug. *Nature*, **263**, 126–129.

Durston, A. J., & Vork, F. 1979. A cinematographical study of the development of vitally stained *Dictyostelium discoideum*. *J. Cell Sci.*, **36**, 261–279.

Early, V. E., & Williams, J. G. 1988. A *Dictyostelium* prespore-specific gene is transcriptionally repressed by DIF *in vitro*. *Development*, **103**, 519–524.

Esch, R. K., & Firtel, R. A. 1991. cAMP and cell sorting control the spatial expression of a developmentally essential cell-type-specific *ras* gene in *Dictyostelium*. *Genes & Dev.*, **5**, 9–21.

Feit, I. N., & Sollitto, R. L. 1987. Ammonia is the gas used for the spacing of fruiting bodies in the cellular slime mold *Dictyostelium discoideum*. *Differentiation*, **33**, 193–196.

Feit, I. N., Bonner, J. T., & Suthers, H. B. 1990. Regulation of the anterior-like cell state by ammonia in *Dictyostelium discoideum*. *Dev. Gen.*, **11**, 442–446.

Fong, D., & Rutherford, C. L. 1978. Protease activity during cell differentiation of the cellular slime mold *Dictyostelium discoideum*. *J. Bacteriol.*, **134**, 521–527.

Fukushima, S., & Maeda, Y. 1991. Whorl formation in *Polysphondylium violaceum*: relevance to organization by cyclic AMP. *Dev. Growth Differ.*, **33**, 525–533.

Gerisch, G., Fromm, H., Huesgen, A., & Wick, U. 1975. Control of cell-contact sites by cyclic AMP pulses in differentiating *Dictyostelium* cells. *Nature*, **255**, 547–549.

Gregg, J. H., & Davis, R. W. 1982. Dynamics of cell redifferentiation in *Dictyostelium mucoroides*. *Differentiation*, **21**, 200–205.

Gregg, J. H., & Karp, G. C. 1978. Patterns of cell differentiation revealed by L-[^{3}H]fucose incorporation in *Dictyostelium*. *Exp. Cell Res.*, **112**, 31–46.

Gross, J. D., Bradbury, J., Kay, R. R., & Peacey, M. J. 1983. Intracellular pH and the control of cell differentiation in *Dictyostelium discoideum*. *Nature*, **303**, 244–245.

Haberstroh, L., & Firtel, R. A. 1990. A spatial gradient of expression of a cAMP-regulated prespore cell-type-specific gene in *Dictyostelium*. *Genes & Dev.*, **4**, 596–612.

Inouye, K. 1989. Control of cell type proportions by a secreted factor in *Dictyostelium discoideum*. *Development*, **107**, 605–609.

Insall, R., Nayler, O., & Kay, R. R. 1992. DIF-1 induces its own breakdown in *Dictyostelium. EMBO J.*, **11**, 2849–2854.

Jermyn, K. A., & Williams, J. G. 1991. An analysis of culmination in *Dictyostelium* using prestalk and stalk-specific cell autonomous markers. *Development*, **111**, 779–787.

Jermyn, K. A., Berks, M., Kay, R. R., & Williams, J. G. 1987. Two distinct classes of prestalk-enriched mRNA sequences in *Dictyostelium discoideum. Development*, **100**, 745–755.

Jermyn, K. A., Duffy, K. T., & Williams, J. G. 1989. A new anatomy of the prestalk zone in *Dictyostelium. Nature*, **340**, 144–146.

Kay, R. R. 1982. cAMP and spore differentiation in *Dictyostelium discoideum. Proc. Nat. Acad. Sci. USA*, **79**, 3228–3231.

Kay, R. R., & Jermyn, K. A. 1983. A possible morphogen controlling differentiation in *Dictyostelium. Nature*, **303**, 242–244.

Kesbeke, F., Van Haastert, P. J. M., & Schaap, P. 1986. Cyclic AMP relay and cyclic AMP-induced cyclic GMP accumulation during development of *Dictyostelium discoideum. FEMS Lett.*, **34**, 85–90.

Kopachik, W. J. 1982. Size regulation in *Dictyostelium. J. Embryol. Exp. Morphol.*, **68**, 23–35.

Kopachik, W. J., Dhokia, B., & Kay, R. R. 1985. Selective induction of stalk-cell-specific proteins. *Differentiation*, **28**, 209–216.

Krefft, M., Voet, L., Gregg, J. H., Mairhofer, H., & Williams, K. L. 1984. Evidence that positional information is used to establish the prestalk-prespore pattern in *Dictyostelium discoideum* aggregates. *EMBO J.*, **3**, 201–206.

Kumagai, A., Pupillo, M., Gundersen, R., Miake-Lye, R., Devreotes, P. N., & Firtel, R. A. 1989. Regulation and function of Gα protein subunits in *Dictyostelium. Cell*, **57**, 265–275.

MacLeod, C. L., & Loomis, W. F. 1979. Biochemical and genetic analysis of a mutant with altered alkaline phosphatase in *Dictyostelium discoideum. Dev. Gen.*, **1**, 109–121.

MacWilliams, H. K. 1982. Transplantation experiments and pattern mutants in cellular slime mold slugs. In *Developmental Order: Its origin and regulation*, Subtelny, S. (ed). pp. 463–483, New York: Alan R. Liss.

McDonald, S. A. 1986. Cell-cycle regulation of center initiation in *Dictyostelium discoideum. Dev. Biol.*, **117**, 546–549.

McDonald, S. A., & Durston, A. J. 1984. The cell cycle and sorting behaviour in *Dictyostelium discoideum. J. Cell Sci.*, **66**, 195–204.

Mehdy, M. C., & Firtel, R. A. 1985. A secreted factor and cyclic AMP jointly regulate cell-type-specific gene expression in *Dictyostelium discoideum. Mol. Cell. Biol.*, **5**, 705–713.

Mehdy, M. C., Ratner, D., & Firtel, R. A. 1983. Induction and modulation of cell-type-specific gene expression in *Dictyostelium. Cell*, **32**, 763–771.

Morris, H. R., Taylor, G. W., Masento, M. S., Jermyn, K. A., & Kay, R. R. 1987. Chemical structure of the morphogen differentiation inducing factor from *Dictyostelium discoideum. Nature*, **613**, 811–814.

Morrissey, J. H., Devine, K. M., & Loomis, W. F. 1984. The timing of cell-type-specific differentiation in *Dictyostelium discoideum. Dev. Biol.*, **103**, 414–424.

Newell, P. C. 1982. Cell surface binding of adenosine to *Dictyostelium* and inhibition of pulsatile signalling. *FEMS Lett.*, **13**, 417–421.

Newell, P. C., & Ross, F. M. 1982. Inhibition by adenosine of aggregation centre initiation and cyclic AMP binding in *Dictyostelium. J. Gen. Microbiol.*, **128**, 2715–2724.

Noegel, A., Gerisch, G., Stadler, J., & Westphal, M. 1986. Complete sequence and transcript regulation of a cell adhesion protein from aggregating *Dictyostelium* cells. *EMBO J.*, **5**, 1473–1476.

Oohata, A., & Takeuchi, I. 1977. Separation and biochemical characterization of the two cell types present in the pseudoplasmodium of *Dictyostelium discoideum. J. Cell Sci.*, **24**, 1–9.

Oyama, M., & Blumberg, D. D. 1986. Interaction of cAMP with the cell-surface receptor induces cell-type specific mRNA accumulation in *Dictyostelium discoideum. Proc. Nat. Acad. Sci. USA*, **83**, 4819–4823.

Oyama, M., Okamoto, K., & Takeuchi, I. 1983. Proportion regulation without pattern formation in *Dictyostelium discoideum. J. Embryol. Exp. Morph.*, **75**, 293–301.

Peters, D. J. M., Cammans, M., Smit, S., Spek, W., Van Lookeren Campagne, M. M., & Schaap, P. 1991. Control of cAMP-induced gene expression by divergent signal transduction pathways. *Dev. Gen.*, **12**, 25–34.

Pogge-Von Strandmann, R., & Kay, R. R. 1990. Position-dependent regulation of the prestalk-prespore pattern in *Dictyostelium* slugs. *Dev. Gen.*, **11**, 447–453.

Raper, K. B. 1940. Pseudoplasmodium formation and organization in *Dictyostelium discoideum. J. Elisha Mitchell Sci. Soc.*, **56**, 241–282.

Richardson, D. L., Hong, C. B., & Loomis, W. F. 1991. A prespore gene dd51, expressed during culmination of *Dictyostelium discoideum. Dev. Biol.*, **144**, 269–280.

Schaap, P. 1986. Regulation of size and pattern in the cellular slime molds. *Differentiation*, **33**, 1–16.

Schaap, P., & Spek, W. 1984. Cyclic-AMP binding to the cell surface during development of *Dictyostelium discoideum. Differentiation*, **27**, 83–87.

Schaap, P., & Van Driel, R. 1985. Induction of post-aggregative differentiation in *Dictyostelium discoideum* by cAMP. Evidence of involvement of the call surface cAMP receptor. *Exp. Cell Res.*, **158**, 388–398.

Schaap, P., & Wang, M. 1984. The possible involvement of oscillatory cAMP signaling in multicellular morphogenesis of the cellular slime molds. *Dev. Biol.*, **105**, 470–478.

Schaap, P., & Wang, M. 1986. Interactions between adenosine and oscillatory cAMP signaling regulate size and pattern in *Dictyostelium. Cell*, **45**, 137–144.

Schaap, P., Konijn, T. M., & Van Haastert, P. J. M. 1984. cAMP pulses coordinate morphogenetic movement during fruiting body formation of *Dictyostelium minutum. Proc. Nat. Acad. Sci. USA*, **81**, 2122–2126.

Schindler, J., & Sussman, M. 1977. Ammonia determines the choice of morphogenetic pathways in *Dictyostelium discoideum. J. Mol. Biol.*, **116**, 161–169.

Siegert, F., & Weijer, C. J. 1992. Three-dimensional scroll waves organize *Dictyostelium* slugs. *Proc. Nat. Acad. Sci. USA*, **89**, 6433–6437.

Spek, W., Van Drunen, K., Van Eijk, R., & Schaap, P. 1988. Opposite effects of adenosine on two types of cAMP-induced gene expression in *Dictyostelium* indicate the involvement of at least two different intracellular pathways for the transduction of cAMP signals. *FEBS Lett.*, **228**, 231–234.

Sternfeld, J., & David, C. N. 1981. Cell sorting during pattern formation in *Dictyostelium. Differentiation*, **20**, 10–21.

Sternfeld, J., & David, C. N. 1982. Fate and regulation of anterior-like cells in *Dictyostelium* slugs. *Dev. Biol.*, **93**, 111–118.

Takeuchi, I. 1963. Immunochemical and immunohistochemical studies on the development of the cellular slime mold *Dictyostelium mucoroides. Dev. Biol.*, **8**, 1–26.

Theibert, A., & Devreotes, P. N. 1984. Adenosine and its derivatives inhibit the cAMP signaling response in *Dictyostelium discoideum. Dev. Biol.*, **106**, 166–173.

Traynor, D., Kessin, R. H., & Williams, J. G. 1992. Chemotactic sorting to cAMP in the multicellular stages of *Dictyostelium* development. *Proc. Nat. Acad. Sci. USA*, **89**, 8303–8307.

Tsang, A., & Bradbury, J. M. 1981. Separation and properties of prestalk and prespore cells of *Dictyostelium discoideum. Exp. Cell Res.*, **132**, 433–441.

Van Lookeren Campagne, M. M., Aerts, R. J., Spek, W., Firtel, R. A., & Schaap, P. 1989. Cyclic-AMP-induced elevation of intracellular pH precedes, but does not mediate, the induction of prespore differentiation in *Dictyostelium discoideum. Development*, **105**, 401–406.

Wang, M., & Schaap, P. 1989. Ammonia depletion and DIF trigger stalk cell differentiation in intact *Dictyostelium discoideum* slugs. *Development*, **105**, 596–574.

Wang, M., Van Haastert, P. J. M., & Schaap, P. 1986. Multiple effects of differentiation-inducing factor on prespore differentiation and cyclic-AMP signal transduction in *Dictyostelium. Differentiation*, **33**, 24–28.

Wang, M., Aerts, R. J., Spek, W., & Schaap, P. 1988a. Cell cycle phase in *Dictyostelium discoideum* is correlated with the expression of cyclic AMP production, detection, and degradation. *Dev. Biol.*, **125**, 410–416.

Wang, M., Van Driel, R., & Schaap, P. 1988b. Cyclic AMP-phosphodiesterase induces differentiation of prespore cells in *Dictyostelium discoideum* slugs: evidence that cyclic AMP is the morphogenetic signal for prespore differentiation. *Development*, **103**, 611–618.

Wang, M., Roelfsema, J. H., Williams, J. G., & Schaap, P. 1990. Cytoplasmic acidification facilitates but does not mediate DIF-induced prestalk gene expression in *Dictyostelium discoideum*. *Dev. Biol.*, **140**, 182–188.

Weijer, C. J., & Durston, A. J. 1985. Influence of cAMP and hydrolysis products on cell type regulation in *Dictyostelium discoideum*. *J. Embryol. Exp. Morphol.*, **86**, 19–37.

Weijer, C. J., Duschl, G., & David, C. N. 1984. Dependence of cell-type proportioning and sorting on cell cycle phase in *Dictyostelium discoideum*. *J. Cell Sci.*, **70**, 133–145.

Williams, J. G., Ceccarelli, A., McRobbie, S. J., Mahbubani, H., Berks, M. M., Kay, R. R., & Jermyn, K. A. 1987. Direct induction of *Dictyostelium* prestalk induction by DIF provides evidence that DIF is a morphogen. *Cell*, **49**, 185–192.

Williams, J. G., Duffy, K. T., Lane, D. P., McRobbie, S. J., Harwood, A. J., Traynor, D., Kay, R. R., & Jermyn, K. A. 1989. Origins of the prestalk-prespore pattern in *Dictyostelium* development. *Cell*, **59**, 1157–1163.

Yamamoto, A., Maeda, Y., & Takeuchi, I. 1981. Development of an autophagic system in differentiating cells of the cellular slime mold *Dictyostelium discoideum*. *Protoplasma*, **108**, 55–69.

27. MODELLING ACTIN FILAMENT ALIGNMENT

Jonathan A. Sherratt[1] **and Julian Lewis**[2]

[1]Centre for Mathematical Biology
Mathematical Institute
24–29 St. Giles'
Oxford OX1 3LB, UK

[2]ICRF Developmental Biology Unit
Department of Zoology
South Parks Road
Oxford OX1 3PS, UK

27.1 INTRODUCTION

Intracellular actin filaments tend to align with an imposed stress field. This phenomenon has been the subject of a number of experimental studies, using micromanipulation (Chen 1981; Kolega 1986), magnetic particle ingestion (Valberg & Albertini 1985) and flow chamber techniques (Franke *et al.* 1984; Wechezak *et al.* 1989). However, in previous mechanical models of morphogenesis, the alignment phenomenon has almost always been neglected (see Oster & Odell 1984 for discussion); a notable exception is the recent computer model of Weliky *et al.* (1991) for notochord morphogenesis. We discuss a new model that enables stress alignment to be incorporated into continuum mechanical models such as the Oster–Murray model (Oster *et al.* 1983; Murray & Oster 1984; Murray *et al.* 1988).

In the absence of a detailed understanding of the mechanisms underlying actin filament alignment, we base our model on simple, intuitively plausible assumptions. In particular, we assume that alignment occurs as a direct response to the ratio of the principal components of stress. We denote by $G_1(\mathbf{r})$ and $G_2(\mathbf{r})$ the 'actin filament density components' along the principal axes of the stress tensor σ, and our model takes the following form:

$$\begin{aligned}
G_1(\mathbf{r}) &= G_0(\mathbf{r}) \cdot f(\sigma_{11}/\sigma_{22}) \\
G_2(\mathbf{r}) &= G_0(\mathbf{r}) \cdot f(\sigma_{22}/\sigma_{11}) .
\end{aligned} \tag{27.1}$$

Here $G_0(\mathbf{r})$ is the total actin filament density, and the function f represents the alignment induced by anisotropy in the stress field. We have previously presented a detailed heuristic argument which gives the following functional form for f:

$$f(\rho) = \frac{(\pi/2)\rho^p}{\rho^p + \pi/2 - 1} \tag{27.2}$$

(Sherratt & Lewis 1992). The parameter p reflects the sensitivity of the actin filament network to changes in the stress field.

Experimental and Theorietcal Advances in Biological Pattern Formation,
Edited by H.G. Othmer *et al.*, Plenum Press, New York, 1993

27.2   ACTIN ALIGNMENT IN WOUND HEALING

A change in the alignment of intracellular actin filaments can be caused by a change in external conditions, and as an example of this we consider the initial response of embryonic epidermis to wounding. In both chick and mouse embryos, Martin & Lewis (1991); Martin & Lewis (1992) have recently shown that a surgical cut in the epidermis induces marked alignment of actin filaments, resulting in a thick cable of filamentous actin around almost all of the wound margin, localized within the leading row of basal cells. By contrast, the free edge in an adult epidermal wound induces increased proliferation and lamellipodial crawling (Winter 1962; 1972; Sherratt & Murray 1990; 1991; 1992a; 1992b).

To investigate the role of stress-induced actin filament alignment in this system, we consider the mechanical balance of forces in the epidermal cell sheet. Following wounding, there are rapid cytoskeletal changes, and the epidermis quickly evolves to a quasi-equilibrium state in which an actin cable runs around the wound margin. The contraction of this cable appears to cause the wound to close, but over a considerably longer time scale; here we restrict attention to the initial response to wounding, and specifically to the initial formation of the actin cable. We represent the post-wounding quasi-equilibrium state using a continuum mechanical model of Oster–Murray form, but with the crucial difference that we include the effects of actin filament alignment using (27.1) and (27.2); the mathematical details of the model are presented elsewhere (Sherratt 1991; Sherratt & Lewis 1992; Sherratt *et al.* 1992). In particular, we consider only the case in which the sensitivity parameter p has the value 1, for mathematical simplicity. For a wide range of parameters, the model solutions predict a

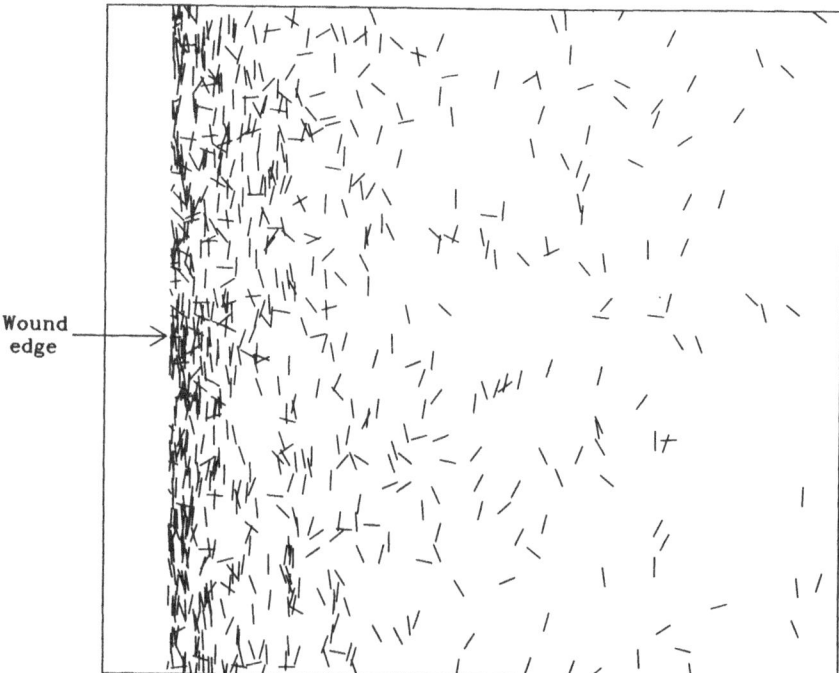

Figure 27.1. An illustration of the model prediction of the initial response of the actin filament network to wounding in embryonic epidermis. We plot a number of small line segments according to the predicted actin filament density and orientation. Details of the model equations and parameter values are given elsewhere (Sherratt *et al.* 1992)

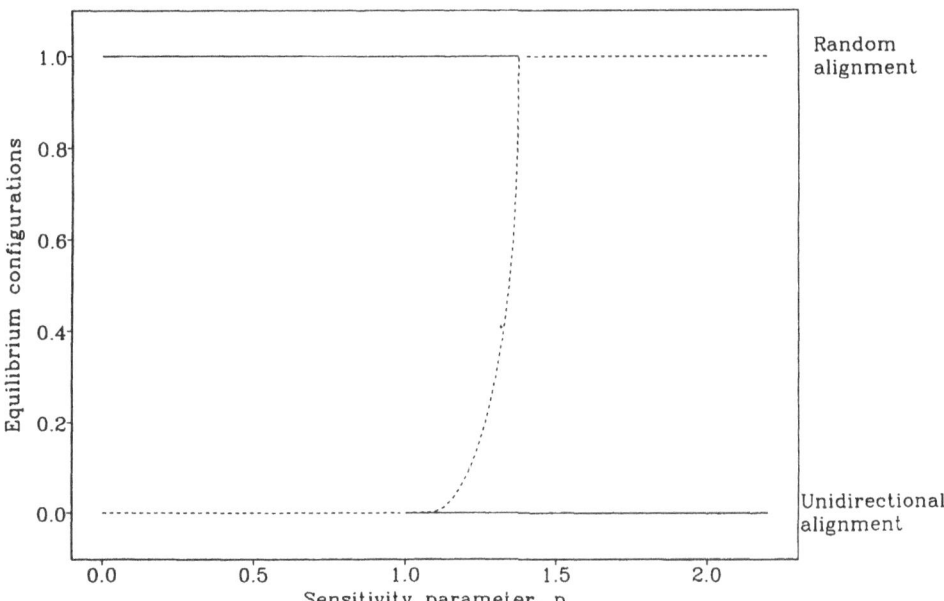

Figure 27.2. The model prediction of the equilibrium configurations of the actin filament network in a cell sheet at rest, as a function of the parameter p, which reflects the sensitivity of actin filament alignment to stress anisotropy. The stable configurations are denoted by full lines, and the unstable configurations by broken lines. The configurations are represented quantitatively by G_1/G_2, which is the ratio of the effective actin filament densities in the directions of least and greatest alignment. Note that for $1 < p < 1.38$, the random and unidirectional configurations are both stable.

sharp increase in the density of actin filaments aligned with the wound edge, corresponding to the actin cable observed experimentally (Figure 27.1). Comparison of model solutions with $p = 1$ and $p = 0$ suggests that the alignment phenomenon is crucial to the formation of the actin cable. Moreover, our model predicts that the formation of the cable can be explained simply as a by-product of the post-wounding quasi-equilibrium in the epidermis.

27.3   AUTO-ALIGNMENT OF ACTIN FILAMENTS

In the absence of external influences, stress forces are exerted by actin filaments, and the resulting stress field controls filament alignment. Intuitively, we expect that this auto-regulation of actin alignment may destabilize a random filament mesh-work, and we now use the model (27.1)-(27.2) to investigate this in the simple case of a cell sheet at equilibrium, with no strain at any point. Following a change in actin filament configuration, the resulting stress field alters almost instantaneously. By contrast, actin filaments align much more slowly following changes in the stress field: experiments suggest a time scale of about 10 seconds (Kolega 1986; Chen 1981). Our stability calculation is based on this difference in time scales.

In the absence of strain, the stress field is due entirely to cell traction, and we take this dependence to be a simple proportionality. Thus at equilibrium, $\sigma_{11}/\sigma_{22} = G_1/G_2 = f(\sigma_{11}/\sigma_{22})/f(\sigma_{22}/\sigma_{11}) \equiv F(\sigma_{11}/\sigma_{22})$, where the 1- and 2-axes are in arbitrary orthogonal directions. This nonlinear equation for the actin filament configuration G_1/G_2 has multiple solutions, whose stability depends on the sign of $F'(G_1/G_2) - 1$ (Sherratt & Lewis 1992). As expected intuitively, the solutions and their stability depend crucially on the sensitivity parameter p (Figure 27.2). When p is low, so that actin filament alignment is not very sensitive to stress anisotropy, the stable configuration is a random filament mesh-work. However, for larger values of p, this random configuration is driven unstable, and the stable state has all the filaments aligned in a single, random direction. Such spontaneous filament alignment could play an important role in morphogenetic processes such as convergent extension (Gerhart & Keller 1986; Lewis, Jesuthasan and Sherratt, (In preparation)).

ACKNOWLEDGMENTS JAS was supported by a Junior Research Fellowship at Merton College, Oxford.

REFERENCES

Chen, W. 1981. Mechanism of retraction of the trailing edge during fibroblast movement. *J. Cell Biol.*, **90**, 187–200.

Franke, R. P., Grafe, M., Schnittler, H., Seiffge, D., Mittermayer, C., & Drenckhahn, D. 1984. Induction of human vascular endothelial stress fibres by fluid shear stress. *Nature*, **307**, 648–649.

Gerhart, J., & Keller, R. 1986. Region-specific cell activities in amphibian gastrulation. *Ann. Rev. Cell Biol.*, **2**, 201–229.

Kolega, J. 1986. Effects of mechanical tension on protrusive activity and microfilament and intermediate filament organization in an epidermal epithelium moving in culture. *J. Cell Biol.*, **102**, 1400–1411.

Martin, P., & Lewis, J. 1991. The mechanics of embryonic skin wound healing – limb bud lesions in mouse and chick embryos. In *Fetal Wound Healing: a Paradigm of Tissue Repair*, Adzick, N. S., & Longaker, M. T. (eds). New York: Elsevier.

Martin, P., & Lewis, J. 1992. Actin cables and epidermal movement. in embryonic wound healing. *Nature*, **360**, 179–183.

Murray, J. D., & Oster, G. F. 1984. Generation of biological pattern and form. *IMA J. Math. Appl. Med. Biol.*, **1**, 51.

Murray, J. D., Maini, P. K., & Tranquillo, R. T. 1988. Mechanochemical models for generating biological pattern and form in development. *Phys. Rep.*, **171**, 59–84.

Oster, G. F., & Odell, G. M. 1984. Mechanics of cytogels I: oscillations in *Physarum*. *Cell Motil. Cytoskel.*, **4**, 469–503.

Oster, G. F., Murray, J. D., & Harris, A. K. 1983. Mechanical aspects of mesenchymal morphogenesis. *J. Embryol. exp. Morphol.*, **78**, 83–125.

Sherratt, J. A. 1991. Actin aggregation and embryonic epidermal wound healing. *J. Math. Biol.* (In press).

Sherratt, J. A., & Lewis, J. 1992. Stress-induced alignment of actin filaments and the mechanics of cytogel. *Bull. Math. Biol.* (In press).

Sherratt, J. A., & Murray, J. D. 1990. Models of epidermal wound healing. *Proc. R. Soc. Lond.*, **B241**, 29–36.

Sherratt, J. A., & Murray, J. D. 1991. Mathematical analysis of a basic model for epidermal wound healing. *J. Math. Biol.*, **29**, 389–404.

Sherratt, J. A., & Murray, J. D. 1992a. Epidermal wound healing: a theoretical approach. *Comments Theor. Biol.*, **2**, 315–333.

Sherratt, J. A., & Murray, J. D. 1992b. Epidermal wound healing: the clinical implications of a simple mathematical model. *Cell Transplantation*, **1**, 365–371.

Sherratt, J. A., Martin, P., Murray, J. D., & Lewis, J. 1992. Mathematical models of wound healing in embryonic and adult epidermis. *IMA J. Math. Appl. Med. Biol.* (In press).

Valberg, P. A., & Albertini, D. F. 1985. Cytoplasmic motions, rheology, and structure probed by a novel magnetic particle method. *J. Cell Biol.*, **101**, 130 140.

Wechezak, A. R., Wight, T. N., Viggers, R. F., & Sauvage, L. R. 1989. Endothelial adherence under shear stress is dependent upon microfilament reorganisation. *J. Cell. Physiol.*, **139**, 136–146.

Weliky, M., Minsuk, S., Keller, R., & Oster, G. 1991. Notochord morphogenesis in *Xenopus laevis*: simulation of cell behavior underlying tissue convergence and extension. *Development*, **113**, 1231–1244.

Winter, G. D. 1962. Formation of the scab and the rate of epithelialization of superficial wounds in the skin of the young domestic pig. *Nature*, **193**, 293–294.

Winter, G. D. 1972. Epidermal regeneration studied in the domestic pig. In *Epidermal Wound Healing*, Maibach, H. I., & Rovee, D. T. (eds). Chicago: Year Book Med. Publ.

28. THE MOTILE BEHAVIOR OF AMOEBAE IN THE AGGREGATION WAVE IN *DICTYOSTELIUM DISCOIDEUM*

David R. Soll, Deborah Wessels, and Andrew Sylwester

Department of Biological Sciences
University of Iowa
138 Biology Building
Iowa City, IA 52242-1324, USA

28.1 INTRODUCTION

Pattern develops during the aggregation process in the cellular slime mold *Dictyostelium discoideum*, and although aggregation appears to represent a more dynamic process than the genesis of a more static morphogenetic field, it shares several of the same basic characteristics. As hypothesized for some forms of positional information (Wolpert 1971), aggregation information is in the form of a signal emanating from a source and in this case is amplified through a relay system between cells. Cells respond individually to the signal with directional movement towards the source (Alcantara & Monk 1974), which is the aggregation center. Because of the ease of experimental manipulation, gene cloning, and targeted mutagenesis, the process of signal genesis, signal transduction, and the motile response in *D. discoideum* is rapidly being elucidated and warrants careful scrutiny by researchers in the field of pattern formation. In this mini-review, we will focus on the behavioral response of cells to the naturally propagated cAMP wave. We will consider a wave paradox which has been explained by a temporal mechanism for chemotaxis (Varnum *et al.* 1985; Soll 1990), and then describe methods for assessing the role of particular cytoskeletal elements in the motile response to the natural wave.

28.2 THE CHEMOTACTIC WAVE PARADOX

When amoebae of the cellular slime mode *D. discoideum* deplete their food source or, in a laboratory situation, are washed free of a food source, they aggregate after a lag period of several hours (Gerisch 1968; Soll 1979). The aggregation process depends upon chemotaxis to move more than 100,000 cells in a large aggregation territory to a single aggregation center. In the aggregation process, precocious cells in the population begin to release the chemoattractant cAMP in a pulsatile fashion (Bonner 1947; Shaffer 1957; Konijn *et al.* 1968; Tomchik & Devreotes 1981) at intervals initially greater than 10 minutes. Cells in the immediate vicinity of a precocious cell respond to the pulse of cAMP by in turn releasing a pulse of cAMP, and by moving in a directed fashion towards the original source of the wave, the aggregation center. After releasing cAMP, cells enter a refractory period

Experimental and Theorietcal Advances in Biological Pattern Formation,
Edited by H.G. Othmer *et al.*, Plenum Press, New York, 1993

325

during which they are unable to release cAMP in response to a pulse of cAMP (Shaffer 1975; Devreotes & Steck 1979). A phosphodiesterase on the cell surface (Malchow *et al.* 1972) then destroys the signal. Cells centripetal to these cells in turn progress through the same relay cycle. Therefore, an initial pulse of cyclic AMP released from a precocious cell in an aggregation territory is relayed outwardly as a non-dissipating wave of attractant. These chemotactic waves have several characteristics. Firstly, they are released at a lower frequency at the beginning of aggregation, with a periodicity greater than 10 minutes, and increase in frequency as aggregation progresses (Durston 1974). The average period of a wave is roughly 7 minutes. Secondly, the peak and trough concentration values of a wave are $10^{-6}M$ and less than $10^{-8}M$, respectively (Devreotes *et al.* 1983). Finally, the speed at which a wave moves outwardly from the center is roughly 300 μm per minute (Alcantara & Monk 1974; Gross *et al.* 1976). Waves appear to be symmetrical (Gross *et al.* 1976; Devreotes *et al.* 1983).

As noted previously, the role of the chemotactic wave is to orient cell movement towards the aggregation center. However, the mechanism underlying the chemotactic response has not been unequivocally demonstrated. Two alternative mechanisms have been considered. Firstly, cells may use a spatial mechanism in which they assess the direction of a cAMP gradient across the cell body by the difference in receptor occupancy between anterior and posterior ends (Devreotes & Zigmond 1988; Fisher *et al.* 1989). Secondly, cells may use a temporal mechanism (Gerisch & Keller 1981; Varnum-Finney *et al.* 1987a; Varnum-Finney *et al.* 1987b; Soll 1990) in which they assess direction by the change in receptor occupancy with time as they move through the gradient. A cell will experience an increase in receptor occupancy with time as it moves up a spatial gradient and a decrease in receptor occupancy with time as it moves down a spatial gradient. Bacteria use a temporal mechanism exclusively in chemotaxis (Boyd & Simon 1982; Brown & Berg 1974; Silverman & Simon 1974). Although a spatial mechanism has been considered the more likely mechanism in *D. discoideum*, (*e. g.* Devreotes & Zigmond 1988; Fisher *et al.* 1989) the results from our laboratory support a temporal mechanism. Indeed, if *D. discoideum* employ a spatial mechanism exclusively, a paradox arises which is related to the dynamics of the natural wave (Soll 1990) (Figure 28.1).

Since a cell is challenged by a positive spatial gradient in the direction of the source in the front of the wave for the same amount of time (*i. e.* 3.5 minute) that it is challenged by a positive spatial gradient in the direction away from the source in the back of the wave (Figure 28.1), and since a *D. discoideum* amoeba can turn in a matter of seconds (Futrelle *et al.* 1982; Swanson & Taylor 1982), each amoeba in an aggregation territory should move towards the source and away from the source for the same amounts of time if a purely spatial mechanism were employed. This would lead to no net progress towards the aggregation center. That is not the way it works. Cells surge periodically towards the source in a natural wave, generating rings of translocating cells around the aggregation center (Alcantara & Monk 1974). At the end of aggregation, most of the cells in an aggregation territory have entered the aggregate.

28.3 CELLS READILY READ TEMPORAL GRADIENTS IN THE ABSENCE OF SPATIAL GRADIENTS: PARTIAL RESOLUTION OF THE PARADOX

If a cell could assess the direction of a temporal wave, it could translocate only when the temporal gradient was increasing (*i. e.* in the front of each passing wave of cAMP). It could then use a spatial or temporal mechanism for assessing the correct direction of translocation. In the former case, it would employ the gradient of receptor occupancy across its cell body, translocating in the direction of higher receptor occupancy. In the latter case,

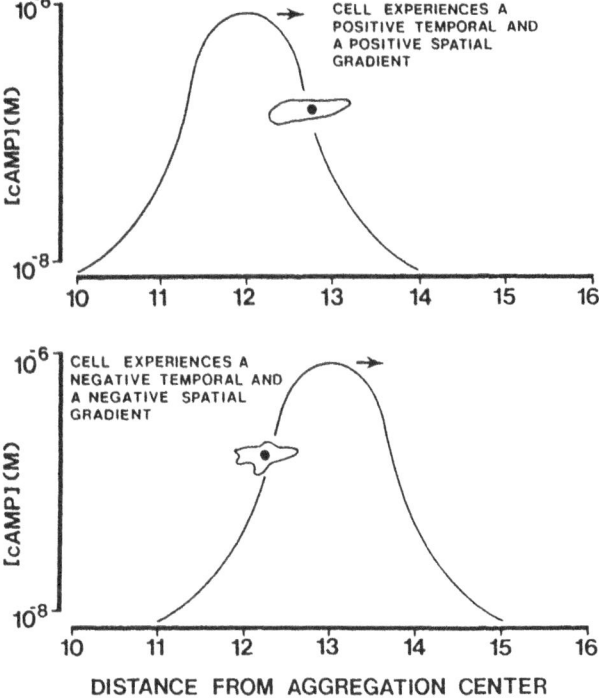

Figure 28.1. The wave paradox.

pseudopod extension up the spatial gradient (*i. e.* in the direction of the source) would result in an increase in receptor occupancy with time, and would signal the correct direction of translocation, while pseudopod extension down the spatial gradient would result in a decrease in receptor occupancy with time, and signal incorrect direction and, in turn, retraction of the pseudopod.

If cells use a temporal mechanism for assessing the direction of a wave, then they must be able to discriminate an increasing from a decreasing temporal gradient in the absence of a spatial gradient. This prediction was tested by analyzing the motile behavior of cells in a perfusion chamber in which the temporal dynamics of waves with 7 minute periodicity were imitated without establishing increasing or decreasing spatial gradients (Varnum *et al.* 1985). In these experiments, it was demonstrated that cells translocated rapidly and persistently during the increasing phase of the temporal wave (see upper portion of Figure 28.3 below). It was also demonstrated that (i) at the beginning of the increasing phase of the wave, cells formed more pseudopodia with less directionality, (ii) in the middle of the increasing phase of the wave, cells were elongated, formed fewer lateral pseudopodia and translocated rapidly, (iii) at the peak of the wave, cells were relatively inactive, and (iv) in the decreasing phase of the wave, cells formed pseudopodia but were less elongated or polar, and therefore made little net progress in any direction (Varnum-Finney *et al.* 1987a). This behavioral cycle could explain why there was periodic motility, or rings of motile cells, in an aggregation territory, and why there was net progress towards the aggregation center. These results definitively demonstrated that cells could assess the direction of a temporal gradient in the absence of a spatial gradient. However, these results did not explain how cells know which way to go in the front of the wave.

28.4   VERIFYING THE TEMPORAL BEHAVIOR CYCLE IN THE NATURAL WAVE

The behavior of a cell in simulated temporal waves of cAMP can explain, at least in part, how amoebae make net progress towards the aggregation center. However, we must first verify that the behaviors observed in a simulated wave occur in a natural territory. We therefore developed methods (Figure 28.2) for monitoring the behavior of single cells at high magnification in natural territories (Wessels *et al.* 1992). Aggregation-competent amoebae are

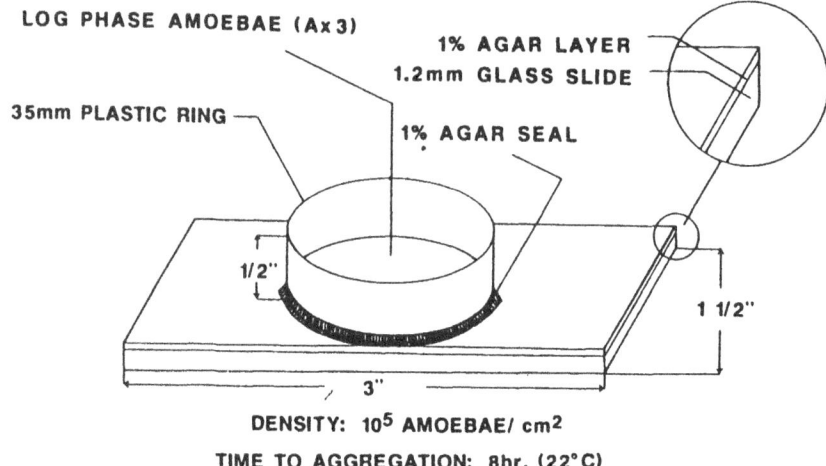

Figure 28.2. The development of natural aggregation under conditions amenable to high resolution videoanalysis.

evenly distributed on a thin agar film on a glass slide within the confines of a ring inserted into the agar film. The ring is capped to produce a humidity chamber. Cell behavior is then video recorded and analyzed with computer-assisted dynamic image analyzing systems (Soll 1988; Soll *et al.* 1987; Soll *et al.* 1988). Aggregation-competent amoebae dispersed on the agar films begin to aggregate after 30 minutes. Independent cells move in a periodic manner towards aggregation centers, and plots of instantaneous velocity for cells exhibiting a periodicity of approximately 7 minutes are indistinguishable from plots of aggregation competent cells challenged with simulated temporal waves (Figure 28.3). In addition, changes in polarity, pseudopod extension and cell shape exhibit the same cycles in natural aggregation territories as they do in simulated waves. These results demonstrate that major aspects of the natural behavior cycle are cued by the temporal characteristics of the wave. In addition, because the concentration dynamics of the simulated wave are experimentally controlled and therefore known, the similarity in behaviors allows us to deduce the dynamics of the natural wave, as diagrammed in Figure 28.3. The front and back phases of the natural wave are deduced as follows. Firstly, the beginning of a wave is deduced by normalizing the beginnings of the increasing phase of the instantaneous velocity plots in simulated and natural waves. Secondly, the deduced trough-to-trough period is halved to provide the beginning of the back of the wave. If the period of the natural wave is 7 minutes, as in the example in Figure 28.3, then the front and back of the natural wave are 3.5 minutes in length, the same as the simulated wave. In

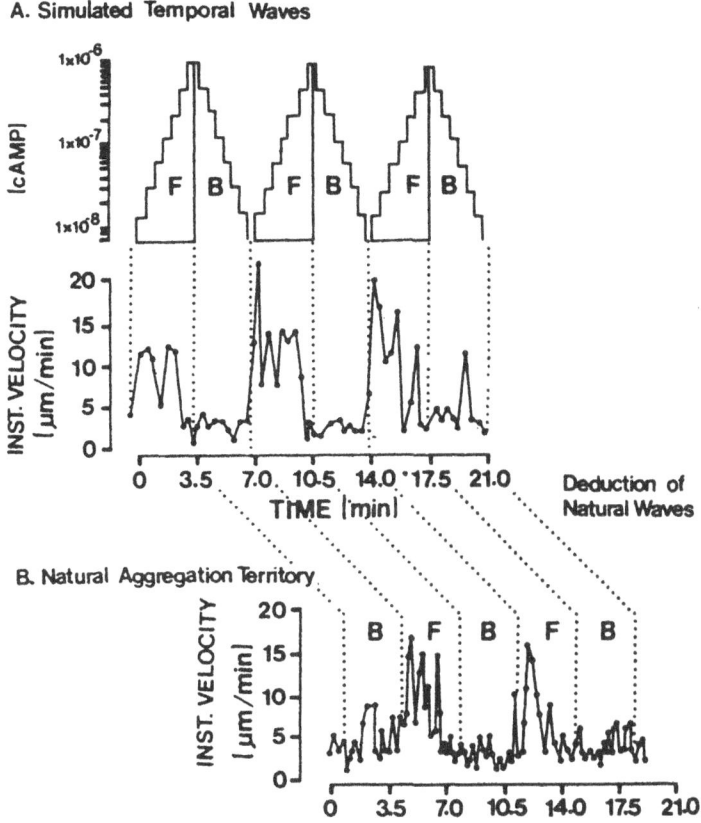

Figure 28.3. The method for deducing the concentration dynamics of a natural wave by comparing cell behavior in a simulated temporal wave.

Figure 28.4, the instantaneous velocity plots are presented for 3 individual amoebae within 60 μm of each other in a natural wave. The cells are almost exactly in phase, demonstrating that they are responding to the same natural waves emanating from a single source. The deduction of wave dynamics provides us with the capacity to characterize motile behavior in the different portions of the wave. Rapid and persistent cellular translocation occurs during the first 150 seconds of the front of a wave with 7 minute periodicity. This compares favorably with the 100 second period of movement estimated by Alcantara & Monk (1974) for cells with a periodicity of 5 minutes. Our estimate of the translocation phase represents 36% of a 7 minute period, while Alcantara and Monk's estimate represents 33% of a 5 minute period. During this translocation phase, cells move on average towards the aggregation center. However, a third of the cells analyzed veer away from the center by more than 45°, but are again on track at the beginning of the next translocation phase. Directional adjustment appears to occur at the onset of the translocation phase. The depression of lateral pseudopod extension, the increase in instantaneous velocity, and the decrease in directional change during this phase suggest that once on track, cells move in a relatively blind fashion.

    During the last 30 seconds of the front of the wave and the first 30 seconds of the back of the wave, cells appear to freeze. During this peak period of the wave, cells exhibit a dramatic decrease in translocation, lateral pseudopod formation, and intracellular particle movement. This "freeze" is analogous to, but apparently of longer duration than, the freeze

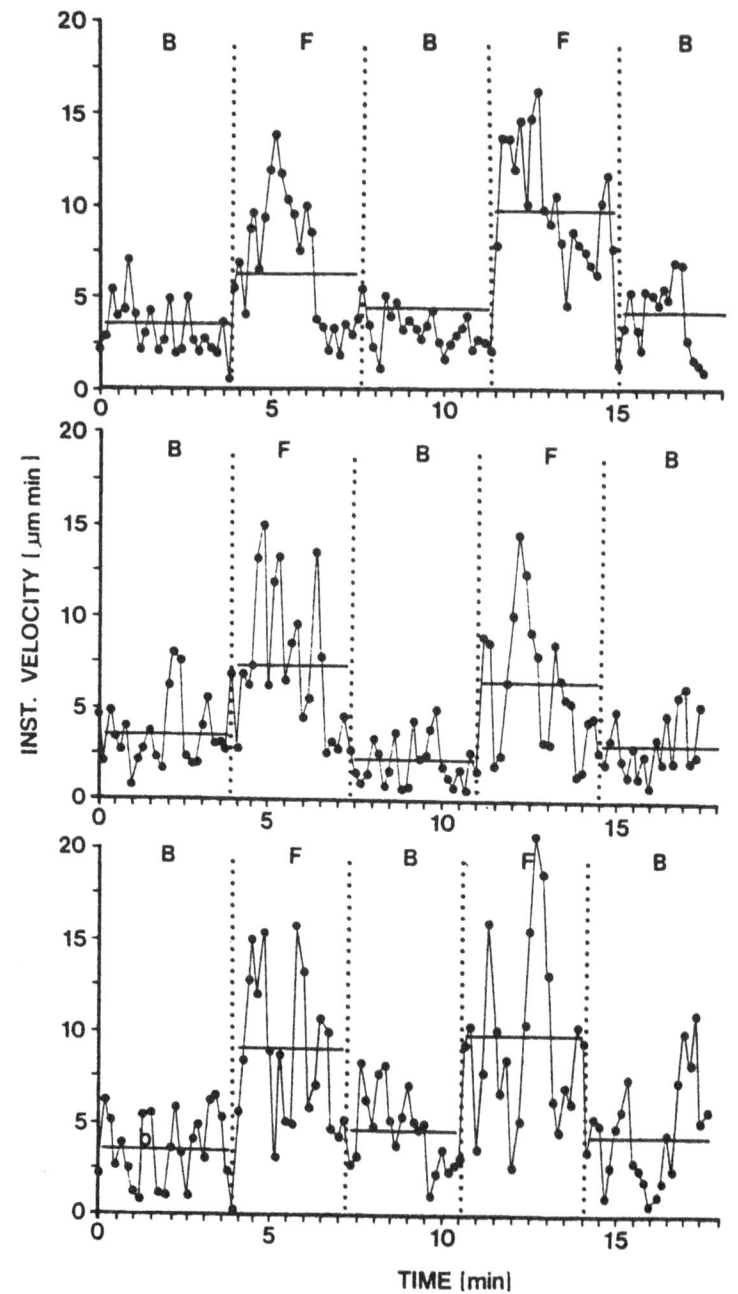

Figure 28.4. Instantaneous velocity of three individual amoebae within 60 μm of each other in a natural aggregation territory.

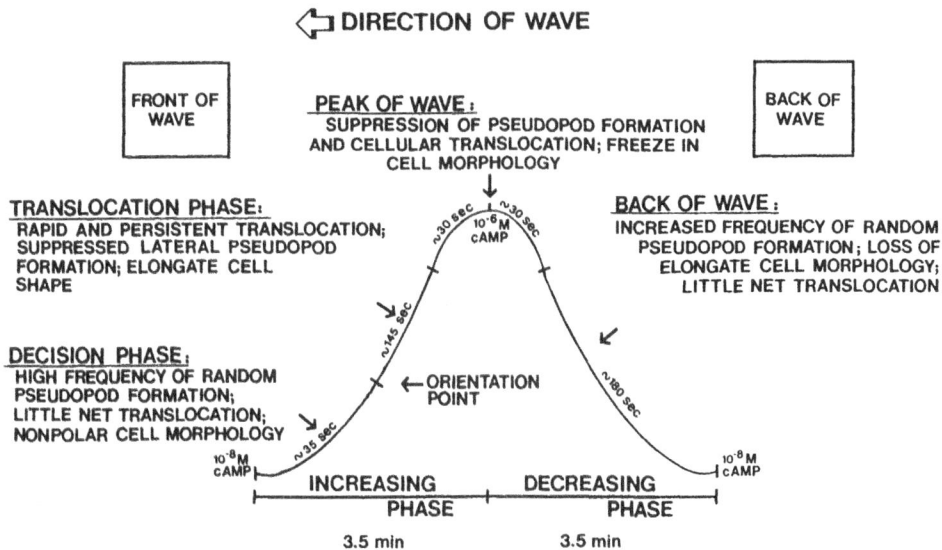

Figure 28.5. Wave behavior interpreted from studies in natural waves and in simulated temporal waves

caused by the rapid addition of 10^{-6}M cAMP to a perfusion chamber containing rapidly translocating amoebae (Wessels *et al.* 1989). During the final 180 seconds of the back of a wave, pseudopod formation is reinitiated, but cells do not regain polarity, and exhibit a very high degree of directional change. Therefore, during this period, cells do not exhibit persistent and directed translocation.

The behavior of cells in natural waves with 7 minute periodicity is similar to that of cells in simulated temporal waves in every respect but directionality. As noted, cells appear to adjust direction at the onset of the front of the natural wave. In simulated waves, where the onset of the wave can be more precisely estimated, there appears to be a short period of increased pseudopod formation followed by the translocation phase (Varnum *et al.* 1985). We believe that cells realign polarity in the direction of the source sometime during this initial period, then move in a relatively blind fashion during the translocation phase. Elucidating the mechanism involved in the decision for subsequent direction represents a major goal in our present research.

28.5 ASSESSING THE INVOLVEMENT OF SPECIFIC CYTOSKELETAL ELEMENTS IN THE AGGREGATION PROCESS

Chemotaxis and aggregation in a population of *D. discoideum* amoebae is a carefully orchestrated process involving signal development and relay, and individual cell response. There are three components of the cellular response (Loomis 1982): (1) secretion of the signal; (2) directed cell motility; and (3) the regulation of gene expression in the program of differentiation and morphogenesis. All of these components are receptor-mediated and involve sensory transduction pathways (Devreotes *et al.* 1987; Newell & Europe-Finner 1990). *D. discoideum* is uniquely suited for analyzing the different responses to receptor occupancy not only because of the ease of experimental manipulation, but also because of the capacity to knock out any gene of interest by site-specific recombination (*e. g.* De Lozanne & Spudich

1987). Since *D. discoideum* is haploid, mutations are immediately dominant. In the case of the genes for all cytoskeletal elements related to actin-mediated events, mutations have not been lethal. Rather, they have resulted in modifications in behavior which have been identified by use of computer-assisted dynamic image analysis (Wessels & Soll 1990; Wessels *et al.* 1991; Wessels *et al.* 1988; Cox *et al.* 1992; Titus *et al.* 1992). For instance, it has been demonstrated that in the absence of conventional myosin II, cells lose polarity, elongate shape, the capacity to respond rapidly to the addition of 10^{-6}M cAMP, and the capacity to translocate rapidly and persistently (Wessels *et al.* 1988; Wessels & Soll 1990). Although they still generate pseudopodia at normal frequencies, the pseudopodia form randomly around the cell periphery, and are smaller. In the absence of either myosin IA (Titus *et al.* 1992) or IB (Wessels *et al.* 1991), members of the myosin I gene family, cells form lateral pseudopodia roughly 3 times as frequently as control cells. Therefore, anterior pseudopod dominance is diminished and cells zig-zag while translocating in buffer or when chemotaxing up a spatial gradient of cAMP. In the absence of the actin cross-linking protein ABP-120, cells form pseudopodia at a reduced rate (Cox *et al.* 1992). Interestingly, in all of the cytoskeletal mutants but the myosin II null mutants which we have tested, cells are capable of some translocation and chemotaxis, but the efficiency and precision of these behaviors are diminished.

However, all previous characterizations of mutant phenotypes suffer from two major flaws. The first involves responsiveness to chemoattractant. In no case has cell behavior been tested in a natural wave. Since all pathways in sensory transduction, signal relay, and the motile response have evolved within the confines of the dynamics of the natural wave, the impact of a mutation, particularly in a cytoskeletal gene, must be assessed in the natural wave. However, the mutation may affect the dynamics of the wave generated by a population of mutant cells. Therefore, we have developed a simple test system (Sylwester *et al.* 1992). Mutant cells are labeled with *Di*I, a vital membrane fluorescent stain, and mixed among wild-type cells in a natural aggregation territory at a ratio of 1 to 10. In this combination, wild-type cells dominate wave development. Therefore, each mutant cell on average is surrounded by control cells, and videorecordings are made of mutant and neighboring control (wild-type) cells as they respond to natural waves (Figure 28.6). This technique has been used to characterize myosin IA deficient cells, and the results suggest that these cells zig-zag as a result of increased pseudopod formation, during the translocation phase. There is little doubt that this technique will be the acid-test for future studies of mutant phenotypes in *D. discoideum.*

28.6   3-DIMENSIONAL ANALYSIS OF CELLULAR BEHAVIOR IN THE NATU-RAL WAVE

The second flaw in all previous studies of wild-type and mutant cells involves dimensionality. To date, the behavioral responses of amoebae to the rapid addition of cAMP, to spatial gradients of cAMP, to a simulated temporal wave, and to the natural wave have been studied in 2-dimensions. This is of course reasonable since natural aggregation territories are generated in 2D. However, as recently demonstrated for polymorphonuclear leukocytes (Murray *et al.* 1992), cellular behavior is 3-dimensional and involves a 3D motility cycle which is not immediately evident using 2D dynamic image analysis technology. Therefore, a complete description of *D. discoideum* behavior in the natural wave must be performed in 3D, and the ultimate characterization of the behavioral phenotype of a mutant cell will also have to be performed in 3D. We have therefore developed a 3D dynamic image analyzing system,

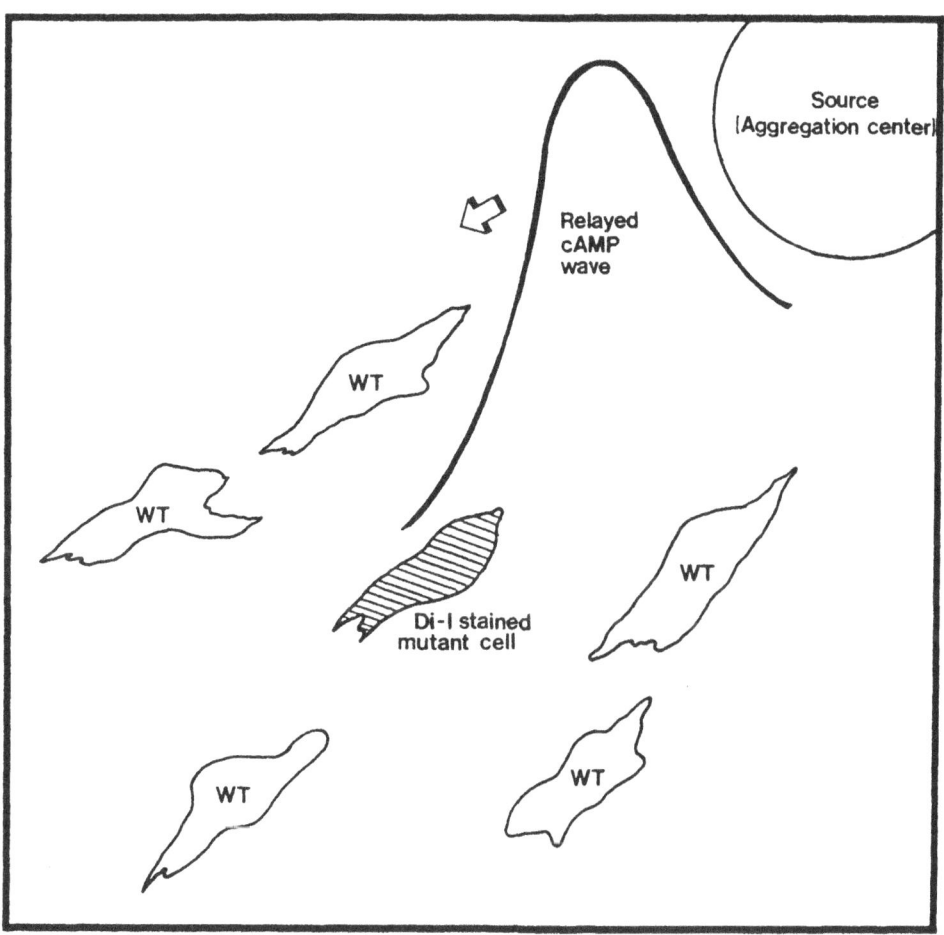

Figure 28.6. Characterizing the behavior of a mutant cell in natural waves generated by wild-type cells

3D-DIAS (Soll *et al.* 1993), which builds on the software of the 2D dynamic image analyzing systems DMS (Soll 1988; Soll *et al.* 1987; Soll *et al.* 1988) and the next generation 2D system 2D-DIAS. For 3D-DIAS, cells are optically sectioned within a 2 second time window using differential interference contrast microscopy at intervals of 5 second. The sections are then reconstructed into a plate image which is used to generate a faceted and then smoothed 3D-image The plated, faceted or smoothed images can then be viewed in 3-D through the polarizing screen of a Tektronix 3-D workstation stereo-monitor. 3D-DIAS software can then calculate over 100 parameters of motility and morphology based upon the 3D path of the centroid and the 3D contour of the cell (Murray *et al.* 1992; Soll *et al.* 1993).

When applied to the basic behavior of a *D. discoideum* amoeba moving on a glass surface in buffer, several new aspects of pseudopod formation and centroid translocation were evident (Wessels *et al.* 1993). Firstly, we discovered that the majority of new lateral pseudopods (70%) were formed off the substratum, and that pseudopods formed off the substratum were usually retracted, while pseudopods formed on the substratum in almost all cases resulted in a turn. Secondly, we found that each *D. discoideum* amoeba moving in rapidly perfused

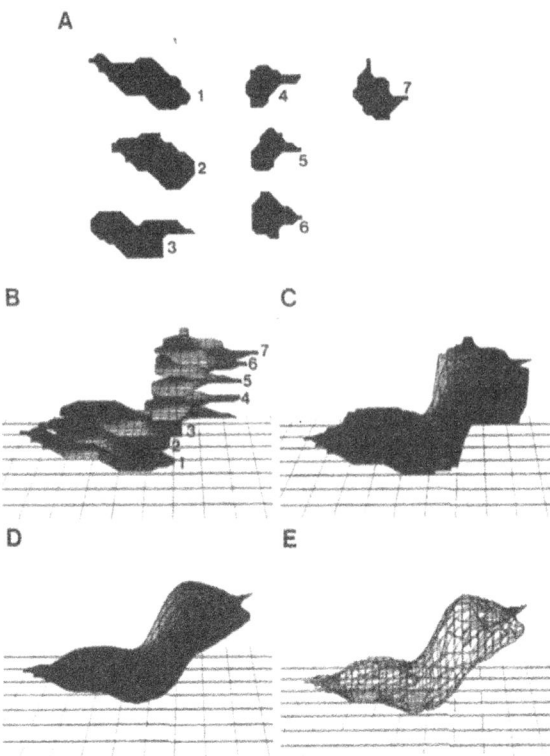

Figure 28.7. Optically sectioning and reconstructing cell morphology in 3D using 3D-DIAS.

buffer in the absence of chemoattractant exhibited a motility cycle of roughly 1.5 minute, similar to the cycle demonstrated by human polymorphonuclear leukocytes (Murray *et al.* 1992). In Figure 28.8, examples are presented of two cycling parameters, 3D instantaneous velocity of the cell centroid and the 3D volume of the expansion zone, for two individual amoebae. The cycle is evident. We are only now applying this new technology to amoebae in natural waves and to cytoskeletal mutants. Our first results with the myosin IA mutant indicate that the increase in lateral pseudopod formation causes a disruption of the normal cycle Figure 28.9

28.7 CONCLUSION

The signal relay system in *D. discoideum* establishes a pattern of rings of translocating amoebae. We have demonstrated that the behavior of amoebae is regulated to a large extent by the temporal dynamics of the chemotactic waves. By comparing behavior in simulated temporal and natural waves, we have found that the individual phases of behavior including rapid translocation in the front of the wave, the freeze at the peak of the wave, and the loss of both directionality and the capacity to move in a persistent fashion in the back of the wave are shared. However, we have not elucidated the mechanism by which cells decide on direction at the beginning of the front of the wave. This crucial decision point might still be based upon a spatial mechanism, although our intuition tells us that a temporal mechanism is involved at the level of pseudopod extension.

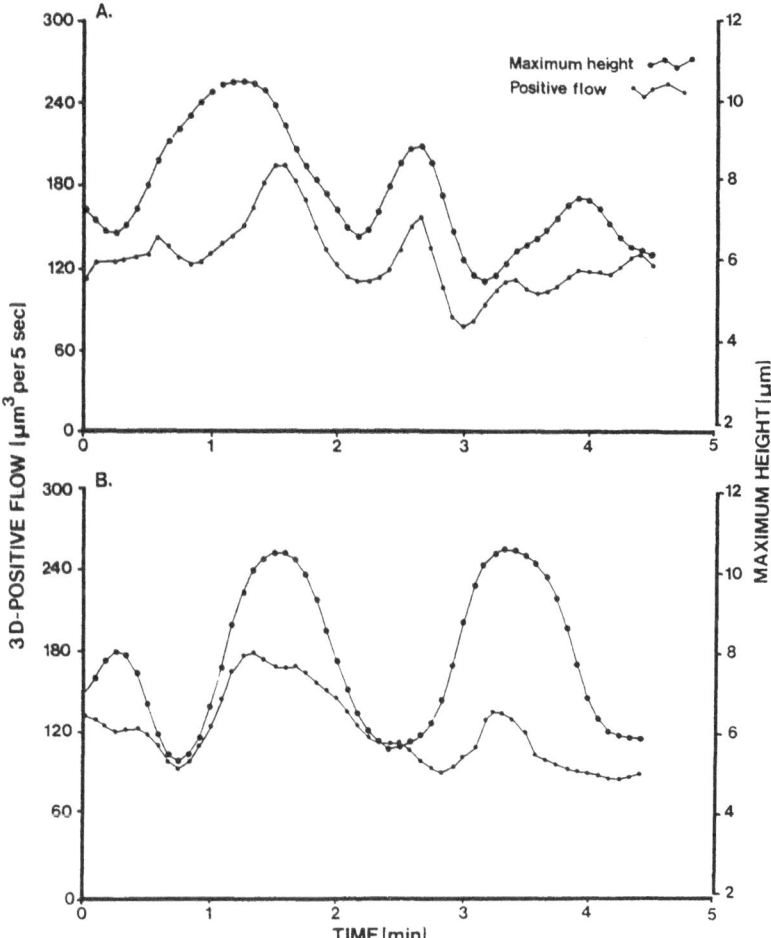

Figure 28.8. The 3-dimensional motility cycle of *wild-type D. discoideum* amoebae translocating in buffer.

We have also begun to look at behavior in 3 dimensions, and have discovered motility cycles in both PMNs and *D. discoideum* amoebae. We believe that the motility cycle represents the basic behavior of a cell, but we do not know how this behavior is manipulated by the natural wave. Finally, we have criticized our previous studies of cytoskeletal mutants for two reasons. Firstly, they were not performed under conditions which tested the responsiveness of cells to natural waves, and secondly, they were performed in 2 dimensions. Techniques for examining mutant cells in 3 dimensions in a natural wave have now been developed in our laboratory, and will be applied to a number of cytoskeletal and sensory transduction mutants.

ACKNOWLEDGMENTS The authors are indebted to E. Voss, H. Vawter-Hugart, and J. Murray for their participation in different aspects of these studies. This work was supported by grant HD18577 and by training grant HD07216 from the National Institutes of Health and by a grant from the Iowa Economic Development Commission.

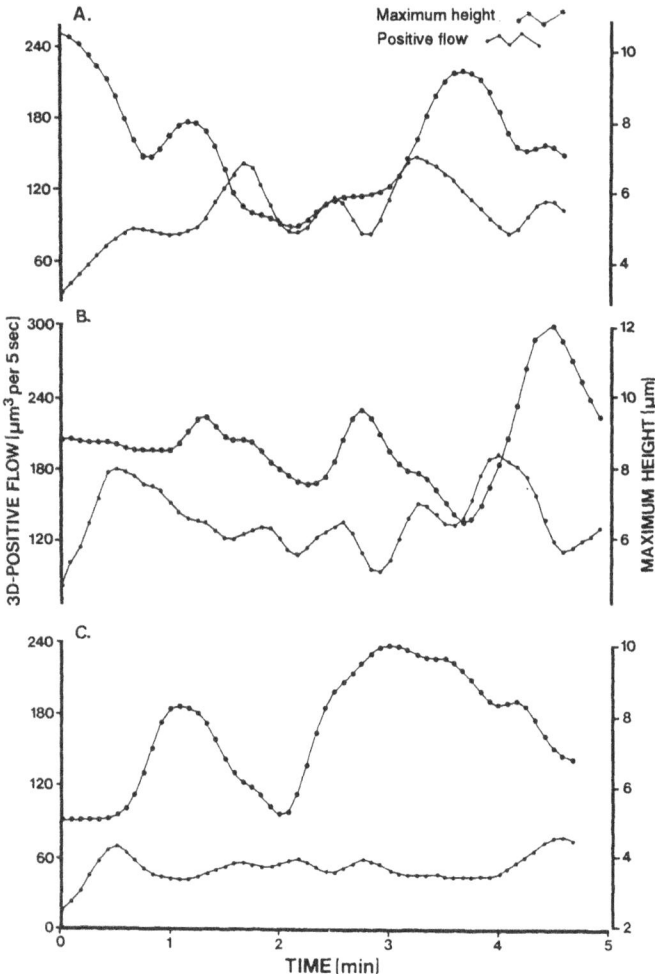

Figure 28.9. Disruption of the 3-dimensional motility cycle in a myosin IA mutant of *D. discoideum*.

REFERENCES

Alcantara, F., & Monk, M. 1974. Signal propagation in the cellular slime mould *Dictyostelium discoideum*. *J. Gen. Micro.*, **85**, 321–324.

Bonner, J. 1947. Evidence for the formation of cell aggregates by chemotaxis in the development of the slime mould *Dictyostelium discoideum*. *J. Exp. Zool.*, **106**, 1–26.

Boyd, A., & Simon, M. 1982. Bacterial chemotaxis. *Ann. Rev. Physiol*, **44**, 501–517.

Brown, P., & Berg, H. 1974. Temporal stimulation of chemotaxis in *Escherichia coli*. *Proc. Natl. Acad. Sci.: (USA)*, **71**, 1388–1392.

Cox, D., Condeelis, J., Wessels, D., Soll, D., Kern, H., & Knecht, D. 1992. Targeted disruption of the ABP-120 gene leads to cells with altered motility. *J. Cell Biol.*, **116**, 943–955.

De Lozanne, A., & Spudich, J. A. 1979. Disruption of the *Dictyostelium* myosin heavy chain gene by homologous recombination. *Science*, **236**, 1086–1091.

Devreotes, P. N., & Steck, T. 1979. Cyclic AMP relay in *Dictyostelium discoideum*. II. Requirements for the initiation and termination of the response. *J. Cell Biol.*, **80**, 300–309.

Devreotes, P. N., & Zigmond, S. H. 1988. Chemotaxis in eukaryotic cells: a focus on leukocytes and *Dictyostelium. Ann. Rev. Cell Biol.*, **4**, 649–686.

Devreotes, P. N., Potel, M., & MacKay, S. 1983. Quantitative analysis of cyclic AMP waves mediating aggregation in *Dictyostelium discoideum. Dev. Biol.*, **96**, 405–415.

Devreotes, P. N., Fontana, D., Klein, P., Scherring, J., & Theibert, A. 1987. Transmembrane signaling in *Dictyostelium. Methods Cell Biol.*, **28**, 299–331.

Durston, A. 1974. Pacemaker activity during aggregation in *Dictyostelium discoideum. Dev. Biol.*, **37**, 225–235.

Fisher, P., Merkl, R., & Gerisch, G. 1989. Quantitative analysis of cell motility and chemotaxis in *Dictyostelium. J. Cell Biol.*, **92**, 807–821.

Futrelle, R., Traut, J., & Mckee, W. 1982. Cell behavior in *Dictyostelium discoideum*: preaggregation response to localized cAMP pulses. *J. Cell Biol.*, **92**, 807–821.

Gerisch, G. 1968. Cell aggregation and differentiation in *Dictyostelium discoideum. Curr. Top. Devl. Biol.*, **3**, 157–197.

Gerisch, G., & Keller, H. 1981. Chemotactic reorientation of granulocytes stimulated with micropipettes containing fMET-Leu-Phe. *J. Cell. Sci.*, **52**, 1–10.

Gross, J., Peacey, M., & Trevan, D. 1976. Signal emission and signal propagation during early aggregation in *Dictyostelium discoideum. J. Cell Sci.*, **22**, 645–656.

Konijn, T., Barkley, D., Chang, Y., & Bonner, J. 1968. Cyclic AMP: a naturally occurring acrasin in the cellular slime molds. *Am. Nat.*, **102**, 225–233.

Loomis, W. F. 1982. The Development of *Dictyostelium discoideum*. New York: Academic Press.

Malchow, D., Nagele, B., Schwarz, H., & Gerisch, G. 1972. Membrane-bound cyclic AMP phosphodiesterase in chemotactically responding cells of *Dictyostelium discoideum. Eur. J. Biochem*, **28**, 136–142.

Murray, J., Vawter-Hugart, H., & Soll, D. R. 1992. Three-dimensional motility cycle in leukocytes. *Cell Motil. Cytoskel*, **22**, 211–223.

Newell, P. C., & Europe-Finner, G. N. 1990. Signal transduction for chemotaxis in *Dictyostelium* amoebae. *Sem. Cell Biol.*, **1**(2), 105–114.

Shaffer, B. M. 1957. Aspects of aggregation in the cellular slime molds. 1. orientation and chemotaxis. *Nature*, **91**, 19–35.

Shaffer, B. M. 1975. Secretion of cAMP-induced by cAMP in the cellular slime mold *Dictyostelium discoideum. Nature*, **255**, 549–552.

Silverman, M., & Simon, M. 1974. Flagellar rotation and the mechanism of bacterial motility. *Nature*, **249**, 73–74.

Soll, D. R. 1979. Timers in developing systems. *Science*, **203**, 841–849.

Soll, D. R. 1988. DMS, a computer-assisted system for quantitating motility, the dynamics of cytoplasmic flow, and pseudopod formation: Its applicating to *Dictyostelium* chemotaxis. *Cell Motil. Cytoskel.*, **10**, 91–106.

Soll, D. R. 1990. Behavioral studies into the mechanism of eukaryotic chemotaxis. *J. Chem. Ecol.*, **16**, 133–150.

Soll, D. R., Voss, E., & Wessels, D. 1987. Development and application of the "Dynamic Morphology System" for the analysis of moving amoebae. *Proc. SPIE*, **832**, 821–830.

Soll, D. R., Voss, E., Varnum-Finney, B., & Wessels, D. 1988. The "Dynamic Morphology System": A method for quantitating changes in shape, pseudopod formation and motion in normal and mutant amoebae of *Dictyostelium discoideum. J. Cell Biochem*, **37**, 177–192.

Soll, D. R., Vawter-Hugart, H., & Voss, E. 1993. "3D-DIAS": A computer-assisted method for quantitating the three-dimensional motion parameters of motile cells. (In preparation).

Swanson, J., & Taylor, D. L. 1982. Local and spatially coordinated movements in *Dictyostelium discoideum* amoebae during chemotaxis. *Cell*, **28**, 225–232.

Sylwester, A. W., Wessels, D., & Soll, D. R. 1992. Myosin-IA cells among aggregating wild-type cells exhibit aberrant chemotactic behaviors. (In preparation).

Titus, M., Wessels, D., Spudich, J., & Soll, D. R. 1992. *The unconventional myosin encoded by the myoA gene plays a role in* Dictyostelium *motility. Molec. Biol. Cell.* (In press).

Tomchik, K., & Devreotes, P. 1981. cAMP waves in *Dictyostelium discoideum*: Demonstration by a novel isotope dilution fluorography technique. *Science*, **212**, 433–446.

Varnum, B., Edwards, E., & Soll, D. 1985. *Dictyostelium* amoebae alter motility differently in response to increasing versus decreasing temporal gradients of cAMP. *J. Cell Biol.*, **101**, 1–5.

Varnum-Finney, B., Edwards, K., Voss, E., & Soll, D. 1987a. Amoebae of *Dictyostelium discoideum* respond to an increasing temporal gradient of the chemoattractant cAMP with a reduced frequency of turning: evidence for a temporal mechanism in ameboid chemotaxis. *Cell Motil. Cytoskel*, **8**, 7–17.

Varnum-Finney, B., Voss, E., & Soll, D. 1987b. Frequency and orientation of pseudopod formation of *Dictyostelium discoideum* amoebae chemotaxing in a spatial gradient: further evidence for a temporal mechanism. *J. Cell Motil. Cytoskel*, **8**(18-26).

Wessels, D., & Soll, D. R. 1990. Myosin II heavy chain null mutant of *Dictyostelium discoideum* exhibits defective intracellular particle movement. *J. Cell Biol.*, **111**, 1137–1148.

Wessels, D., Soll, D. R., Knecht, D., Loomis, W. F., DeLozanne, A., & Spudich, J. A. 1988. Cell motility and chemotaxis in *Dictyostelium* amoebae lacking myosin heavy chain. *Dev. Biol.*, **128**, 164–177.

Wessels, D., Schroeder, N., Voss, E., Hall, A., Condeelis, J., & Soll, D. R. 1989. cAMP-mediated inhibition of intracellular particle movement and actin reorganization in *Dictyostelium*. *J. Cell Biol.*, **109**, 2841–2851.

Wessels, D., Murray, J., Jung, G., J.A. Hammer, III, & Soll, D. R. 1991. Myosin IB null mutants of *Dictyostelium* exhibit abnormalities in motility. *Cell Motil. Cytoskel.*, **20**, 301–315.

Wessels, D., Murray, J., & Soll, D. R. 1992. Behavior of *Dictyostelium* amoebae is regulated primarily by the temporal dynamic of the natural cAMP wave. *Cell Motil. Cytoskel*, **23**, 145–156.

Wessels, D., Murray, J., Sylwester, A., Vawter-Hugart, H., & Soll, D. R. 1993. The three dimensional dynamics of pseudopod formation and turning during the motility cycle of *Dictyostelium*. (Submitted).

Wolpert, L. 1971. Positional information and pattern formation. *Curr. Top. Dev. Biol.*, **6**, 183–224.

29. RETINOIC ACID: AN AUTOCATALYTIC MORPHOGEN

Dennis Summerbell

The National Institute for Medical Research
The Ridgeway, Mill Hill
London NW7 1AA

29.1 DEVELOPMENTAL MECHANISMS

Human development starts with the production of sperm and egg and ends with death. In between lie the embryo, the fetus, the child and the adult. Most developmental biologists are interested primarily in a short period of twelve weeks lying between fertilisation and fetus, The period of embryogenesis. During embryogenesis the fertilised egg cell develops into an immature form of the normal adult organism. A single relatively unspecialised cell will divide into 25×10^9 cells, comprising a few hundred different cell types, arranged in a complex three dimensional pattern. During this period of human development three main cellular activities are involved. Firstly, the cells have to divide. Initially this is simple and repetitive, but gradually a complicated pattern of varying cell cycle times will emerge. Secondly, the cells must differentiate in different ways so as to produce the diversity of cell types. This involves both transcription of new genes and a progressive restriction of a cell's potential differentiative paths. Thirdly, the different cell types must be arranged in an appropriate pattern so as to produce a functional embryo. Pattern formation can involve both decision making that leads to differential transcription and rearrangement of cells that have already differentiated. If we compare human development with common laboratory models it is clear that they consist of similar cells arranged in different patterns. A human embryo comprises much the same cell types as a monkey, a mouse, or even a magpie. The difference lies in the details of the three dimensional pattern as much as in the gene products. My main interest lies in factors that control this pattern.

29.1.1 Pattern Formation

In principle there are two ways of controlling pattern formation. First, the fertilised egg may be laid out as a map of the final form. As the cells divide, the egg cytoplasm becomes sub-divided into smaller units. Eventually each nucleus becomes surrounded by a unique fraction of the original cytoplasm. The cytoplasmic constituents influence the control of transcription in the nucleus and determine the cells terminal differentiated state. The final pattern is a fixed function of the initial pattern of cytoplasm. This view of development was initially supported by the experiments of Roux (1988). He used a hot needle to kill one of the first two blastomeres of the frog egg. The surviving cell produced only half of the embryo. This type of development Roux called "mosaic", or "self-differentiation".

Experimental and Theorietcal Advances in Biological Pattern Formation,
Edited by H.G. Othmer *et al.*, Plenum Press, New York, 1993

The second mechanism supposes that the products of division cycles are intrinsically equipotential. The identical cells require external stimuli to determine how they will differentiate. These developmental signals can come from the external environment, or from other cells. The evidence supporting this type of mechanism was originally provided by Driesch (1892) who showed that if one separated blastomeres of sea urchins, each cell went on to produce the entire embryo. He styled this type of development as regulative.

In the 100 years since Driesch's paper we have come a long way, yet we still cannot say that we understand development. We have learned that both mechanisms are involved in normal development. For example, cytoplasmic determinants play a critical role during early development of amphibian embryos. However, long range signals seem to be of greater importance in many later stages of development. In this paper we will examine in detail one such signalling system: the function of retinoic acid during pattern formation in the developing vertebrate limb.

29.2   THE CHICK EMBRYO

Initially the fertilised egg is transformed into a multi-cellular layered structure (comprising ectoderm, mesoderm and endoderm) with a head, a tail and bilateral symmetry about the mid-line. This is sometimes known as the formation of the primary body plan. At this stage we would describe the whole embryo as comprising a single embryonic field (Huxley & de Beer 1934). This first field maps out the positions of secondary fields and creates the initial conditions which control their subsequent development. The limb bud is one of these secondary fields. It arises from an oval shaped patch of cells in the lateral plate. It comprises a thin peripheral layer of ectoderm enclosing a loosely packed mass of mesoderm. The limb bud mesoderm cells divide faster than the cells of the surrounding flank tissue so that the limb bud starts to bulge out. Cell division remains high at the distal tip of the bud but decreases proximally. Thus the mass of cells is expanding most rapidly at the tip. The limb unfolds with proximal structures appearing from the undifferentiated tip region before the more distal structures. Wolpert (1969) described the limb as a three-dimensional structure mapped along three orthogonal coordinate axes: dorso-ventral, antero-posterior and proximo-distal. He expected positional information to be provided independently for each axis so that every cell would know its position within the structure.

29.2.1   Dorso-ventral and proximo-distal axes

We know comparatively little about control of development across the dorso-ventral axis. We know rather more about the proximo-distal axis. Distally it is regulative, proximally it is mosaic. It has a distal organiser (the apical ectodermal ridge or AER) and at least two, as yet unidentified, medium range (approximately 300μm) signals. One of these signals maps out the region at the distal tip known as the progress zone. Cells in the progress zone are able to respond to changes in environmental signals, modifying their developmental potential. This includes responding to signals controlling the antero-posterior axis.

29.2.2   The Antero-Posterior Axis

When tissue from the posterior lateral margin of a limb bud is grafted to the anterior margin it induces a mirror-image duplication of the limb pattern (Saunders & Gasseling 1968). The graft (the zone of polarising activity, or ZPA) organises host cells to form the additional structures (Summerbell 1981). It has been suggested that the ZPA produces a morphogen that diffuses across the limb to form a monotonic graded concentration profile. The local concentration of the morphogen determines the pattern of differentiation of the responding limb cells (Tickle *et al.* 1975). The chief candidate for the hypothetical morphogen is all-trans retinoic acid (RAt). When a local source of RAt is implanted at the anterior margin it mimics the effect of a ZPA graft (Summerbell 1983; Tickle *et al.* 1982). Furthermore, RAt is present in normal limbs and is relatively more abundant in the posterior (ZPA containing) region (Thaller & Eichele 1987). These and numerous other observations have led to the hypothesis that RAt is the natural morphogen controlling pattern formation across the AP axis.

29.3 RETINOIC ACID

Retinoic acid (RA) is a lipophilic molecule of about 300 daltons belonging to a class of vitamin A derivatives, the retinoids. They are involved in a wide range of biological phenomena. These include differentiation, cell communication, proliferation, cell movement, vision and nutrition. They are of considerable clinical significance in the treatment of acne, skin cancer and differentiative diseases of the skin. It is of increasing interest to the cosmetics industry because of its ability to reverse photo-aging of skin. On the down side it is a proven potent teratogen and exposure is strongly contra-indicated for all females of child bearing age.

RA is synthesised from retinol (vitamin A) via retinaldehyde and is removed by oxidation into relatively hydrophilic analogs. It is carried round the body bound to a serum binding protein and is present in the cytoplasm of cells bound to cellular retinoic acid binding protein (CRABP). RA also binds to a family of nuclear receptor proteins (retinoic acid receptors or RARs) that are members of the steroid/thyroid hormone receptor group. Each receptor protein in the family binds strongly to its specific hormone (or ligand). The resulting ligand-receptor complex is able to bind to a specific response element in the DNA. This up-regulates or down-regulates expression of the gene(s) controlled by that response element.

29.3.1 Alternative hypotheses

The basic working hypothesis is that retinol, the precursor, is converted to RA by the cells in the ZPA. The RA then diffuses across the limb bud to form a graded monotonic concentration profile: high in the posterior margin and low in the anterior margin. At high concentrations the RA-RAR ligand-receptor complex binds to appropriate response elements modifying the pattern of gene transcription across the limb. During a ZPA graft one is moving a group of RA producing cells into the anterior margin. They continue to produce RA setting up a symmetrical gradient and therefore inducing additional limb elements in mirror image symmetry. Implantation of a local source of RA into the anterior margin provides a passive reservoir of RA sufficient to set up and maintain a similar concentration profile.

An alternative hypothesis is that RA is not the product of the ZPA but rather that RA acts to induce the formation of a ZPA from normal limb cells. When a local source of RA is implanted in the anterior margin of the limb bud, tissue next to the implant acquires functional ZPA-like abilities within 18-24 hours (Summerbell & Harvey 1983). The *de novo* ZPA cells are able to induce mirror-image reduplication when grafted to a secondary host.

Indirect methods suggest that the additional structures produced are not the result of passive transfer of RA. Various strategies have been devised to show that the tissue adjacent to the graft cannot contain sufficient RA to cause the induction of additional digits (Bryant *et al.* 1991; Summerbell & Waterson 1991; Tickle 1991; Wanek *et al.* 1991).

It is also possible to measure RA directly in the "induced" ZPA and in the implant using high performance liquid chromatography (HPLC). The figures are disappointing as they fail to give unequivocal support for the data provided using indirect bio-assays. Tissue adjacent to an RA implant contains significantly less RA than is needed in an implant to ensure a full reduplication (approximately 10 times less). Yet it contains much more RAt than is present endogenously in a normal ZPA (approximately 100 times more). That same block of tissue, if grafted to a secondary host, would normally give a full reduplication. The direct measurements lie awkwardly in a range that fails to support either hypothesis, (Summerbell & Waterson 1991).

29.3.2   Autocatalysis

The abundance of RA in the tissue adjacent to the implant suggests the possibility that the implant is stimulating the adjacent tissue to produce its own retinoic acid. This has been tested using two functional synthetic analogs that cannot be metabolised to RA.

Both analogs induce the synthesis of copious amounts of RA in the experimental limbs (10-100 fold normal levels). The induced retinoic acid is distributed in a monotonic concentration profile across the limb, high towards the implant and low away from the implant. RA itself also produces a graded concentration profile across the experimental limb but at a significantly lower level. Retinol (the metabolic precursor of RA) has a weak ability to induce additional digits. Direct measurements using HPLC show that Retinol does induce synthesis of RA, but at a much lower level and is not in a gradient. The amount of RA induced per analog correlates well with the known potency of inducing extra digits, CD367 = TTNBP > RAt >> Rol (Maden *et al.* 1991; Maden & Summerbell 1986).

29.3.3   Other retinoids

Retinoic acid is not the only retinoid that has been proposed as a possible natural morphogen. Retinol (Rol), a natural precursor of retinoic acid, is normally present at a concentration about 10-fold higher than RAt. 3,4-di- dehydroretinoic acid (ddRA), an en- dogenous functional analog, is normally present in similar concentrations to RA and has been proposed as a possible alternative morphogen (Thaller & Eichele 1990). It binds efficiently to RARs and has similar morphogenetic potency. Levels of both are enhanced following induction, but neither produce a consistent spatial or temporal pattern. 9-cis retinoic acid has been suggested more recently but as yet no data is available on its ability to induce production of itself or RA.

29.4   DISCUSSION

Numerous theoretical models have been proposed that attempt to elucidate the under- lying mechanisms controlling the antero-posterior axis of limb development. The simplest have been based on a variation of the source-sink diffusion model (Crick 1970; Wolpert 1969). If the cells at one extremity act as a source and all other cells destroy the morphogen then one obtains a concentration profile that is exponential even at equilibrium (Tickle *et al.* 1975). This has proved a very robust model though it has the disadvantage that it only works in semi-infinite fields where the size of the field is greater than the effective range of diffusion of the morphogen (Summerbell 1979). The model accounts for almost all of the experimental data obtained for the antero-posterior axis of the chick limb.

The most widely discussed alternatives to source-sink diffusion models are based on Turing models (Turing 1952). An example is the reaction-diffusion model (Gierer & Meinhardt 1972). Here, an array of cells is assumed to produce two substances. The activator (A) diffuses slowly from cell to cell across the field. The inhibitor (I) diffuses more rapidly. Activator stimulates production of both activator and inhibitor whereas inhibitor only inhibits production of activator. Initially, an array of cells may have homogeneous, low concentrations of both activator and inhibitor. However, this is unstable; chance local fluctuations will in places elevate activator concentration to a level where autocatalysis amplifies its own production and it starts to rise rapidly. At the same time it also amplifies production of inhibitor. Because inhibitor is free to rapidly diffuse away from the site of peak production it represses any rise in activator in adjacent cells, but the initial site of activation escapes from inhibition. The array of cells reaches an equilibrium with areas of stable high concentration and areas of stable low concentration. Peaks, dependent on various other parameters, tend to be evenly spaced. The model is at its best when explaining repeating patterns such as coat colour on skin (Murray 1981). However if one looks at the special case where the array of cells is the same length as the wavelength of the peaks then one has a formal situation difficult to distinguish from some source-sink diffusion models.

The observation that retinoic acid may be autocatalytic must tilt our thinking towards Turing models. Our next step will be to search for direct evidence for the catalytic mechanism. It seems unlikely that the synthetic analogues can act directly in the metabolic pathway of RAt synthesis. The obvious route of action is therefore via transcriptional control mechanisms. Both TTNBP and CD367 (Bailly *et al.* 1990) bind strongly to known RAt nuclear receptor proteins (NRP's), three of which have already been identified (RAR-α, RAR-β, RAR-γ). The holoprotein (RAt-NRP) is in turn expected to bind to appropriate response elements on the DNA so as to regulate transcription of their target gene (see review, Tabin 1991). Response elements have already been identified in the promoter region of genes coding for RAR-β and a cytoplasmic retinol binding protein (CRBP-1). If we could demonstrate a positive feedback from either of these proteins on synthesis of RAt we would have closed the loop and provided a definitive mechanism for RAt autocatalysis.

REFERENCES

Bailly, J., Delescluse, C., Bernardon, J. M., Charpentier, B., Martin, B., Pilgrim, W. R., Shroot, B., & Darmon, M. 1990. Differentiation of F9 embryonal carcinoma cells by synthetic retinoids: amplitude of plasminogen activator production does not depend on retinoid potency or affinity for F9 nuclear retinoic acid receptors. *Skin Pharmacol.*, **3**, 256–267.

Bryant, S. V., Hayamizu, T., Wanek, N., & Gardiner, D. M. 1991. Position dependent properties of limb cells. In *Developmental Patterning of the Vertebrate Limb*, Hinchliffe, J. M. Hurle J. R., & Summerbell, D. (eds). pp. 133–142, New York: Plenum Press.

Crick, F. H. C. 1970. Diffusion in embryogenesis. *Nature*, **225**, 420–422.

Driesch, H. 1892. The potency of the first two cleavage cells in echinoderm development, experimental production of partial and double formation. In *Foundations of Experimental Embryology*, Willier, B. H., & Oppenheim, J. M. (eds). pp. 38–50, New York 1964: Prentice Hall.

Gierer, A., & Meinhardt, H. 1972. A theory of biological pattern formation. *Kybernetik*, **12**, 30–39.

Huxley, J. S., & de Beer, G. R. 1934. *The Elements of Experimental Embryology*. Cambridge University Press.

Maden, M., & Summerbell, D. 1986. Retinoic-acid-binding protein in the chick limb bud: identification at developmental stages and binding affinities of various retinoids. *J. Embryol. Exp. Morphol.*, **97**, 239–250.

Maden, M., Summerbell, D., Maignan, J., Darmon, M., & Shroot, B. 1991. The respecification of limb pattern by new synthetic retinoids and their interaction with cellular retinoic acid-binding protein. *Differentiation*, **67**, 49–55.

Murray, J. D. 1981. On pattern formation mechanisms for lepidopteran wing patterns and mammalian coat markings. *Phil. Trans. R. Soc. Lond.*, **B295**, 473–496.

Roux, W. 1988. Contributions to the developmental mechanics of the embryo. In *Foundations of Experimental Embryology*, Willier, B. H., & Oppenheim, J. M. (eds). pp. 20–37, New York 1964: Prentice Hall.

Saunders, J. W., & Gasseling, M. T. 1968. Ectodermal-mesenchymal interactions in the origin of limb symmetry. In *Epithelial-Mesenchymal Interactions*, Fleischmajer, R., & Billingham, R. (eds). pp. 78–97, Baltimore: Williams & Wilkins.

Summerbell, D. 1979. The zone of polarising activity: Evidence for a role in normal chick limb morphogenesis. *J. Embryol. exp. Morph.*, **50**, 217–233.

Summerbell, D. 1981. The control of growth and the development of pattern across the antero-posterior axis of the chick limb bud. *J. Embryol. exp. Morph.*, **63**, 161–180.

Summerbell, D. 1983. The effect of local application of retinoic acid to the anterior margin of the developing chick limb. *J. Embryol. exp. Morph.*, **78**, 269–289.

Summerbell, D., & Harvey, F. 1983. Vitamin A and the control of pattern in developing limbs. In *Limb Development and Regeneration*, Fallon, J. F., & Caplan, A. I. (eds). pp. 109–118, New York: A. R. Liss.

Summerbell, D., & Waterson, N. 1991. Does retinoic acid organise a limb or induce a ZPA? In *Developmental Patterning of the Vertebrate Limb*, Hinchliffe, J. M. Hurle J. R., & Summerbell, D. (eds). pp. 151–155, New York: Plenum Press.

Tabin, C. 1991. Retinoids, homeoboxes, and growth factors - toward molecular-models for limb development. *Cell*, **66**, 199–217.

Thaller, C., & Eichele, G. 1987. Identification and spatial distribution of retinoids in the developing chick limb bud. *Nature*, **327**, 625–628.

Thaller, C., & Eichele, G. 1990. Isolation of 3,4-didehydroretinoic acid, a novel morphogenetic signal in the chick wing bud. *Nature*, **345**, 815–819.

Tickle, C. 1991. Retinoic acid and limb patterning and morphogenesis. In *Developmental Patterning of the Vertebrate Limb*, Hinchliffe, J. M. Hurle J. R., & Summerbell, D. (eds). pp. 143–149, New York: Plenum Press.

Tickle, C., Summerbell, D., & Wolpert, L. 1975. Positional signalling and specification of digits in chick limb morphogenesis. *Nature*, **254**, 199–202.

Tickle, C., Alberts, B., Wolpert, L., & Lee, J. 1982. Local application of retinoic acid to the limb bond mimics the action of the polarizing region. *Nature*, **296**, 564–565.

Turing, A. M. 1952. The chemical basis of morphogenesis. *Phil. Trans. R. Soc. Lond.*, **B237**, 37–72.

Wanek, N., Gardiner, D. M., Muneoka, K., & Bryant, S. V. 1991. Conversion by retinoic acid of anterior cells into ZPA cells in the chick wing bud. *Nature*, **350**, 81–83.

Wolpert, L. 1969. Positional information and the spatial pattern of cellular differentiation. *J. Theor. Biol.*, **25**, 1–47.

30. THE HARDWARE AND SOFTWARE OF MORPHOGENESIS: STUDIES WITH THE *DICTYOSTELIUM DISCOIDEUM* SLUG, A SIMPLE TISSUE

Keith L. Williams and Gregory H. Joss

School of Biological Sciences
Macquarie University
Sydney, NSW 2109, Australia

30.1 INTRODUCTION

In this article we consider morphogenesis as involving two different aspects which we call "hardware" and "software". By "hardware" we mean things that exist and would be found on an inventory of the bits that make up an organism. We refer to "software" as features of the system that involve use of hardware in the process of building the organism.

Experimentalists usually set out to discover hardware features of morphogenesis; *i. e.* they study components of the system, at least in the first instance. Theoreticians are more likely to seek the software, the key processes enabling morphogenesis. We propose that it is useful for both groups to think about morphogenesis in terms of parts and processes, and to link the two when they do experiments. We suspect that there is a significant difference in the complexity of hardware and software. Hardware probably involves many components and much redundancy, while software is likely to be simpler. If there were no master rules (simple software), it is hard to see how such a complex structure as a three dimensional tissue could form. The rules for laying out *Drosophila* development indicate this simplicity (Nüsslein-Volhard 1991).

Once developmental biologists studied morphogenesis with a scalpel and good ideas. The explosion of technological advances in recombinant DNA, monoclonal antibody technology, microscopy and computer based micro-instrumentation, means that experimentalists now have more tools than any one person can hope to master. The stage is set for an explosion of information; the challenge is to make sense of it. There has been no better time for people with different skills and perspectives to collaborate in order to understand what comprises an organism and how it is assembled. While it is unusual for experimentalists and theoreticians to sit in the same laboratories, mixing these different world views makes for lively and productive science. Subtle items of experimental information are often filtered out of the literature, which may lead to incomplete views of experimental findings. In our laboratory we have a multidisciplinary approach to our experiments, employing theoreticians and collaborating widely. A flavour of the fruits of such an approach is given in this report.

Experimental and Theorietcal Advances in Biological Pattern Formation,
Edited by H.G. Othmer *et al.*, Plenum Press, New York, 1993

30.2 HISTORY OF STUDIES ON MORPHOGENESIS

Development of multicellular animals remains one of the major challenges that confronts biologists in the 1990's, despite the fact that morphogenesis is a field of study that dates back to the very beginning of experimental biology. Until recently it largely involved description or crude surgical manipulations (Child 1941; Wolpert 1992). This tradition of descriptive morphology has been continued at the molecular level using monoclonal antibodies in both light and electron microscope studies. Gene probes have also been used to study the expression of specific genes in developing tissues, and in a few cases small molecules have been traced in developing tissues (Devreotes 1989). Analysis of mutants has always had a role in studies of morphogenesis, although the impact of this work has been less than might have been expected due to redundancy in developmental pathways (Witke *et al.* 1992). Recent studies on tumour cell lines have been instructive with respect to cell-cell adhesion (Takeichi 1991). The advent of transgenic organisms with specific gene disruptions (the best developed multicellular system being *Dictyostelium discoideum* (De Lozanne & Spudich 1987)) has opened up new possibilities for definitive identification of key elements in morphogenetic cascades.

The preceding paragraph briefly summarises the study of morphogenesis by experimentalists; it is detailed and full of facts (hardware oriented). However, there is another tradition of study that is theoretical which largely derives from the studies of Turing (1952) (see also Meinhardt 1982). It involves the application of non-linear systems and catastrophe theory to biology. In essence this is a minimalist approach to understanding the key rules upon which morphogenesis is based (software focus). Our view is that these processes are not clearly defined, and we consider both software and hardware when designing experiments and interpreting the results.

30.3 *DICTYOSTELIUM DISCOIDEUM* AS AN EXPERIMENTAL SYSTEM

Some years ago we decided to choose the simplest possible "complex" experimental system in order to allow a multidisciplinary approach to understanding morphogenesis. We sought an organism that allowed the full range of approaches involving theory, biology, morphology, genetics, biochemistry and molecular biology. We chose *Dictyostelium discoideum*. It is an eukaryote, with alternating unicellular and multicellular phases. Many key features of morphogenesis are exhibited in its development and the fact that many laboratories study *D. discoideum* from different viewpoints is a bonus. Of particular note are the studies on the single cell phase (Soll 1987; Gerisch 1986), that provide great precision in understanding individual cells, the basic building blocks of a tissue.

D. discoideum enables examination of morphological development and differentiation essentially without the complication of cell division. Morphogenesis is triggered by starvation and involves a number of steps: aggregation of initially dispersed solitary amoebae as a result of cell-cell signaling; formation of streams of cells, which involves both cell signaling and cell-cell contacts; construction of a progression of increasingly complex three dimensional multicellular structures known to the experimentalist as the "mound", "tipped aggregate" and "standing finger" stages; subsequent formation of the migratory "slug", which has polarity and at least two differentiated cell types, which communicate via signal molecules, cell-cell contact and an extracellular matrix. Finally the asexual fruiting body, comprising a spore head on a cellular stalk which is inserted into a cellular basal disc, is formed by movements of cells in a fashion reminiscent of gastrulation in animal systems (Raper 1984).

These morphological stages encompass many of the features seen in more complicated developmental systems, and they are accessible to experimental study. Cell-cell signaling can be studied in a chemotactic field; cell signalling and cell-cell contact can be examined in 2-dimensional sheets of cells in aggregating streams, or 3-dimensional arrays in later developmental stages. This conversion of 2-dimensional layers of cells to 3-dimensional mounds is correlated with the appearance of an extracellular matrix. As well as being accessible to study using morphological, biochemical and molecular biology techniques, mutants formed by classical genetic techniques or targeted gene disruption are available for study (De Lozanne & Spudich 1987).

30.4 WHICH EXPERIMENTS?

Our focus for experiments has been the "slug" stage, which can be maintained as a motile organism on a petri dish for long periods (days). In the past, the *D. discoideum* slug, a simple 3-dimensional tissue, has been thought of essentially as a "bag of cells" (Odell & Bonner 1986). However, it is well known that the amoebae differentiate into two basic types of cells (prestalk and prespore) in the process of becoming organised to form the slug, so at the very least there are two types of cells in the bag (Odell & Bonner 1986; Williams *et al.* 1986).

The migrating slug exhibits complex behaviour in both space and time. Although the accessibility of this system to experimentation means that this organism is extremely well understood, how shape is determined and how the organism becomes mobile remains obscure. These are the questions that preoccupy us.

Our results suggest that the slug is organised in a much more sophisticated way than previously thought. Here we briefly summarise seven experiments that illustrate new features of morphogenesis in *D. discoideum*. These experiments involve techniques such as surgical manipulation of the organism, various kinds of microscopy, time lapse filming, flow cytometry and analysis of mutants formed by gene disruption. The diversity of approach leads to a mosaic picture of morphogenesis in *D. discoideum*. We are now conscious of the difference between "hardware" and "software" aspects of the developing organism, and each experiment is discussed in relation to this idea. Under the heading "hardware" are items that constitute the nuts and bolts of the system, while under "software" we consider those aspects that involve instructions on how to build an organism. We expect that with this duality in mind, new experiments will be designed that have a better focus.

30.4.1 Experiment 1: Surgical Manipulation to Understand Length Regulation (software)

Observations of time lapse films of emerging slugs of the rapidly developing wild strain WS380B have shown that slug formation is staged in time. In this strain the tip forms and commences migration before the rear of the slug emerges from the aggregate; effectively the front of the organism forms before the rear. There is a period when the slug elongates as the tip migrates. This is followed by a period when there is rapid advancement of the rear to define the adult compact rear zone; this involves shortening of the slug and establishment of the mature length, which is essentially unchanged for several days of migration (Breen *et al.*

1992). Bisection experiments indicate that the slug length is regulated and, if some cells are removed, the remaining cells change shape and/or position to maintain the original length.

Although the importance of length as a variable in morphogenesis was emphasised very early (Morgan 1905), it has generally been ignored (with the exception of Othmer & Pate (1980)). So far we have recognised the importance of length, but this gives no clues about the molecule(s) involved in its definition or their instructions. It is very clear that length regulation involves software rather than hardware (e. g. instructions about counting cells, etc.).

It is interesting to think about the experiments on length establishment and regulation in terms of a multiplicity of signaling pathways. Traditionally, pattern formation is considered as cells sensing gradients of morphogens and differentiating accordingly (Wolpert 1969). The process is often thought of as responding to a single gradient. From studies on *D. discoideum* it is clear that several informational systems are present and they may act in a time dependent sequence. Firstly, there is polarity, or specification of where the tip (organiser) will form. Then there is proportioning to apportion cells to alternate differentiation states. Finally, in the light of the experiments on length regulation, there is definition of the overall shape (length) of the organism. These different features of the organism are not only spatially distinct, but also temporally separate. In this way morphology becomes more complex by layering of different components.

30.4.2   Experiment 2: Microscopic Studies on the Peripheral Cells Reveal an Epithelium (hardware with software input)

Several groups (Schaap *et al.* 1985; Fuchs *et al.* 1992) have improved techniques for preserving the *D. discoideum* multicellular stages for light and electron microscopic observations. Using electron microscopy and freeze substitution techniques, we have recently observed close contact between the cells of the slug and organisation of the peripheral cells into a primitive epithelium (Fuchs *et al.* 1993). At the anterior and ventrum of the slug, the epithelium consists exclusively of prestalk or anterior-like cells; on the dorsal posterior region of the slug, the epithelium consists of a mixture of prespore and prestalk cells.

The epithelium is a boundary layer of cells that defines the extent of the organism. Only these cells have direct contact with the extracellular matrix of the slug. Most work on epithelia has concerned very complicated systems (e. g. skin, cornea, gut lining, etc.). We contend that the formation of an epithelium is integral to the construction of multicellular tissues. By studying the delicate tissue of the *D. discoideum* slug we have shown that an epithelium does occur, although it is fragile and similar to that seen in some keratin mutants of animals (Coulombe *et al.* 1991). A simple tissue like *D. discoideum* can cope with this simple arrangement, whereas an animal needs better stitching. Studies on *D. discoideum* will help us to define this basic epithelium.

The results of our microscopic studies leave us with a set of images (hardware). What do they mean? By thinking about the results, some ideas of the control elements (software) in the system can be deduced. Formerly the extracellular matrix has been thought of as possibly providing shape information (*i. e.* like a sausage skin), although this seemed unlikely. Clearly the epithelial layer is a candidate for a major role in the determination of slug shape.

The rule for epithelium formation in *D. discoideum* allows for different classes of cells to take this role. So in this organism the epithelium is not strictly defined by differentiated state. While most epithelial cells are prestalk or prestalk-like, some on the dorsum are not. Such flexibility is probably only possible in a simple organism such as *D. discoideum*, whose tissues are only loosely organised.

30.4.3 Experiment 3: Use of Flow Cytometry to Demonstrate Gradients in Cell Surface Molecules Along the Slug (hardware with software implications)

The concept of diffusible morphogens dominates ideas about cell patterning, yet morphogens are poorly defined. In animals, only retinoic acid is seriously considered as a possible morphogen, and even in this case there is controversy as retinoic acid specifies a region of influence rather than the influence itself (Noji *et al.* 1991; see other papers in this volume). In *Drosophila* development, where early stages occur in a plasmodium, there is evidence that two maternal proteins, bicoid and dorsal, are present in a gradient which defines the body axis (St. Johnston & Nusslein-Volhard 1992).

By contrast, development in *D. discoideum* is cellular (like mammalian systems) and several morphogens which diffuse between cells are well understood (Firtel 1991; Williams 1988). Here the morphogens are small molecules (*e. g.* cAMP, DIF) rather than gene products.

We have asked the question as to whether there are gradients in cell surface molecules along the *D. discoideum* slug. The answer is yes. Our best studied molecule is PsA, a major surface glycoprotein on prespore cells at the slug stage. By dissecting slugs along their anterio-posterior axis and analyzing, using flow cytometry, the levels of PsA on the surface of prespore cells after disaggregation of the slug sections, a gradient of PsA was observed (Browne & Williams 1993a). PsA is the first cell surface molecule to be shown to be present in a gradient in an emerging tissue. We suspect that this gradient is not instructive, but rather it reflects underlying morphogen gradients (possibly cAMP as pspA, the gene encoding PsA, is cAMP regulated). The gradient may make the prespore cells more cohesive at the rear than the front of the slug. This would mean that the prespore zone is more compacted at the rear, thus preventing loss of cells.

30.4.4 Experiment 4: Use of Flow Cytometry to Examine Redevelopment Characteristics of Separated Slug Cells (software)

We have recently studied the fate of cells disaggregated from the *D. discoideum* slug and sorted into pure populations of prespore and prestalk cells by flow cytometry (Browne & Williams 1993b). It has long been known that cells disaggregated from a slug can reaggregate and form a new slug (Raper 1984). We have shown that the prestalk cells behave in this way, but that the prespore cells dedifferentiate on separation, before regaining aggregation competence. Our interpretation of this is that the prestalk cells are still effective in isolation after disaggregation from the tissue. The prespore cells are much more differentiated as they are preparing to form spores, the long-term-survival stage. They probably act as "cargo" in the slug, where the prestalk cells do most of the work of moving the organism.

In more complex systems, cells divide in response to signals. Recently it has been suggested that cells in tissues only stay alive if they get messages to do so (Raff 1992). We suggest that in the slime mould slug, the prespores maintain their differentiation as long as they are signaled to do so, and that this only occurs in the tissue of the slug. The challenge is to identify this putative signaling molecule (possibly cAMP?). These considerations emphasise the importance of dynamic signaling in morphogenesis.

30.4.5 Experiment 5: Use of Immunohistochemistry to Understand the Role of Myosin II in Morphogenesis (hardware)

With the availability of monoclonal antibodies, it is now possible to study the distribution of specific macromolecules during morphogenesis using antibody labelling techniques at both light and electron microscope levels. One such critical molecule is myosin II, which many expected would be essential for cell survival (De Lozanne & Spudich 1987). We examined the localisation of myosin II in slugs of *D. discoideum*, and showed that it is located at the periphery of cells in the epithelial layer (Eliott *et al.* 1991), suggesting a role for these cells in moving the slug.

Mutants lacking myosin II grow, divide, and aggregate to form mounds of cells; further development is blocked and the aggregate is stationary rather than consisting of cells which move in a spiral fashion (Eliott *et al.* 1993). Therefore myosin II is required for the formation of the slug stage.

The studies on the myosin II deficient mutants show that this molecule acts differently in growth and 2-dimensional development compared to its role in 3-dimensional development. There is sufficient redundancy in the system to allow survival, growth and development of sheets of cells in a relatively normal fashion; the 3-dimensional tissue is abnormal due to lack of myosin II, so myosin II is essential for morphogenesis to proceed beyond the mound stage.

30.4.6   Experiment 6: Extracellular Matrix Contact Zones Revealed by Immunohisto-chemistry (hardware with hints about software)

An extracellular matrix (slime sheath) is secreted around the anterior of the slug, which then moves through it. This material is left behind the slug as a collapsed trail, which allows examination of its molecular features. In the past the extracellular matrix was thought to be composed of a homogeneous mixture of cellulose and protein (Hohl & Jehli 1973). We have used monoclonal antibodies to study the location of specific glycoproteins, and calcofluor to identify cellulose (Vardy, Blanton and Williams, In preparation). Cellulose is almost exclusively located on the ventral extracellular matrix in zones of contact where the projecting tip touches down on the substratum. The patterns of cellulose deposition indicate that it is deposited as an outline around the position of cells in contact with the extracellular matrix at the time of initial contact with the substratum. A group of glycoproteins called the sheathins is co-deposited with cellulose in these contact zones (Zhou-Chou 1992).

The biochemical and mechanical features of this "contact zone" suggest that it is an area of strengthening on the ventrum of the sheath and that it is a sufficient platform to allow aerial projection of the slug. None of these findings was predicted; we merely set out to locate particular molecules which were thought to be components of the extracellular matrix. We are very interested in the biochemistry of the cell print regions, as they indicate a coordinated and rapid co-secretion of cellulose and a family of glycoproteins. Hence, further studies on the hardware side of morphogenesis are in progress. More importantly, these hardware discoveries have led us directly to consider the software issue of how the slug moves (Williams *et al.* 1986).

30.4.7   Experiment 7: Time Lapse Imaging and Computer Analysis of Slug Movement (software)

Using personal computer equipment, our laboratory has developed a specialized image processing system (IPS). We record and analyze the time history of shape and position of both the whole organism and individual cells in various multicellular stages of *D. discoideum*. At the whole organism level the shape and allometry of the migrating slug has been studied. Individual cells have been tracked in aggregating streams, the surface layer of mounds and on the ventral layer in contact with the substratum in migrating slugs.

Studies on movement indicated that slugs move in a pulsatile fashion, with pronounced aerial projection of the tip or shuffling involving a series of small projections (Breen *et al.* 1987). The cell prints discovered in Experiment 6 coincide with the regions where the slug tip touches the substratum after its aerial projection.

Our initial conclusion from the discovery of cell prints in the slug extracellular matrix (see Experiment 6) was that cells in the touchdown zone remain stationary and project the inner cells forward (Vardy *et al.* 1986). However, measurements using the IPS have shown that the ventral layer of cells in a migrating slug is in constant movement. Therefore, the cells can only stop for a very short period upon touchdown of the slug tip. This is a good example of the complementarity between different techniques, which leads to a new view of a phenomenon. The cell prints reflect the secretion of cellulose and sheathin glycoproteins by cells at touchdown, but this process must be rapid, as the cells move on very soon afterwards. We conclude from these and other experiments that the cell print zones reflect areas of strengthening of the extracellular matrix which form the base for subsequent launching of the tip into the air. Any new theories of slug movement must consider how cells at the tip are projected aerially, what happens at the contact zone, and how cells in the rear of the slug are moved forward.

30.5   WHAT DOES IT ALL MEAN?

The seven experiments described above all add some information to the jigsaw puzzle that is morphogenesis. Most of the findings have not been predicted by theory. For example, theories to explain slug movement did not consider the pulsatile aerial projections by which slugs advance, nor was specialisation of the extracellular matrix at the touchdown zone envisaged. These discoveries came from careful observation of the movement process using time lapse filming techniques and immunohistochemical analysis (Experiment 6). Subsequent experiments have shown that the touchdown zone is an area of strengthening rather than a region where cells adhere (Experiment 7). In most of the experiments, technical advances were critical to the new discoveries. Hence our finding of specialisation of the peripheral cells as a primitive epithelium (Experiment 2) was only convincing when we used freeze substitution techniques to prepare the material. The role for the epithelium in morphogenesis is still unclear, but we now know that specialisation of the cells at the periphery contributes to the boundary definition and hence presumably the shape of the slug. The next question is what makes the epithelial cells take up particular attitudes? Flow cytometry (Experiments 3 and 4) has made possible new discoveries about surface molecules within the slug tissue and controls on their maintenance. The availability of gene disruption mutants lacking myosin II has helped us to understand that molecules affecting cell shape become critical in building an organism only when the cells are piled on top of each other to form a 3-dimensional structure (Experiment 5). Finally, Experiment 1 indicated that even simple biological experiments involving description and surgical manipulation of the tissue still have a role in uncovering new features of morphogenesis.

Our aim when doing these experiments was to describe the hardware of slug movement, yet along the way they have helped us develop new views on the software. How can we focus our experiments in future to learn more about the underlying controls of morphogenesis? It is apparent from Systems Theory that simple systems can exhibit complex behaviour (*e. g.* Turing machines, catastrophe and chaos theory, fractals; (May 1976; Gleick 1987; Mandelbrot 1983; Goldbeter 1981)), and yet on the other hand, that stable behaviour or morphology of a system is obtained by adding complexity in the form of redundancy, interaction between components, and feedback control.

It is becoming clear that there is much redundancy, interaction and feedback control in the biochemical mechanisms by which cells and organisms live, grow and reproduce. We do not believe that the control of morphological development could operate effectively if it were complexly interactive or redundant. During development there is considerable switching between modes, and such changes would be resisted, or only gradual, if the control system(s) were complexly interwired. Simplicity in control wiring seems essential for the dramatic changes observed in animal development. The explanation of morphological development lies in the structure of the system and the behavioural processes the structure permits.

While DNA contains the specifications for the ingredients of the system, and as experimentalists we are inevitably involved in discovering and describing these components, the morphogenetic recipe requires more than merely a list of the ingredients. The description of how to generate morphology is likely to be much simpler than a description of resultant morphologies. Understanding the control circuitry is more conceptual in the first instance, and the proof of ideas about this requires different approaches than those generally used by experimentalists. The challenge is to separate and extract an understanding of those aspects of the biological system structure which form the controlling process of morphological development, from those which maintain the viability of the particular stage of the life cycle in which the life form is currently invested.

We think that this view of the world may help those interested in understanding morphogenesis, be they theoreticians or experimentalists, find a way forward. The search goes on!

ACKNOWLEDGMENTS Our research is supported by a Program Grant to KLW from the Australian Research Council and grants from the Australian NH&MRC. We thank the students and staff of the Dicty lab at Macquarie University for their enthusiasm and productivity, and Jean Joss and Brenda Goodman for helpful comments.

REFERENCES

Breen, E. J., Vardy, P. H., & Williams, K. L. 1987. Movement of the multicellular slug stage of *Dictyostelium discoideum*, an analytical approach. *Development*, **101**, 313–321.

Breen, E. J., Eliott, S., Vardy, P. H., White, A., & Williams, K. L. 1992. Length regulation in the *Dictyostelium discoideum* slug is a late event. *J. Exp. Zool.*, **262**, 299–306.

Browne, L. H., & Williams, K. L. 1993a. Gradients in the expression of cell surface glycoproteins in a simple tissue, the *Dictyostelium discoideum* slug. *J. Gen. Microbiology*. (In press).

Browne, L. H., & Williams, K. L. 1993b. Pure populations of *Dictyostelium discoideum* prespore and prestalk cells obtained by flow cytometry have different redevelopment characteristics at their cell surfaces. *Cytometry*. (In press).

Child, C. M. 1941. *Patterns and Problems of Development*. Chicago: University of Chicago Press.

Coulombe, P. A., Hutton, M. E., Letai, A., Hebert, A., Paller, A. S., & Fuchs, E. 1991. Point mutations in human keratin 14 genes of epidermolysis bullosa simplex patients: genetic and functional analyses. *Cell*, **66**, 1301–1311.

De Lozanne, A., & Spudich, J. A. 1987. Disruption of the *Dictyostelium* myosin heavy chain gene by homologous recombination. *Science*, **236**, 1086–1091.

Devreotes, P. 1989. *Dictyostelium discoideum*: a model system for cell-cell interactions in development. *Science*, **245**, 1054–1058.

Eliott, S., Vardy, P. H., & Williams, K. L. 1991. Myosin - a key molecule for multicellular development; distribution in the slug of *Dictyostelium discoideum*. *J. Cell Biol.*, **115**, 1267–1274.

Eliott, S., Joss, G. H., Spudich, A., & Williams, K. L. 1993. Patterns in *Dictyostelium discoideum*: the role of myosin II in the transition from the unicellular to the multicellular phase. *J. Cell Sci.*, **104**. In press.

Firtel, R. A. 1991. Signal transduction pathways controlling multicellular development in *Dictyostelium*. *Trends in Genetics*, **7**, 381–388.

Fuchs, M., Jones, M. K., & Williams, K. L. 1992. Freeze-substitution for the preservation of cell-cell contact regions in the multicellular *Dictyostelium discoideum* slug. *J. Comp. Asstd. Microscopy*, **4**. (In press).

Fuchs, M., Jones, M. J., & Williams, K. L. 1993. Characterisation of an epithelium-like layer of cells in the multicellular *Dictyostelium discoideum* slug. (Submitted).

Gerisch, G. 1986. Inter-relation of cell adhesion and differentiation in *Dictyostelium discoideum*. *J. Cell Sci. Suppl.*, **4**, 201–219.

Gleick, J. 1987. *Chaos, Making a New Science*. New York: Viking.

Goldbeter, A. 1981. Bifurcations and the control of developmental transitions: Evolution of the cyclic AMP signalling system in the Slime Mold *Dictyostelium discoideum*. In *Mathematical Biology*, Burton, T. A. (ed). pp. 79–95, Pergamon Press.

Hohl, H. R., & Jehli, J. 1973. The presence of cellulose microfibrils in the proteinaceous slime track of *Dictyostelium discoideum*. *Arch. Microbiol.*, **92**, 179–187.

Mandelbrot, B. B. 1983. *The Fractal Geometry of Nature*. New York: W.H. Freeman.

May, R. M. 1976. Simple mathematical models with very complicated dynamics. *Nature*, **261**, 459–467.

Meinhardt, H. 1982. *Models of Biological Pattern Formation*. London: Acad. Press.

Morgan, T. H. 1905. Regeneration of heteromorphic tails in posterior pieces of *Planaria simplicissima*. *J. Exp. Zool.*, **1**, 385–393.

Noji, S., Nohno, T., Koyama, E., Muto, K., Ohyama, K., Aoki, Y., Tamura, K., Ohsugi, K., Ide, H., Taniguchi, S., & Saito, T. 1991. Retinoic acid induces polarizing activity but is unlikely to be a morphogen in the chick limb bud. *Nature*, **350**, 83–86.

Nüsslein-Volhard, C. 1991. Determination of the embryonic axes of *Drosophila*. *Development 1991 Supplement*, **1**, 1–10.

Odell, G. M., & Bonner, J. T. 1986. How the *Dictyostelium discoideum* grex crawls. *Phil. Trans. R. Soc. B London*, **312**, 487–525.

Othmer, H. G., & Pate, E. F. 1980. Scale invariance in reaction-diffusion models of spatial pattern formation. *Proc. Natl. Acad. Sci. U.S.A.*, **77**, 4180–4184.

Raff, M. C. 1992. Social controls on cell survival and cell death. *Nature*, **356**, 397–400.

Raper, K. B. 1984. *The Dictyostelids*. Princeton: Princeton University Press.

Schaap, P., Pinas, J. E., & Wang, M. 1985. Patterns of cell differentiation in several cellular slime mold species. *Dev. Biol.*, **111**, 51–61.

Soll, D. 1987. Methods for manipulating and investigating developmental timing in *Dictyostelium discoideum*. *Meth. Cell Biol.*, **28**, 413–431.

St. Johnston, D., & Nusslein-Volhard, C. 1992. The origin of pattern and polarity in the *Drosophila* embryo. *Cell*, **68**, 201–219.

Takeichi, M. 1991. Cadherin cell adhesion receptors as a morphogenetic regulator. *Science*, **251**, 1451–1455.

Turing, A. M. 1952. The chemical basis of morphogenesis. *Phil. Trans. R. Soc. Lond.*, **B237**, 37–72.

Vardy, P. H., Fisher, L. R., Smith, E., & Williams, K. L. 1986. Traction proteins in the extracellular matrix of *Dictyostelium discoideum* slugs. *Nature*, **320**, 526–529.

Williams, J. G. 1988. The role of diffusible molecules in regulating the cellular differentiation of *Dictyostelium discoideum*. *Development*, **103**, 1–16.

Williams, K. L., Vardy, P. H., & Segel, L. A. 1986. Cell migrations during morphogenesis: some clues from the slug of *Dictyostelium discoideum*. *BioEssays*, **5**, 148–152.

Witke, W., Schleicher, M., & Noegel, A. A. 1992. Redundancy in the microfilament system : abnormal development of *Dictyostelium* cells lacking two F-actin cross-linking proteins. *Cell*, **68**, 53–62.

Wolpert, L. 1969. Positional information and the spatial pattern of cellular differentiation. *J. Theor. Biol.*, **25**, 1–47.

Wolpert, L. 1992. The shape of things to come. *New Scientist*, June 27, 38–42.

Zhou-Chou, Alice Ti. 1992. *Characterization of cell surface and extracellular matrix glycoproteins in* Dictyostelium discoideum. Ph.D. thesis, Accepted.

31. FACTORS CONTROLLING THE DIRECTIONALITY OF MESODERM CELL MIGRATION IN THE *XENOPUS* GASTRULA

Rudolf Winklbauer, Martina Nagel, and Andreas Selchow

Max-Planck-Institut für Entwicklungsbiologie
Spemannstraße 35
7400 Tübingen, Germany

31.1 INTRODUCTION

The goal-directed migration of cells is a wide-spread phenomenon in the development of organisms. Examples range from the aggregation of *Dictyostelium* amoeba during slug formation to the pathfinding of axonal growth cones during development of the nervous system. These examples typically share the characteristic that it is a population of single, isolated cells (or motile parts of cells as in the case of growth cones) which moves toward a pre-specified target. In contrast to this, the directional migration of the mesoderm during *Xenopus* gastrulation involves the movement of a multilayered coherent cell aggregate on a planar substrate.

In *Xenopus*, as in other amphibians, the mesoderm provides the main driving force for the process of gastrulation. Being initially part of the blastocoel wall, the mesoderm involutes at the blastopore lip, becomes apposed to the inner surface of the blastocoel roof (BCR), and moves on this substrate directionally toward the animal pole of the embryo (Figure 31.1) (Keller 1986; Keller & Winklbauer 1992, for review). The mesoderm forms a continuous ring around the equator of the embryo, and involution starts on the dorsal side, progresses laterally and eventually occurs also on the ventral side of the gastrula. Therefore, in sagittal sections through the gastrula, mesoderm is seen to involute and to migrate toward the animal pole both on the dorsal and the ventral side (Figure 31.1).

Throughout gastrulation, the mesoderm behaves as a coherent cell mass. Thus, a piece of mesoderm can be excised from the embryo and moved around as a whole without disintegrating into individual cells. Also, on artificial substrates or on the BCR, mesoderm explants move as coherent aggregates, with single cells rarely leaving the main cell mass (Winklbauer 1990; Winklbauer & Nagel 1991; Winklbauer *et al.* 1991; Keller & Winklbauer 1992). This cohesion is a necessary condition for the movement of the multilayered mesoderm as a whole on a planar substrate: only the basal cells are able to migrate actively by exerting traction on the substrate, while those in the upper layers have to be transported passively by the basal ones. This requires some minimum degree of mutual cell adhesion within the mesoderm. It may be expected that these specific conditions are reflected in the mechanisms involved in directing mesoderm translocation.

In this article, we review some of our work on directional mesoderm movement during *Xenopus* gastrulation. We show that two independent mechanisms are involved in guiding

Experimental and Theorietcal Advances in Biological Pattern Formation,
Edited by H.G. Othmer *et al.*, Plenum Press, New York, 1993

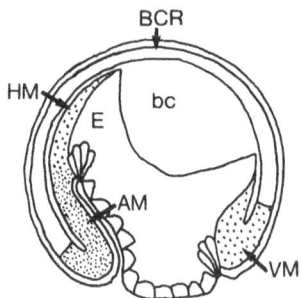

Figure 31.1. Schematic drawing of a sagittal section through a *Xenopus* middle gastrula. Future dorsal side to the left, animal pole to the top. bc, blastocoel; BCR, blastocoel roof; HM, prospective head mesoderm, forming the leading anterior part of the dorsal mesoderm; AM, prospective axial mesoderm; VM, prospective ventral mesoderm; E, prospective endoderm apposed to the migrating mesoderm. From Winklbauer *et al.* (1991).

mesoderm cells to the animal pole. Firstly, a substrate-dependent guidance mechanism ensures that all cells in contact with the BCR substrate can migrate directionally. Secondly, an intrinsic tissue polarity of the mesoderm is also able to determine the direction of mesoderm movement independent of external, substrate-dependent cues. In the embryo, both mechanisms support movement in the same direction, toward the animal pole.

31.2   GUIDANCE CUES IN THE EXTRACELLULAR MATRIX OF THE BLASTO-COEL ROOF

The inner surface of the BCR, which serves as the substrate for the migration of the mesoderm, is covered by a delicate network of extracellular matrix fibrils. As shown by Nakatsuji & Johnson (1983), this matrix can be transferred to a plastic surface by culturing the BCR with its inner surface down on the bottom of a tissue culture dish. When the BCR explant is eventually removed, the fibrillar matrix is left behind (Figure 31.2). Such conditioned substrate supports the attachment and migration of mesoderm cells (Nakatsuji & Johnson 1983; Shi *et al.* 1989; Johnson *et al.* 1990; Winklbauer & Nagel 1991). Moreover, the movement of mesoderm explants on conditioned substrate is directional. The positions of the blastopore lip and the animal pole can be marked on the bottom of the dish during conditioning (Figure 31.2(a)). When a piece from the leading, anterior part of the dorsal mesoderm of *Xenopus*, the prospective head mesoderm (HM), is placed on conditioned substrate either in normal orientation, or rotated 180°, it always migrates toward the animal pole, regardless of whether its original leading edge points toward the animal pole, as in the embryo (Figure 31.2b), or away from it (Figure 31.2(c); Figures 31.5(a,b) in Winklbauer & Nagel 1991; Winklbauer *et al.* 1991). Obviously, the conditioned substrate contains cues which are sufficient to determine the direction of migration of the anterior mesoderm. Such substrate-dependent guidance cues have also been demonstrated for *Ambystoma* (Nakatsuji & Johnson 1983; Johnson *et al.* 1992) and *Pleurodeles* (Shi *et al.* 1989; Riou *et al.* 1990) embryos.

The guidance cues present in the extracellular matrix of the BCR are not specifically recognized by mesoderm cells only. Other cell types from the *Xenopus* gastrula, like prospective endoderm cells which normally do not come in contact with the extracellular matrix of the BCR, or aggregates of cells from the animal cap, also migrate directionally toward the animal

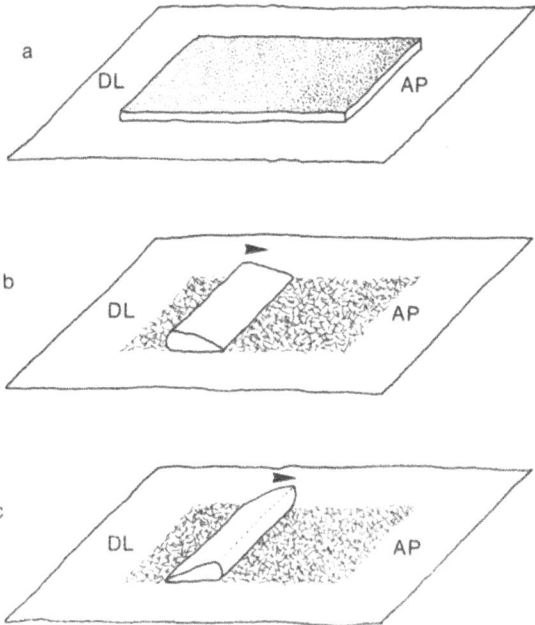

Figure 31.2. Mesoderm migration on conditioned substrate. (a) A piece of BCR (stippled), with known orientation (DL, dorsal blastopore lip; AP, animal pole), is cultured for 2 hours with its inner surface down to transfer its extracellular matrix to the bottom of the culture dish. (b) After removal of the BCR explant, anterior dorsal mesoderm (prospective head mesoderm) is placed in normal orientation (tapering anterior end toward AP, as in the embryo) on the conditioned substrate (dashes). (c) Prospective head mesoderm on conditioned substrate in reverse orientation, with tapering anterior end toward DL. Mesoderm explants move to the AP in (b) and (c) (arrowheads).

pole position on conditioned substrate (Figures 31.8c,d Winklbauer & Nagel 1991). Thus, the mechanism of action of these guidance cues will probably not involve highly specific recognition processes as, for example, in chemotaxis. A mechanism relying on more general features of the substrate would be haptotaxis (Carter 1965). It has been shown *in vitro* that migrating cells will often exhibit a preference to move up a gradient of adhesiveness of the substrate, and this could in principle be used to guide cell migration.

Adhesion of mesoderm cells to conditioned substrate is mediated by the extracellular matrix protein fibronectin (FN) (Shi *et al.* 1989; Winklbauer & Nagel 1991). Each of the two nearly identical subunits of FN consists of a linear array of domains which mediate the interaction with other matrix constituents and with the cell surface (Hynes 1985). The main cell binding site of FN is characterized by an arg-gly-asp (RGD) sequence of amino acids (Pierschbacher & Ruoslahti 1984; Yamada & Kennedy 1984; Hynes 1987). This site is recognized by cellular receptors belonging to the integrin family of transmembrane proteins (Hynes 1987). As many other cell types, *Xenopus* mesoderm cells adhere to a substrate of immobilized FN via this RGD site within the cell binding domain of FN (Smith *et al.* 1990; Winklbauer 1990). The attachment of mesoderm cells to conditioned substrate can be completely inhibited by an excess of synthetic RGD peptide in the incubation medium, or by antibodies to FN (Winklbauer & Nagel 1991). RGD peptide competitively inhibits the cellular FN receptor, whereas functional FN antibodies render the RGD site of FN inaccessible to the FN receptor. This shows that the RGD cell binding site is normally responsible for the attachment of mesoderm cells to conditioned substrate.

If haptotaxis were involved in guiding mesoderm migration, the FN-mediated adhesiveness of conditioned substrate should increase in the direction of migration, *i. e.* from the blastopore lip to the animal pole. This could simply be achieved by a gradual increase in FN density along the axis of migration. However, there is no indication of a significant increase in FN density toward the animal pole (unpublished results). Although suggestive, this is not conclusive evidence against haptotaxis. The adhesiveness of the FN substrate could be modulated otherwise, *e. g.* by the chemical modification of FN, or by the presence of factors that do not mediate attachment by themselves, but are able to regulate the strength of adhesion. Therefore, the adhesiveness of conditioned substrate between blastopore lip and animal pole was tested directly (Winklbauer & Nagel 1991).

For this purpose, isolated mesoderm cells were seeded onto conditioned substrate extending from the blastopore lip to the animal pole. After having allowed the cells to attach to the substrate, the incubation dish was turned upside down in a tank containing incubation medium, gently moved to remove non-attached cells, and reverted to its normal position. Cells were photographed shortly before and after inversion of the dish, and the difference in cell density at a given region was taken as a measure of the adhesiveness of that part of the substrate. As can be seen from Figure 31.3, no preferential adhesion of mesoderm cells at the animal pole region can be observed.

These experiments were also evaluated quantitatively, by dividing the conditioned substrate into three zones along the axis of mesoderm movement (dorsal lip, intermediate, animal pole, Figure 31.3), and counting cell numbers in each field before and after inversion of the dish. Moreover, these tests were performed under a variety of conditions, with high or low overall adhesion, to exclude the possibility that spatial differences in adhesiveness were obscured by weakly adherent cells being still attached strongly enough to resist detachment in our assay. Nevertheless, in all of these experiments, a significant increase in adhesive capacity of the substrate toward the animal pole was never observed (Winklbauer & Nagel 1991). Also other tests, for example measuring the degree of spreading of mesoderm explants at different positions on conditioned substrate, gave no indication of reproducible spatial differences in the adhesiveness of substrate (unpublished results). This makes it, in our view, very unlikely that haptotaxis plays a role in guiding mesoderm cell migration in the *Xenopus* gastrula.

To promote mesoderm cell adhesion and random migration, a diffuse layer of FN molecules adsorbed to an appropriate surface is sufficient (Nakatsuji 1986; Smith *et al.* 1990; Winklbauer 1990). On conditioned substrate, which supports directional movement of mesoderm explants, FN can be seen by immunofluorescence staining with antibodies to *Xenopus* FN to form a loose two-dimensional network of fibrils. It was Nakatsuji & Johnson (1983) who proposed that these FN fibrils may have a polar structure, which could be recognized by mesoderm cells and used for orientation. To test this hypothesis, we examined whether intact FN fibrils are indeed necessary for the directional movement of the mesoderm on conditioned substrate.

FN fibril formation occurs on the cell surface and requires the interaction of secreted FN with a cellular integrin receptor via the RGD cell binding site of FN (Darribere *et al.* 1990; Yost 1992). Thus, FN fibril formation, but not FN secretion and deposition, should be inhibited by RGD peptide present in the incubation medium during substrate conditioning with explanted BCR. This is indeed observed (Winklbauer & Nagel 1991; Winklbauer *et al.* 1991). Figure 31.4(a) shows the network of FN fibrils on the inner surface of the BCR, as visualized by immunofluorescence staining with antibody to FN. This network of fibrils can be transferred to the bottom of a tissue culture dish (Figure 31.4(b)). Substrate conditioned in the presence of RGD peptide shows only diffuse staining with FN antibody, and no fibril network (Figure 31.4(c)). However, mesoderm cells attach to that substrate and migrate,

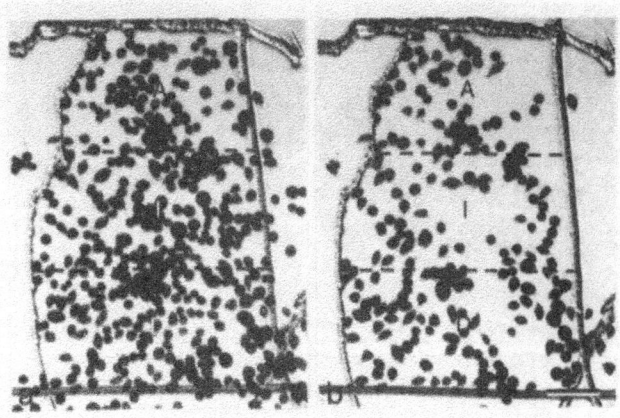

Figure 31.3. Adhesion of mesoderm cells to different regions of the conditioned substrate. Dissociated HM cells were incubated on conditioned substrate for 10 minutes and photographed (a). After turning the incubation dish upside down in a tank containing incubation buffer (Modified Barth's Solution, MBS) to remove nonattached cells, the remaining cells were photographed again (b). For quantitative analyses, conditioned substrates were divided into a region close to the dorsal blastopore lip (D), an intermediate region (I), and an animal pole region (A) (dashed lines). Bar = 200 μm. From Winklbauer & Nagel (1991).

and adhesion can be inhibited by RGD peptide or antibodies to FN (Winklbauer & Nagel 1991; Winklbauer *et al.* 1991). FN fibril formation can also be inhibited by conditioning substrate in the presence of cytochalasin B, which disrupts the actin cytoskeleton of the cell. Again, immunofluorescence staining shows that no extensive network of FN fibrils forms (Figure 31.4(d)), but mesoderm cells attach to the substrate in an RGD-dependent manner (Winklbauer & Nagel 1991; Winklbauer *et al.* 1991). Thus, with both inhibitors, FN is secreted and deposited on the substrate, but not assembled into fibrils. When mesoderm explants are now placed on such non-fibrillar conditioned substrates, they attach, spread and migrate, as expected from the presence of FN. However, the interesting point is that mesoderm translocation is no longer toward the animal pole: explants move in all directions equally well (Figures 31.5(c,d); Winklbauer & Nagel 1991; Winklbauer *et al.* 1991). This suggests that intact FN fibrils are required for the directional migration of the mesoderm, and it is consistent with the hypothesis that the guidance cues of the BCR extracellular matrix are related to a polarized structure of these fibrils.

To demonstrate the existence of guidance cues in the BCR extracellular matrix of *Xenopus* gastrulae, whole pieces of dorsal anterior mesoderm (HM) were used as responding tissue. Since the mesoderm forms a coherent cell mass in the embryo, this may represent a reasonable approximation to the *in situ* condition. To see whether artificial separation of cells has any effect on the directionality of mesoderm translocation, we observed the migration of dissociated, isolated cells on conditioned substrate and compared it to the movement of cells in aggregates (Winklbauer *et al.* 1991; Winklbauer *et al.* 1993).

A striking feature of the migratory behavior of single cells is that movement is highly non-persistent. Cells frequently change their direction of movement abruptly, such that the net distance covered during a given period of time is significantly shorter than the total path length (Figure 31.6(a); Winklbauer *et al.* 1991; Winklbauer & Selchow 1992). This makes translocation of isolated mesoderm cells inefficient. An even more severe limitation arises

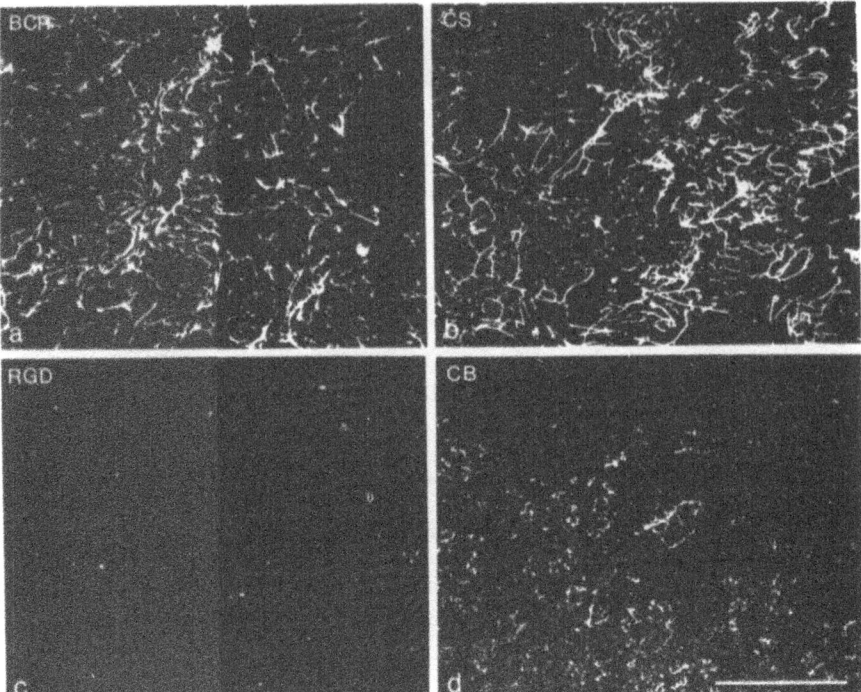

Figure 31.4. Immunofluorescence staining of BCR whole mounts and conditioned substrate with antibodies to *Xenopus* plasma FN. Inner surface of BCR (a), normal conditioned substrate (b), substrate conditioned in the presence of 4mg/ml of GRGDSP peptide (c), or of 20 μg/ml of cytochalasin B (d). Bar = 50 μm. From Winklbauer & Nagel (1991).

from the fact that movement is no longer directional. On conditioned substrate or on the BCR, isolated mesoderm cells move in all directions equally well (Keller & Hardin 1987; Winklbauer 1990; Winklbauer *et al.* 1991; 1993). In contrast, even small aggregates of 10-20 cells allow mesoderm cells to move on conditioned substrate persistently, on straight pathways, and directionally toward the animal pole (Figure 31.6(b); Winklbauer *et al.* 1991; 1993). Thus, aggregation of mesoderm cells into a coherent cell mass is essential for the efficient translocation of these cells and for their ability to follow the guidance cues provided by the substrate.

One of the primary effects of aggregate formation seems to be on cell morphology. Isolated mesoderm cells are bi- or multipolar on the BCR, on conditioned substrate or on artificial FN substrates (Figures 31.7(a-c); Winklbauer 1990; Winklbauer *et al.* 1991; Winklbauer & Selchow 1992). Two or more cytoplasmic lamellae extend simultaneously from the cell body and pull in diverging directions. Movement usually occurs when one of the mutually antagonizing lamellae is retracted and the cell body is pulled rapidly in the direction of the remaining lamellae. This generates the non-persistent, intermittent type of movement described above (Winklbauer & Selchow 1992). In contrast, mesoderm cells moving directionally as part of a larger explant on conditioned substrate or in the embryo appear more or less unipolar when viewed from the substrate side (Figure 31.7(d); Winklbauer & Nagel 1991; Winklbauer *et al.* 1991; Winklbauer & Selchow 1992). Cytoplasmic lamellae extend from the cell body only in the direction of mesoderm movement, *i. e.* toward the animal pole. Moreover, the posterior part of a cell is typically underlapped by the anterior part of the cell

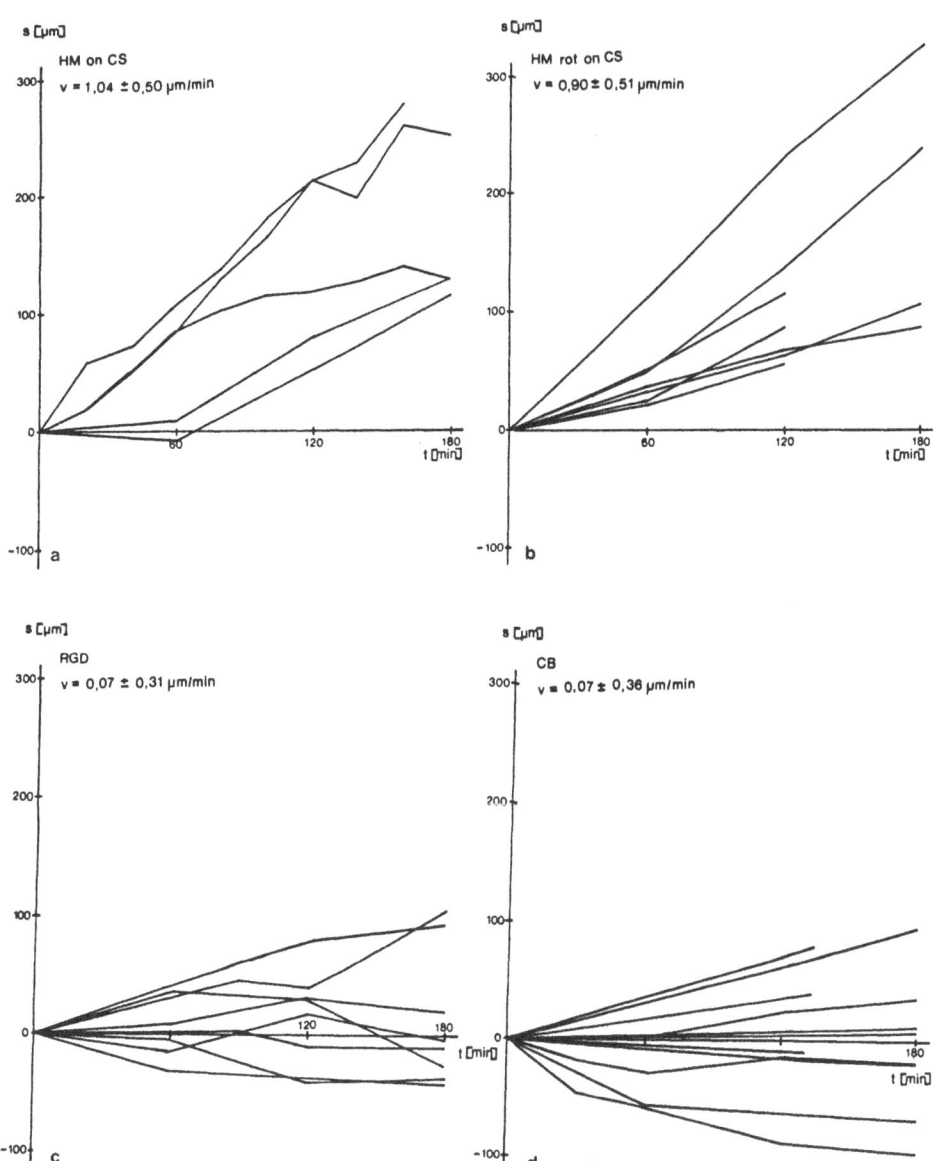

Figure 31.5. Dorsal anterior (head) mesoderm (HM) explant movement on conditioned substrate. HM was placed either in normal orientation (a) or rotated 180° (b), on conditioned substrate close to the position of the dorsal blastopore lip. Similarly, HM was placed between the dorsal lip and the animal pole on substrate conditioned in the presence of 4 mg/ml of GRGDSP peptide (c), or of 20 µg/ml of cytochalasin B (d). Horizontal axis represents time in minutes, vertical axis indicates the distance of the explant from the starting position. A positive slope indicates movement away from the dorsal blastopore lip, toward the animal pole. Average velocities and the respective standard deviations are indicated. Each line represents a separate explant. From Winklbauer *et al.* (1991).

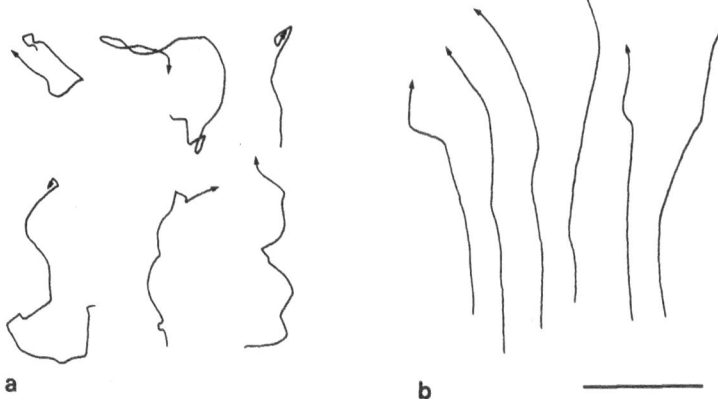

Figure 31.6. Persistence of mesoderm cell migration. Cell trails of isolated HM cells moving on conditioned substrate (a), or of HM cells moving as part of an explant on conditioned substrate (b) were reconstructed from video time-lapse recordings. Cells were followed for 2 hours. Bar = 100 μm.

behind it, so that only the anterior edge of any cell is in contact with the substrate. This shingle arrangement, and the unipolar structure of cells in aggregates, is consistent with their directional migration and their high persistence of movement under these conditions.

31.3   TISSUE POLARITY OF THE MESODERM

In all the experiments described above, where a piece of mesoderm migrates on conditioned substrate toward the animal pole, regardless of its orientation, explants from the anterior margin of the dorsal mesoderm (prospective head mesoderm, HM) were employed (see Figure 31.1). The result of such an experiment differs, however, in an interesting way if more posterior dorsal mesoderm, the prospective axial mesoderm (AM, Figure 31.1), is used (Winklbauer & Nagel 1991). Axial mesoderm placed on conditioned substrate in its normal orientation migrates to the animal pole position, as it would in the embryo (Figure 31.8(a)). However, when rotated 180°, such that its leading edge points away from the animal pole, it no longer moves preferentially toward its normal target, as anterior mesoderm does, but it also does not move consistently toward the blastopore lip (Figure 31.8(B)). This demonstrates two points. Firstly, the axial mesoderm responds to the guidance cues of the substrate, since it recognizes the orientation of the substrate and translocates only when its own anterior-posterior axis aligns with the corresponding axis of the substrate. Secondly, the axial mesoderm obviously possesses an intrinsic tissue polarity which allows movement in the direction of its anterior end, but precludes movement in the direction of its posterior edge. In the embryo, the guidance cues of the substrate and the polarity axis of the mesoderm point in the same direction, and mesoderm translocation can occur. When under experimental conditions these two directing influences are brought to point in opposite directions, they seem able to balance, or neutralize each other, such that no appreciable net movement occurs.

In the above experiments, post-involution axial mesoderm was used which had already been in contact with the oriented BCR substrate (see Figure 31.1). Therefore, the axial mesoderm could have become polarized after involution, through interaction with the guidance cues on the BCR surface which are known for their ability to orient cell movement. However,

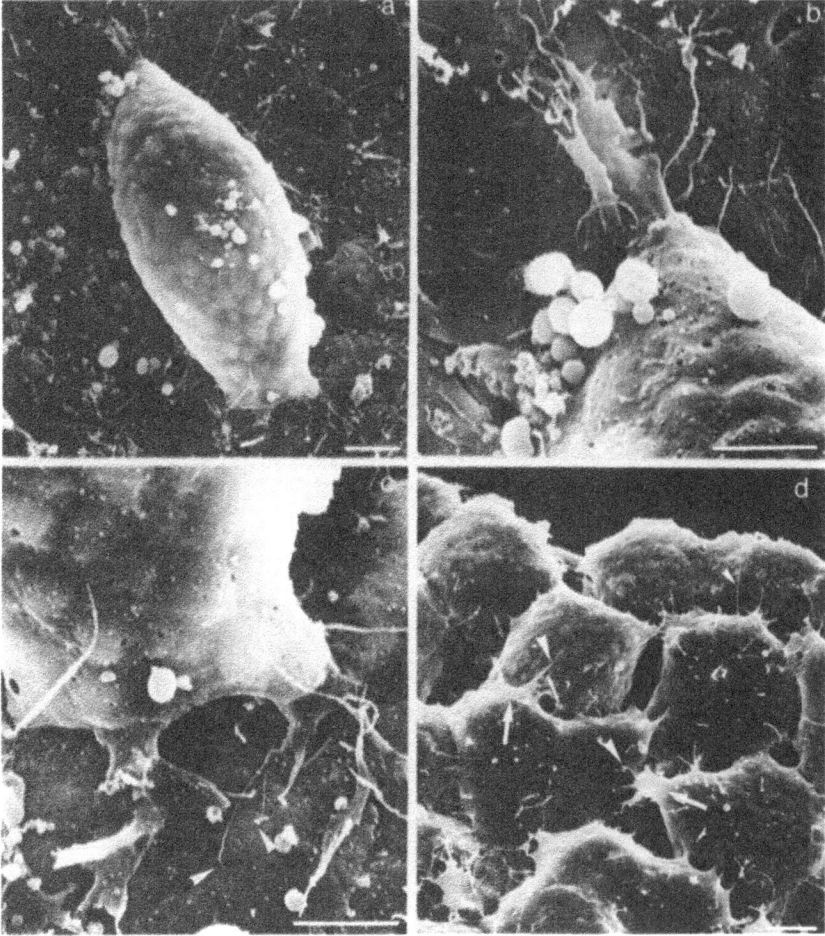

Figure 31.7. Morphology of mesoderm (HM) cells in the scanning electron microscope. (a) Isolated HM cell after 1 hour of incubation on the inner surface of the BCR. Higher magnification of cytoplasmic protrusions at the upper (b) and lower (c) end of the same cell. Close association of filiform protrusions with FN fibrils on the BCR substrate is indicated by the arrowheads in (c). (d) Leading edge of dorsal mesoderm from middle gastrula. The side of the mesoderm cell mass which had been in contact with the substrate is exposed after removal of the BCR after fixation. Direction of migration is to the top. Filiform processes (large arrowheads) extend from lamelliform protrusions (arrows) or directly from the cell body (small arrowheads). Bars = 10 μm (a,d) or 5 μm (b,c). From Winklbauer & Selchow 1992.

when a strip of pre-involution axial mesoderm is explanted onto an artificial non-oriented FN substrate, it assumes a polar morphology and moves in the direction of its well-spread anterior edge, as it would in the embryo after involution (Figures 31.9(c,d); Winklbauer 1990). This demonstrates that an intrinsic tissue polarity which is able to determine the direction of movement is present in the axial mesoderm before this tissue starts to move across the BCR surface.

The nature of this tissue polarity which is defined through its effects on mesoderm translocation is totally obscure. At present, we know only a single, potentially interesting correlation between this polarity and a parameter of cell behavior. The morphology of a migrating mesodermal slug on FN suggests that adhesion to FN may be graded along its

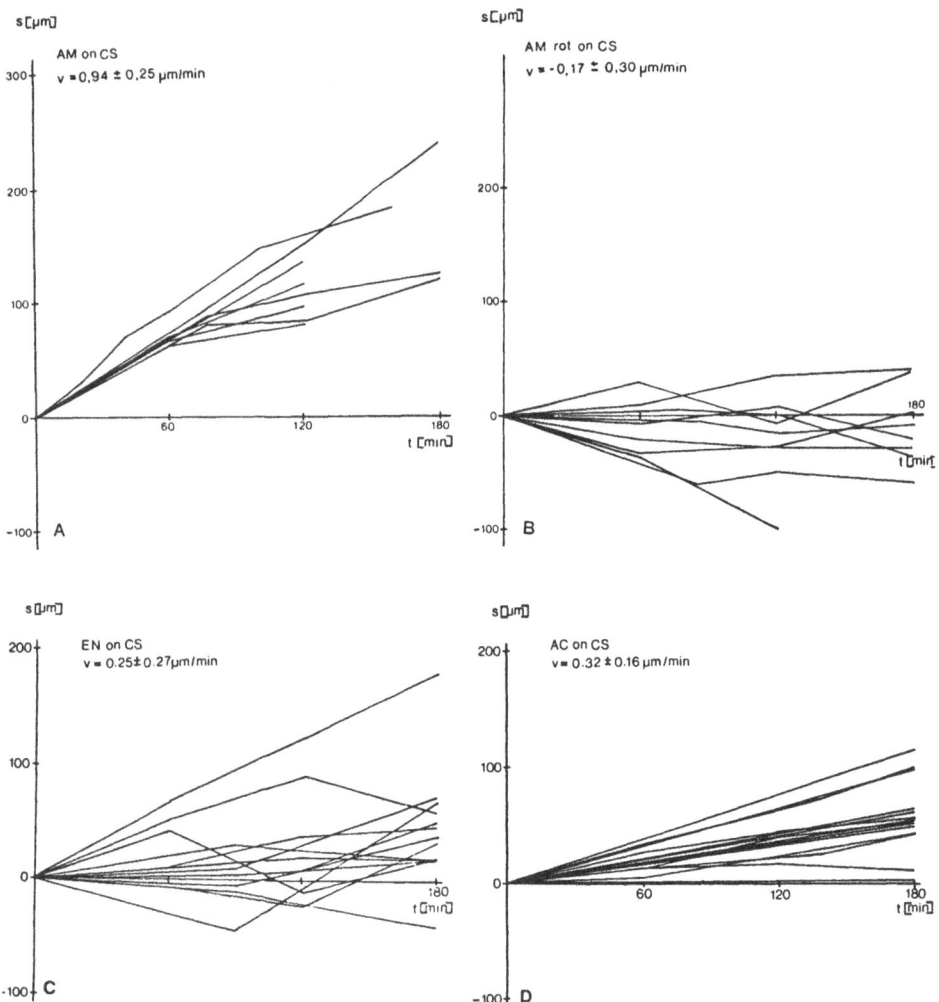

Figure 31.8. Movement of axial mesoderm, endoderm, and the inner layer of the animal cap on conditioned substrate. Axial mesoderm was placed on conditioned substrate between the dorsal blastopore lip and the animal pole in normal orientation (A) or rotated 180° (B). Endoderm from the center of the vegetal yolk mass (C) and aggregates from cells of the inner layer of the animal cap (D) were placed in random orientation on conditioned substrate near the blastopore lip region. Average velocities and the respective standard deviations are indicated. Movement toward the animal pole is statistically significant, *i. e.* the mean velocity is significantly greater than zero, for the axial mesoderm in normal orientation, for the endoderm, and for the animal cap explants (significance level $\alpha = 0.001$ in all cases). The average velocity of rotated axial mesoderm is not significantly higher than zero ($\alpha = 0.05$), and it is significantly lower than that of normally oriented axial mesoderm ($\alpha = 0.001$). From Winklbauer & Nagel (1991).

anterior-posterior axis: the elongated mesoderm explant spreads strongly at its anterior end, but more and more weakly toward its posterior end (Figure 31.9(c,d); Winklbauer 1990). In fact, a gradient of adhesiveness to FN could be demonstrated in the dorsal mesoderm of *Xenopus*. Adhesiveness decreases continually from the anterior head mesoderm to the posterior part of the axial mesoderm (Winklbauer 1990). If this were correlated with tissue polarity, it would suggest that this polarity is not only expressed in the axial mesoderm,

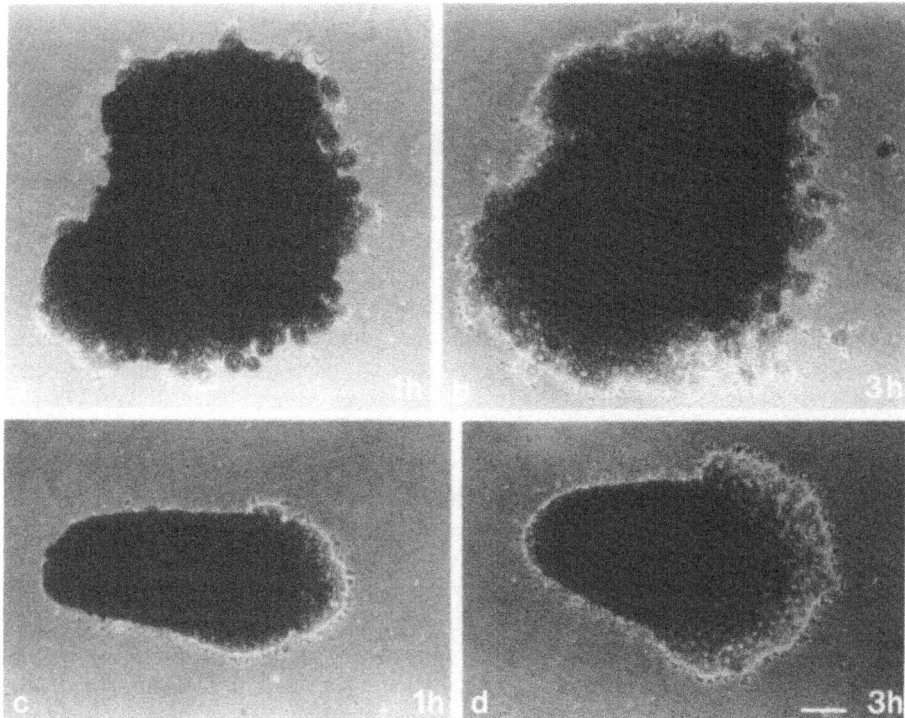

Figure 31.9. Mesoderm explants on fibronectin substrate. (a,b) Involuted prospective head mesoderm including blastopore lip mesoderm of early gastrula 1 and 3 hour after explantation on FN-coated tissue culture plastic. (c,d) Pre-involution axial mesoderm including part of the prospective neuroectoderm of early gastrula under the same condition. Movement of explant relative to debris (arrows) can be observed. (a-d) same magnification, bar = 100 μm. From Winklbauer (1990).

but also in the mesoderm anterior to it. On a non-oriented FN substrate, a weak tendency of the anterior mesoderm to spread out in the anterior direction can indeed be observed (Figure 31.9(a,b); Winklbauer 1990). It is conceivable that the expression of polarity in this region is just not strong enough to prevent rotated explants from moving toward the animal pole on conditioned substrate. However, interpretation of this gradient of adhesiveness to FN is complicated by the fact that it extends continuously beyond the posterior margin of the mesoderm, into the adjacent prospective ectoderm, up to the animal pole (Winklbauer 1990), and it is not clear whether such a gradient of responsiveness to FN which encompasses virtually the whole embryo is of any significance in directing mesoderm translocation.

31.4 CONCLUSIONS

In the *Xenopus* gastrula, two factors have been identified which are able to direct the migrating mesoderm. Firstly, guidance cues are present on the substrate of mesoderm migration, the extracellular matrix of the BCR (Winklbauer & Nagel 1991; Winklbauer *et al.* 1991). These cues are not specific for the mesoderm, but can also be recognized by other cell types from the embryo. The cues do not consist of a gradient of adhesiveness of the substrate, and haptotaxis is most likely not the mechanism by which the mesoderm is directed toward the animal pole. On the other hand, extracellular matrix fibrils consisting of the adhesion protein fibronectin (FN) are necessary for mesoderm guidance by the substrate (Winklbauer & Nagel 1991). Thus, one should examine the hypothesis that these FN fibrils possess a polar structure, due *e. g.* to a regular arrangement of subunits with chiral symmetry, as in microtubules or actin microfilaments.

The second factor involved in mesoderm guidance is an intrinsic polarity of the mesoderm which reveals itself in the strong preference of a mesoderm explant to move in the direction of its anterior end, independently of the substrate. This polarity is strongly expressed in the axial mesoderm, and perhaps weakly in the anterior mesoderm (Winklbauer 1990; Winklbauer & Nagel 1991). In the gastrula, both factors independently guide the mesoderm in the same direction, thus mutually reinforcing their effects. When, under experimental conditions, the substrate-dependent cues and the polarity of the mesoderm point in opposite directions, their effects may balance each other, and movement may become arrested (Winklbauer & Nagel 1991).

An aspect common to both guidance mechanisms is tissue polarity. Whatever the exact nature of the substrate-dependent cues may be, their local effects on the direction of mesoderm movement have to be coordinated globally to ensure continuous, direct movement toward the animal pole from all positions on the BCR (Winklbauer & Nagel 1991). This suggests that the BCR cell sheet possesses an anisotropic structure which can be translated into a globally coordinated orientation of the extracellular matrix. Thus, both the coherent cell sheet which provides the substrate for migration, and the coherent, migratory mesodermal cell mass express a tissue polarity which is relevant for morphogenetic processes during gastrulation. We are at present totally ignorant about the cellular or molecular basis of this polarity, and even its phenomenology has yet to be established. For example, it is not known whether the *Xenopus* gastrula, which basically consists of a single coherent cell mass, shows a continuous polar organization which encompasses the whole embryo and can be used as a frame of reference for locally orienting cellular activities, or whether different parts of the gastrula develop tissue polarity independently and by completely different means.

REFERENCES

Carter, S. B. 1965. Principles of cell motility: The direction of cell movement and cancer invasion. *Nature*, **208**, 1183–1187.

Darribere, T., Guida, K., Larjava, H., Johnson, K. E., Yamada, K. M., Thiery, J.-P., & Boucaut, J.-C. 1990. *In vivo* analyses of integrin ß1 subunit function in fibronectin matrix assembly. *J. Cell Biol.*, **110**, 1813–1823.

Hynes, R. O. 1985. Molecular biology of fibronectin. *Ann. Rev. Cell Biol.*, **1**, 67–90.

Hynes, R. O. 1987. Integrins: A family of cell surface receptors. *Cell*, **48**, 549–554.

Johnson, K. E., Darribere, T., & Boucaut, J.-C. 1990. Cell adhesion to extracellular matrix in normal *Rana Pipiens* gastrulae and in arrested hybrid gastrulae *Rana pipiens* (m.) and *Rana esculenta* (f.). *Dev. Biol.*, **137**, 86–99.

Johnson, K. E., Darribere, T., & Boucaut, J.-C. 1992. *Ambystoma maculatum* gastrulae have an oriented, fibronectin-containing extracellular matrix. *J. exp. Zool*, **261**, 458–468.

Keller, R., & Hardin, J. 1987. Cell behaviour during active cell rearrangement: Evidence and speculations. *J. Cell Sci. Suppl.*, **8**, 369–393.

Keller, R., & Winklbauer, R. 1992. The cellular basis of amphibian gastrulation. *Current Topics in Dev. Biol.*, **27**, 39–89.

Keller, R. E. 1986. The cellular basis of amphibian gastrulation. In *The Cellular Basis of Morphogenesis*, Browder, C. W. (ed). pp. 241–327, New York/London: Plenum. Development Biology, Vol. 2.

Nakatsuji, N. 1986. Presumptive mesoderm cells from *Xenopus laevis* gastrulae attach to and migrate on substrata coated with fibronectin or laminin. *J. Cell Sci.*, **86**, 109–118.

Nakatsuji, N., & Johnson, K. E. 1983. Conditioning of a culture substratum by the ectodermal layer promotes attachment and oriented locomotion by amphibian gastrula mesodermal cells. *J. Cell Sci.*, **59**, 43–60.

Pierschbacher, M. D., & Ruoslahti, E. 1984. Cell attachment activity of fibronectin can be duplicated by small synthetic fragments of the molecule. *Nature*, **309**, 30–33.

Riou, J.-F., Shi, D.-L., Chiquet, M., & Boucaut, J.-C. 1990. Exogenous tenascin inhibits mesodermal cell migration during amphibian gastrulation. *Dev. Biol.*, **137**, 305–317.

Shi, D.-L., Darribere, T., Johnson, K. E., & Boucaut, J.-C. 1989. Initiation of mesodermal cell migration and spreading relative to gastrulation in the urodele amphibian *Pleurodeles waltl.* *Development*, **105**, 351–363.

Smith, J. C., Symes, K., Hynes, R. O., & DeSimone, D. 1990. Mesoderm induction and the control of gastrulation in *Xenopus laevis*: The role of fibronectin and integrins. *Development*, **108**, 229–238.

Winklbauer, R. 1990. Mesodermal cell migration during xenopus gastrulation. *Dev. Biol.*, **142**, 155–168.

Winklbauer, R., & Nagel, M. 1991. Directional mesoderm cell migration in the *Xenopus* gastrula. *Dev. Biol.*, **148**, 573–589.

Winklbauer, R., & Selchow, A. 1992. Motile behavior and protrusive activity of migratory mesoderm cells from the *Xenopus* gastrula. *Dev. Biol.*, **150**, 335–351.

Winklbauer, R., Selchow, A., Nagel, M., Stoltz, C., & Angres, B. 1991. Mesodermal cell migration in the *Xenopus* gastrula. In *Gastrulation: Movements, Patterns and Molecules.* New York: Plenum Press.

Winklbauer, R., Selchow, A., Nagel, M., & Angues, B. 1993. Cell interaction and its role in mesoderm cell migration during Xenopus gastrulation. *Dev. Dyn.* (In press).

Yamada, K. M., & Kennedy, D. W. 1984. Dualistic nature of adhesive protein function: Fibronectin and its biologically active peptide fragments can autoinhibit fibronectin function. *J. Cell Biol.*, **99**, 29–36.

Yost, H. J. 1992. Regulation of vertebrate left-right asymmetries by extracellular matrix. *Nature*, **357**, 158–161.

32. ANALYSIS AND MODELLING OF THE ORGANIZATION OF THE MAMMALIAN CEREBRAL CORTEX

Malcolm P. Young and Jack W. Scannell

University Laboratory of Physiology
Parks Road
Oxford OX1 3PT, UK

32.1 INTRODUCTION

The functional organization of an information-processing device, such as the brain, is related to the connectional, rather than spatial, organization of its processing elements. To give a simple example, one can change the spatial relationship of the components of a personal computer (by moving the screen, keyboard and disk drives to different positions on the desk) without having to worry about changing the way the computer works, because the connectional organization of the electronic processing elements remains unchanged. At the rather more complex level represented by the brain, a typical mammalian central nervous system (CNS) contains about 10^{11} neural processing elements, and each one of these has about 10^4 connections, or synapses, with other neurons. Hence, the brain's connectional architecture - the pathways in which sensory information and motor signals may flow - is defined by about 10^{15} points of possible transmission between cells: a number roughly comparable to the number of stars in 10,000 galaxies. It is this connectional architecture, rather than the gross morphology, that determines the functional organization of the mammalian brain.

If there were no gross regularities with which to simplify the task of describing the connection scheme, the processing system of the brain could only be described by characterizing all 10^{15} synapses. Happily, though, neurons with similar patterns of connections are clustered together and these clusters correspond roughly to 'areas' of the cerebral cortex and to the subcortical nuclei (though in detail, cortical areas and subcortical nuclei may be further subdivided into connectionally and functionally distinct cell populations). Figure 32.1 shows the spatial distribution of areas on the cortical sheet of the cat brain. Connections between these larger brain structures may be composed of tens of thousands or even millions of individual nerve fibres.

At the level of the connections linking cortical areas and subcortical nuclei, a considerable degree of complexity remains. In the monkey and cat, for example, there are about 70 cortical areas, linked by 1000 or so connections. This network is further complicated by the fact that connections between cortical areas originate and terminate in different layers of the cortex. The organization of networks of this scale is difficult to understand without some simplifying analysis. In this chapter we consider analyses of the area to area, and the layer to layer patterns of connectivity which contribute to the functional organization of the

Experimental and Theoretical Advances in Biological Pattern Formation,
Edited by H.G. Othmer *et al.,* Plenum Press, New York, 1993

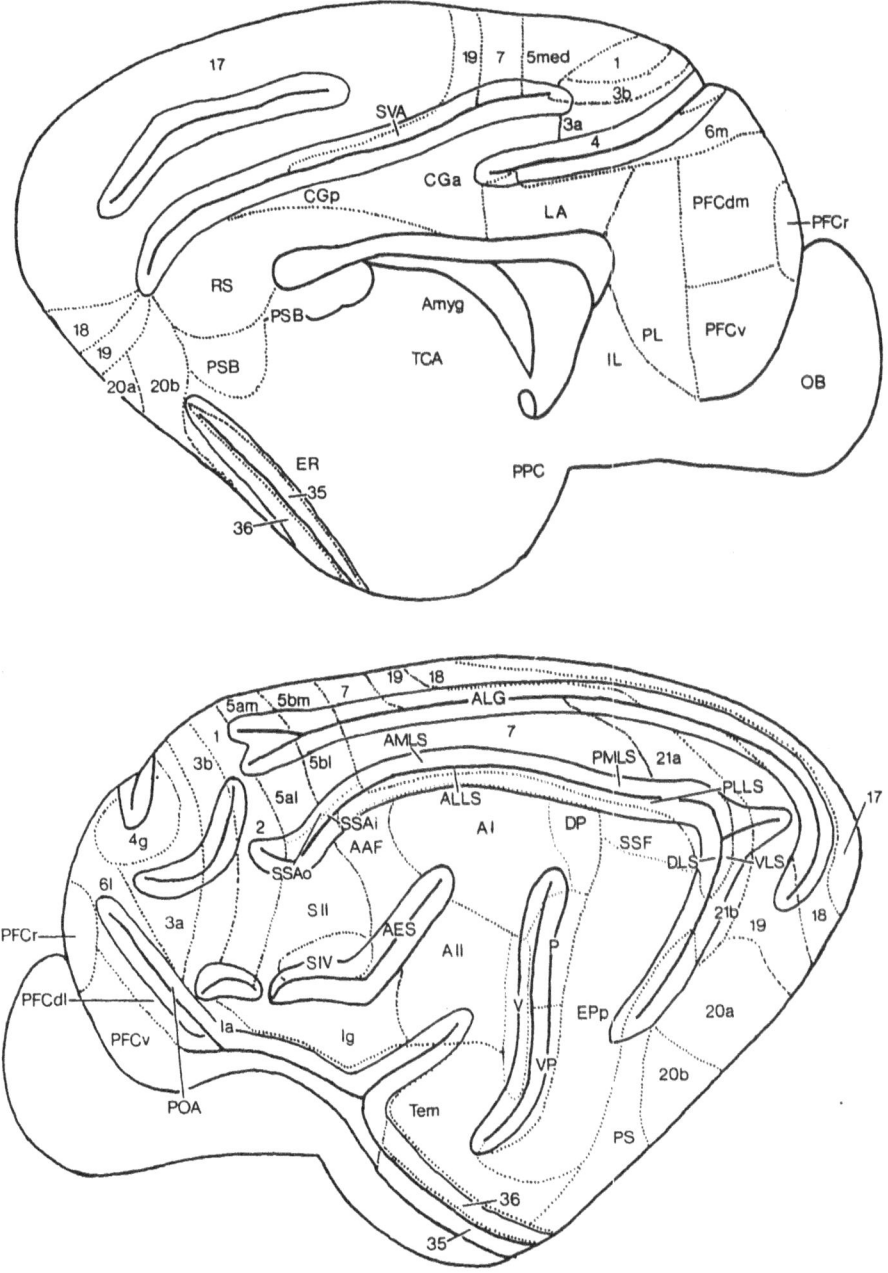

Figure 32.1. A diagrammatic representation of the medial (top) and lateral (bottom) views of a mammalian brain, in this case that of the cat, showing the division of the cortical sheet into areas. The abbreviations stand for the neuro-anatomical names of the areas (*e. g.* AMLS: the anterior medio-lateral suprasylvian area). These names are not crucial to the considerations of this chapter, but they typically refer to the spatial position of an area, particularly relative to gross morphological landmarks such as the convex gyri and concave sulci.

mammalian cerebral cortex, and explore some of the mechanisms that may be responsible for the formation of such connection patterns.

32.2 REPRESENTING THE ORGANIZATION OF NEURAL SYSTEMS

Thousands of man years of labour have been spent tracing connections between brain structures. Several approaches have been used to try to gain insights into the organization of the system lying behind the now huge catalogue of connections. Unfortunately, the most prevalent has been to carry out no analysis of the connection data whatsoever, and then to articulate various informal speculations supported only by sheer force of personality. Recently, however, more systematic approaches have been developed. "Hierarchical analysis", for example, exploits information about the cortical layers in which connections originate and terminate. By defining certain patterns of origin and termination as "ascending", "descending", or "lateral", it is possible to arrange interconnected groups of cortical areas into reasonably consistent hierarchies. We will return to this approach later.

A third approach begins with a matrix of connections between various brain structures, like that in Table 32.1. The entries in this matrix can be ones and noughts, indicating whether a connection between two brain structures has been identified, or can be ranks which categorise the connection as dense and strong (3), moderate (2), sparse and weak (1), or unreported (0). In either case, the entries in a connection matrix are *proximities* that describe the relations between points representing the brain structures in some high-dimensional multivariate "connection" space. The configuration of points in this high-dimensional space perfectly represents the connectional topology of the brain structures: strongly interconnected structures are positioned very close together, moderately strongly connected structures not quite so close, sparsely interconnected structures less close still, and unconnected structures far apart. Humans cannot, however, understand spatial relations in more than three dimensions, so the task of making the connectional organization of a set of brain structures understandable is just the problem of dimensional reduction: flattening the high-dimensional configuration that perfectly reflects the connectional topology of the areas into a low-dimensional configuration that fits the high-dimensional one as closely as possible. A routine statistical tool, non-metric multidimensional scaling (MDS) (Kurskal 1964; Shepherd 1962; Shepherd 1980; Takane *et al.* 1977; Young 1978; Young & Harris 1990), performs exactly this operation, and we have employed it to analyze the organization of cortical brain systems. We call the resulting representations of cortical connectivity *topological* structures. We think that the topological models of the nervous system provide some previously unobtainable insights into the large scale connection patterns in the brain, and that the patterns revealed by this quantitative analysis give us new opportunities to explore the mechanisms of the formation of cortical connection patterns.

32.3 THE TOPOLOGICAL ORGANIZATION OF THE MAMMALIAN CEREBRAL CORTEX

32.3.1 The Primate Visual System

We will use the primate visual system, one of the most studied cortical networks, to illustrate the results of the objective dimension-reduction technique.

Examination of Figure 32.2, which shows the results of applying MDS to Table 32.1, reveals several features of the connectional topology of the cortical visual system. The two dimensions of the structure approximately correspond to the posterior-anterior (left to right in Figure 32.2) and to the dorsal-ventral (top to bottom) spatial distribution of the areas in the brain. For example, areas of the posterior parietal cortex appear concentrated in the top

Table 32.1. The matrix of connections between areas of the macaque visual cortex. The cortical parcellation and connections are exactly as Felleman & Van-Essen (1991) with the exception of MIP and MDP, which have been excluded because of uncertainty concerning their relation to, or identity with, an area called 7m. Connections coded as '2' are reciprocal, those coded as '1' are one-way projections (whose directions are indicated in Figure 32.1) and those coded '0' are either projections which have been explicitly tested for and found absent or connections which are not presently known. No information about the spatial position of the areas in the brain, the laminar termination patterns of the connections, the continuity or patchiness of the distribution of cells giving rise to a projection or about the relative density of projections is represented in the matrix. The information represented concerns only the existence of a connection between two areas, and is therefore the coarsest and most reliable that can be extracted from neuroanatomical studies.

	V1	V2	V3	Vp	V3a	V4	VOT	V4t	MT	MSTd	MSTl	FST	PITd	PITv	CITd	CITv	AITd	AITv	STPp	STPa	TF	TH	PO	PIP	LIP	VIP	DP	A7a	FEF
V2	2																												
V3	2	2																											
Vp	0	2	2																										
V3a	2	2	2	2																									
V4	2	2	0	2	2																								
VOT	2	0	0	0	0	2																							
V4t	2	2	2	2	0	2	0																						
MT	2	2	2	2	2	2	2	2																					
MSTd	0	2	0	0	0	0	0	0	2																				
MSTl	0	2	2	2	0	2	2	2	2	2																			
FST	0	2	0	2	0	2	0	2	2	2	0																		
PITd	0	0	0	0	0	2	0	0	0	0	0	0																	
PITv	0	0	0	0	0	2	1	0	0	0	0	0	2																
CITd	0	0	0	0	0	2	1	0	0	0	0	0	2	2															
CITv	0	0	0	0	0	0	0	0	0	0	0	0	2	2	2														
AITd	0	0	0	0	0	0	0	0	0	0	0	0	2	0	2	2													
AITv	0	0	0	0	0	0	0	0	0	0	0	0	0	2	0	2	2												
STPp	0	0	0	0	0	0	0	0	2	2	2	2	0	0	2	2	0	2											
STPa	0	0	0	0	0	0	0	0	2	0	2	2	0	0	2	2	2	2	2										
TF	0	2	2	2	2	2	0	0	2	0	0	0	2	2	2	2	0	0	2	2									
TH	0	0	0	0	0	0	0	0	0	0	0	0	0	0	0	0	0	0	2	2	2								
PO	0	2	2	2	2	0	0	0	2	0	0	0	0	0	0	0	0	0	0	0	0	0							
PIP	0	0	2	2	2	0	0	0	0	0	0	0	0	0	0	0	0	0	0	0	0	0	2						
LIP	0	2	0	2	2	2	0	0	2	2	2	2	0	0	2	2	0	0	2	2	0	0	2	2					
VIP	0	2	0	0	2	2	0	0	2	2	2	2	0	0	0	0	0	0	2	2	0	0	2	2	2				
DP	0	0	0	2	2	2	0	0	2	2	0	0	0	0	2	0	0	0	0	2	0	0	2	2	2	2			
A7a	0	0	0	0	0	0	0	0	2	2	2	2	0	0	2	2	2	2	2	2	2	2	2	2	2	2	0		
FEF	0	2	0	0	2	2	0	2	2	2	2	2	0	0	2	2	2	2	2	2	2	0	0	2	2	2	2	2	
A46	0	0	0	0	0	0	0	0	0	2	2	0	0	0	0	0	0	0	2	2	0	0	0	0	2	2	0	0	0

part of the diagram, while areas of the inferior temporal cortex are located in the lower part. Because no information regarding the spatial position of the areas entered the analysis (only information concerning the areas to which each area is connected) this feature suggests that the spatial position of an area in the brain is a good predictor of the areas to which the area is likely to be connected, and that nearby areas tend to innervate one another (see below).

Visual signals from the eyes enter the primary visual cortex (V1), which is located at the far left of Figure 32.2. Cells in V1 have small receptive fields (RFs) and are influenced by simple events like local textural discontinuities in the visual field. Visual signals pass from V1 to a cluster of prestriate areas (V2, V3, VP, V4t, V3A, MT, and PIP) which are somewhat hierarchically organized, with areas V3A and MT being topologically less peripheral than other members of the group. MT is distinguished topologically by its one-way projections to frontal cortex. Cells in MT have slightly larger RFs, and are mainly concerned with whether something is moving in a particular direction within the RF. Every area of the "prestriate group" sends output connections to a further cluster of areas comprising FST, MSTd, MSTl, VIP, PO, LIP and DP. The projections from the "prestriate group" to this group appear highly redundant, which might account for the fact that partial damage to these prestriate areas does not seriously disrupt spatial vision (Ungerleider & Mishkin 1982). Cells in some of these areas are typically responsive to wide-field movements, such as expansion or contraction, that might result from motion of the viewer. One of the other areas of this group, LIP, contains cells that may "decompute" the viewer's eye, head and body motions from the signals they receive to build up a stable representation of the position of objects in the world. Signals from the "MST/posterior parietal complex" then pass to the frontal eye fields, area 7a, the posterior part of the superior temporal polysensory area (STPp), and eventually to area 46 and the anterior STP (STPa).

Moving ventrally from V1, signals are relayed via V4 and VOT into the "inferior temporal (IT) complex". The "IT complex" appears to be hierarchically organized, in the sense that more anterior stations are topologically further from the periphery, and is associated with parahippocampal areas TF and TH. The topologically "higher" areas of IT, where some cells respond with high specificity for particular visual patterns, such as faces, hands and even learned fractal patterns (Gross *et al.* 1972; Miyashita & Chang 1988; Perrett *et al.* 1982; Young & Yamane 1992) project to area 46 and to STPa.

Connections between the dorsal and ventral streams are much less dense than those within each stream, and opportunities for cross-talk do not exist at every station. Both streams, however, project selectively to area 46 and to STPa. Area 46, for example, receives signals which presumably concern what an object is (from IT), where it is (area 7a, LIP), its movement in visual space (MT), its colour (V4) and its relation to movements of the eyes (FEF).

These results suggest that, despite the enormous complexity of the cortical visual system, at this gross level it may be organized according to four principles. (i) It is dichotomised into two streams, (ii) both streams are hierarchies, (iii) the streams re-converge in area 46 and STPa, and (iv) neighbouring areas tend to innervate one-another (Young 1992).

32.3.2 The Global Cortical System Of The Primate

Figure 32.3 represents 834 connections between 72 cortical areas of the visual, auditory, somatosensory-motor, frontal and limbic cortex of the primate. Despite the complexity of the figure, features of the gross organization of the system are apparent. Elements within each of the visual, auditory and somatosensory-motor systems cluster together and each sensory cluster is topologically distant from the other sensory clusters. Visual cortical areas are to the left, with striate and prestriate areas located toward the bottom and progressively higher visual areas located progressively toward the top left of the structure. The two "streams" of the visual system are separated, with elements of the "dorsal stream" (*e. g.* FEF, LIP, 7a) positioned relatively further toward the centre of the structure, because of their greater connectivity with

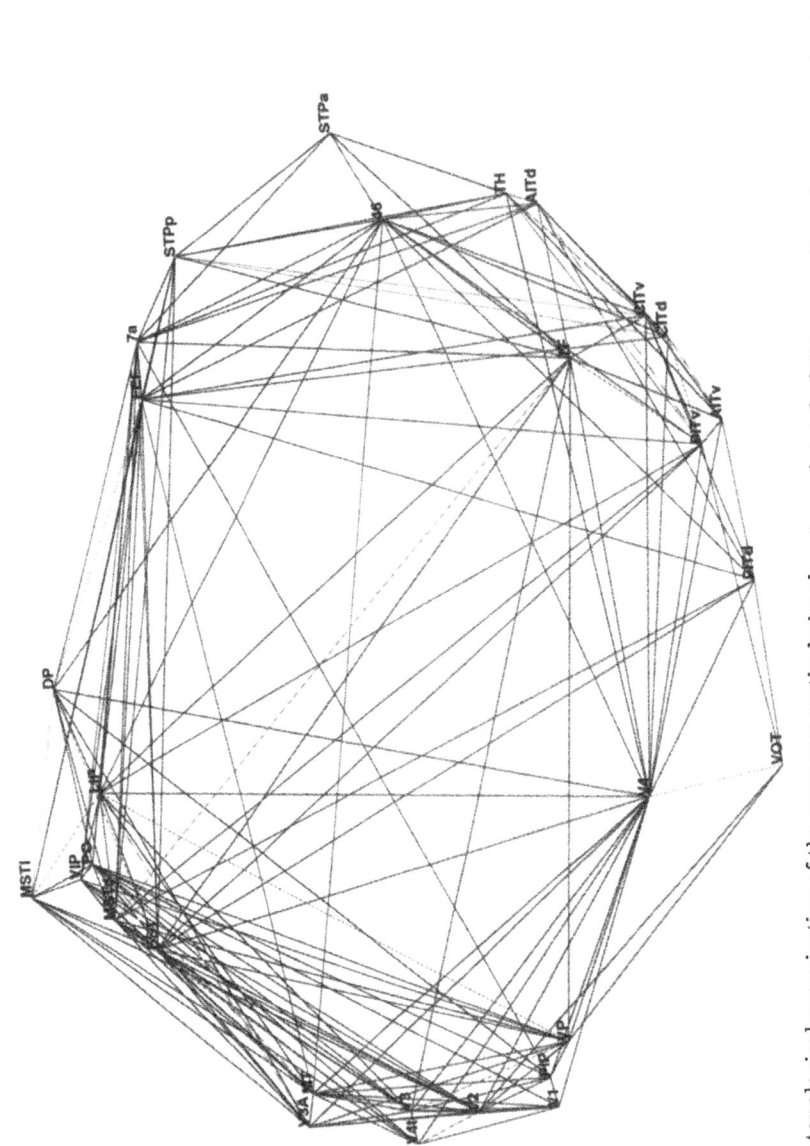

Figure 32.2. The topological organization of the macaque cortical visual system. A total of 301 connections is represented, of which 62 are one-way. The structure was derived by submitting the proximity matrix in Table 32.1 to non-metric multidimensional scaling. This structure was derived with an ordinal level of measurement in 2 dimensions, and the configuration of points (60 parameters) accounted for 40% of the variability in Table 32.1 (435 parameters). Reproduced with permission from Young 1992

the somatosensory-motor system. Auditory cortical areas are at the top right, with primary and secondary areas furthest right and higher auditory cortex located progressively further toward the top left of the structure. The somatosensory-motor cortex is located at the bottom right of the structure, with primary areas at the very bottom. The primary motor cortex (area 4) is in the center of this group of areas and is furthest from the center of the structure. Area 7b, as well as being strongly connected to the other areas of the somatosensory-motor system, is interconnected with the other sensory systems and has been pulled nearer the centre of the structure. Figure 3 near here - topology of global system

Areas at the top of each sensory system, where the most elaborated sensory signals are processed, are connected with a cluster of areas at the top left of the structure. This cluster is composed of the limbic system (entorhinal, perirhinal and cingulate cortex and the hippocampus) in association with some areas of the frontal and prefrontal cortex (*e. g.* areas 13, 10 and 12). The only limbic structure not placed within the fronto-limbic complex is the amygdala, which is drawn into the centre of the structure by its very rich output connectivity: the monkey amygdala projects to all but eight of the cortical areas. The fronto-limbic complex, at the top left of the structure, is the topologically most central group of areas, being furthest from the periphery of the cortical processing system which is represented at the bottom and right-hand edge of the structure. This confirms the topological centrality of the limbic system, but also shows the association of limbic structures with elements of frontal cortex. Frontal cortex, however, may not be a topologically homogeneous grouping (in contrast to occipital cortex) since some frontal cortical areas are associated with the limbic system, while others are more associated with one of the sensory modalities (*e. g.* areas 8 (FEF) and 46 with vision, area 45 with the somatosensory-motor system). STPa may be the least peripheral area in the cortical processing system, and this may be related to the elaborate stimulus selectivities of cells in that area (Perrett *et al.* 1987; Young & Yamane 1992).

The somatosensory-motor system is located between the other two sensory modalities: it is closer topologically to the visual and auditory systems than vision and audition are to one-another. Cross-talk between the auditory and visual systems, except at the highest level, seems limited to interactions mediated through the frontal eye fields (FEF). Interactions between the somatosensory-motor system and audition are more abundant, particularly involving connections of Areas 7b and Tpt. Interactions of the visual and somatosensory-motor systems are also rich, especially those mediated by areas 7b, FEF, 46 and LIP.

32.3.3   The Global Cortical System Of The Cat

For comparison with the global organization of the primate cortical systems, Figure 32.4 shows the global organization of the cortical systems of the cat, as revealed by an analysis of the 1134 connections reported to pass between its 64 cortical areas (Scannell & Young 1992). The figure shows that, like the macaque, the cat cortex has four main clusters of areas, corresponding to the visual, auditory, somatosensory-motor system and the fronto-limbic complex. There are, however, in contrast to the primate, prolific connections linking the four systems so that cross-talk between the sensory systems is not limited to connections between higher areas. The early visual areas are at the bottom of Figure 32.4, and these areas appear to give rise to two visual streams. One, the "visual-somatosensory-motor stream" (AES, SSF, 7, POA and 5bl) passes towards the somatosensory-motor system and the fronto-limbic complex. The other, the "visual-frontolimbic-auditory stream" (EPp, 20a, 20b, and PS) passes towards the fronto-limbic complex and the auditory system.

The auditory system occupies the left of the structure and is broadly hierarchical: a peripheral-to-central axis is discernible, in which AI, AAF, V and VP are the most peripheral processing areas, while DP, AII and the Temporal field occupy more central positions. The somatosensory-motor system is at the right side of Figure 32.4. A clear progression from

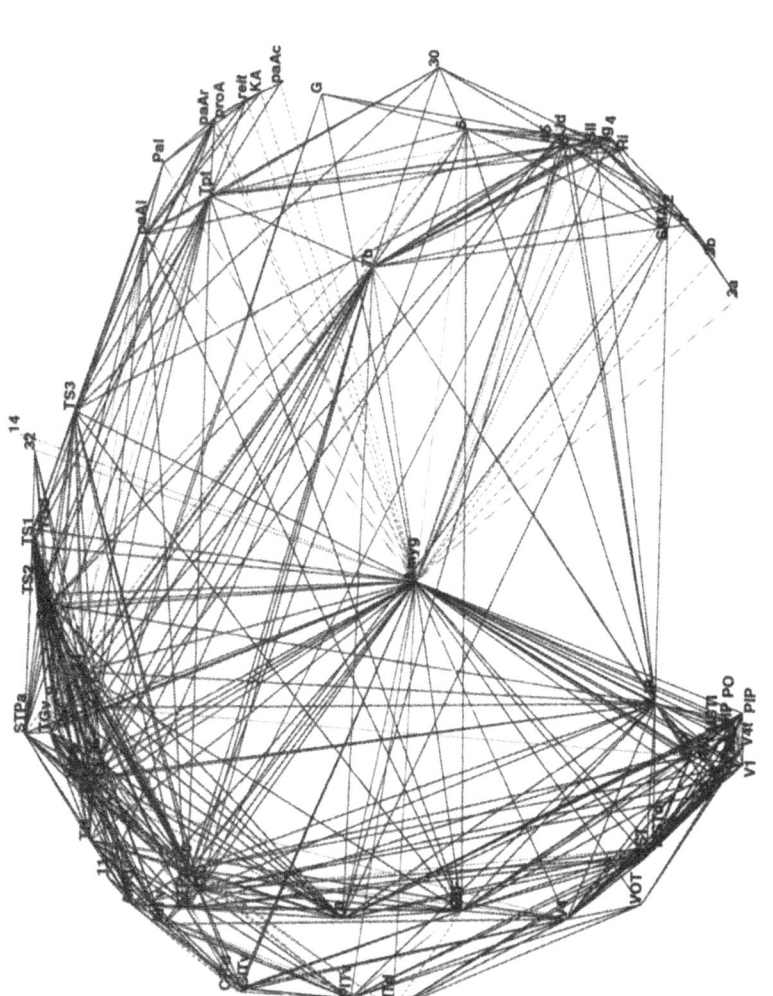

Figure 32.3. The topological organization of the entire macaque cortical processing system. A total of 758 connections between the 73 areas is represented, of which 136 (18%) are one-way. This connectivity represents 15% of the possible connections between these areas. This non-arbitrary structure represents in a spatial framework the organizational structure of the network of cortico-cortical connections of this animal. This configuration of points (144 parameters) accounted for 25% of the variability in Table 32.1 (2556 parameters).

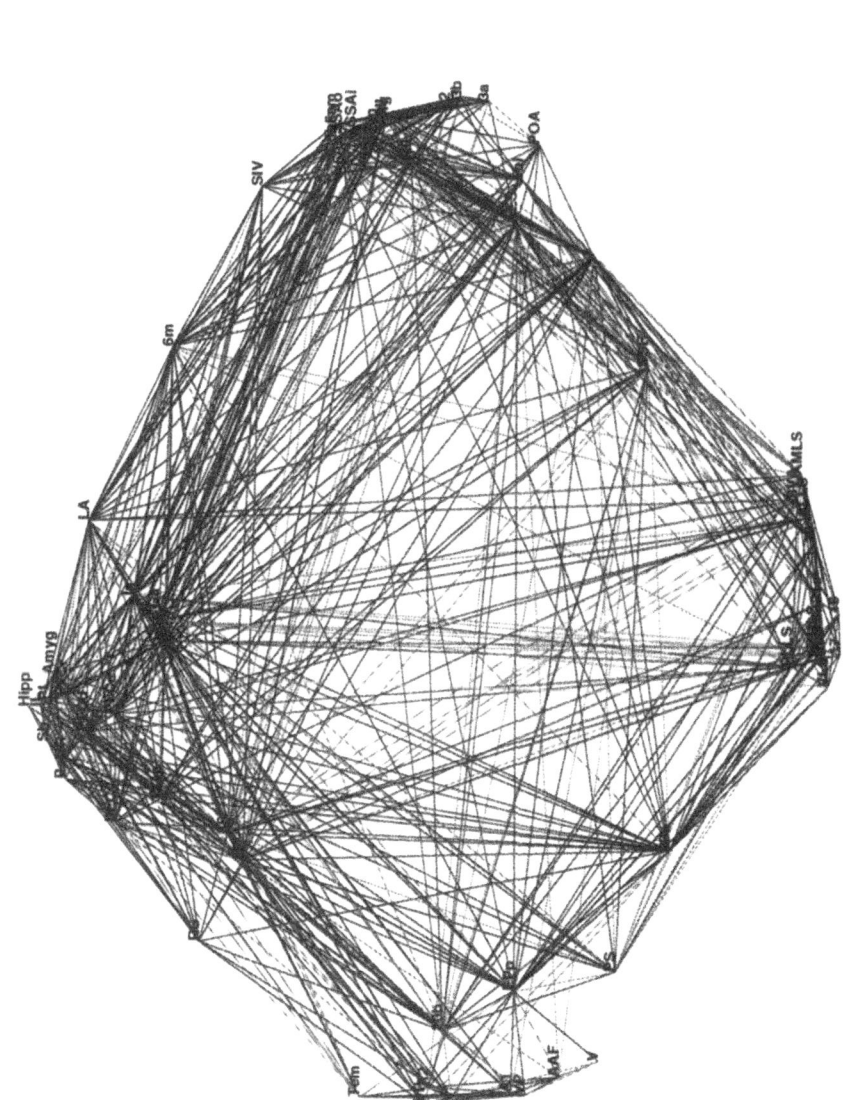

Figure 32.4. The topological organization of the cat cerebral cortex. A total of 1134 connections between 64 areas are represented, of which 321 are one-way. This represents 27% of the possible connections between these areas. The configuration of points (130 parameters) accounted for 34% of the variability in the input matrix (4225 parameters).

the primary somatosensory areas (3a, 3b, 1, and 2) to the high somatosensory motor areas (*e. g.* SIV and 6m), leading towards the fronto-limbic complex, is particularly noticeable.

The fronto-limbic system forms a topological group at the top of Figure 32.4. Areas topologically close to the sensory systems are 35 and 36, Ia, Ig, CGa, CGp, LA, and RS, which form the anatomical rim of the fronto-limbic complex. They provide the major gateway between the sensory systems and the remaining frontal and limbic areas. The cortical areas most distant from the periphery include the prefrontal cortex, hippocampus, subiculum, presubiculum, the amygdala, the entorhinal cortex and the pre- and infra-limbic areas.

Several differences are apparent between the topological organization of the cat and macaque cortical systems. Firstly, there is a higher degree of connectivity between the sensory systems in the cat. Secondly, the relation between the cortical systems differs in the two animals. The cat visual system, for example, diverges because of the differential connectivity of the higher areas with the somatosensory-motor and auditory systems, while both the "dorsal" and "ventral" streams of the macaque visual system have strong connections with the fronto-limbic complex but interact weakly with the other sensory systems. Similarly, the extrinsic connections of the fronto-limbic complex of the monkey are predominately made with the visual system, whereas the cat visual system does not dominate the extrinsic connectivity of the cat fronto-limbic complex. Thirdly, the topological position of individual structures, such as the amygdala, differs markedly between the two species, indicating that the input-output relations, and hence functions, of apparently homologous, morphological structures may be different in these two mammalian species.

These differences notwithstanding, there are some suggestive similarities between the topological organizations of the cat and monkey. The division of the cortex, at this gross level, into four major topological clusters, corresponding to the visual, auditory, somatosensory-motor systems, and the fronto-limbic complex appears in the cat, macaque, and rat. In all studied animals, the fronto-limbic complex is situated at the furthest remove from the sensory/motor periphery, being connected predominately with the higher areas of each sensory system. These common aspects may represent general features of mammalian cortical organization.

32.4   THE HIERARCHICAL ORGANIZATION OF THE MAMMALIAN CEREBRAL CORTEX

The connectional structures considered so far can be derived from the crudest, most reliable neuroanatomical data: simply the presence or absence of connections between cortical areas. Cortical areas are not, however, internally homogeneous. They contain substructures, such as six distinct layers, which may have particular patterns of intra- and extra-areal connectivity. Experiments on the primate visual system have shown that connections from the primary visual cortex to the "higher" visual areas tend to originate in the superficial cortical layers and terminate around the cell-rich layer IV of their target areas. The projections back from the "higher" areas to the primary visual cortex tend to originate in the deep layers and terminate in the deep and superficial cortical layers (Rockland & Pandya 1979). This has led to the suggestion that the former pattern of connections reflected a system for feed-forward information-processing, and the latter a system for feedback processing (Rockland & Pandya 1979; Felleman & Van-Essen 1991).

The rules for classifying the direction of connections have been extended. In recent formulations, connections originating in the superficial, or both the deep and superficial, layers and terminating in cortical layer IV are taken to be "ascending". Connections originating in both the deep and superficial layers and terminating in a columnar manner throughout the depth of the cortex are "lateral", and connections originating in the deep, or the deep and superficial layers, and terminating in the superficial and/or deep layers are "descending" (Felleman & Van-Essen 1991).

We see three chief limitations to the hierarchical approach to the analysis of cortical networks. Firstly, detailed data on the layers from which connections originate and/or terminate is necessary to determine the hierarchical organization of a cortical network, but it is considerably more difficult to localise projections to the level of individual cortical layers than it is to obtain the cruder "area to area" data. In fact, there is no information on laminar origin or termination for the majority of the connections that are known to exist, so these connections cannot be classified as ascending, descending or lateral. The cat visual system, shown in Figure 32.5, contains 193 connections of which only about 50 may be classified by data on their laminar origins and terminations.

The second problem arises from hierarchical inconsistencies: connections whose direction, as defined by origin and termination patterns, do not fit with the rest of the hierarchy. About 10% of connections in the macaque visual system (Felleman & Van-Essen 1991), and at least 9 of connections in the cat visual system fall into this group.

The third major limitation is that cortical hierarchies are necessarily unidimensional. While this single dimension may indicate the direction or order of signal flow in the network, any other possible organizational features of the system are inevitably under-represented. Topological models, on the other hand, can be derived in an arbitrary number of dimensions, which can accommodate many organizational features. Hierarchical and topological analyses are therefore complementary.

The "state of the art" in hierarchical analysis is Felleman and Van Essen's (1991) treatment of the primate visual system, which includes 32 visual areas linked by over 300 connections. Approximately 80 of the connections they show have data on both their origin and termination. Less extensive hierarchies have been produced for the macaque somatosensory-motor system (Felleman & Van-Essen 1991), and the cat auditory system (Rouiller *et al.* 1991).

32.5 MODELS OF TOPOLOGICAL PATTERN FORMATION

The complexity and apparent order of the area-to-area and layer-to-layer patterns of neural connectivity prompt the question of how such patterns are formed.

We have pursued two approaches to begin to investigate possible mechanisms in the development of the area-to-area connectivity patterns. The first approach is to examine the relations between real and artificial connection matrices, in which the artificial matrix is ordered according to some simple wiring rule.

We noted, above, that the topological analysis of the monkey visual system corresponds approximately to the spatial organization of the areas in the brain. Since the analysis received information about connectivity but no information about the spatial position of the areas, this result suggests that the spatial position of an area is a good predictor of the areas with which it will be connected. This suggests, in turn, that neighbourhood wiring rules may underlie the connectivity patterns. We investigated this possibility by constructing matrices analogous to Table 32.1 for the primate visual system. The first matrix was derived by scoring hypothetical connections between areas which share a common border as "1" and all other possible connections as "0": this is the "nearest-neighbour" wiring model (all connections were assumed to be reciprocal). This nearest-neighbour wiring matrix accounted for 61 out

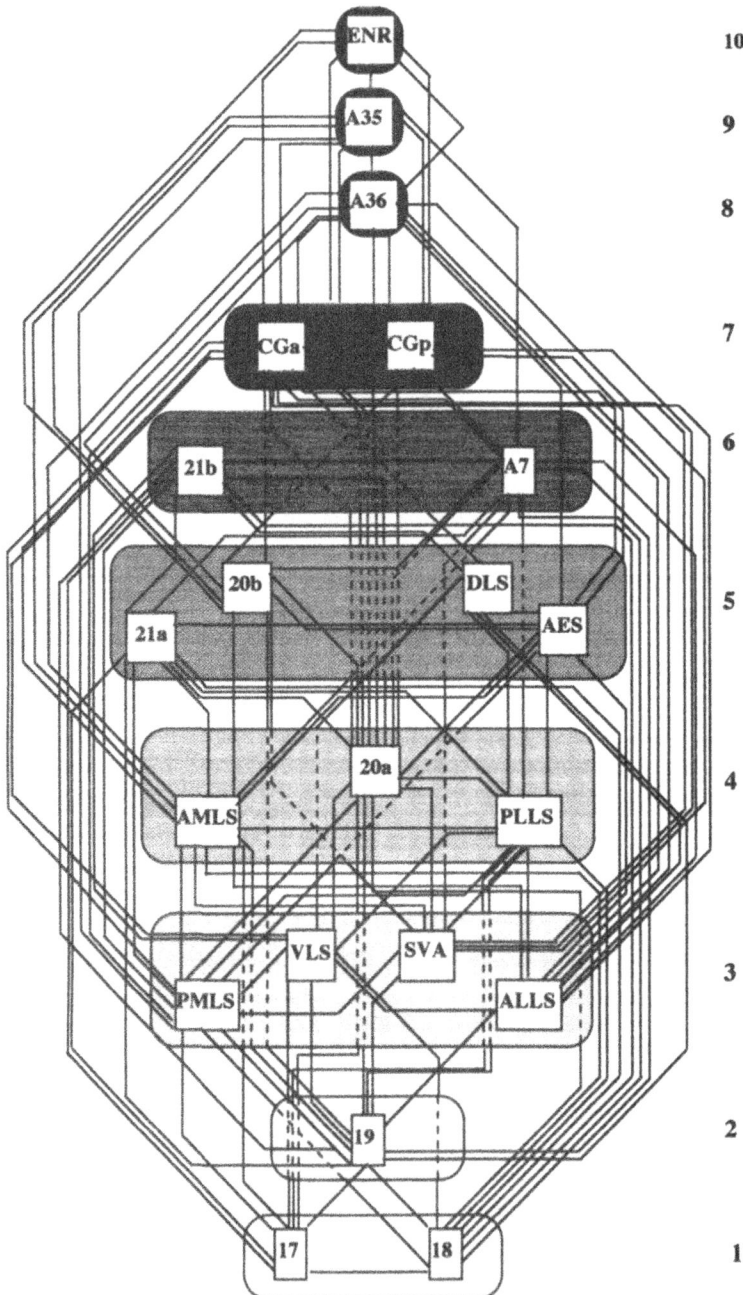

Figure 32.5. The hierarchical structure of a cortical visual system, in this case that of the cat. The figure shows 193 connections, of which 166 are reciprocal, between 21 cortical areas occupying 10 levels (indicated in the right margin). Areas placed at the same level are enclosed within a panel to show that left-to-right placement of areas is not meaningful: only higher-than and lower-than assignments are constrained by the analysis.

of the 301 connections of the visual system (27%), and so is inadequate as an account of the wiring of the system.

A second matrix was derived by scoring possible connections as a "1" if the areas share a common border or if they are separated by only one intervening area which abuts both areas: this is the "nearest-neighbour or next-door-but-one" model. Of the 301 connections in Table 32.1, 169 (55%) were connections between nearest-neighbour or next-door-but-one areas. Of those connections that were not accounted for by the nearest-neighbour-or-next-door-but-one wiring rule, the majority were between areas that were clustered into four spatial groups. The clustering of such areas implies that a small number of "fascicles" (*e. g.* 4), long-distance connection tracts, together with the nearest-neighbour-or- next-door-but-one wiring rule, would be able to account for the observed connectivity.

The nearest-neighbour-or-next-door-but-one wiring rule, however, would produce too many connections: it predicts that some connections will exist between areas that are not, in fact, connected in the real brain. It is possible, though, that this over production of connections may be related to a perinatal period during which exuberant neonatal connections are "pruned", possibly by an activity-dependent mechanism, to yield the adult connectivity pattern. We are investigating whether the particular exuberant projections predicted by the nearest-neighbour-or-next-door-but-one wiring rule are indeed those that are deleted during this phase.

In a second approach, we used the mathematical tractability of the structure in Figure 32.2 to investigate quantitatively whether the topological organization of the visual cortex reflects neighbourhood wiring. The nearest-neighbour and nearest-neighbour-or-next-door-but-one wiring matrices were submitted to MDS, and the resulting configurations were compared with Figure 32.2 by a regression-like procedure, Procrustes rotation. The statistical rarity of each comparison was assessed by approximate randomisation.

The two models which represented border relations between areas explained about the same amount of variability (30%) and were related to the structure derived for the real cortical visual system at a level which would not be expected by chance (less than 1 in 600 probability). Neighbourhood wiring is, therefore, a feature of the wiring of the primate cortical visual system. Table 32.2 shows the results of exactly similar analyses for all systems of the cat cerebral cortex.

These analyses show that neighbourhood wiring may be an important factor determining the connectivity of the cortex as a whole, and that in some cases such a wiring rule can account for much of the observed connectivity pattern. This feature might represent a parsimonious developmental mechanism, or an efficient minimum-wiring solution to cortical connectivity. Alternatively it might be that new areas evolve from functional heterogeneity within a given area, followed by "divergent evolution".

32.6 MODELS OF HIERARCHICAL PATTERN FORMATION

The hierarchical organisations of the cat and macaque visual systems, as shown by analysis of laminar origin and termination patterns, correlate with the distance from primary visual cortex along the posterodorsal-anteroventral and posterior-anterior axes. The simplest account of the formation of these hierarchies might therefore be to invoke the "maturational gradient" of neocortical development, and a recently-discovered time-dependent "stop" signal which is expressed in the middle layers of the cortex at a particular developmental stage (Molnar & Blakemore 1991).

Table 32.2. Quantitative comparisons between border relation models and connectional topology for cat cerebral cortex. The table summarises the results of the quantitative comparisons made between the topology of the visual, auditory, somatosensory-motor, and frontolimbic systems, and the entire cortex (global), with the border relation models. Comparisons are also shown between subsets of the cortical areas and their respective border relation models. "Neocortex"; all cortical areas except the frontolimbic system, "Fron/Aud/Som"; all cortical areas except the visual system, "Fron/Aud/Vis"; all cortical areas except the somatosensory system, "Fron/Som/Vis"; all cortical areas except the auditory system. The * symbol marks comparisons that were statistically significant. "% Real Connections Predicted", shows the percentage of anatomically demonstrated connections which are predicted by the border relation model. The "% Model Connections Covered" shows the percentage of the models' connections which overlap with real connections.

Set of cortical areas	Nearest Neighbour Model			Next Door But One Model		
	Variance explained	% Real Connections Predicted	% Model Connections Covered	Variance explained	% Real Connections Predicted	% Model Connections Covered
Visual	.10*	26.1	77.2	.16*	56.6	61.1
Auditory	.73*	27.9	70.6	.68*	61.6	64.6
Som/motor	.13	18.2	75.6	.16*	44.4	65.9
Frontolimbic	.26*	20.2	81.8	.29*	50.2	77.9
Global	.14*	20.1	69.8	.31*	47.8	51.0
Neocortex	.64*	25.8	69.0	.70*	56.0	48.5
Fron/Aud/Som	.32*			.52*		
Fron/Aud/Vis	.31*			.32*		
Fron/Som/Vis	.37*			.62*		

The neocortex, lateral to the rhinal sulcus, develops along an anteroventral to posterodorsal temporal gradient (Bayer & Altman 1991), in which the anteroventral regions mature earlier than posterodorsal ones. It could be supposed that during the development of cortico-cortical connections, a particular piece of cortex, wiring according to a neighbourhood wiring rule, would innervate more mature anterior cortex in which the "stop" signal is being expressed in layer IV and so give rise to "ascending" connections terminating in layer IV. Innervation of less mature posterior cortex would be unaffected by the "stop" signal, which would not yet be expressed, causing projections from more anterior cortex to terminate in the superficial and deep cortical layers characteristic of "descending" projections. This wiring pattern would yield a hierarchical organization running from posterior areas being "low" to anterior areas being "high" as observed for the visual systems of the cat and monkey.

Our consideration of other sensory systems, however, shows that a unidimensional developmental gradient model cannot account for the hierarchical organization of all cortical systems. In the cat auditory and monkey somatosensory-motor systems, for example, the hierarchical level of cortical areas is correlated with their distance from their respective area of primary sensory cortex, but not to their position along the presumed cortical developmental axis. An alternative rule that accounts well for the hierarchical level of cortical areas in all the systems so far considered is "hierarchical level increases with distance from primary sensory cortex".

At least two mechanisms could account for this feature. The first is the production of diffusible messenger substances by the primary sensory areas. Axons growing up the concentration gradient would persist in the adult as "descending" connections, those going down the concentration gradient would become "ascending" connections, and axons growing across the gradient would become "lateral" connections.

The second mechanism depends on axons from the primary sensory areas carrying a surface marker. Axons growing in the opposite direction to the axons from the primary cortex would become descending, those growing in the same direction would become ascending, and axons growing orthogonally to axons from the primary cortex would become lateral connections. Exuberant connections early in development could act as a scaffold, so that the hierarchical organization is established in areas to which the primary sensory cortex does not project in the adult, because of activity-dependent deletion mechanisms.

ACKNOWLEDGMENTS This work was supported by the Royal Society, MRC, SERC, Wellcome Trust, and the Oxford McDonnell-Pew Centre for Cognitive Neuroscience.

REFERENCES

Bayer, S. A., & Altman, J. 1991. *Neocortical development*. New York: Raven Press.

Felleman, D. J., & Van-Essen, D. R. 1991. Distributed hierarchical processing in the primate cerebral cortex. *Cerebral Cortex*, **1**, 1–47.

Gross, C. G., Rocha-Miranda, C. E., & Bender, D. B. 1972. Visual properties of neurons in the inferotemporal cortex of the macaque. *J. Neurophys*, **35**, 96–111.

Kurskal, J. B. 1964. Multidimensional scaling by optimizing goodness of fit to a nonmetric hypothesis. *Psychometrika*, **29**, 115–129.

Miyashita, Y., & Chang, H. S. 1988. Neuronal correlate of pictorial short term memory in the primate temporal cortex. *Nature*, **331**, 68–70.

Molnar, Z., & Blakemore, C. 1991. Lack of regional specificity for connections formed between thalamus and cortex in coculture. *Nature*, **351**, 475–477.

Perrett, D. I., Rolls, E. T., & Caan, W. 1982. Visual neurones responsive to faces in the monkey temporal cortex. *Exp. Brain Res.*, **47**, 329–342.

Perrett, D. I., Mistlin, A. J., & Chitty, A. J. 1987. Visual neurons reponsive to faces. *Trends Neurosci*, **10**, 358–364.

Rockland, K. S., & Pandya, D. N. 1979. Laminar origins and terminations of cortical connections of the occipital lobe in the rhesus monkey. *Brain Res.*, **179**, 3–20.

Rouiller, E. M., Simm, G. M., Villa, A. E. P., de Ribaupierre, Y., & de Ribaupierre, F. 1991. Auditory corticocortical interconnections in the cat: evidence for parallel and hierarchical arrangement of the auditory cortical areas. *Exp Brain Res.*, **86**, 483–505.

Scannell, J. W., & Young, M. P. 1992. Organization of the cortico-cortical connections of the cat: I: Qualitative aspects of topology and hierarchy. *(In preparation)*.

Shepherd, R. N. 1962. Multidimensional scaling with an unknown distance function. *Psychometrika*, **27**, 125–140.

Shepherd, R. N. 1980. Multidimensional scaling, tree fitting, and clustering. *Science*, **210**, 390–398.

Takane, Y., Young, F. W., & de Leeuw, P. 1977. Non-metric individual differences multidimensional scaling: an alternative least squares method with optimal scaling features. *Psychometrika*, **42**, 7–67.

Ungerleider, L. G., & Mishkin, M. 1982. Two cortical visual systems. In *Analysis of Visual Behavior*, Goodale, M. A., Ingle, D. G., & Mansfield, R. J. Q. (eds). pp. 549–586, Cambaridge, MA: MIT.

Young, F. W. 1978. *Multidimensional Scaling: History, Theory and Applications*. New Jersey, 1987: Hillside.

Young, F. W., & Harris, D. F. 1990. *SPSS Base System User's Guide*. Chicago: SPSS inc.

Young, M. P. 1992. Objective analysis of the topological organization of the primate cortical visual system. *Nature*, **358**, 152–155.

Young, M. P., & Yamane, S. 1992. Sparse population coding of faces in the inferotemporal cortex. *Science*, **256**, 1327–1331.

INDEX

The manufacturer's authorised representative in the EU is Springer
Nature Customer Service Centre GmbH, Europaplatz 3, 69115 Heidelberg,
Germany. If you have any concerns regarding our products, please
contact ProductSafety@springernature.com

Printed and bound by CPI Group (UK) Ltd, Croydon, CR0 4YY

23/04/2026

02095607-0017